Dynamic Knowledge Representation in Scientific Domains

Cyril Pshenichny
ITMO University, Russia

Paolo Diviacco
Istituto Nazionale di Oceanografia e di Geofisica Serimentale, Italy

Dmitry Mouromtsev
ITMO University, Russia

IGI Global
DISSEMINATOR OF KNOWLEDGE

A volume in the Advances in Web Technologies and Engineering (AWTE) Book Series

Published in the United States of America by
IGI Global
Engineering Science Reference (an imprint of IGI Global)
701 E. Chocolate Avenue
Hershey PA, USA 17033
Tel: 717-533-8845
Fax: 717-533-8661
E-mail: cust@igi-global.com
Web site: http://www.igi-global.com

Library of Congress Cataloging-in-Publication Data

Names: Pshenichny, Cyril, 1973- editor. | Diviacco, Paolo, editor. | Mouromtsev, Dmitry, editor.
Title: Dynamic knowledge representation in scientific domains / Cyril Pshenichny, Paolo Diviacco, and Dmitry Mouromtsev, editors.
Description: Hershey PA : Engineering Science Reference, [2018] | Includes bibliographical references.
Identifiers: LCCN 2017038363| ISBN 9781522552611 (hardcover) | ISBN 9781522552628 (ebook)
Subjects: LCSH: Communication in science. | Communication of technical information. | Information technology--Social aspects.
Classification: LCC Q223 .D95 2018 | DDC 501/.4--dc23 LC record available at https://lccn.loc.gov/2017038363

This book is published in the IGI Global book series Advances in Web Technologies and Engineering (AWTE) (ISSN: 2328-2762; eISSN: 2328-2754)

British Cataloguing in Publication Data
A Cataloguing in Publication record for this book is available from the British Library.

For electronic access to this publication, please contact: eresources@igi-global.com.

Advances in Web Technologies and Engineering (AWTE) Book Series

Ghazi I. Alkhatib
The Hashemite University, Jordan
David C. Rine
George Mason University, USA

ISSN:2328-2762
EISSN:2328-2754

Mission

The **Advances in Web Technologies and Engineering (AWTE) Book Series** aims to provide a platform for research in the area of Information Technology (IT) concepts, tools, methodologies, and ethnography, in the contexts of global communication systems and Web engineered applications. Organizations are continuously overwhelmed by a variety of new information technologies, many are Web based. These new technologies are capitalizing on the widespread use of network and communication technologies for seamless integration of various issues in information and knowledge sharing within and among organizations. This emphasis on integrated approaches is unique to this book series and dictates cross platform and multidisciplinary strategy to research and practice.

The **Advances in Web Technologies and Engineering (AWTE) Book Series** seeks to create a stage where comprehensive publications are distributed for the objective of bettering and expanding the field of web systems, knowledge capture, and communication technologies. The series will provide researchers and practitioners with solutions for improving how technology is utilized for the purpose of a growing awareness of the importance of web applications and engineering.

Coverage

- Knowledge structure, classification, and search algorithms or engines
- Data analytics for business and government organizations
- Software agent-based applications
- Mobile, location-aware, and ubiquitous computing
- Data and knowledge validation and verification
- Radio Frequency Identification (RFID) research and applications in web engineered systems
- Case studies validating Web-based IT solutions
- Human factors and cultural impact of IT-based systems
- IT readiness and technology transfer studies
- IT education and training

IGI Global is currently accepting manuscripts for publication within this series. To submit a proposal for a volume in this series, please contact our Acquisition Editors at Acquisitions@igi-global.com or visit: http://www.igi-global.com/publish/.

The Advances in Web Technologies and Engineering (AWTE) Book Series (ISSN 2328-2762) is published by IGI Global, 701 E. Chocolate Avenue, Hershey, PA 17033-1240, USA, www.igi-global.com. This series is composed of titles available for purchase individually; each title is edited to be contextually exclusive from any other title within the series. For pricing and ordering information please visit http://www.igi-global.com/book-series/advances-web-technologies-engineering/37158. Postmaster: Send all address changes to above address.

Titles in this Series

For a list of additional titles in this series, please visit: www.igi-global.com/book-series

Global Perspectives on Frameworks for Integrated Reporting Emerging Research and Opportunities
Ioana Dragu (Babes-Bolyai University, Romania) Adriana Tiron-Tudor (Babes-Bolyai University, Romania) and Szilveszter Fekete Pali-Pista (Babes-Bolyai University Romania)
Business Science Reference • ©2018 • 160pp • H/C (ISBN: 9781522527534) • US $135.00

Novel Design and the Applications of Smart-M3 Platform in the Internet of Things Emerging Research and Opportunities
Dmitry Korzun (Petrozavodsk State University (PetrSU), Russia) Alexey Kashevnik (St. Petersburg Institute for Informatics and Automation of the Russian Academy of Sciences (SPIIRAS), Russia & ITMO University, Russia) and Sergey Balandin (FRUCT Oy, Finland & St. Petersburg State University of Aerospace Instrumentation (SUAI, Russia)
Information Science Reference • ©2018 • 150pp • H/C (ISBN: 9781522526537) • US $145.00

Developing Metadata Application Profiles
Mariana Curado Malta (Polytechnic of Oporto, Portugal & Algoritmi Center, University of Minho, Portugal) Ana Alice Baptista (Algoritmi Center, University of Minho, Portugal) and Paul Walk (University of Edinburgh, UK)
Information Science Reference • ©2017 • 248pp • H/C (ISBN: 9781522522218) • US $170.00

Game Theory Solutions for the Internet of Things Emerging Research and Opportunities
Sungwook Kim (Sogang University, South Korea)
Information Science Reference • ©2017 • 221pp • H/C (ISBN: 9781522519522) • US $130.00

Design Solutions for Improving Website Quality and Effectiveness
G. Sreedhar (Rashtriya Sanskrit Vidyapeetha (Deemed University), India)
Information Science Reference • ©2016 • 423pp • H/C (ISBN: 9781466697645) • US $220.00

Handbook of Research on Redesigning the Future of Internet Architectures
Mohamed Boucadair (France Télécom, France) and Christian Jacquenet (France Télécom, France)
Information Science Reference • ©2015 • 621pp • H/C (ISBN: 9781466683716) • US $345.00

Artificial Intelligence Technologies and the Evolution of Web 3.0
Tomayess Issa (Curtin University, Australia) and Pedro Isaías (Universidade Aberta (Portuguese Open University), Portugal)
Information Science Reference • ©2015 • 422pp • H/C (ISBN: 9781466681477) • US $225.00

IGI Global
DISSEMINATOR OF KNOWLEDGE

701 East Chocolate Avenue, Hershey, PA 17033, USA
Tel: 717-533-8845 x100 • Fax: 717-533-8661
E-Mail: cust@igi-global.com • www.igi-global.com

Editorial Advisory Board

Table of Contents

Detailed Table of Contents

This chapter presents a further step of the formalization of the event bush method by means of graphs and graph homomorphisms. First, the authors revise an analysis of the structure of event bushes by means of bipartite structure graphs that are equipped with a multi-flow structure. They consider event bushes as a compact description of a set of alternative scenarios of directed changes of environments. They give a first formalization of the concept of scenario. A snapshot describes the state of an environment, which evolves in accordance with a given scenario at a certain moment in time. A sequence of snapshots describes how an environment may evolve over time. Cartoons are maximal sequences of snapshots. The authors discuss, define, and exemplify the concepts scenario, snapshot, and cartoon.

The aim of this chapter is to delve into the issues related to the formalization and transmission of knowledge within the scope of collaborative scientific research and to propose a new approach to address such difficulties. Analyzing methods and practices of collaborative research, the authors highlight that observation and reasoning are systematically prone to flaws, so that theorization is made highly conjectural. To gain reliability, points of views and visions need then a support from a community; in other words, they become public. To allow convergence to take place, conceptualizations need to be understood by people with possibly different cognitive models. Therefore, the authors propose using artifacts that can be strongly structured in individual use while weakly structured in common use. These artifacts take place generally as graphic representations, and as in the case of the arts, they can be realistic or abstract, they can emphasize, hide, or allow different, contrasting and concurrent interpretations of the exposed knowledge. Keywords: Collaborative Research, Scientific reasoning, Knowledge Representation, Knowledge Formalization, Boundary objects.

Formation and evolution of ice sheets is one of the "hot" problems of modern geosciences, as it has direct implication on the issues of climate change and sea level rise. Different methods of measurement or computing the mass balance of modern ice sheets based on various physical models sometimes give conflicting results. To understand them, one should first reconcile the models they are based on. This, in turn, requires one to decipher the vision different researchers have on the generation and evolution of ice sheets. This vision is initially qualitative. Hence, a qualitative model is desired that would reconcile various, and sometimes conflicting, physical models. This chapter proposes this model.

Formation of dikes is typical for most of the volcanic centers. Different types of eruptive centers tend to create different patterns of dike distribution by strike. Looking at this distribution, one may suggest, by purely geometrical reasons, where the vent should be (if buried) or should have been (if removed by denudation or new eruptions). However, wide application of this technique raises the issue of interpretation of the results in the context of the geological history of a particular volcano. Impartial and universal interpretation may be possible in a framework of knowledge of eruptive center evolution that can be built by means of the event bush method. Such a framework, however, may have wide application in volcanology extending far from the initial task it has been developed for.

The chapter addresses the mechanism of growth of druses of faden quartz in a tectonically deformed sandstone. A peculiar feature of this type of quartz, the so-called "white tape" represented by fine subparallel cracks with fluid inclusions, appears highly informative about the genesis of the mineral and tectonic regime of its growth. Two stages of formation of druses of faden quartz are recognized. The suggested mechanism is checked for contradictions by means of the event bush method. The proposed event bush model appears to describe a wider range of quartz formation environments and therefore may serve as a conceptual framework for various models of quartz growth in sedimentary rocks.

The chapter reports the state of the art in the modeling of coastal environments by the method of event bush. This method appears to be an efficient alternative to highly subjective conceptualizations used in the databases that provide data for probabilistic assessment of evolution of the coast. The event bushes themselves may be used for Bayesian computation, but there emerge complications that pose intriguing theoretical tasks. Still, their application requires deep conceptual rethinking of the field of knowledge including both the terminology and concepts traditionally accepted in it.

Purely descriptive knowledge appears to have a structure that can be formalized using the newly developed tool of dynamic knowledge representation, the event bush. In its framework, a singular biography can be transformed into a knowledge-based system with an opportunity of creation of a database. However, the very approach to the database design may undergo quite a change thereafter. This change may concern the concept of primary key, composition of data scheme, and other fundamental issues of database design.

Many manmade systems, from tea clippers to banking database management tools, exhibit an important common feature that a useful impulse is produced and functionates in them, being affected by favorable or unfavorable circumstances. While some of these systems are highly formalized, others remain operated or investigated quite intuitively. Their structural similarity urges one to think of conveying of experience of formalization from most to least formalized systems, but this process itself needs formal grounds to avoid the errors of inadequate formalization. In this chapter, a solution in the dynamic knowledge representation method and underlying theory of multitudes are sought.

Two alternative methods of knowledge representation, Petri net and event bush, have been applied to represent the same environments of alternative directed changes. The chapter examines the interrelation between the two methods and considers the ways of their mutual optimization and co-application. It discusses the balance between conceptual thoroughness and formal transparency from one side and readability, representation power, and common-sense clarity from the other. The interrelation and interplay of these two methods may bring fruitful revelations about how to treat dynamic knowledge, what strategy to choose to represent scenarios depending on the task in mind. This is seen as an important contribution to a new field of research, representation of dynamic knowledge, which may have wide application in a variety of fields, from history to technical design.

Foreword

This volume presents a radical new approach to knowledge engineering (and information processing, and data formalisation). Developed by an international community of researchers led by Cyril Pshenichny, Uwe Wolter, Paolo Diviaccio and co-workers, it has found its first applications in the geosciences but potentially is of far wider applicability, as it develops some fundamentally new mathematical concepts. The basis of this mathematics is explained in the earlier chapters of the book (Section 1), and a selection of applications illustrated by case studies in the subsequent chapters (Section 2) in geosciences, and (Section 3) in other fields of science and the humanities).

I first met Cyril Pshenichny, then a young researcher in the Institute of Mathematical Geology (IMAG) which had been founded by the great Andrey Vistelius, on a visit to St Petersburg in 2000. Vistelius was a pioneering Russian geologist who was one of the founders of the field of mathematical geology, but sadly had died 5 years previously.

We shared a common interest in "geo-reasoning" - my own, from experience with relational database management systems and particularly the inadequacy of predicate logic to capture the full information content of geological observations, and Cyril's, researching a broader and more fundamental problem with the use of formal logic in understanding and interpreting the history of geological systems. Even at that stage he had shown the seeds of his "event bush" ideas, which have now matured into a grand body of theory which is expounded in this book. This has been greatly enriched by his subsequent experience internationally, in varied fields of geoscience, and with a large number of co-workers in Russia and elsewhere, two of whom are included as editors of this volume.

They have taken on a great challenge. Many geologists studied the subject because of an aversion to (or even a fear of) mathematical rigour, with the result that mathematical geology is a relatively young branch of the science, with its own international association formed only 50 years ago. Much (indeed most) of the innovation in the field comes from mathematicians applying their methods to this new science: it is rare for geologists to make contributions of such importance to mathematics. The "Theory of Multitudes" which is the subject of this book is, however, one such contribution. It is clear that it has great practical potential in the geosciences and in other observational sciences. Its significance as a purely mathematical theory has yet to be fully realised and recognised by the mathematical community. Furthermore, in my own opinion, this theory could well be one of the keys to future development of complex knowledge-based systems which will be needed for "artificial intelligence" as it provides a set of tools for formalisation of knowledge structures and relationships under uncertainty - in other words, in the real world.

Although it is a radical new approach, requiring, as it does, even a new terminology to express its concepts (with some words from the Russian language, unfamiliar to most western readers, but necessarily so because there is no precise English translation of the concepts they encapsulate), nevertheless it follows logically from much previous work such as the mathematical modelling of geological processes pioneered by Vistelius, and by the more widely known work of Edgar Codd, a researcher with IBM who developed the relational database management concept based on predicate logic. In his later years Codd recognised that simple two-valued logic is insufficient to represent the full meaning of even comparatively simple data sets. The Theory of Multitudes goes beyond the axioms of logic and set theory to provide a general-purpose methodology for representation, structuring, and modelling knowledge in the real world.

Stephen Henley
Resources Computing International Ltd., UK

Preface

Human eye and reasoning use to distinguish things that change and things that do not. The latter are commonly expressed as relations between concepts. Relations and concepts, generally defined in the semantic science as objects, have become a key issue of knowledge representation and are well tractable by computer. They represent the "anatomy" of a subject area, whatever it be – a factory, an ecosystem, the Internet or philosophical theory. For instance, the concepts "production unit" and "raw material" would be linked by the relation "receive" in a "Factory" subject area, concepts "plants" and "solar energy", by the relation "consume" in "Ecosystem", concepts "people" and "webpages" – by the relation "open" in the "Internet", concepts "positivism" and "experiment", by the relation "favor" in "Philosophical Theory". Correspondingly, static knowledge is that in which the relations between concepts are considered fixed – i.e., once asserted, are never negated. This assumption has proven to be efficient for a wide range of tasks like building data models and search algorithms or description of those domains in which things, their properties and relations are considered stable. Static, or "fixed", relations are perfectly captured by the well-known formalisms of ontologies and conceptual graphs, as well as by numerous knowledge representation tools: formal concept analysis, ontology web language, knowledge interchange format, resource description framework, entity-relationship diagrams and many others, recently grouped under an umbrella term "object-based methods". Formally and visually, all of them can be represented by a network graph with a semantics that includes two types of object: concepts in the nodes and relations put strictly between them (either as arcs or as special type of nodes). Sometimes, if data are accurately presented, a detailed and formalized description of the phenomenon in question may bring an illusion that the static part of scientific research gives its full explanation. Still, such description lacks understanding of this phenomenon across time and under different conditions.

Thus there comes to play the other kind of relations, those able to change in time, space or any other concern, thus reflecting the well-known change of things in the world by external or internal reasons. Exactly these reflect the "physiology" of the subject area, or the functioning of the phenomenon that is mirrored by it. The knowledge about changing relations is regarded as dynamic.

The division of knowledge into static and dynamic is natural for many sciences, e.g., anatomy and physiology (of plants, animals and human beings), planetary science and cosmogony. "Static" (classification) branches are clearly seen in history, while the main body of its knowledge is certainly "dynamic". Importantly, dynamic does not mean "editable", "extensible" or "incomplete" as is it is understood in the so-called dynamic ontologies or dynamic graphs, which describe essentially static knowledge. "Dynamic" here relates not to the way the knowledge is represented but to the phenomenon of reality this knowledge describes.

Such "unstable" relations, being treated from various perspectives as events, (changeable) states, transformations, processes, scenarios and the like, remain a controversial philosophical, linguistic and scientific issue and have gained notably less systematic attention by knowledge engineers. Methods of their representation, recently defined as event-based – Petri nets, event trees, Bayesian belief networks and trees, UML behavior diagrams, sequence diagrams and others – have quite loose semantics of nodes and arcs. There is no clarity in how to express the change of relation formally; however, what looks most likely is that not relations between the objects but relations between their combinations should be analyzed, otherwise each assertion or negation of the same relation would ruin the existing network graph and urge to create a new. Hence, these combinations should be placed in the nodes of a graph, while arcs, likely, should mean cause-effect relation, which may be well or poorly understood. In practice, however, nodes in such graphs may include just parameters (e.g., in influence diagrams or, occasionally, belief networks), questions or one-word phrases. Nevertheless, accurate consideration of their meaning is likely to reveal hidden relations between concepts.

Furthermore, methods that describe events, processes and scenarios have been scattered in many disciplines, from computer science to geology, and their discovery in various fields of science and practice is an important task itself.

In addition to the above, there has appeared a method addressing the changing relations on an elaborated syntactic and semantic basis, the event bush. This method was suggested specially for the volcanology and tested by a number of tasks in this science. Later it was extrapolated to many other domains, from physics to history, but geoscience remains its main "testbed". This can be explained by the importance of timing in the explanation of geologic phenomena. Earth science in fact investigates and explains causation in the natural word, much as historians of human history investigate and explain the development of mankind.

The event bush method not only structures the knowledge in given domain but also constrains it semantically and allows, inter alia, conversion in Bayesian belief network or direct probabilistic computation, quantitative assessment of similarity of modeled phenomena, qualitative "parsing" and optimization of physical models, building of expert elicitation protocols and online collaborative work. Still, formal (logical and mathematical) foundations of the event bush still remain largely unclear, and all the encouraging results of its application should be referred rather to lucky intuition.

The book you are holding in your arms aims to present a snapshot of the cutting-edge achievements in the dynamic knowledge representation and engineering. After numerous journal papers and separate chapters, this is the first book entirely dedicated to this issue. Judging by the response to the scientific meetings, at which it was sound (annual KESW, MSEPS workshop in Mumbai in 2012, KEMG 2012, 2015, and 2016), as well as by the regular reads and downloads of related content at the ResearchGate, the topic is of a long-lasting interest for the international scientific community.

The main challenges of research in this field include, though are not limited to, the following dimensions.

1. Development of a formalism to underlie the dynamic knowledge representation. Methodologies of static knowledge representation are firmly grounded in the classical logic and set theory. Changeability of relations in dynamic knowledge urges us to look for other, but not less strict, ways to manage concepts and statements.
2. Adaptation of this formalism to the psychological, sociological and cultural peculiarities of the scientific research. Every scientist possesses his/her own cultural background and shares numerous societal and corporative values, which influence, and perhaps even determine, his/her vision of a

domain of knowledge and his/her own research. To accommodate a new methodology in a scientific community, one needs to make it user-friendly – both in terms of general whys and wherefores and in terms of software usability.

3. Application of the dynamic knowledge representation in particular domains of knowledge brings a theoretical claim to rethink the whole body of knowledge in these domains, i.e., untangle the threads of inference, accurately select arguments to support or refute hypotheses, compare models, evaluate expert judgments and re-comprehend the field of knowledge in its entirety. As most fields of dynamic knowledge are largely descriptive and language-dependent (history, geology, psychology and others), this is a real challenge that suggests, first of all, the need to structure the knowledge and, wherever possible, semantically constrain it.

This book contributes to all these avenues of research. The first two issues, development of basic formalism and its adaptation to the scientific community, will be considered in the first section of the book. The third issue, the application in particular domain(s) of science, will be dealt with in the second and third sections.

In these domains, the book will present a way to model in a strict way what has been least formalized – verbal descriptions and meanings of graphic objects. Geoscientific applications are grouped in the second section of the book, and applications in the other fields, spanning from engineering to humanities, in the third section.

The three sections of the book aim to attract each its own audience. The first section may be of interest for computer scientists, information scientists, philologists and linguists, mathematicians (especially those interested in the set theory) and logicians. Chapter 4 of Section 1 should also attract the attention of philosophers of science, psychologists and sociologists. The second section is intended to be read by geoscientists. In addition, some chapters of the Section 2 will be hopefully valuable for applied physicists and mathematicians working in the field of modeling of geological, geophysical and other processes. Section 3 may attract attention of specialists in various fields (economics, history, again psychology and sociology and many others) who suffer of poor organization of theoretical dynamic knowledge in their field and are looking for an example of a formalized descriptive science, more or less close to their area of research. Finally, the whole book will be interesting for knowledge engineers, designers of knowledge-based systems and developers of intellectual software tools.

Likewise, the authors of the book belong to a variety of fields: computer science, logic, mathematics, database design, geosciences, engineering, archivistics. Section 1 describes the theoretical foundations of the suggested approach. In the first chapter, Cyril Pshenichny presents a new formal theory, the *Theory of Multitudes*. This theory may be equally well related to the mathematics, logic, or linguistics. In mathematics, with further development, it may pretend to play nearly the same role as the set theory or category theory, i.e., a universal ground to represent mathematical theories. In logic, it offers another way of encoding general knowledge than does the classical predicate logic, and this way may appear more friendly to many natural languages and natural human thinking. The same time, it remains open to non-classical (multivalued, and even fuzzy) interpretations. In linguistics, it enters a wide pool of the theories of meaning, ranging at least from Wittgenstein to Grice.

The second chapter, also authored by Cyril Pshenichny, develops the ideas put forth in the first one up to the state of instrument – importantly, both qualitative and quantitative – to process knowledge and data both about static and changing world. For the qualitative reasoning, the formalism of Novorossian logic is suggested, and for quantitative computation, the Novorossian mathematics. The Novorossian logic gives ground for numerous tools of representation and engineering of both static and dynamic knowledge. Still, while static knowledge has been well addressed by methods based on traditional predicate logic (for instance, by ontologies or conceptual graphs), the dynamic one was lacking an appropriate semantic-based formalism, though many existing methods like UML behavior diagrams could act as such with a due semantic basement, and a special method of event bush was suggested solely to operate with the dynamic knowledge. The Novorossian mathematics is fully compliant with the Novorossian logic but, the same time, may serve as a "bridge" to many mathematical approaches.

The key concept that unites the two is scenario, which is a succession of transitions from some states (events) to some others. The presence of scenarios is the virtue of dynamic knowledge. This feature is thoroughly studied by Uwe Wolter in the third chapter of the book by the example of the "mainstream" method of dynamic knowledge engineering, the event bush. U. Wolter considers the evolution of any context, or environment, described by the event bush, as a "cartoon" consisting of snapshots, each being determined by a corresponding step (event) of all involved scenarios.

These theoretical constructs may look compelling or not per se, but they immediately encounter methodological problems when being applied to any domain of knowledge. These problems commonly arise at application of any formal approach. Paolo Diviacco in Chapter 4 stresses one such problem, which seems to be the most urgent for the theory of multitudes (both Novorossian logic and Novorossian mathematics): the incompatibility of efficient formalization and representation. While formalization requires explicit disclosure of all relevant pieces of knowledge, the representation is strong enough only if addresses the knowledge hidden in personal or collective backgrounds and never explicated. The solution suggested by P. Diviacco is boundary objects shared by scientific communities or individual scientists, present but not structured in the formalization and the same time deeply structured at representation by each of these parties (while the individual/group contexts of these parties are deliberately left beyond formalization).

Section 2 of the book includes case stories of dynamic knowledge representation in the domain, for which it was first designed, the geoscience. Six chapters of Section 2 cover diverse areas within it, such as glaciology, volcanology and tectonics, mineralogy, coastal geomorphology, regional geology and tectonics, geographical topology and theoretical geomorphology.

Kirill Khvorostovsky and coauthors turned to the dynamic knowledge representation aiming to conceptualize the accumulation of ice in the ice sheets of Greenland and Antarctic. Their ultimate goal is to use this conceptualization to constrain the quantitative modeling of mass balance variation in ice sheets. However, this goal appeared to be too specific and requiring highly detailed conceptual model, which can be developed only on top of a basic one. Still, such basic model itself gives enough food for meditation relating the qualitative formalization of the entire field of ice sheet behavior.

On the contrary, Sophie Koneva and Cyril Pshenichny achieve their goal at least principally (see Chapter 6) yet from the very beginning. Their task, to conceptualize the formation of dikes in volcanoes and other eruptive centers, appears to be, with some additional knowledge, a reformulation of the question of evolution of eruptive centers. This is one of those fundamental issues of volcanology, on top of which the formalization of any other subject area in this field should rest. However, the model developed to capture this issue appears so large that only a part of it may be presented in visual form in the publication format. This is another problem of dynamic knowledge engineering.

However, in some lucky cases the problem appears compact enough to be captured entirely by a graph of reasonable size and complexity. An example of such problem is presented by Oleg Krisak and Vyacheslav Bezrukov in Chapter 7. The initial motivation for their research was the phenomenon of faden quartz growth in sedimentary rocks. Considering a model of its formation suggested in the literature, they queried, what other forms of quartz may be produced if to change the conditions in the proposed model. Having constructed an event bush based solely on one scenario, the one that leads to faden quartz formation, and using some pieces of general mineralogic and tectonic knowledge as rules, they built an event bush, in which other scenarios of quartz formation in sedimentary rocks readily follow, pointing to peculiar diagnostic features of these species of quartz. One of the most useful features of the event bush technique is that it may start from bare depicting of one scenario and naturally lead to the full group of scenarios that may take place in the considered environment.

This environment may be already addressed by some preexisting methods of knowledge engineering. However, absence of strict semantic basement makes these results vulnerable. Victoria Shterkhun et al. (Chapter 8) reexamined a Bayesian model of marine coast evolution. Building an event bush for the environment in question allowed them to revise the Bayesian model and substantially improve its semantics. Coupling of various methods of dynamic knowledge engineering (event bush, Bayesian belief network and others) in studying the same environment will be also performed in a few chapters below.

In the study presented by Biju Longhinos with coauthors, the environment, in fact, has not only thematic but also spatial localization. This is a first attempt of application of dynamic knowledge engineering (in particular, the event bush method) to some specific area of land. The authors are trying to solve a geological problem that has been solved either strictly, but based on a very much limited set of parameters that may not give an adequate picture of reality, or based on real, dominantly qualitative, geological data, but purely intuitively. This problem is: whether the existing geological data support or refute given tectonic hypothesis? In the event bush framework, this problem has found a reliable, though laborious, solution based on a strict and formal treatment of the geological data.

Spatial localization of any environment is one of the most important methodological peculiarities of the geosciences. Spatial properties themselves are described by formalisms, which, obviously, are potentially influential on any research in the geosciences, but especially on the geomorphology. If these properties change in time, special formalisms are needed, able to account for the dynamic character of this knowledge. Such formalisms are reviewed by Jayanthi Ganapathy and Uma Vijayasundaram in Chapter 10. These authors outline an urgent goal of the geospatial research in the framework of the theory of multitudes – to reinterpret these formalisms in terms of the said theory.

Section 3 presents a selection of case stories of dynamic knowledge representation in various domains of knowledge, from technical disciplines to humanities, storytelling and everyday reasoning, including some cross-disciplinary parallels and theoretical insights inspired by the tasks addressed.

Maria Mandrik and Sergey Arustamov perform a research in archivistics aiming at optimizing the thematic search of documents (Chapter 11). They show that, similar to an event bush describing the growth of faden quartz and delivering para gratis other possible scenarios of evolution of the environment, a biography of a singular person may give birth to an event bush that would disclose a much wider historical context. Studying this feature of the event bush, the authors come to a brand new vision of database design that may be applied in a variety of fields, extending far from archivistics.

Andrey Naumov with coauthors in Chapter 12 demonstrate that dynamic knowledge representation may help consciously reuse formal models developed for one case (e.g., banking database management design systems), in many others, thus increasing the value of these models. The authors study two competing methods of dynamic knowledge representation, the evenet bush and activity diagrams.

Vyacheslav Rogalchuk et al. (Chapter 13) also studied the interrelation of event bush and another method of knowledge engineering, the Petri nets, by strikingly diverse examples of information technologies and modern politics that exhibit surprising structural similarity. V. Rogalchuk et al. demonstrate universal applicability of both methods and look for an efficient strategy of their co-application.

Uwe Wolter et al. (Chapter 14), on the contrary, focused on the singular method of event bush, but addressing complex issues, although borrowed from everyday reasoning, and thus requiring composite, multi-bush structures. These structures were analyzed in the light of the graph and category theories, and some general suggestions were formulated for the syntax of the event bush method.

Finally, Pavel Komar and coauthors (Chapter 15) focused on a case when one theme may have different aspects belonging to different fields of knowledge. P. Komar and coauthorss addressed geological drilling that has, along with geological, also technical, physical, economic and other aspects, described, correspondingly, by various knowledge representation constructs. Mind that in all these aspects the knowledge may well be static and dynamic. The authors studied two aspects of drilling, geological and physical, analyzed how physical models can be parsed and optimized by the event bush and how thematically diverse event bushes may relate to each other if describe the same issue.

The selection of chapters that makes up this book characterizes not just the state-of-the-art but, better say, the state of the dawn of the dynamic knowledge representation. At this stage, full formalization of the theory of multitudes is still ahead, different authors may be using somewhat different versions of the event bush and following different classifications of environments, but what can be definitely said is sthat a common semantic ground already exists for a wide spectrum of independent knowledge representation and engineering techniques able to capture the changes of our vivid and versatile world. Each of the chapters included in Sections 2 and 3, in fact, marks a gateway to a new field of research in the domain of knowledge it relates to, be it geoscience or something else. We hope that this book would foster the research and stimulate the interest of practitioners worldwide. Moreover, we believe that building extended semantic models will, perhaps in a distant future, shift the accent from representation of knowledge to penetration into the world this knowledge is about.

Cyril Pshenichny
ITMO University, Russia

Paolo Diviacco
Istituto Nazionale di Oceanografia e di Geofisica Serimentale (OGS), Italy

Dmitry Mouromtsev
ITMO University, Russia

Section 1

Theory of Multitudes and Related Methodological Issues

Chapter 1
Theory of Multitudes as an Alternative to the Set Theory

Cyril Pshenichny
ITMO University, Russia

ABSTRACT

The initial goal of the research reported in this chapter was to develop an extension of the set theory able to handle the meanings of concepts. However, this purpose led not to extension of the said theory but to development of a new one, and this chapter presents an outline of a newly constructed theory of multitudes, which nevertheless seems to share some of the features of the set theory. Still, its approach is to clearly separate the qualitative and quantitative concerns, and first formalize the qualitative one, based on the vision communicated by natural languages. Though still not entirely formal, the proposed theory is seen to provide a common ground for a wide spectrum of independent knowledge engineering and representation techniques.

INTRODUCTION

Since the times of Cantor the classical set theory has evolved a lot and now, in tight connection with the classical predicate logic, is the ground for formalization of many theories of mathematics, as well as other fields of knowledge. Modern tools of knowledge engineering and representation essentially based on these two formalisms allow building ontologies, conceptual graphs and other formalized conceptualizations widely used in the science, education, industry and business. Various extensions of the set theory allow to account for fuzzy membership of an element in a set (Zadeh, 1965), multiple occurrence of an element in a set as "copies" (multiset theory of Blizard, 1989), ordering of elements in a set (posets in order theory – see, e.g., Davey and Priestley, 2002, or trees in descriptive set theory – see Kechris, 1995) and other important peculiarities of sets.

Nonetheless, there remains an issue that is of vital importance for knowledge representation but, in our view, may not be correctly addressed by the classical set theory in its present form, nor by any existing extension. This is the issue of meaning of concepts that underlies ontology design and related techniques.

DOI: 10.4018/978-1-5225-5261-1.ch001

Along with personality in psychology or field in physics, meaning is an entity as widely used as poorly understood. The abundance of theories of meaning, from competing semantic views of Frege, Russel and Tarsky to linking it to syntax by Chomsky, skepticism of Wittgenstein and, in addition, intentional-behaviorist views of Grice, Donaldson and many others (see, e.g., a review of Speaks, 2016, and references therein) fairly well illustrate this fact. This urged us, considering the meaning, develop our own vision of the subject, mainly based on actual use of this term in the knowledge engineering and representation.

As might be concluded from the cornerstone publications and numerous application reports, concepts in ontologies (Gruber, 1993), conceptual graphs (Sowa, 2006), entity-relationship diagrams (Chen, 1976) and, presumably, other knowledge representation constructs are commonly treated as potentially non-empty sets. Still, the adequacy of this approach looks debatable. For instance, compare two cases: (i) only green apples were delivered to the supermarket *X* and (ii) apples can be green, yellow or red. In the former case, one surely may treat the concept "green apples" as a set, while in the latter there is no confidence if apples ever exist. If to substitute apples with flying hills, and colors, with "disintegrating", "aggregating" and "stable", we get empty sets, but the logic of relations remains the same, and meanings, however unexpected, are still there. Furthermore, "atomic" has been considered in the science as a synonym to "further indivisible". However, whatever atom to take, in the real world it appears not only splittable into pieces (e.g., apple can be sliced) but, yet more painful, often consisting of notably different parts (e.g., apple consists of skin, flesh, and seeds). Perhaps only quarks in the physics are considered absolutely atomic, but these, by the same theory, may not occur alone, i.e., may not be treated separately. In addition, "atoms" recognized in the real world are not warranted to avoid coalescence, disintegration or complete disappearance.

Moreover, we have serious problems identifying "elements" (i.e., smallest indivisible parts) of water, air, or any continuous medium, as well as energy, light, and the like. Are these molecules/atomic particles or "elementary units"? If the former, their properties may have nothing to do with the properties of the whole thing (as different are the micro- and macroworld), and if the latter, how big these units are? Unlike apples, here the choice of an element is absolutely provisional. Obviously, taking "apples", "water", "air", "energy", "light" for sets, we have to admit that they are filled with some other "type of existence" than elements.

Finally, saying that courage may be caused by the sense of duty or psychopathy, we should exhibit the highest degree of creativity to imagine "elements" of courage, sense of duty and psychopathy. Even if to reformulate this notion and say, "in every person courage may be caused by the sense of duty or psychopathy" to identify some hypothetic persons *X*, *Y*, *Z*, …, as elements, we do not solve the problem because once courage, sense of duty and psychopathy are predicates, we consider them concepts, and if they are concepts, they must be also regarded as sets, and the question is back: what are the elements of these sets? – The facts of occurrence of the said properties in a particular *X*, *Y* or *Z*? But these properties never occur alone, in every particular case may be accompanied by vast many other features, and objective criteria of their occurrence are inexistent. In fact, here we have to operate with meanings only and do not deal with elements at all.

These and other imperfections of the set theory were summarized by Simons (2005) who analyzed them from mathematical, philosophical, linguistical and common-sense perspectives. He was supported by Smith (2005) who related the flaws of the set theory to the widely accepted view that "the key to the ontological structure of reality is captured syntactically in the 'F*a*' (or, in more sophisticated versions, in the 'R*ab*') of first-order logic, where 'F' stands for what is general in reality and '*a*' for what is individual. Hence "fantology". Because predicate logic has exactly two syntactically different kinds of referring expressions – (F), (G), (R), etc., and (*a*), (*b*), (*c*), etc. – so reality must consist of exactly two correspondingly different kinds of entity: the general (properties, concepts) and the particular (things, objects), the relation between these two kinds of entity being revealed in the predicate-argument structure of atomic formulas in first-order logic" (p. 1-2). Nevertheless, Smith advocates the predicate logic, saying that this doctrine stems not from the predicate logic per se but from its philosophical ("fantological") (mis)interpretation suggesting that all generality and all meaning is grasped by *F* and nothing left for *a*.

Opposing this, inevitably, though implicitly, Smith returns to the Aristotelian understanding of categorical proposition as a subject-predicate structure, *S* – *P*, in which *S* is the subject, *P*, the predicate, and both – absolutely equal concepts (e.g., "Some apples are green"). However, this also brings a trouble – the equality of subject and predicate allows one to change their places that in many cases obviously leads to confusion ("Green is (some) apples") .

Therefore, a theory is desired that would (i) account for different types of "fill" of a concept (including, but not limited to elements) and (ii) handle meanings and their relations, with or without any "fill". This seems to be a departure from the set-element vision, as well as from predicate or subject-predicate ones, far enough to think of an alternative theory rather than extension of the set theory.

To do this, it looks necessary to:

- Put forth basic terms and axioms;
- Introduce basic relations and operations;
- Look how these may work in a natural language and
- Discuss their relationship to the existing mathematical and logical framework.

These tasks will be addressed in the following sections of the chapter.

BASIC TERMS AND AXIOMS OF THE THEORY OF MULTITUDES

Undefined Terms

Undefined terms of the theory are: *meaning*, *language*, *name*, *fill*, *multitude*, *to be discernible part of* and *to be indiscernible part of*.

We do not define these terms but will comment them. Meaning has got various definitions and understandings in the science and humanities. Recent studies, pretending to suggest yet deeper concept than meaning, namely "smysl", were performed by Akimova-Shterkhun and Pshenichny (2007) and Elsukov et al. (2015). Here we generally follow their line of reasoning and abstract from communicative (pragmatic) aspect of meaning. Any thing and any property in the material or non-material world necessarily have meaning, and any meaning refers either to a property or to a thing. Every meaning is unique, and therefore meanings may not be similar (or synonymic) to each other, and if two words or other denotations are synonymic, they bear at least one same meaning. For the same reason, meaning may not enter any relation as a whole and as a part of something (unlike the set theory where a set and its subset may well be members of the same superset). Meaning may refer to a property, only if it is in some relation to another meaning that refers to a thing. For instance, we may say that something is wooden if there is such thing as wood. Similarly, something is white if there is whiteness (white color) per se in the world, and someone is running only if we assume that running exists as an entity.

We believe that meanings exist apart from any material or lingual world more or less like Plato's ideas in the World of Forms, in which meanings themselves are things and properties. Still, meanings are captured, better or worse, by different *languages*. We mean languages in the broadest sense, from programming languages to the language of music, the "languages" of animal tamers and others, however intuitive, ones – but will be concerned mainly with the so-called natural languages. For instance, in some Slavonic languages the meaning that unites "light" (svet) and "all" (vse, vsyo) is eminent, while in English one can easily determine another meaning that binds "light" with "ease to carry" and then, in turn, with "air" ("airy nothing" of Shakespeare). All European languages inherit meanings grasped by Greek and Latin, and some of these may have yet deeper, Arian or other, demarcation.

Meanings reside in texts, sentences, phrases, words and their parts. Moreover, meanings can be preserved in some auxiliary parts of speech (e.g., prepositions), in poetic rhymes and just any kind of similarity between words, as well as in the parts of words. In the latter case, while base of the word is an obvious container of meaning, prefix, suffix or flexion may also bear one, and the inheritance from other languages may be especially pronounced in them. For instance, English flexion "-tion" bears the same meaning (hardly definable, but well felt) as Spanish "-cion" or Italian "-zione", and this meaning was first caught in Latin.

However, it should be noted that in any language along with expressions that represent meanings may well exist those correctly built (or grammatically correct) expressions that do not – i.e., those deemed to have hidden controversy, or absurd (e.g., e.g., "green colorless ideas" or "dry water") or fatal uncertainty ("un-neurotic courage" of Kurt Vonnegut). This fully applies to the formal language developed below.

For this reason, absurd will never start making sense, and meaning will never become absurd. This should not be confused with the case when there are meanings that have, or had, or will have, no examples, like flying hills in the beginning of the chapter, alcohol-free beer in the 1970s or good manners now. Absurd name may not relate to a meaning. If a name is known to relate to a meaning, neither this name nor other names known to be its synonyms are absurd.

Importantly, even absurd presents an affirmation (an affirmation of some controversy, e.g., "Water is dry" if we assume that water cannot be such). The name that is complete denial ("Not bad", "Not Ivan", "Neither new nor interesting") is bare expression in some language, neither meaning, nor even absurd.

Fill is understood as bare materialization ("fulFILLment") of meaning, an existence per se, considered apart from particular properties. Even the indices (e.g., "no. 1", "no. 2" and so forth) are a kind of meaning, as will be shown below in the text. Fill may be nameless and numberless pieces of anything (more or less like urelements in the set theory), or an integer "lump", or a continuous "flow" or "wave" of being. Fill does not necessarily imply material existence – for instance, the psycho-semantic theories in psychology study how various words of a natural language and underlying concepts are being filled with human emotions. One may say that, to exist in any sense (even in someone's imagination), thing must be filled with being in this very sense.

Multitude is an object that consists of one meaning and possibly fill. In other words, multitude is any "real object", or "real thing", both material and non-material. No other part than fill or meaning may be in a multitude. Meaning must be generally affirmative and integral. We postulate that once an object is said to be multitude, there is a meaning that cover it entirely and does not overwhelm it. Meaning is intentional, and fill, extensional parts of the multitude.

Furthermore, meaning is the *discernible part of* the multitude, and fill (if present), is its *indiscernible part*. Henceforth, they will be rephrased for brevity: *to be indiscernible part of* as *to belong to* (or *to be a fill of*), and *to be a discernible part of*, just as *to be a part of*. To be discernible part reserves an opportunity of being a whole for something else (i.e., having its own parts called submeanings), while indiscernible part has no such privilege. For instance, "green apple" is a part of meaning of "apple", and "psychopathic courage", as well as "sense of duty", are parts of the meaning "courage". In all these cases, the multitudes that include parts are not necessarily filled if the multitude that includes the whole is filled – i.e., apples are not necessarily green, and courage – psychopathic. In another case, part may be always "present" in the whole, e.g., the meaning "hunger" is a part of the meaning "appetite", and the corresponding multitude defined by the meaning "hunger" is always filled when the multitude defined by the meaning "appetite" is filled (i.e., no hunger – no appetite). However, "to be discernible" may be understood rather freely, from "to have at least some distinction" to "to have nothing in common".

The fill may not be part of meaning, and vice versa. We believe that mereology per se (see, e.g., Keet, 2006) focuses exactly on the type of discernible part-whole relation, though many points advocated by her do not apply to the theory reported here.

These terms – multitude, meaning, fill, to be discernible part of and to be indiscernible part of – will be used in the formalism developed below.

Multitudes are denoted by bold Greek capital letters in italics (***A***, ***B***, ***Γ***, ***Δ***, ...), possibly numbered ($\boldsymbol{A}_1$, $\boldsymbol{A}_2$, $\boldsymbol{A}_3$, $\boldsymbol{A}_4$, ...).

Definition 1: Multitude is described as

A:=(<*Meaning*>|<*Fill*>),

with the *Fill* slot possibly empty:

A:=(<*Meaning*>|∅).

If a multitude has non-empty fill slot, let us call it *filled*.

Meanings are denoted by capital Cyrillic letters in italics: *А, Б, В, Г*, ..., possibly numbered: $А_1$, $А_2$, $А_3$, $А_4$, ..., or their correctly built combinations, however complex (see below). If ***A*** is a multitude and *Б* is the meaning of ***A***:

A:=(*Б*|<*Fill*>) or ***A***:=(*Б*|∅),

we put *Б*<***A***, where < denotes the discernible part-whole relation. The fill, if present, is denoted depending on its type considered below.

Indiscernible part-whole relation is denoted by ⊂: <*Fill*>⊂***A*** (this expression will be put more precisely below, at consideration of types of the fill).

Finally, if there is null in place of meaning in the description of a multitude due to absurd or uncertainty, it is postulated that there is also no fill in such multitude, so the multitude is *absolutely empty*:

C:=(∅|∅).

This literally means: what has no meaning, may not be exemplified. Therefore, all meaningless expressions of a language may be taken one for another, and "green colorless ideas" is synonymic to "green colorless ideas that furiously sleep" and the same time to "dry water" and also to "un-neurotic courage".

Multitudes have no relations with each other. Rather, their meanings do. These are described in the next section.

Formal Relations Between Meanings

Formal relations between meanings, as well as meanings themselves, are necessarily specified in a language. One meaning may be denoted by a few expressions of language, which are considered synonymic. This applies also to the formal language below. In this language, the sign = will be used to denote synonymic expressions of the same meaning.

Subordination

Meaning may have proper parts. Parts of the meaning are also meanings.

Definition 2: Proper parts of a meaning are termed submeanings, and their relation to the whole is termed *subordination*:

А∠Б,

where *Б* is meaning, *A*, submeaning, and ∠, subordination. For instance, the meaning "melon" is subordinate to (is a part of) meaning "berry".

Axiom 1: Subordination is transitive relation for meanings. If *A∠Б* and *Б∠B*, then *A∠B*.

Importantly, this applies to meanings only and not to the multitudes ("things") that include them. It will be considered below how the part-whole relation in the "real world" is captured in this theory.

Negation

Definition 3: If *A∠Б*, then the rest of the meaning *Б* that is not *A* is termed *negation* of *A* in *Б*: ~ *ᴮA*.
Axiom 2: ~*A*=∅ (negation is not a meaning) if there are no other submeanings in *Б*.

Thus, introducing subordination, one narrows the meaning to its proper part (e.g., the meaning of berry, to melon).

Compatibility and Incompatibility

Definition 4: If two meanings have common submeaning, they are *compatible*. (e.g., meanings "berry" and "food"). Otherwise they are *incompatible*, e.g., meanings "berry" and "ball for play":

A𝔎Б.

Axiom 3: Any two meanings may have not more than one common submeaning.

Still, this axiom does not forbid this submeaning to have its own submeaings.

If the negation of meaning A is meaning itself, by definition of negation, it is incompatible with A:

$A\text{Ꞩ}{\sim}A$ if not ${\sim}A=\varnothing$.

Union

Suppose that meaning $Б$ has parts (submeanings) $Б_1, Б_2, \ldots, Б_n$: $Б_1\angle Б, Б_2\angle Б, \ldots, Б_n\angle Б$ and nothing else, and these submeanings are mutually incompatible: $Б_1\text{Ꞩ}Б_2\text{Ꞩ}\ldots\text{Ꞩ}Б_n$.

Definition 5: Meaning $Б$ is the *union* of these submeanings:

$Б=(Б_1\cup Б_2\cup\ldots\cup Б_n)$.

Intersection

Definition 6: If there are two meanings A and $Б$ and meaning B that is proper part of both A and $Б$: $B\angle A$ and $B\angle Б$, then A and $Б$ are *intersecting*: $A\cap Б$. The multitude B defined by this submeaning is termed *intersection* of the two multitudes:

$B=A\cap Б$.

By the Axiom 3, no other submeanings but B may be common for A and $Б$.

Axiom 4: Intersection with a negation is not meaning ($A\cap{\sim}Б=\varnothing$) if, and only if, this negation is not meaning itself (${\sim}Б=\varnothing$).

Devotion

Definition 6: Let A, $Б$ and B be meanings. *Devotion* of $Б$ and B to A is a ternary relation:

$A–Б, ({\sim})B$,

such that $A\angle Б$ and $:A\cap B$, and $({\sim})$ stands for "with or without negation".

Definition 7. The subordinating meaning $Б$ is termed *universum* for A and any its part.

Let us denote the meaning resulting from the intersection as $Г$: $Г=A\cap B$. Given the devotion, both $Г$ and ${\sim}Г$ are meanings and are expressed as:

$Г=(A–Б, B); {\sim}Г=(A–Б, {\sim}B)$.

As both of them are meanings, they are incompatible: $\Gamma \mathfrak{K} \sim\Gamma$, and both are parts of A: $\Gamma\angle A$ and $\sim\Gamma\angle A$, their union gives A:

$$A=(\Gamma\cup\sim\Gamma).$$

This may be also formulated as

$$(A–Б, B)\angle A;\ (A–Б, \sim B)\angle A;\ A=((A–Б, B)\cup (A–Б, \sim B)).$$

$Б$ is the universum for the devotion but not for B.

This relation can be extended to n-ary: let there be l meanings of the $Б$ type and m meanings of the B type, so that $l+m=n-1$ (as A id the n-th member of the relation):

$A\angle Б_1,$

$A\angle Б_2,$

…,

$A\angle Б_l,$

$A\cap B_1,$

$A\cap B_2,$

…,

$A\cap B_m,$

and the whole relation would take the following form:

$$A–Б_1, Б_2, …, Б_l, (\sim)B_1, (\sim)B_2, …, (\sim)B_m.$$

Correspondingly, there will be a list, or family, of p submeanings Γ that differ by presence or absence of negation at B:

$\Gamma_1=(A–Б_1, Б_2, …, Б_l, \sim B_1, \sim B_2, …, \sim B_m),$

$\Gamma_2=(A–Б_1, Б_2, …, Б_l, B_1, \sim B_2, …, \sim B_m),$

$\Gamma_3=(A–Б_1, Б_2, …, Б_l, B_1, B_2, …, \sim B_m),$

…,

$Г_p=(А–Б_1, Б_2, ..., Б_l, В_1, В_2, ..., В_m)$,

the number of strings p being determined by the binomial coefficient C_m^{m-1}. Then, the meaning A is the union of $Г_1, Г_2, ..., Г_p$:

$А=(Г_1 \cup Г_2 \cup ... \cup Г_p)$

or

$А=((А–Б_1, Б_2, ..., Б_l, \sim В_1, \sim В_2, ..., \sim В_m) \cup (А–Б_1, Б_2, ..., Б_l, В_1, \sim В_2, ..., \sim В_m) \cup ... \cup (А–Б_1, Б_2, ..., Б_l, В_1, В_2, ..., В_m))$.

Like subordination, devotion is transitive given the fixed universa.

For instance, consider a devotion

$(А–Б_1, Б_2, ..., Б_l, В_1, В_2, ..., В_m) \angle А$.

Now let us add new meanings $(Д_1, Д_2, Д_3, ..., Д_r)$ intersecting with A to any meaning $Г_i$ (e.g., to $Г_p$):

$(А–Б_1, Б_2, ..., Б_l, В_1, В_2, ..., В_m, Д_1, Д_2, Д_3, ..., Д_r)$,

$(А–Б_1, Б_2, ..., Б_l, В_1, В_2, ..., В_m, \sim Д_1, Д_2, Д_3, ..., Д_r)$,

$(А–Б_1, Б_2, ..., Б_l, В_1, В_2, ..., В_m, Д_1, \sim Д_2, Д_3, ..., Д_r)$,

$(А–Б_1, Б_2, ..., Б_l, В_1, В_2, ..., В_m, \sim Д_1, \sim Д_2, Д_3, ..., Д_r)$

and so forth.

Each of these newly formed submeanings is a part of the initial one $(Г_p)$:

$(А–Б_1, Б_2, ..., Б_l, В_1, В_2, ..., В_m, Д_1, Д_2, Д_3, ..., Д_r) \angle (А–Б_1, Б_2, ..., Б_l, В_1, В_2, ..., В_m)$,

$(А–Б_1, Б_2, ..., Б_l, В_1, В_2, ..., В_m, \sim Д_1, Д_2, Д_3, ..., Д_r) \angle (А–Б_1, Б_2, ..., Б_l, В_1, В_2, ..., В_m)$,

$(А–Б_1, Б_2, ..., Б_l, В_1, В_2, ..., В_m, Д_1, \sim Д_2, Д_3, ..., Д_r) \angle (А–Б_1, Б_2, ..., Б_l, В_1, В_2, ..., В_m)$

$(А–Б_1, Б_2, ..., Б_l, В_1, В_2, ..., В_m, \sim Д_1, \sim Д_2, Д_3, ..., Д_r) \angle (А–Б_1, Б_2, ..., Б_l, В_1, В_2, ..., В_m)$

and further on.

Then, the meaning $Г_p=(А–Б_1, Б_2, ..., Б_l, В_1, В_2, ..., В_m)$ with the "first generation" of added meanings, $В$, can be represented as the union of submeanings with the "second generation" of added meanings, $Д$:

$$Г_p=(А–Б_1, Б_2, \ldots, Б_l, В_1, В_2, \ldots, В_m)=(А–Б_1, Б_2, \ldots, Б_l, В_1, В_2, \ldots, В_m, Д_1, Д_2, Д_3, \ldots, Д_r)\cup(А–Б_1, Б_2, \ldots, Б_l, В_1, В_2, \ldots, В_m, \sim Д_1, Д_2, Д_3, \ldots, Д_r)\cup(А–Б_1, Б_2, \ldots, Б_l, В_1, В_2, \ldots, В_m, Д_1, \sim Д_2, Д_3, \ldots, Д_r)\cup(А–Б_1, Б_2, \ldots, Б_l, В_1, В_2, \ldots, В_m, \sim Д_1, \sim Д_2, Д_3, \ldots, Д_r)\cup \ldots \cup(А–Б_1, Б_2, \ldots, Б_l, В_1, В_2, \ldots, В_m, \sim Д_1, \sim Д_2, \sim Д_3, \ldots, \sim Д_r).$$

The same refers to any other $Г_i$ and, hence, their union, the meaning *A*, can be represented as a union of new submeanings including the newly introduced members ($Д_1, Д_2, Д_3, \ldots, Д_r$).

If the meanings are introduced but not specified, they are just indicated by names (see below) or simply numbered. Then the name or natural number (or index: $Б_1, Б_2, \ldots$) may be understood as an indication of a meaning.

Types of Meaning: Meanings of Things and Meanings of Properties

Any meaning refers either to a property or to a thing.

Definition 8: The meaning that refers to the whole thing and denoted by a name not divided into parts is termed *predmet*, *S*, and that referring to a thing but denoted by a name consisting of further indivisible sub-names, *event*, *E*.

Correspondingly, predmet is assumed to be denotable by a synonymic name further divisible into parts. Meaning that refers to a thing and is denoted by a name consisting of sub-names, which, in turn, may be divided, has no special term in the theory of multitudes.

Definition 9: The meaning of a property is termed *primeta*, *P*.

(Letters *S* and *P* are suggested to remind, though not exactly refer to, the traditional logical concepts of subject and predicate; although, despite the analogies, there is quite a difference between subject and predmet, predicate and primeta. Where it is important to emphasize the roles of meanings, these symbols will be used.) Meanings of properties may not have parts. Also, they may not define (i.e., be part of) multitudes. Any multitude is defined by predmet or event. Properties resulting from the formal relations between meanings specified above (subordination, negation, (in)compatibility, intersection and devotion), as well as equality (synonymy) of lingual expressions, are not expressed as primetas.

One meaning of a thing ("father" meaning, e.g., whiteness) per se may "produce" only one meaning of a property ("daughter" meaning, e.g., white) because otherwise we will have to admit that there may be two absolutely similar meanings, but such "twins" are not admitted. Symmetrically, each meaning of a property may, and must, be "produced" by only one meaning of a thing but may not relate to it further, to avoid tautological collisions like "liquid water" or "military militant". If predmet includes events, it may not produce primetas anymore, but the events may.

Though, one may suppose that one predmet may "give birth" to various primetas, e.g., "irony" may lead to "ironic" and "ironizing". Still, only one primeta, "ironic", would be the "daughter" of "irony", while any other ("ironizing", "too ironic" and whatever else) would be a "daughter" of events that are in some relation to the predmet "irony" (see below).

Each predmet or its events may appear in many universa. This reflects the well-known case that the same thought can be approached by different people from purely different sides.

Some more points should be stressed herewith:

1. Predmet and primeta may not be the same, meaning of a thing is not the same as meaning of a property;
2. One predmet must, and two may not, enter definition of an event;
3. One primeta may not enter twice the same definition of event (though may be a part of another primeta that also enters this definition – however, this case is not considered at present);
4. All primetas in a definition of event are bent with the predmet by the same relation;
5. Only finite number of primetas may be present in a definition of event;
6. No relation between two or more primetas is possible, and transitivity does not apply to relations of primetas and predmets where it may lead to direct relation between primetas;
7. If predmet is a part of universum, all its events are parts of this universum too;
8. If predmet (as a whole) produces a primeta, it may not intersect with other primetas; alternatively, not the predmet but its events may produce primetas.

In addition to the above requirements, the meaning should

9. Cover the entire multitude (applies to predmets and events only),
10. Be generally affirmative and
11. Be integral.

The requirement to cover the entire multitude has been commented. To be generally affirmative means that negation may be only partial in every submeaning. This also implies that the relation between predmet and primeta may not be negative, otherwise nothing will be affirmed in the submeaning. A complication may arise herewith that in some cases negation of simple meaning may be meaningless (e.g., "not a lion"), but in others, quite sense-making (e.g., "not a liar" or "not a female"). If so, either one of the counterparts is implicitly meaningless (e.g., liar is the one who says *not* truth, is *not* honest and behaves *not* fine), or there is no dichotomy. The latter may mean that either there are more than two options (e.g., one has no sex at all – e.g., an embryo or a cripple, the one is female and not male or, otherwise, male and not female, or one is both – hermaphrodite; in this case "male" and "female" are just independent meanings, both opposing "to have no sex at all") – or the options are two, but both have independent meanings and are not just refusal of each other – e.g., subject is that acts, object is that accepts action, and tertium non datur.

Absurd names may originate from: (i) senseless words that mean nothing, (ii) primetas incompatible with the predmets they are related to in a phrase in given language, or (iii) primetas incompatible with each other or each other's negation.

The requirement of integrity sets out that all submeanings in a composite meaning must have something common that gives ground to unite them into an integral whole (otherwise we are at risk of uniting something completely different that has no general meaning).

Formally, the three types of meaning are formally described by the devotion relation. Given

$$A\text{–}Б_1, Б_2, \ldots, Б_p, (\sim)В_1, (\sim)В_2, \ldots, (\sim)В_m,$$

and

$$Г_1=(A\text{–}Б_1, Б_2, \ldots, Б_p, \sim В_1, \sim В_2, \ldots, \sim В_m),$$

$Г_2=(A–Б_1, Б_2, ..., Б_l, В_1, \sim В_2, ..., \sim В_m)$,

$Г_3=(A–Б_1, Б_2, ..., Б_l, \sim В_1, В_2, ..., \sim В_m)$,

$Г_4=(A–Б_1, Б_2, ..., Б_l, В_1, В_2, ..., \sim В_m)$,

...,

$Г_p=(A–Б_1, Б_2, ..., Б_l, В_1, В_2, ..., В_m)$,

A is predmet, *Б* and *В* are primetas, and $Г_i$, $1 \leq i \leq p$, events:

$S:=A$, $P:=(Б, В)$, $E:=Г$.

The same primeta may intersect with one predmet or event (i.e., be its universum) and subordinate another.

Therefore, in terms of types of meaning, the relation of devotion looks as follows:

$S–P^u_1, ..., P^u_m, (\sim)P_{m+1}, (\sim)P_{m+2}, ..., (\sim)P_n,$

where $P^u_1, ..., P^u_m$ are universa (finite in number), and $(\sim)P_{m+1}, ..., (\sim)P_n$ are non-universum primetas, also finite in number.

Running through the interval of natural values from 1 to *n*, this formula, obviously, produces a number of mutually exclusive cases:

$E_1=(S–P^u_1, ..., P^u_m, \sim P_{m+1}, \sim P_{m+2}, ..., \sim P_n)$,

$E_2=(S–P^u_1, ..., P^u_m, P_{m+1}, \sim P_{m+2}, ..., \sim P_n)$,

$E_3=(S–P^u_1, ..., P^u_m, \sim P_{m+1}, P_{m+2}, ..., \sim P_n)$,

$E_3=(S–P^u_1, ..., P^u_m, P_{m+1}, P_{m+2}, ..., \sim P_n)$,

...,

$E_s=(S–P^u_1, ..., P^u_m, P_{m+1}, P_{m+2}, ..., P_n)$,

These cases form the *family of events, E*.

Definition 9: Family of events is the full list of incompatible events having the same predmet.

Obviously, the family of events is equal to the predmet that forms it, *S*:

$S=E=(E_1 \cup E_2 \cup ... \cup E_p)$,

and each event of the family is subordinate to the family:

$E_i < E,\ 1 \le i \le p.$

In this study, we assume for simplicity that primetas involved in one family of events are independent of each other, that is, none of them is produced by an event from this family. However, in the future, this limitation should be removed and the full spectrum of variants considered.

Definition 10: Exclusion is an expression of the same form as event that does not make sense.

If primetas do not make sense with the given predmet (even if their negation does), they should not appear in the event. Besides, there is also a possibility that primetas do not make sense with each other or each other's negation – e.g., "Bob pays a visit and is lying in bed with fever", "Water evaporates in the air and infiltrates in the ground" or, to exclude fuzzy vision and simultaneously greet the memory of Julio Cortazar, "Cronopios are subject to gravitation and are not physical bodies". In this case, the expression that does not make sense (i.e., may not have meaning) must be *excluded* from those denoting a family of events. Such exclusion can be done only axiomatically, because this meaningless expression is formally perfect and does not differ from others. For instance, if there is an axiom $(E_k=(S–P_1, P_2, P_3, \ldots, \sim P_{n-1}, \sim P_n))=\varnothing$ (that reads, $S–P_1, P_2, P_3, \ldots, \sim P_{n-1}, \sim P_n$ does not mean an event), then the family of events will be:

$$E=((S–P_1,(\sim)P_2,(\sim)P_3, \ldots,(\sim)P_{n-1}, (\sim)P_n); (S–(\sim)P_1,P_2,(\sim)P_3, \ldots,(\sim)P_{n-1}, (\sim)P_n); \ldots; E=(S–(\sim)P_1,(\sim)P_2,(\sim)P_3, \ldots,(\sim)P_{n-1}, P_n))EXCL(S–P_1, P_2, P_3, \ldots, \sim P_{n-1}, \sim P_n),$$

where *EXCL* is exclusion.

Exclusion can be expressed formally as subordination with indication of absurd: $E\ EXCL\ E_k=E\angle E_k$, given $E_k=\varnothing$. This literally means that the excluded should be a part of the subordinating meaning unless it is absurd. In other words, exclusion says that the expression that denotes this absurd is very much alike those which denote real submeanings of the same meaning (i.e., consists of the same sub-expressions that mean real submeanings). In fact, exclusion is an option of a "spy" or "betrayer".

One particular type of event is $S–P_1^u, P_2^u, \ldots, P_n^u$, where P_i^u, $1 \le i \le n$, are universa (e.g., human beings (S) can reason (P_1^u) and permanently live on the Earth (P_2^u)". This event, denoted as E^u and termed *universum event*, would be equal to its predmet: $E^u=S–P_1^u, P_2^u, \ldots, P_n^u=S$.

Definition 11: Uiniversum event is an event such that all its primetas are universa.

From now on, universum event is NOT meant when the term "event" is used, unless explicitly stated otherwise.

Natural question arises, whether the same predmet can be involved in several families of events. If there were no universum and families were not equal to the predmet, then obviously they may differ from each other, coexisting in the same predmet and "occupying" different parts of its meaning. However, with a universum, such families are readily merged in one, equal to the predmet.

Nevertheless, even if two predmets, S_i and S_j, have the same primetas, $P^u, P_1, P_2, P_3, \ldots, P_n$, there is no ground to consider them the same. In any case, they are different: $S_i \ne S_j$.

Virtually everything we take for predmets in our every-day or scientific reasoning represents, in fact, events in some larger area, simply the primetas that form these events are for some reason (lack of knowledge, irrelevance, fear and so forth) ignored by us. And reversely, putting down events of an event family, it is clear that, narrowing the area of our interest, one may take any of the events of this family for an independent predmet, like any point at the wave front is the source of a wave itself, by Huygens. Then, for purely applied reasons, let us introduce the concept of *predmet event*.

Definition 12: Predmet event means a meaning considered as predmet in an artificially narrowed area but representing an event in a real, large area.

Such predmet events are all, or almost all, predmets we actually deal with.

There may be predmets in a universum that do not intersect with some primetas. If all primetas, with whom the predmets in the universum ever intersect, are known and enumerated, for such predmets one may put: $S_i – P^u$, $\sim P_m$ or $S_j – P^u$, $\sim P_k$, $\sim P_l$ and so forth (keeping in mind that P^u is always affirmed as universum). Then we may consider this as a "complementary negative definition" of the universum which states what is prohibited in it, though not necessarily these negative relations really indicate the nature of P^u or its "father predmet".

Then, let us postulate that if primeta is universum (P^u), it is allowed to include *n* predmets or events (i.e., these predmets/events are proper parts of it – e.g., any eater and any sleeper are breathing, so to be breathing may be a universum for these two predmets). Thus there appear systems of interrelated meanings within a universum.

Moreover, being a primeta, universum requires a "father" predmet. Hence, however big, universum is still not all-encompassing. Rather, there necessarily is something outside of it, namely the predmet that is its "father" (the simplest example: Father God and the Universe He produced). The predmets, the universa each of these predmets belongs to, the "father" predmets of all these universa and "wife" and "daughter" primetas of these predmets constitute the *context*. Although, the predmets that are parts of one universum may be "fathers" for other primetas, which could be universa for other predmets perhaps including the "father" of the present one – let us call this "cross-universum". This gives ground to further investigate possible structures of contexts. Some of their features are quite straightforward.

Thus, incompatibility plays an important role in organization of a universum in a context. Incompatible in the universum are any two premdets:

$$S_i \Re S_j$$

and any two events in an event family and ever:

$$E_i \Re E_{j'}$$

as well as the universum and its "father" predmet or event:

$$S_p{}^u \Re P^u,\ E_p{}^u \Re P^u.$$

Likewise, incompatible is father predmet/event with daughter primeta (including universum):

$S_i \mathfrak{K} P_{Si}$, $E_i \mathfrak{K} P_{Ei}$.

Also, incompatible may be predmet with some primetas, as well as some primetas with each other (though the latter may be established only via a predmet or event).

If an event family is a part of several universa, its semantics would have as many non-deniable pairs as the number of universa.

As seen from the above, requirements to the universum include, though not necessarily are limited to: (i) universum must comprise at least one predmet; (ii) universum must comprise at least one other primeta; (iii) every primeta in the universum must have "father" predmet and be "married" to (intersect with) at least one predmet in the universum; (iv) "father" predmet of every considered primeta may be within or outside of the universum, but the "father" of the universum itself must be outside of it; (v) all families of events in the universum are equal to the predmets they include; (vi) Primeta that is entirely a part of predmet may not be universum; (vii) Primeta that is a universum for one predmet, may intersect with another one, but for the latter it would not be universum; (iix) Different universa may intersect but may not be equal or subordinate to each other..

These constraints show that the universum, being a specific type of primeta, (i) has internal structure and (ii) opens a few promising opportunities for reasoning about its parts, such as:

- Exhaustingly disclose the predmet via the event family (including the case of denying all primetas except the universum) and
- Follow all part-whole relations through events.

The internal structure of the universum represents ensembles, or systems, of primetas intersecting each with at least one predmet, while predmets may be linked only via common primetas.

Thus the universum should be understood not as a world but as a "home" in the world ("kept" by primeta who bears the surname of her "father").

To round up, in any context the following binary relations may take place:

- Predmet and predmet – always incompatibility;
- Predmet S and primeta P – intersection $S \cap P$, if S and P join one event and P is not the universum; subordination $S \angle P$, if P is the universum for S; incompatibility $S \mathfrak{K} P$ if S and P do not join one event including the case that P is universum and S, its "father" predmet;
- Predmet S and event E' – subordination $E' \angle S$, if the event E' is based on the predmet S, and incompatibility $S \mathfrak{K} E'$ otherwise;
- Predmet S and family of events E – equality $S=E$, if this family is based on that predmet, and incompatibility $S \mathfrak{K} E$ otherwise;
- Primeta P and event E' – subordination $E' \angle P$ or incompatibility $E' \mathfrak{K} P$, if primeta intersects with the predmet on which the event E' is based; subordination $E' \angle P$ if primeta is universum; incompatibility $E' \mathfrak{K} P$, if neither;
- Primeta P and family of events E – the same as predmet S and primeta P;
- Event E' and event E'' – always incompatibility;
- Family of events and family of events – always incompatibility $E' \mathfrak{K} E''$;
- Event E' and family of events E – subordination $E' \angle E$, if the event is in this family (i.e., based on the same predmet), and incompatibility $E' \mathfrak{K} E$ otherwise;

- Primeta P_i and primeta P_j – relations are set indirectly, via the relation of each primeta to the same predmet or event.

In addition, primeta may not be subordinate to predmet, even if our intuition tells us so. For instance, the meaning "red" seems to be primeta subordinate to the predmet "color". Nevertheless, it is comprehended as event "red color" originating from intersection of the primeta "red" and the predmet "color". Query, what else in the universe except color may ever be red? Still, we never know everything, and poets may readily deem of, say, "red voice" or "yellow dream", insisting for some reason, yet unknown to them, the dreams are exactly yellow, or voice, red. Theory of multitudes is capable to capture even such highly intuitive meanings.

These regularities are universal; however, additional constraints may lead to various types of contexts fitting different fields of knowledge and different theoretical and applied tasks. Among these, that of predmet event looks the most approprpiate: as event acts as a predmet in a context, one may suppose that the predmet one has is, in fact, an event in some unknown event family in a larger context. In practice, virtually all predmets we may think of should be regarded as such.

Contrary to the subject and predicate of traditional logic, the roles of predmet and primeta are fixed for any meaning, given the universum. What is left of "–" in any formula, would never occur right, and vice versa. Contrary to the predicate logic, predmet, primeta and event may not be comprehended as predicates $P(x)$, $S(x)$ or $E(x)$, correspondingly, because, first, they do not ever imply any individual variable x, and then, because their roles are not equal. Predmet (this word is masculine in the Slavonic languages) and event may be "fathers" for primetas (that are, correspondingly, feminine and can be "daughters" of predmets or events, "born" by these like from the Adam's rib, "wives" for other predmets or events and "mothers" for new events), and only primetas may be universa. Correspondingly, predmets may be fathers and husbands only, and events, fathers, husbands, and sons. With this in mind, *concept* in the classical logic and the knowledge engineering/representation can be paralleled with a meaning in the multitude theory. We believe that ontologies and other knowledge models that use concepts can be well built based on the meanings.

Types of Fill

Fill is a thing, and meaning of fill is a predmet in the universum "to be in the world". Besides, in the same universum we recognize, inter alia, such predmets as "integer", "grain" (pixel, quark, quant, urelement) and "field". They form corresponding primetas "to be *integer*", "to be *grainy*", "to be *field*". So far, these primetas are not defined but apprehended intuitively. Still, their definition will be required in the future for parameter-based and quantitative computations based on the theory of multitudes. These primetas intersect with the said predmet "fill" and form the following events: "integer fill", "grainy fill" and "field fill". Their combinations, as well as negation of all of them, are excluded by an exclusion axiom, so that no other events are present in this family and the union of these three is equal to the predmet "fill".

Then, the indiscernible part of a multitude ***A*** is denoted either as *<Integer Fill >*⊂***A***, or as *<Grainy Fill >*⊂***A***, or *<Field Fill >*⊂***A***.

Grainy fill includes grains, or atoms, denoted x, y, z, …, or x_1, x_2, x_3, … Then one should put, e.g., x⊂ ***A***. If an atom does not belong to the multitude, the notation is y⊄***A***.

Philosophically (or physically) speaking, the case of infinite number of elements may be also understood as a continuous flow (or field) with indiscernible elements. In this case, an artificial quantitative measure is often used instead of real elements – e.g., values of speed, discharge, energy and so forth.

Atom is considered here as something that has absolutely no meaning, possesses no properties, bears no name and is just a "bit of existence". In no way such atom may be regarded as a multitude itself. Rather, it should be understood as "absolute element", or urelement, in terms of the classical set theory.

Axiom 5: Fill always belongs to a multitude formed by a meaning of thing; moreover, only one.

In other words, if fill belongs to an event, the predmet this event is formed by is not considered multitude. Another corollary from this axiom is that fill may not be thought "independent", apart from any multitude.

Axiom 6: Every multitude has only one type of fill (integer, grainy, or field) and may not change it.

Mind that grainy fill may include only one atom but it is not the same as integer fill. This difference will be commented below in the present book.

On the other extreme, given the events and the grainy fill, if atoms in the predmet are countless (i.e., their number is infinite), then the infinite number of atoms would also be allocated at least to some of the events, because dividing an infinity by a finite number (e.g., the number of events in the family) or subtracting any finite number from it, one would get just a "lesser" infinity but not a finite number. Vice versa, infinite number of atoms in an event automatically implies that they are infinite in the predmet. Furthermore, once atoms are indiscernible in the predmet, then they would be indiscernible in all its events (if sand is countless in a desert, it is countless in Sahara).

In general, we think that fill is a kind of "omitted primeta" (or primetas, for grainy fill), so that the multitude is a predmet event for "ignored events". When we say that there are three thousand bricks in the wall, we imply that we won't say anything more specific about any of them. Similarly, having said that something singular exists, we are not going to add more primetas to define it. Saying that meaning is a multitude (i.e., is filled), we stop defining it and omit all possible relevant primetas that we have not put.

Different methodologies of mathematics may relate to different types of fill: set theory includes only multitudes with grainy fill (with account of the axioms 4-6), category theory, those with integer, and field theory, with field fill.

MEANINGS AND LANGUAGE

Meanings per se are set by the God or by human intuition and cannot be apprehended or understood, unless shaped up in a language, natural or artificial.

Definition 12: Meta (Russian "mark") is an expression in given language that represents meaning.

To represent a meaning implies to mimic all its part-whole relations, both to its multitude and to other meanings represented by metas, both as part and as whole, and avoid offering new relations, absent "in the original". For instance, "bike (meta of an event) is a means of transportation (meta of the predmet event, which is the whole to the bike) driven by treadles (meta of universum primeta) and may have two or three wheels (metas of other primetas; bike with two wheels and bike with three wheels are parts of the meaning "bike", the joint assertion of primetas "to have two wheels" and "to have three wheels" is excluded, and meanings "means of transportation driven by treadles and having two/three wheels" are events, and hence, parts, of the predmet event "means of transportation".

As an absurd name may not relate to meaning, neither it does to meta. Also, keeping in mind that the name that is complete denial may not be meaning, those negations of a meta that are not correctly-built metas themselves neither have equivalent meaning nor are metas themselves.

All formal expressions given above that relate to meanings do apply to metas, and "–" should be understood as a syntactic relation between metas in given language.

Let us postulate that

- Interrelated meanings always have metas in the same language and
- If the meta has parts, all its parts are in the same language as the meta itself.

Then, for any meaning with all related meanings (if any) there is at least one language, in which they are mirrored by a system of metas and their relations. If this language is artificial (i.e., formal), then this correspondence is expected to be rather transparent. However, for a natural language, some grammatical constraints are likely required to develop the rules for construction of such systems. This is, in fact, a claim for a constrained natural language (see, e.g., Boyd et al., 2005; Swanson et al., 2014).

We see two major challenges in construction of such language for the said purpose: to avoid offering new, purely grammatical, part-whole relations and to avoid names that are formally perfect but bear no meaning.

The former problem resides in the fact that simple meaning appears to be not the simplest but, on the contrary, the most general and includes composite meanings as parts, while in the language simple names are parts of composite syntactic constructions (phrases or sentences).

The solution we see is that in the universum simple meanings of things (i.e., predmets) may be disclosed and substituted by equal composite meanings (event families), which explicate all part-whole relations "wrapped" in predmets and are generic for a few other composite meanings.

Here we will suggest some basic principles for constrained natural languages describing meanings in universum. In doing this, we will keep in mind only some European languages (Slavonic, Roman and German), which are more or less familiar to us, but hope our results will be adaptable to other languages too.

Let us put forth that predmets are expressed as *S-metas*, primetas, as *P-metas*, and events, as *E-metas*. *S-metas* may be simple or composite, *P-metas*, only simple, and *E*-ones, only composite.

Grammatically, simple metas (*S*- and *P*-ones) are any content word – noun, adjective, verb, adverb and so forth, possibly with preposition, and in some cases perhaps preposition per se if it bears its own meaning, as happens, e.g., in the prose of Bob Dylan ("&" in his "Tarantula") or James Joyce ("yes" said 749 times in the "Ulysses"), or even part of word like meaningful prefix, suffix or flexion. Moreover, the meaning may hide in phonetic consonance or rhymes that makes the text poetic. *P-metas*, in addition, may be a phrase built of this word and subordinate ones if the "father meaning" of the primeta is not predmet but event. However, the constrained natural language should offer rules of reduction of all this diversity to a limited kit of patterns. E.g., if the prepositions "under" or "in" are considered meaningful, there should be introduced some artificial nouns like "underness" or "inity" to denote corresponding predmets, and then adjectives, e.g., "under" or "ining", for the "daughter" primetas. At best, these reduction rules should be executable by special parsing software interacting with a human user.

E-metas are simple semi-compound sentences ("Means of transportation may carry something useful, are driven by treadles, have two wheels"), and composite *S-metas* are compound sentences ("Means of transportation may carry something useful, are driven by treadles, have two wheels, or means of transportation may carry something useful, are driven by treadles, have three wheels, or means of transportation may carry something useful, are not driven by treadles at all"). The relation of grammar of an *E-meta* sentence to the part-whole structure of a composite meaning is as follows. The subject of this sentence has no subject group of subordinate parts and relates to the predmet, *S*. The predicate in the *E*-meta represents the relation "–". This must be simple verbal predicate, preferably "to be". For instance, quoting Elliott,

"Here we go round the prickly-pear, prickly-pear, prickly-pear,

Here we go round the prickly-pear

At five o'clock in the morning":

S – "going"; "–" – "to be"; P_1 – "around the prickly-pear" ("father" predmet or, possibly, event is "prickly-pear"); P_2 – "at five o'clock in the morning" ("father" event is "five o'clock in the morning"). The *E*-meta would be: "Going is around the prickly-pear and at five o'clock in the morning". Different primetas should be separated by comma or by "and" or "or".

Composite *S*-meta is a compound sentence consisting of *E*-metas divided by a grammatical connective, e.g., "Going is around the prickly-pear and at five o'clock in the morning and/or going is around the prickly-pear and not at five o'clock in the morning".

In some languages, e.g., Russian, predicate may have additional verbal or nominal part (that is termed compound verbal or nominal predicate, correspondingly). Still, there always exists an opportunity to move this addition, perhaps at the expense of style, from the predicate proper to the predicate group. If the relation itself appears to include objects that look like names, e.g., "to be father of", perhaps this indicates that not all submeanings have been picked up when defining a meaning, and yet another submeaning ("father" in the considered example) should be added. The predicate group represents primetas intersecting with the predmet and therefore likely includes similar members. Negation is also moved to the predicate group.

Each subordinate part of the sentence, be it a word or a phrase that can be further split into subordinate subparts, are related to the *i*-th primeta and its "father event". Further research may reveal finer relations between the predicate group and primeta.

Opposite to meanings and metas, names can be synonymic. If more than one names are considered to have, perhaps under some assumption, the same meaning, they are said to be synonyms, or synonymic, or be in relationship of synonymy to each other and all relate to the same meaning and, hence, the same meta and probably multitude.

Finally, if a word or phrase used as a name appears to have sharply different meanings, e.g., a coach as a trainer and a coach as a car, it is considered to represent different meanings, be different metas ("coach 1" and "coach 2") and, if appropriate, pertain to different multitudes.

DISCUSSION

Full theoretical substantiation of the theory of multitudes including the verification for self-consistency is still ahead. Nonetheless, even at the present stage this theory seems to be a tool able to handle many issues of scientific and everyday reasoning. However, prior to announce its wide implementation, one should look at how well it may cope at least with some most obvious cases of reasoning readily emerging both in the science and every-day life.

The conceptual chart depicting the relations between meanings will be henceforth called *predmetka*. Solid circles (ovals) delineate predmets, and dashed, primetas. The relation "–" is indicated by intersection of solid and dashed circles or by inclusion of a solid into dashed one, and "→", by dashed arrows. The relation "=" exists between names, not meanings, and is not shown.

- **Multiple predmets and primetas in a context:** The internal structure of the universum represents ensembles, or systems, of primetas intersecting each with at least one predmet, which may be linked only via common primetas. Examples of such ensemble is given in Figure 1.

In Figure 1 and others, gray are predmets, and white, primetas. Let us denote them, Spirit – S_1, Asket – S_2, Creativity – S_3, Faith – S_4, Art – S_5, Science – S_6, Spiritual – P^u, Asketic – P_1, Creative – P_2. Then, S_1 is the "father" for P^u (S_1=$S_P{}^u$), S_2 – for P_1, and S_3 – for P_2. P^u is the universum for S_4 and S_5, but not for S_1, S_2, S_3 and S_6. She intersects with S_6, but their intersection is not an event, as no event family can be formed if predmet is not covered at least by one universum. However, other primetas, P_1 and P_2, do form event families and, hence, events, with S_4 and S_5. Events here are: $\{(S_4–P^u, P_1, P_2), (S_4–P^u, P_1, \sim P_2), (S_4–P^u, \sim P_1, P_2), (S_4–P^u, \sim P_1, \sim P_2)\}$ (family 1) and $\{(S_5–P^u, P_1, P_2), (S_5–P^u, P_1, \sim P_2), (S_5–P^u, \sim P_1, P_2), (S_5–P^u, \sim P_1, \sim P_2)\}$ (family 2).

- **High-order ("predicate of predicate") relations:** Figure 2 a shows in a simplified way how primeta may be formed by "father" event (not predmet) in an ensemble shown in the previous figure. Colors mean the same as in Figure 1. It is assumed that the same universum P^u includes the same meanings of things (predmets and events) as in Figure 1, and primeta P_2 ("creative") does

Figure 1. An example of ensemble, or system, of events. See comments in the text.

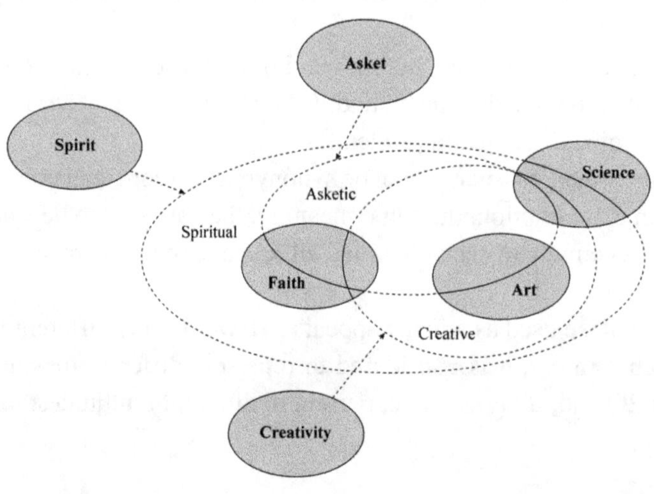

not apply. Then, the event "Faith is asketic" is short formulation in natural language for *(S_4–P^u, P_1, ~P_2)*, and it forms primeta "Asketically faithful", which, intersecting with another predmet, "Monk" (within the same universum), forms events "Monk is spiritual(ly-minded) and asketically faithful" and "Monk is spiritual(ly-minded) yet not asketically faithful". The shortened version of the former is shown in Figure 2a as "Monk is asketically faithful". Predmet or event not necessarily lead to a new primeta, but every primeta must be produced by some predmet or event.

Figure 2b illustrates the same for a somewhat more complicated case of "three in one" ("Eagle flies high in the sky"), and Figure 2c gives an idea of a complex universum ("to be skilled in cooking" for the predmet "women" – assuming that it is totally impossible that woman is not skilled in cooking).

- **Exhausting the universum:** In some cases, universum may be completely exhausted by predmets. This might appear typical for descriptions of fundamentals of a field of knowledge or even fundamentals of the entire human reasoning. For instance, the universum, which may be entitled "Involvement", includes two predmets: object and subject. Object is what is actively involved, and subject, passively (Figure 3).

Time*;* is not always accounted for in the knowledge representation. The understanding of time may be quantitative (continuous), from radioactive decay to cycles in the Universe, as commonly accepted in the natural sciences, or qualitative (discrete), which implies that time changes when something happens in the world, and stops otherwise. In the former understanding, the time flows, in the latter, it jumps and halts. Both concepts of time are fairly well pronounced in geology, the former as radiogenic, and

Figure 2. a, b, c Generation of complex primetas by events. See comments in the text.

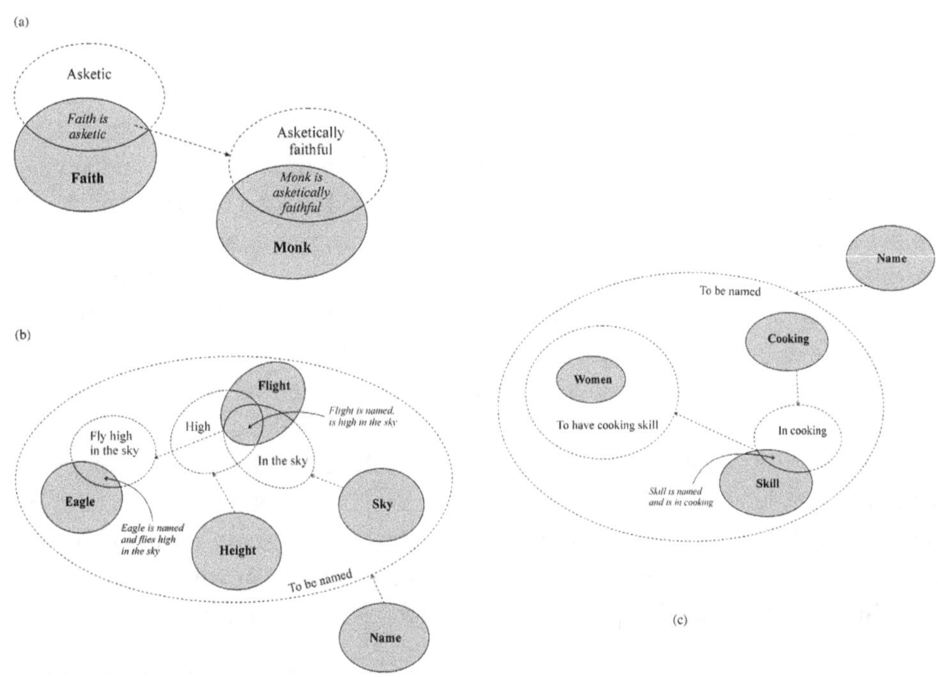

Figure 3. Exhausting of universum ("Do") by predmets ("Object" and "Subject").

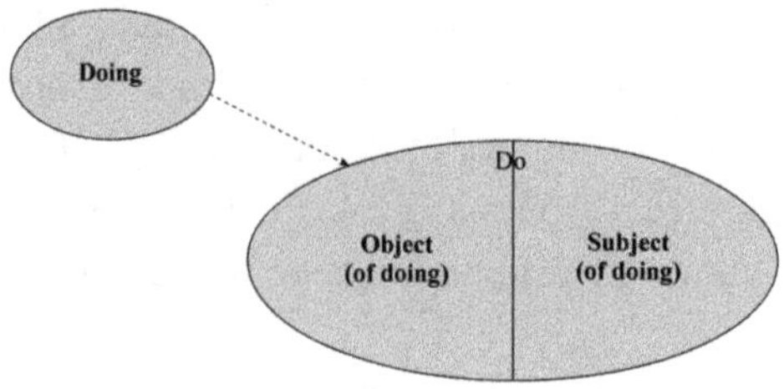

Figure 4. Semi-quantitative representation of time in predmetka

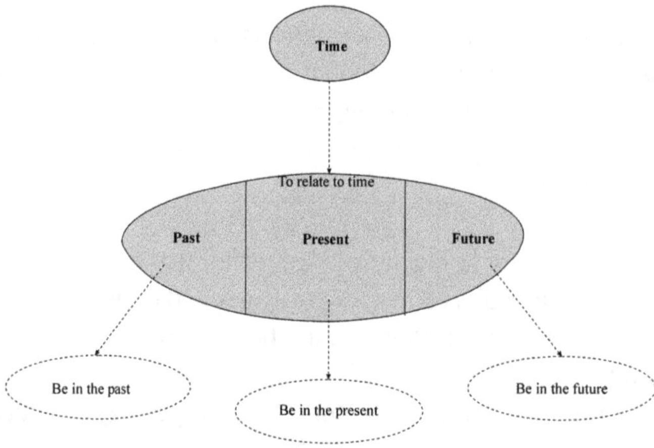

the latter, as biostratigraphic, time. For the purposes of knowledge representation, the main concern is qualitative ("jumping") time. It is commonly present in the event bush but expressed only implicitly, by the very fact of presence of connectives (directed edges) between the events (nodes).

Still, if needed, time may be explicitly embedded in the formulation of meanings in predmetka. This can be done by an exhausted universum considered directly above, as shown in Figure 4.

Binary relations; are an immanent part of virtually any field of knowledge, and any language pretending to express it must cope with them. An example of encoding of a binary relation ("to be brother of") is given in Figure 5. It can be used as a pattern for any binary relation.

Mind that binary relation is expressed as TWO events, each having its own predmet and an opportunity of incorporating different primetas – e.g., "Ivan is named, is a brother of Peter and boxer" and "Peter is named, is a brother of Ivan and the father of Susanna". Mind that the initial predmet is not "brother" but "brotherhood" per se, because considered are not things but their meanings. Also, it is noteworthy that some primetas may allow "slots", or "blanks", denoted as "something" or "somebody", and perhaps these are indicative of relations (rather than properties) of a thing denoted as predmet. For instance, "to enjoy brotherhood" may be easily reformulated as "to enjoy brotherhood with somebody/something".

Likewise, **n-ary relations** are expressed as several independent events, each with its own predmet. Considering a brotherhood of n men, one should plot the primeta "To have brother X" crossing $n-1$ predmets denoting other brothers than X. However, with n big enough this may appear tricky to plot.

Figure 5. Representation of binary relation "Ivan is the brother of Peter" in predmetka

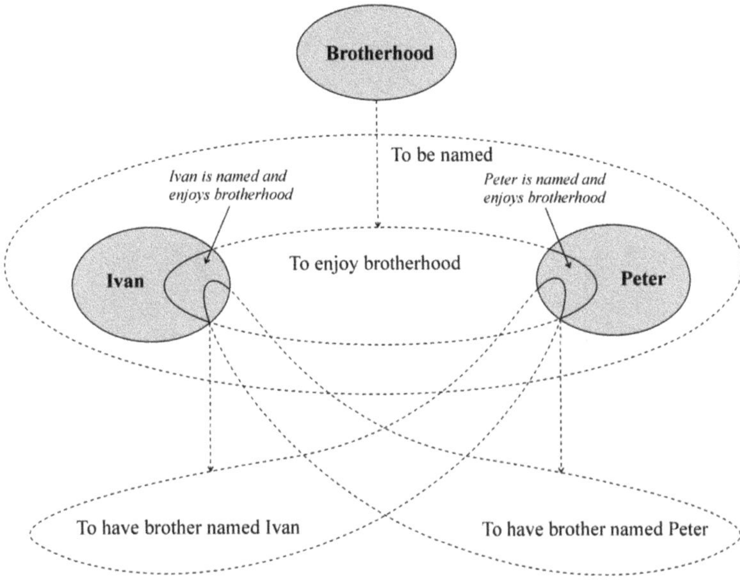

Directed relations are also present in great many fields of knowledge. They are characterized by non-symmetric roles of objects, e.g., if Ivan is the father of Peter, Peter is not father of Ivan. Figure 6 presents a more malignant example, which nonetheless will be used further. Directed binary relations are also expressed in two events.

Omnidirectional relations are a particular case of directed ones, which is often not discerned from non-directed *n*-ary relations. This type of relations exemplified by bidirectional relation "John and Bob fight" is shown in Figure 7. We stress herewith that "to fight" means, in fact, "to beat and be beaten by the same persons" (or roughly the same, if there are many actors). Compare the bidirectional relation in Figure 7 with a binary direction in Figure 5.

In general, an omnidirectional relation (e.g., "*n* fools fight with each other") would require a pair of primetas – e.g., "To beat up/be beaten reciprocally", intersecting with *n* predmets denoting particular fools, with all the concomitant changes to the plot including *4*(n-1)* events for each predmet "Fool *X*".

Auto-action is shown at general conceptual level in Figure 8.

Importantly, looking at this figure, if to abstract from auto-action, one may suppose that the predmet "Action" itself gives primeta "To act", which would, naturally, intersect with the predmet "Subject" giving an event family "Subject is named and acts/does not act". Such predmetka may be actually present in some knowledge bases. However, when merging these bases, one should carefully reexamine all such events and prefer the most detailed one, leaving the names of more general events (e.g., "Subject is named and acts/does not act") as bare expressions of language having no meaning from now on. The general primeta (just "To act") makes no practical sense anymore and should not be plotted, being replaced by more specified ones, "To act on oneself" and "To act on something else".

Representation of part-whole relations between things; The approach of the theory of multitudes is shown in Figure 9. Let us express the knowledge that leg is a part of a body (Figure 9a, b), and toe is a part of the leg (Figure 9b).

Figure 6. Representation of a directed relation "John beats Bob" in predmetka

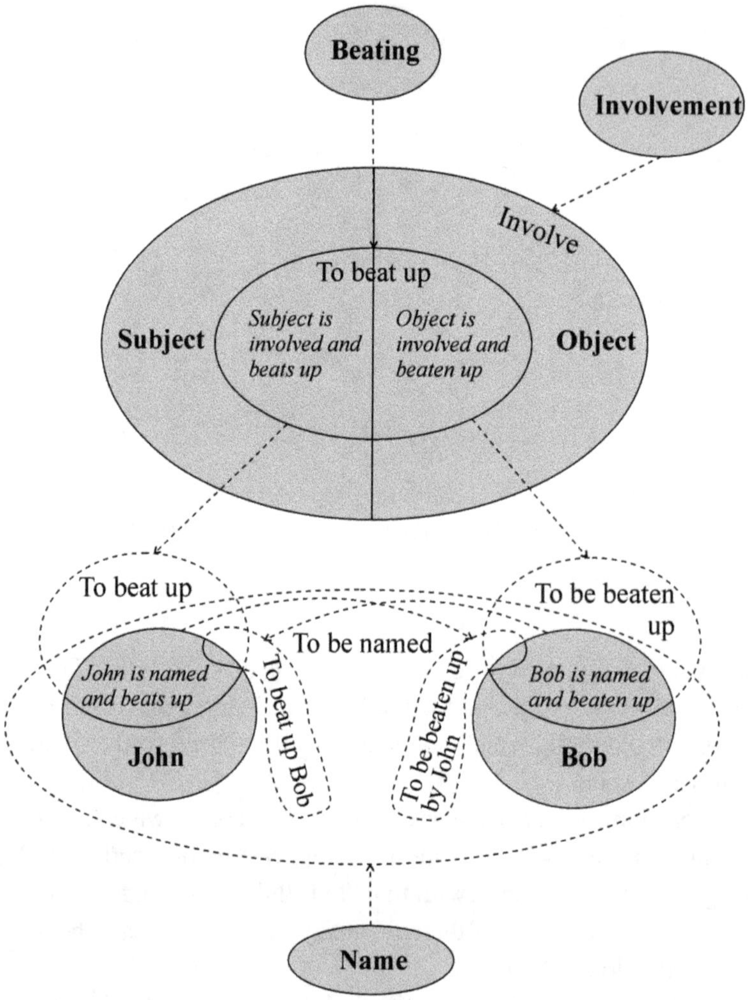

A few observations can be made from this figure. First, it is clearly seen that the part-whole relation here is not transitive. Though Figure 9a is somewhat "nested" in Figure 9b, a notion that toe is a part of the body (that should be expressed as "To be part-body") does not follow from the facts that it is a part of the leg ("is part-leg") and leg is a part of the body ("is part-body"). The same time, primeta "To be whole-body" intersects with the predmet that denotes the creature possessing this body. Otherwise, if it is not whole-body, it does not live in the material world anymore. Likewise, in the case described in Figure 9b, the predmet that intersects with primeta "To be whole-leg" is exactly "Body", which otherwise is cripple.

Also, this relation is good not only for things but for meanings either. Let us replace "Body", "Leg" and "Toe" with "Predmet", "Expression linking predmet to primetas" and "Event", correspondingly. Then the primeta "To be wholepredmet" would intersect with the universa this predmet belongs to (thus admitting that these universa could well exist without this predmet), primeta "To be part-predmet" would be a universum totally exhausted by the predmets "Expression linking predmet to primetas", "To be

Figure 7. Representation of omnidirectional relation "John fights with Bob" in predmetka

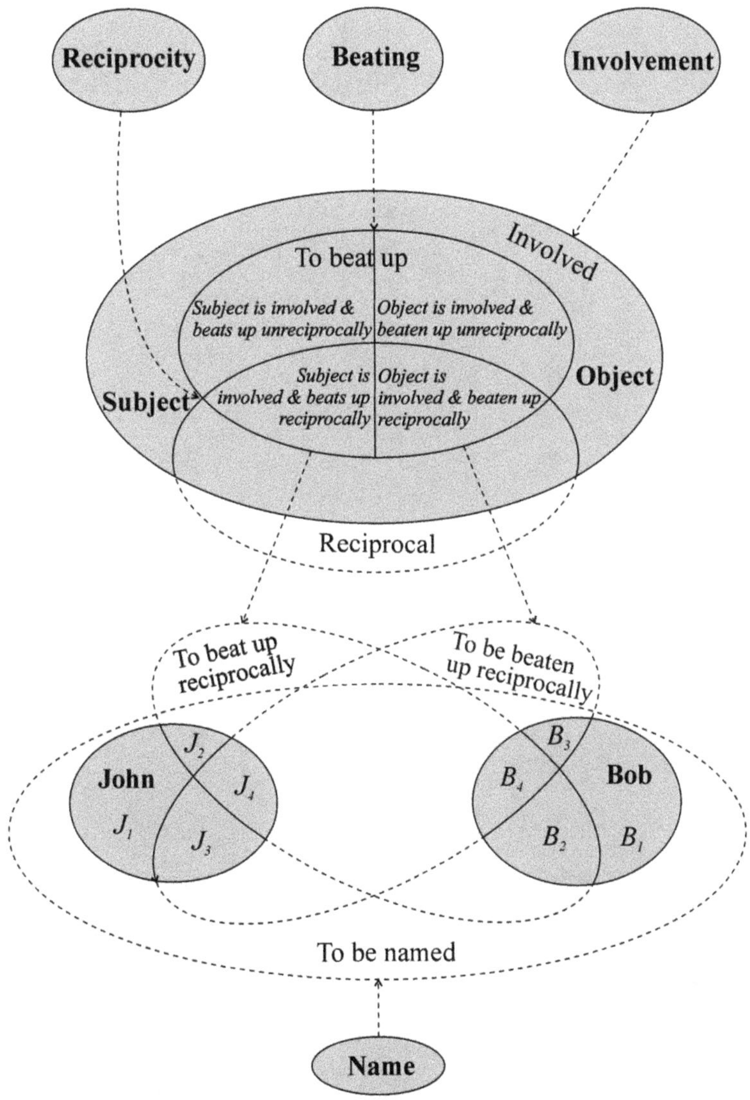

whole-expression..." would intersect with the "Predmet", and "To be part-expression", with "Event". Thus the proposed theory appears at least partly self-descriptive.

Representation of modeling or vision is illustrated by Figure 10. We do not specify what particular theory, philosophical or applied, or what vision, from observation to imagination and from bare eye to a kit of dogmas, is involved. In any case, there would be predmets (*A*, *B*, *C*) that are partly interpreted/seen and may be predmets (*D* and *E*) that are entirely interpreted or seen. The latter are likely the objects of the theory itself or something entirely imagined by a visionary. As for the former, their events "Predmet... is named and interpreted by theory .../seen" are taken for the entire predmets and serve as *models/images* of these (or "symbols" in the initial Greek sense).

Lack of Knowledge: Let us try to formalize the following text. "In our container are three packages. In the first one there is, beside other things, a bottle of vodka. The second one contains only a kilogram

Figure 8. Representation of auto-action in predmetka

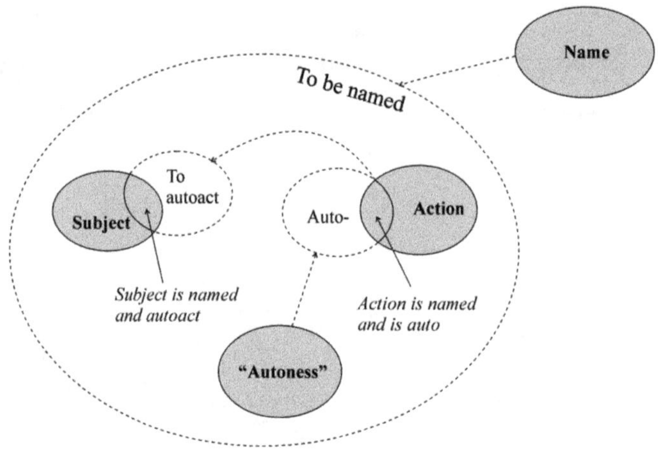

Figure 9. Representation of part-whole relation between things in redmetka

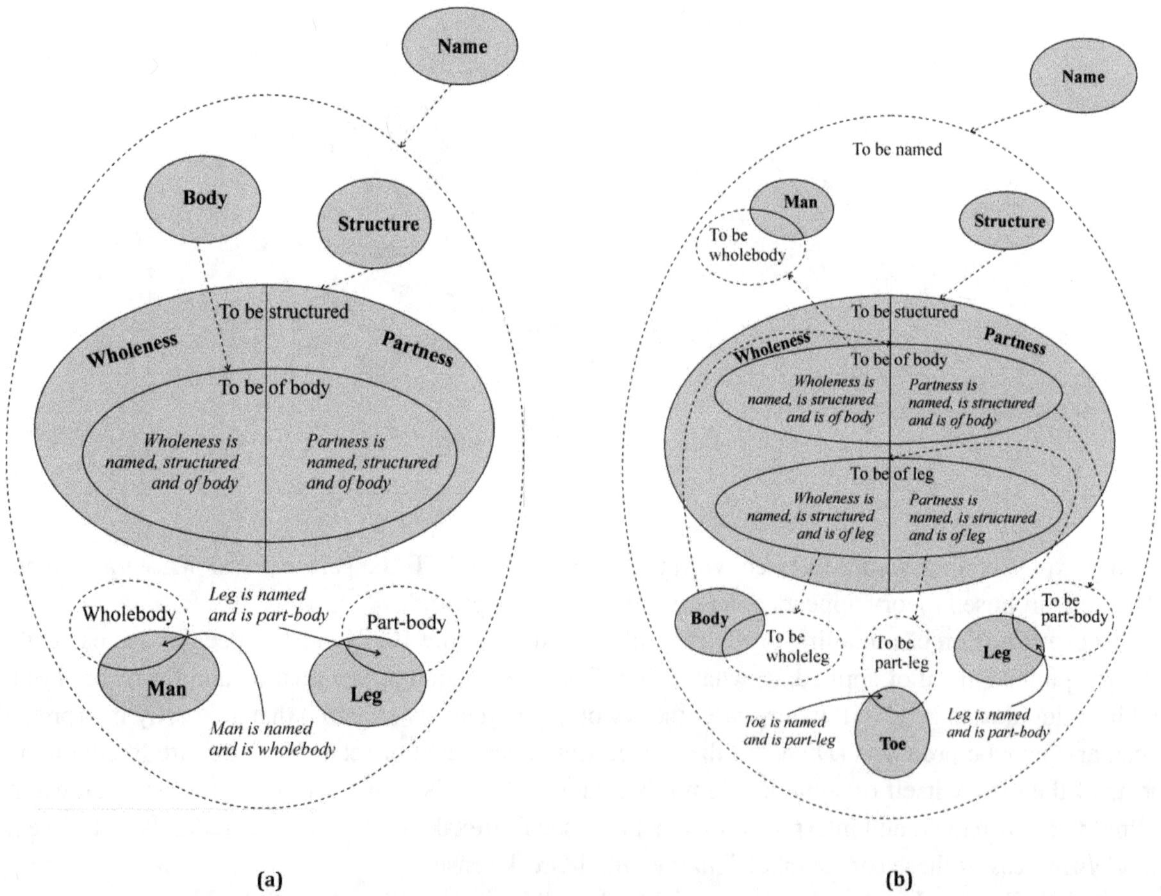

Figure 10. Representation of observation or modeling in predmetka

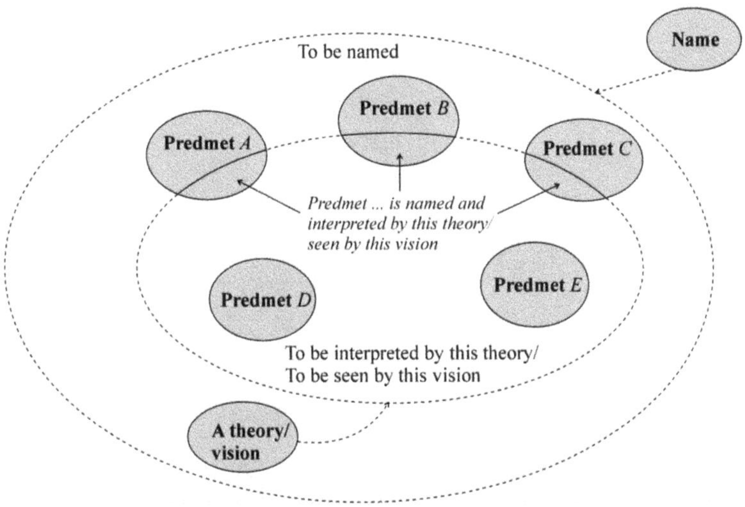

of chocolate. The last one is not empty, but I have no idea what is inside." Unlike the previous examples, it has uncertainty expressed either explicitly ("beside other things"; "I have no idea what is inside") or implicitly, in numbering of packages. Assigning numbers, one ignores properties, therefore numbering is an expression of uncertainty.

The formalization is shown in Figure 11.

The uncertainty of the latter type is expressed via "blank", or unknown, predmets. The "unknown" here is not a name but bare indication of a name, which must exist. As their extent is not known, such predmets are contoured by dotted line. The former type uncertainty is, as is always done by human thinking, expressed in quantities (we avoid the word "number" not to confuse with the number meaning order that is quite a certain named predmet in the predmetka in Figure 11. Quantities are used both to indicate weight (of a chocolate) and number (of a package). Predmets indicating quantities ("1", "2", "3") lie outside of the "To be named" primeta.

The universum "To be named", as well as some other primetas ("To be structured", "To be involved"), appear in many, or even most, of the modeled situations. One may expect that they may, from one side, be used as universal tools of building a context in a subject area and, from the other, have deeper, perhaps philosophical, sense, and demark some vision of the world that appears to be readily expressible by the theory of multitudes.

Within this hypothetic framework, predmetkas are easily built one on/under another and extended "horizontally". Of course, innumerable hindrances may occur herewith, which can hardly be foreseen now. Among those more or less expected, there may appear valuable in some contexts the distinction between the singular and plural. For instance, children may have other properties than child, crowd, other than man, colony of cells, other than individual cell, hair, other than singular hair, and so forth. Perhaps a solution can be another universal predmet "Plurality" producing primeta "To be plural", whose intersections with corresponding predmets or predmet events gives the required meanings as events: "Child is ... (*universa*), being plural" (i.e., children), "Man is ... (*universa*), being plural" (i.e., crowd), "Cell is ... (*universa*), being plural" (i.e., colony), or "Hair is ... (*universa*), being plural" (i.e., hair on the head). Alternatively, to express such plurality as, say, nation, one may put "German" as universum

Figure 11. Representation of uncertainty in predmetka

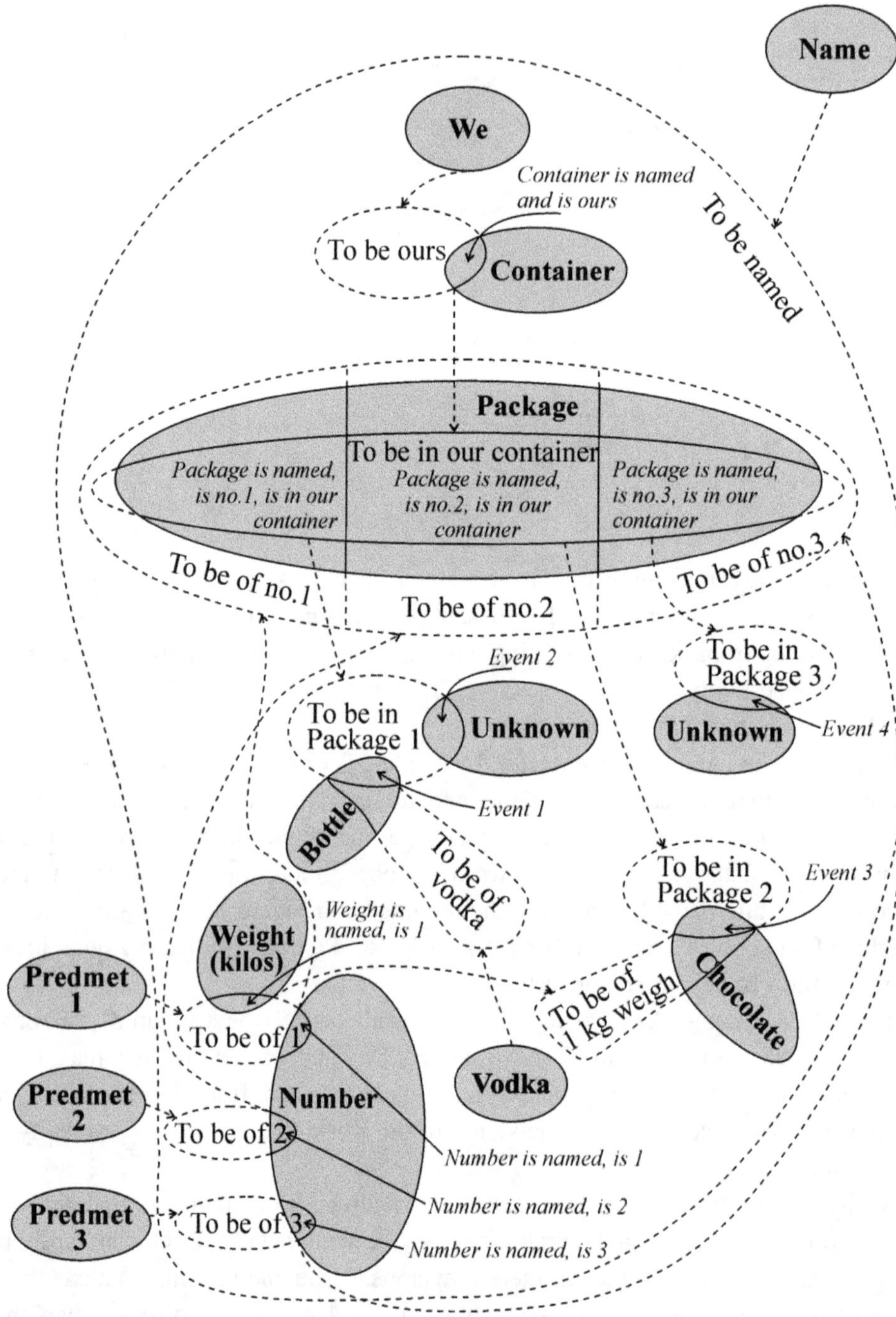

(originating from, say, some "Germanity") and including any individual Herr *X* and Frau *Y*. Otherwise, plurality may be just a matter of fill (namely, fill by atoms), while the meaning is singular – e.g., event "Man is named and is German" would have dozens of millions of atoms but is singular itself.

Figure 12. Relations between different types of fill in predmetka

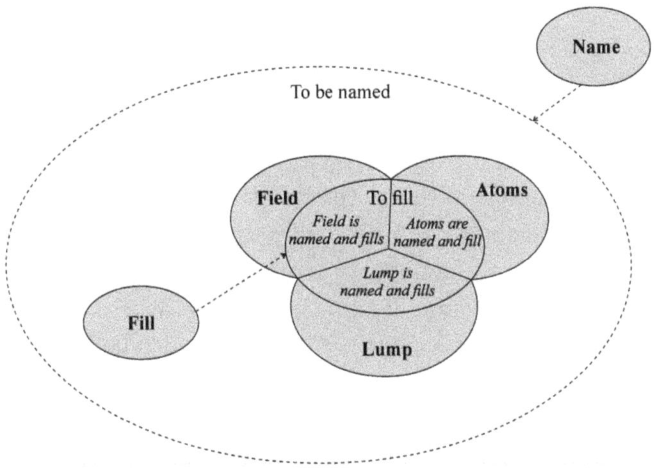

Yet another aspect of plurality can also be covered by fill, though, likely, by the lump type of fill – plurality of small and limited number of parts, like pants, hands, poles of the Earth and so forth. The very concept of part implies plurality, limited number and not necessarily discernible character of things that together belong to, or even form, the whole. Still, if semantic difference can be found, the matter is lump fill, if not, atomic.

In Figure 11, along with the named predmets, there occur predmets that represent bare numbers, which form primetas that intersect either with the predmets "Number" or "Weight" belonging to the universum "To be named", thus producing events meaning order numbers or quantities (or, broadly speaking, numbers). This is a way to express numerical information; however, the properties of predmets that represent numbers should be analyzed further.

When passing to atomic fill, events start to resemble logical expressions with existential quantifier, and predmets in a universum, those with universal quantifier. Still, it is debatable whether the atoms of a multitude may be thought as instances of individual variables of the predicate logic.

In general, the relation of predmet "Fill" to different types of fill is shown in Figure 12, and the relation between type of fill and particular filled meanings (multitudes) fits the pattern presented in Figure 6, with the type of fill acting as subject (filling) and multitude, object (being filled).

An unresolved problem is presence of slots "something"/"somebody" in primetas. They indirectly point to n-ary relations and such primetas likely can all be formalized in the way shown in Figure 6 and 7.

The patterns given in these figures, along with the numerical predmets and primetas and concept of types of fill may hopefully open an opportunity to encode at least some quantitative and qualitative mathematical knowledge. We suppose that mathematical field theory best fits the concept of field fill, set theory, atomic fill, and category theory, lump fill of multitudes. Still, this should be considered a task for distant future. However, the thing to be done next is extension of the reported theory to the case of dependent primetas.

Nevertheless, even in the present state this theory allows some qualitative and quantitative inference and formalization of many subject areas of human knowledge that will be demonstrated in the following chapters.

CONCLUSION

The theory of multitudes proposed above lays down the general mathematical principles for operating with meanings and their fill. Multitudes are defined by meaning and one of three possible types of fill: field, atoms, or lump.

Meanings refer either to things or to properties and, depending on this, may or may not enter formal relations of different types. Meanings of things (predmets or events) may be disclosed via their parts (submeanings). Together the meanings of things and properties (primetas) may form ensembles, or systems, that might be used to model the fields of knowledge along with, or perhaps even instead of, existing tools of engineering/representation of static knowledge such as ontologies, conceptual graphs and the like. With further elaboration, the theory of multitudes also looks promising for formalization of dynamic knowledge engineering. Still, these systems are feasible only within a specified universum, and this allows to overcome the old problem coming from the logical side – reverse relation between the intentional and the extensional components of a concept.

In natural languages, meanings are expressed as names that may be grammatically diverse but reducible, though at the expense of expressive power, to a limited and readily formalized kit of types of word, phrase, and sentence, which would form then a kind of constrained natural language, in which all syntactic relations will be reduced to a limited number of specific types of relations that bijectively map the part-whole relations between and within the corresponding metas. Names that appear to have no meaning are characterized by internal contradiction and denote absolutely empty multitudes, or absurd. By convention, absurd may not be exemplified and, hence, the multitude that has absurd instead of meaning may not have atoms too.

The proposed theory is seen to give common ground for a vast number of independent knowledge engineering and representation approaches and, with further development, maybe serve as a possible ground for some mathematical theories.

ACKNOWLEDGMENT

The author is deeply obliged to Paolo Diviacco and Roberto Carniel for their encouragement and moral support for many years, to Uwe Wolter whose encyclopedic mathematical and philosophical background, acute vision and exceptional tolerance helped organize the author's mind and make it produce something hopefully classifiable as a formalized theory, and to Ivelina Nikolova who helped to better relate this theory to natural language and the common sense.

REFERENCES

Akimova-Shterkhun, V. L., & Pshenichny, C. A. (2007). Quick Extraction of Multi-Format Relevant Volcanological Information by Means of Semantic and Smysl Networks. *Abstracts, IUGG General Assembly XXIV2007*. Retrieved from http://www.iugg2007perugia.it/webbook/

Blizard, W. (1989). Multiset Theory. *Notre Dame Journal of Formal Logic*, *30*(1), 36–66. doi:10.1305/ndjfl/1093634995

Boyd, S., Zowghi, D., & Farroukh, A. (2005). Measuring the expressiveness of a constrained natural language: an empirical study. *13th IEEE International Conference on Requirements Engineering*, 339-349. doi:10.1109/RE.2005.39

Chen, P. P.-S. (1976). The Entity-Relationship Model: Toward a Unified View of Data. *ACM Transactions on Database Systems*, *1*(1), 9–36. doi:10.1145/320434.320440

Davey, B. A., & Priestley, H. A. (2002). *Introduction to Lattices and Order* (2nd ed.). Cambridge University Press. doi:10.1017/CBO9780511809088

Elsukov, M.Yu., Prozorova, V.B., Pshenichny, C.A., & Cherkashin, K.V. (2015). Smysl i prednaznachenie Novorossii (Smysl and Predestination of Novorossia). *MIA Novorossia*. Retrieved from mianews.ru/ru/2015/12/23/smysl-i-prednaznachenie-novorossii

Gruber, T. R. (1993). Toward Principles for the Design of Ontologies Used for Knowledge Sharing. *International Journal of Human-Computer Studies*, *43*(5-6), 907–928. doi:10.1006/ijhc.1995.1081

Kechris, A. S. (1995). *Classical Descriptive Set Theory*. Springer. doi:10.1007/978-1-4612-4190-4

Keet, C. M. (2006). *Introduction to part-whole relations: mereology, conceptual modelling and mathematical aspects*. KRDB Research Centre Technical Report KRDB06-3, Faculty of Computer Science, Free University of Bozen-Bolzano, Italy.

Simons, P. (2005). Against Set Theory. *Proceedings of the 2004 Wittgenstein Symposium*, 143–152.

Smith, B. (2005). Against Fantology. *Proceedings of the 2004 Wittgenstein Symposium*, 153–170.

Sowa, J. (2006). *Semantic Networks*. Retrieved from http://www.jfsowa.com/pubs/semnet.htm

Speaks, J. (2016). Theories of Meaning. In *The Stanford Encyclopedia of Philosophy*. Retrieved from http://plato.stanford.edu/archives/spr2016/entries/meaning

Swanson, B., Yamangil, E., & Charniak, E. (2014). Natural Language Generation and Vocabulary Constraints. *Proceedings of the Ninth Workshop on Innovative Use of NLP for Building Educational Applications*, 124-133. doi:10.3115/v1/W14-1815

Zadeh, L. (1965). Fuzzy sets. *Information and Control*, *8*(3), 338–353. doi:10.1016/S0019-9958(65)90241-X

Chapter 2
Qualitative and Quantitative Formalisms for Knowledge Representation in the Theory of Multitudes

Cyril Pshenichny
ITMO University, Russia

ABSTRACT

The theory of multitudes pretends to be an alternative to virtually all existing versions of the set theory and claims to better handle the knowledge about changing and evolving world. Then, by analogy, one may expect an original logical system based on the theory of multitudes, and within this logic, an authentic calculus. This chapter presents such calculus. Moreover, a new mathematical methodology can be developed on top of it, which together with the underlying logic, should clearly separate qualitative and quantitative, static and dynamic concerns and offer a formal method to proceed from representation of expert knowledge to modeling the world this knowledge is about.

INTRODUCTION

Knowledge engineering implies not only extraction of knowledge from experts and putting into a computer-tractable form, but also making it complete and self-consistent. The latter task necessarily requires inference, which seems to be the only tool to fulfill it. For the traditional (static) knowledge engineering, this tool has been logical calculus, either in its strong (i.e., sensu Gentzen, 1934) or weak (simple modus ponens or modus tollens) form. However, the classical logic is indivisible from the classical set theory, which, as was shown above (Pshenichny, 2017, see Chapter 1 of this book), is crucially different from the theory of multitudes. Therefore, another formalism is needed to provide an inference based on the theory of multitudes. The purpose of the present work is to suggest such a formalism.

To achieve this purpose, we will:

DOI: 10.4018/978-1-5225-5261-1.ch002

- Consider an opportunity of creation of a calculus within the theory of multitudes;
- If a calculus can be created, find out whether it may be efficient at least for some cases;
- Discuss what should be done to extend these cases to known and possible methods of dynamic knowledge engineering.

THE NOVOROSSIAN LOGIC

Let us denote the Novorossian logic as a formal system, which includes

1. The language that has:
   a. Meanings: *S*, *P* and *E* and relations: "~", "–", "→" and "=".
   b. Technical signs: left bracket "*(*", right bracket "*)*", comma "," order subscript indices, $_1$, $_2$, ..., $_i$, $_n$, and dots, "...";
   c. Metasymbols: possible negation, "*(~)*", lists of primetas (Greek capitals *A*, *B*, *Γ*, *Δ*, etc.), universal meaning of thing, *V*, and exclusion sign *EXCL*;
   d. The truth values: truth (*T*) and its opposite false (*F*);
   e. The turnstile: ⇒ or horizontal line;
   f. Indicators of disclosure, *, of reference, ', of being universum, ^{u};
2. The formulae: complete and incomplete;
3. The calculus of meanings and
4. Nothing else.

Meanings and relations have been defined by Pshenichny (2017; see previous chapter). Technical signs are not defined.

Metasymbols include possible negation ("with or without negation"), "*(~)*", lists of primetas (Greek capitals *A*, *B*, *Γ*, *Δ*, etc.), universal meaning of thing, *V*, and an exclusion sign *EXCL*. Possible negation and exclusion sign have been introduced in the quoted chapter. The list of primetas Δ_i can be inductively defined as

$\Delta_1=(\sim)P_1$ if $i=1$;

$\Delta_2=(\sim)P_1, (\sim)P_2$ if $i=2$;

$\Delta_i=(\sim)P_1, (\sim)P_2, ..., (\sim)P_i$ if $i>2$.

List of universa is:

$\Gamma_1= P^u_1$ if $i=1$;

$\Gamma_2= P^u_1, P^u_2$ if $i=2$;

$\Gamma_i= P^u_1, P^u_2, ..., P^u_i$ if $i>2$.

Universal meaning of thing, V (standing for "vesch", the Russian for "thing") is:

$V=S$ if there is expression $V–\Gamma^u_i, \Delta_j$ or $V\to P$ and

$V=E$ otherwise.

Truth values are defined quite simply: expressions in the language of Novorossian logic that mean absurd are always false, others, always true (because they always have a meaning).

Turnstile is used in the inference and is considered below.

Indicators refer to meanings and may say that meaning is disclosed, or is that being given now, or is universum (for primetas only).

Meaning is disclosed if it is:

1. Predmet;
2. Primeta that (i) has a father and (ii) is a mother, or wife, or both;
3. Event that (i) has a father and mother(s) and, excluding universum (ii) has brothers (note: brothers are those events having common father; no brotherhood by mother is defined).

Full (complete) formula of the Novorossian logic, or just Novorossian formula, is one of the following expressions in the language of the Novorossian logic:

$S–P^u_1, \ldots, P^u_m$, or $S–\Gamma^u_m$, where $P^u_1, \ldots, P^u_m$ are universa (finite in number) and Γ^u, list of universum primetas;

$S–P^u_1, \ldots, P^u_m, (\sim)P_{m+1}, \ldots, (\sim)P_n$ $EXCL$ $E_i=\varnothing$, or $S–\Gamma_m, \Delta_n$ $EXCL$ $E_i=\varnothing$, where $P^u_1, \ldots, P^u_m$ are universa (finite in number), and $(\sim)P_{m+1}, \ldots, (\sim)P_n$ are non-universum primetas, also finite in number; notation "$(\sim)$" means, "with or without negation"; Γ_m is list of universa, Δ_n, list of other primetas, and $EXCL$ $E_i=\varnothing$ means exclusion of absurd expressions.

$S\to P$, where S is father predmet and P, daughter primeta;

$E\to P$, where E is father event (possibly universum event) and P, daughter primeta;

$E_i=S–P^u_1, \ldots, P^u_m, (\sim)P_{m+1}, \ldots, (\sim)P_n$, or $E_i=\Gamma_m, \Delta_n$ $EXCL$ $E_i=\varnothing$, with one of the $(\sim)P_k$, $m+1\le k\le n$, fixed as negation or assertion.

Incomplete Novorossian formula is a formula, which has a blank in place of at least one of the meanings. Blank is denoted as three dots.

Incomplete formula is one of the following:

$S–\ldots;$

$S–\ldots, P, \ldots;$

$S–P^u,\ldots;$

$\ldots–P,\ldots;$

$\ldots–P^u,\ldots;$

$\ldots– \ldots, P,\ldots;$

$\ldots– \ldots, \sim P;$

$S\to\ldots;$

$\ldots\to P;$

$\ldots\to P^u.$

In the Novorossian logic formulae, *S* may denote predmet event but in this case it cannot be automatically substituted by event of an upper-level predmet because, given no other information, one may not be certain about which of the universa of this predmet event are also universa for the upper-level predmet, and which are just intersecting with it.

THE CALCULUS OF MEANINGS

Traditional definition of logical inference, dating back to Gentzen (1934), is

$\Gamma\to\Delta,$

where Γ is a sequent (list) of formulae of logic $A_1, A_2, \ldots, A_m$, Δ is a sequent $B_1, B_2, \ldots, B_n$, and $\to$ is the turnstile denoting the logical inference, or entailment: $A_1, A_2, \ldots, A_m \to B_1, B_2, \ldots, B_n$. In terms of propositional logic, this is understandable as $(A_1\&A_2\&\ldots\&A_m)\supset(B_1\vee B_2\vee\ldots\vee B_n)$, where & is conjunction, $\vee$ is disjunction, $\supset$ is implication in terms of the classical logic. The opportunity to put "→", i.e., to infer one sequent from another, is ensured by the rules of inference suggested by Gentzen (1934). These rules may be structural, i.e., changing the position of formulae in a sequent, or logical, introducing a new formula.

We will use this general understanding of inference.

In the Novorossian logic, *sequent* is a sequence of Novorossian formulae. Sequent is not ordered – e.g., if Sequent One is "Formula 1, Formula 2, Formula 3", and Sequent Two, "Formula 2, Formula 3, Formula 1", these sequents are the same.

Raw sequent is a sequence, in which incomplete formulae are present.

Mature sequent is a sequence, in which all formulae are full, no metasymbols are present except the exclusion sign, and all meanings are disclosed.

Inference in the Novorossian logic is replacement of incomplete formulae by complete ones. Each step of inference is singular application (enforcement) of one *rule of inference*. At each step, the turnstile will be denoted as ⇒ or the horizontal line (while → will have another meaning, the "father-daughter" relation between the multitudes). Formulae, complete or incomplete, left of the turnstile or above the line, are termed premises, and those complete ones right of the turnstile or below the line, conclusions. Like in the logical inference, if premises are true, the conclusions are true either.

Rules of inference may be structural (extensional) and semantic (intentional).

Structural rules of inference include:

ST1 (equality $S'=S_i$): $\dfrac{<\text{ FORMULA WITH } S' >}{<\text{ FORMULA WITH } S_i >}$

ST2 (equality $P'=P_i$): $\dfrac{<\text{ FORMULA WITH } P' >}{<\text{ FORMULA WITH } P_i >}$

ST3: $\dfrac{<\text{ FULL FORMULA } >,\ldots}{<\text{ FULL FORMULA } >}$

In the rules ST1-2 formula may be full or incomplete.

Semantic rules of inference include:

SM1: $\dfrac{S-\ldots}{S-P^u,\ldots}$

SM2a: $\dfrac{S-\ldots,\ P_1,\ \ldots}{S-P^u,\ P_1,\ldots;\ S-P^u,\ \sim P_1,\ldots}$

SM2b: $\dfrac{S-\ldots,\ \sim P_1,\ \ldots}{S-P^u,\ P_1,\ldots;\ S-P^u,\ \sim P_1,\ldots}$

SM3a: $\dfrac{S-P^u,\ldots;\ S-\ldots,\ P_1,\ \ldots}{S-P^u,\ P_1,\ldots;\ S-P^u,\ \sim P_1,\ldots}$

SM3b: $\dfrac{S-P^u,\ldots;\ S-\ldots,\ \sim P_1,\ \ldots}{S-P^u,\ P_1,\ \ldots;\ S-P^u,\ \sim P_1,\ \ldots}$

SM4: $\dfrac{\ldots-P^u}{S-P^u}$

SM5a: $$\frac{\ldots - \ldots,\ P_1,\ \ldots}{S - P^u,\ P_1,\ldots;\ S - P^u,\ \sim P_1,\ldots}$$

SM5b: $$\frac{\ldots - \ldots,\ \sim P_1,\ \ldots}{S - P^u,\ P_1,\ldots;\ S - P^u,\ \sim P_1,\ldots}$$

SM6: $$\frac{S \rightarrow \ldots}{S \rightarrow P'}$$

SM7a: $$\frac{\ldots \rightarrow P^u{}_i}{V \rightarrow P^u{}_i}$$

SM7b: $$\frac{\ldots \rightarrow P_i}{V \rightarrow P_i}$$

The relation axiom, $V=S$ or $V=E$, may be introduced in the process of inference.

SM8: $$\frac{S - P^u{}_1,\ldots;\ S - P^u{}_2,\ldots}{S - P^u{}_1,\ P^u{}_2,\ldots}$$

SM9a: $$\frac{S - \ldots,\ P_1,\ \ldots;\ S - \ldots,\ P_2,\ \ldots}{S - \ldots,\ P_1,\ P_2,\ldots;\ S - \ldots,\ \sim P_1,\ P_2,\ldots;\ S - \ldots,\ P_1,\ \sim P_2,\ldots;\ S - \ldots,\ \sim P_1,\ \sim P_2,\ldots}$$

SM9b: $$\frac{S - \ldots,\ \sim P_1,\ \ldots;\ S - \ldots,\ P_2,\ \ldots}{S - \ldots,\ P_1,\ P_2,\ldots;\ S - \ldots,\ \sim P_1,\ P_2,\ldots;\ S - \ldots,\ P_1,\ \sim P_2,\ldots;\ S - \ldots,\ \sim P_1,\ \sim P_2,\ldots}$$

SM9c: $$\frac{S - \ldots,\ P_1,\ \ldots;\ S - \ldots,\ \sim P_2,\ \ldots}{S - \ldots,\ P_1,\ P_2,\ldots;\ S - \ldots,\ \sim P_1,\ P_2,\ldots;\ S - \ldots,\ P_1,\ \sim P_2,\ldots;\ S - \ldots,\ \sim P_1,\ \sim P_2,\ldots}$$

SM9d: $$\frac{S - \ldots,\ \sim P_1,\ \ldots;\ S - \ldots,\ \sim P_2,\ \ldots}{S - \ldots,\ P_1,\ P_2,\ldots;\ S - \ldots,\ \sim P_1,\ P_2,\ldots;\ S - \ldots,\ P_1,\ \sim P_2,\ldots;\ S - \ldots,\ \sim P_1,\ \sim P_2,\ldots}$$

SM10 (exclusion rule): $$\frac{E;\ E_i = \varnothing}{E\ EXCL\ E_i}$$

The exclusion axiom, $E_i=\varnothing$, may be introduced in the process of inference.

Proof is a succession of steps of inference leading to a mature sequent from a raw or from another mature one. The latter case is trivial, as the rules applicable to mature sequents are renamings only, ST1 and ST2. In the former case, the inference proceeds at least until no incomplete formulae are present in the sequent.

When all formulae are complete, meanings are checked for disclosure. If a meaning is disclosed in the sequent, it is marked by the corresponding indicator, *.

Each predmet S_i is marked * anyway once all formulae in the sequent are full. Then, for each primeta P_j the sequent is scanned for a formula $S \rightarrow P_j$ or $E \rightarrow P_j$. If it is found, P_j is marked as disclosed: P_j *, and if not, an incomplete formula $\ldots \rightarrow P_j$ is added to the sequent, which thus becomes raw again. Then, each predmet that includes events is verified for completeness of the event family by its definition with account of exclusion axioms. If all options in the definition $S–\Gamma_m, \Delta_n$ $EXCL$ $E_l = \varnothing$ with and without negation of non-universum primetas are defined as events E_1, E_2, …, E_k, they are considered disclosed and marked *. If some events are missed, they are introduced and marked *. Then the inference is over.

All meanings of a mature sequent make up a context as defined by Pshenichny (2017; see chapter 1 of this book). The relations between meanings in a context are visualized by the conceptual chart called predmetka (Pshenichny 2017; see chapter 1 of this book).

The context represented by a mature sequent has its internal structure. First, it may have predmets only (Figure 1a). Predmets that belong to no universa may produce daughter primetas, which neither subordinate nor intersect with any predmet (i.e., are neither mothers nor wives – Figure 1b). Then, the context may include predmets that belong to no universa and produce daughter primetas, which intersect with (Figure 1c) and/or subordinate (Figure 1d) other predmets. All four ways may be present in one context. However, only the "marriage" (Figure 1c) and "motherhood" (Figure 1d) of primetas provide links between predmets (via primeta). Therefore, if, and only if, solely (c), or (d), or both are present in a context, the latter represents a system of meanings, in which any two meanings are directly or indirectly interconnected by formal relations.

In the first two cases (see Figure 1a, b), there may be recognized isolated structures consisting either of predmets (Figure 1a) or of father predmet and daughter primeta (Figure 1b). In the third and fourth cases (Figure 1c, d), each "building block" of the context includes predmet that may also be family of events (Figure 1c), universum ("mother") primetas, intersecting ("wife") primetas, and predmets or events that are "fathers" of universa and intersecting primetas. Importantly, all primetas may be shared by several (whatever number of) "building blocks" and therefore play a role of "cement". If only these "building blocks" are present in the context, the latter becomes thoroughly interconnected. Each "father – daughter" pair may participate in whatever number of such blocks. If all meanings are interconnected, one may "travel" from any circle to any other circle in this chart (see Figs. 1c, d).

However, in any case, every predmet in a context either (i) has daughter primeta and is not split into events, or (ii) has no daughter and is split into events, or (iii) neither, but never both. Also, the predmet may be covered by universum (universa) or not, and if yes, each of these universa may subordinate and/or intersect with other predmets.

FILLING THE MEANINGS

Properties of Multitudes

The predmetka chart shows merely the relations between meanings, regardless of whether they are multitudes (i.e., are filled). To make it represent even the static knowledge, this scheme must be "inhabited" by some kind of fill (grainy – i.e., atoms, integer – i.e., lump, or field). Inhabited context will be henceforth termed as *environment*. Therefore, context is formed by a mature sequent only; environment is inhabited context and includes multitudes and primetas; then multitudes should be fully disclosed meanings of things from a mature sequent.

Figure 1. a, b, c, d. Types of structure of context. See comments in the text.

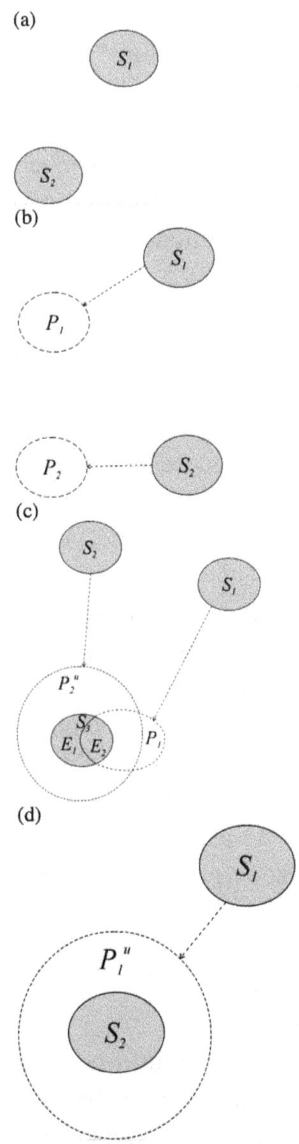

First of all, let us put forth the following axioms.

1. Any fill in environment belongs to one multitude.
2. Only predmets may have fill independently of other meanings.
3. Not a single fill may belong to two predmets (though, this may be regarded as a consequence from incompatibility of any two predmets and of the axiom 1).
4. Primetas may only have the fill that belongs to some predmet. (In other words, primetas may be filled only at intersection with a predmet or, in case of universum, in a subordinate predmet).

Putting $x \subset \boldsymbol{A}$, one implies that atom x belongs to the entire multitude $\boldsymbol{A}$, not to a part of it (e.g., such part as intersection with another multitude). Therefore, one may put $x \subset S$ but may not, $x \subset P$.

5. Primeta may ever be filled (i.e., may share fill with some predmet) only if its own father predmet is filled.
6. Predmets intersecting with primetas (i.e., forming "ordinary", non-universum events), if filled with lump, would produce events, each of them also being filled with lump.

An illustration to the last axiom may be that predmet "democracy" has integer (lump) fill, and its events "Greek democracy", "Novgorodian democracy" or "Western liberal democracy" are also lump-filled.

Thence, we define some properties of multitudes in regard of their serial parts, or fill (atomic, i.e. atoms, integer, i.e. lump, or field).

Bytie (Russian "being") is a property of a meaning to have fill independently of others (i.e., to be multitude). Meaning is said to have bytie if it is filled regardless of type of the fill.

Nebytie is the opposite property, to have no fill.

Probytie is filling a father predmet or event.

Soprobytie is simultaneous filling of a father and husband predmets. It is applicable to the cases shown in Figure 1c, d and implies probytie and *sobytie*. Sobytie is filling of event. If the predmet has wives, i.e., forms a family of events, its sobytie with wife primetas that leads to inhabitation of particular event is considered a property of event as well. Then the event is said to have bytie.

Bytie and nebytie are properties of predmets; primetas have no bytie (i.e., no fill independently of predmets) by axiom, and events, by definition. Every predmet has either just bytie (the only option for the case depicted in Figure 1a), or probytie (predmets in Figure 1b; predmets S_1 and S_2 in Figure 1c; predmet S_1 in Figure 1d) or sobytie with some primeta (predmet S_3 with primetas P^u_2 and P_1 in Figure 1c, and S_2 in Figure 1c; predmet S_2 with primeta P^u_1 in Figure 1d), or soprobytie, which implies sobytie and probytie together (Figs. 1c, d). It should be noted herewith that sobytie is impossible without probytie and therefore, if a predmet has sobytie with some primeta, it has soprobytie with some other predmets as well.

Inhabited predmetka chart gives a static picture of the world, which may be considered a distant analog of ontology sensu Gruber (1993).

Transition of Fill

Dynamic knowledge implies change of things in time, or space, or any other respect. This is comprehended as a transition of fill from one meaning to another. This approach has a lot in common with that suggested by Petri (1973) in the formalism he developed, later entitled Petri nets. An opportunity of building a much wider formalism based directly on the Novorossian logic, which would include the Petri nets as a particular case, will be discussed below.

As can be intuitively concluded from Figure 1, only soprobytie allows transitions. Then, contemplating how fill may move from one predmet to another, two types of transition can be thought of: (i) just moving of fill (atoms or another) from one multitude to another and (ii) their dying in one multitude and appearance in another. Let us call the former, *perebytie*, and the latter, *ubytie* (disappearance in one multitude) and *vozbitye* (appearance in another).

The following axiom is adopted herewith:

- *Fill may not (dis)appear within a predmet.*

In other words, ubytie and vozbytie between events of one family is not allowed. Then, only the perebytie may proceed between events of the same predmet (from E_2 to E_4 in Figure 2a), and u- and vozbytie may do between a predmet/event in a universum and the father of this universum (between E_1 and S_1 in Figure 2b) or between different predmets/events in one universum (between E_2 and E_4 in Figure 2c).

Obviously, within an environment, ubytie and vozbytie are the two sides of the coin and act jointly; however, in a part of the environment, it is possible that only one of them is seen. If to consider the whole physical world, it seems to be a part of some wider context, and a Russian proverb applies to it, "God gives, God takes". However, if the fill is measurable (i.e., in case of atoms or field), there may

Figure 2. Three options of transition in environment. See comments in the text.

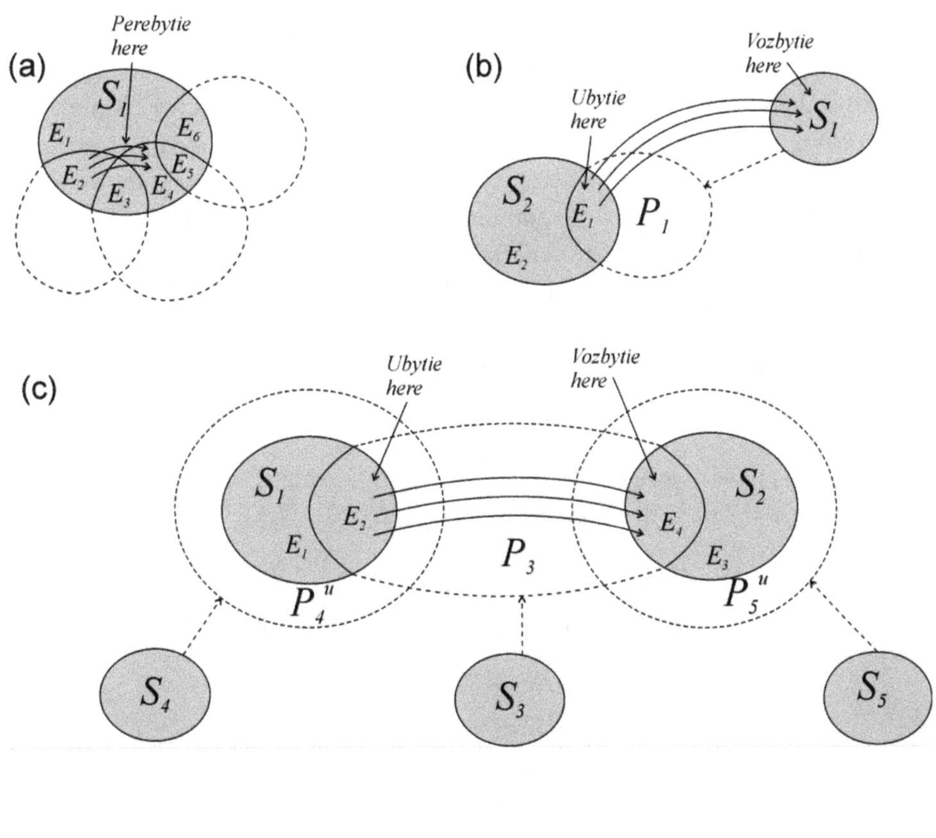

be a proportion of waxing vs. waning fill, not necessarily one to one – in contrast to perebytie, which implies only one-to-one transition.

To formally describe the transition in the theory of multitudes, a concept of *connective* was introduced by Pshenichny and Mouromtsev (2015). With some modification, they defined connective as an operator linking two lists of multitudes:

$$(V_1, V_2, \dots, V_i)\ Connective\ (V_{i+1}, V_{i+2}, \dots, V_j)$$

where $i, j=1, 2, \dots, n$, n being natural number, V is a multitude (i.e., disclosed meaning of thing), $(V_1, V_2, \dots, V_i)$ is the left, or causal list, and $(V_{i+1}, V_{i+2}, \dots, V_j)$, the right, or effect, list. Every V is either a predmet that has daughter primeta (and hence is not divided into events; it may be covered or not by universum), or predmet covered by universum and having no daughters ("singleton"), or event from any family (which may have daughter primetas).

None of these lists may be empty, both of them must be finite; within one list, multitudes may not repeat, but the same multitude may occur both on the left and on the right.

Change of fill in an environment, or just *change*, is a correctly built expression including one connective of any type and the multitudes it refers to, i.e., its left and right lists. One may say that the multitudes of the left list jointly *lead to* the multitudes of the right list.

At the first step, connectives are classified by the number of multitudes in the left and right lists. Four cases are defined as follows and denoted by capital letters of the Old Slavonic alphabet zhivot, shta, yon and izhen (Figure 3).

These letters were chosen because they well represent visually the relation between the causal and effect lists, meaning the causal list above, and effect, below in each letter: "many lead to many" (*zhivot*), "many lead to one" (*shta*), "one leads to many" (*yon*) and "one leads to one" (*izhen*). Old Slavonic (or Early Cyrillic) alphabet can be easily found on the Internet as pixel graphics (see, e.g., Kodeks webpage at the Otto-Friedriech-Universitaet Bamberg website (http://kodeks.uni-bamberg.de/aksl/Schrift/AkslAlphabet.htm), the Omniglot page on Old Slavonic alphabet (http://www.omniglot.com/writing/ocslavonic.htm) or Wikipedia entry for Early Cyrillic alphabet, http://en.wikipedia.org/wiki/Early_Cyrillic_alphabet), though pronunciation of letters may slightly differ depending on the source. The easiest way of using these symbols (followed by the authors of this chapter) is saving a graphic file and copying the symbols from there as graphic objects. Alternatively, one may just put Latin names of these letters in the formulae; exactly this option will be taken further on in this book.

The concept of connective is somewhat wider than that of transition. Transition *sensu stricto* is described by connectives without the same multitudes in the left and right lists. The same multitude may

Figure 3. Main types of connectives of the unified grammar of dynamic knowledge: zhivot (a), shta (b), yon (c) and izhen (d); from Pshenichny and Mouromtsev (2015).

$$\{E_1, E_2, \dots, E_i\}\ \text{Ж}\ \{E_{i+1}, E_{i+2}, \dots, E_j\} \quad \text{(a)}$$

$$\{E_1, E_2, \dots, E_i\}\ \text{Щ}\ E_{i+1} \quad \text{(b)}$$

$$E_1\ \text{Ѭ}\ \{E_2, E_3, \dots, E_i\} \quad \text{(c)}$$

$$E_1\ \text{I}\ E_2 \quad \text{(d)}$$

join both left and right lists. This is needed to express that nothing changes in an environment that allows changes. The following axiom is set:

- *One connective may not describe both transition and absence of transition.*

Absence of transition can be described by *izhen* and *zhivot* connectives. However, it may not be described by *yon* and *shta* because if they do, the multitude V_i that is singular in a list (left in the *yon*, right in the *shta*) would imply, the same time, absence of transition (by putting V_i on each side) and transition of V_i to all others V or that of all others to V_i. Then two independent connectives should be introduced, one V_i *izhen* V_i, and the other, V_i *yon*... or ...*shta* V_i. For the same reason, if there is one multitude repeating in the left and right lists of *zhivot*, there must repeat all the rest, too.

To relate the connectives to the options of transition, another axiom is set:

- *Only one option of transition (either perebytie or u- and vozbytie) may be described by one connective.*

This means that if there is an event in the left (causal) list, either there are (i) other events of the same predmet on the right or (i) any other predmets or events, but not (i) and (ii) simultaneously.

Thence, either a connective describes transitions between events within one predmet (perebyites) or any other transitions (between different predmets and/or events of different families).

With these restrictions in mind, one may classify the connectives each into four types as shown in Figure 4.

Each of these types, henceforth termed *modus of connective*, may be illustrated by examples.

Figure 5 illustrates all modi of the simplest connective, *izhen*, Figure 6, those of the *yon* and Figure 7, of the *shta*. Modi of the *shta* connective are obtained from those of yon by reversing the direction of transition. For simplicity, not more than two events of a predmet are considered.

Modi of the *zhivot* are being formed by combination of modi of the *yon* and *shta* connectives plus one more modus analogous to Modus 1 of the *izhen*, describing no transition, with the left and right lists of meanings being the same.

- Modus 1 (Figure 5a) is the case of no transition.

As stated in Figure 5a, the following sequent should be suggested:

Figure 4. Classification of modi of connectives

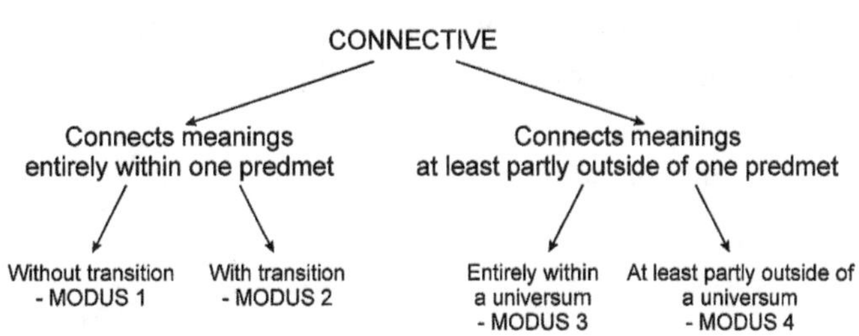

Figure 5a. Examples of modi of the izhen connective, modus 1

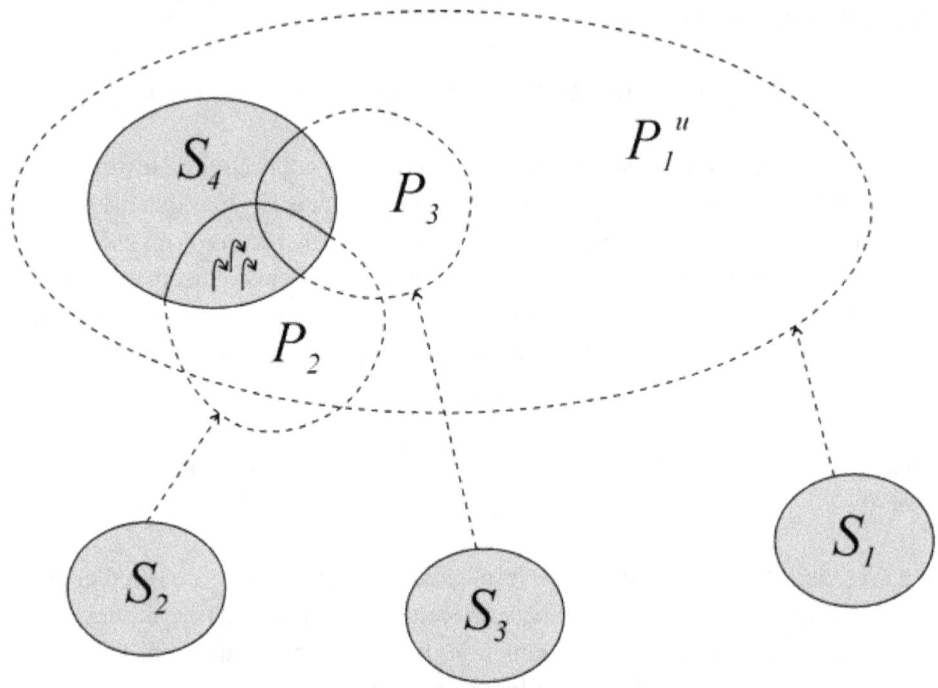

- *henceforth: predmet that must be filled (at least partly, like S_4 here)*

$S_1 \rightarrow P^u_1$; $S_2 \rightarrow P_2$; $S_3 \rightarrow P_3$; $S_4 – P^u_1, P_2, P_3$; $S_4 – P^u_1, P_2, \sim P_3$; $S_4 – P^u_1, \sim P_2, P_3$; $S_4 – P^u_1, \sim P_2, \sim P_3$.

Let us specify the events of S_4:

$E_1 = S_4 – P^u_1, P_2, P_3$; $E_2 = S_4 – P^u_1, P_2, \sim P_3$; $E_3 = S_4 – P^u_1, \sim P_2, P_3$; $E_4 = S_4 – P^u_1, \sim P_2, \sim P_3$.

Then the change would take the form: E_2 *Izhen* E_2.

This change may only take place if S_1, S_2 and E_2 are filled (i.e., are multitudes). S_1 and S_2 must be filled because these are prerequisites for the event E_2 to be filled. However, this condition is not sufficient by itself, so the filling of E_2 per se is required as well.

In any other change considered below, actual transition of fill occurs. Therefore, two states of soprobytie should be specified, before and after the transition. The preceding state includes all preconditions for the transition to occur, and the following state is the result of change of the preceding state made by this transition.

Also, in all subsequent cases two options will be considered: (i) the transition empties the source multitude or (ii) not. In the former case, we will term it full, or complete, transition.

- Modus 2 of the *izhen* (Figure 5b, 5c) represents transition between two events within one predmet.

Figure 5b. Examples of modi of the izhen connective, modus 2 – change of fill from one event to another within a predmet that could be first in a succession of changes.

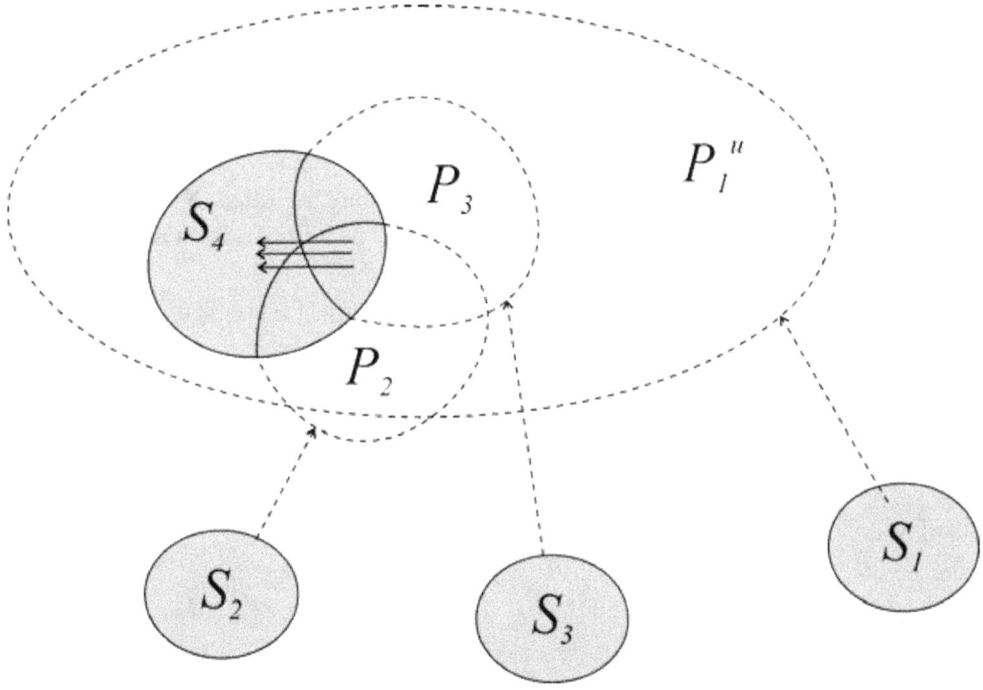

Figure 5c. Examples of modi of the izhen connective, modus 2 – similar change that could follow the change shown in (5b).

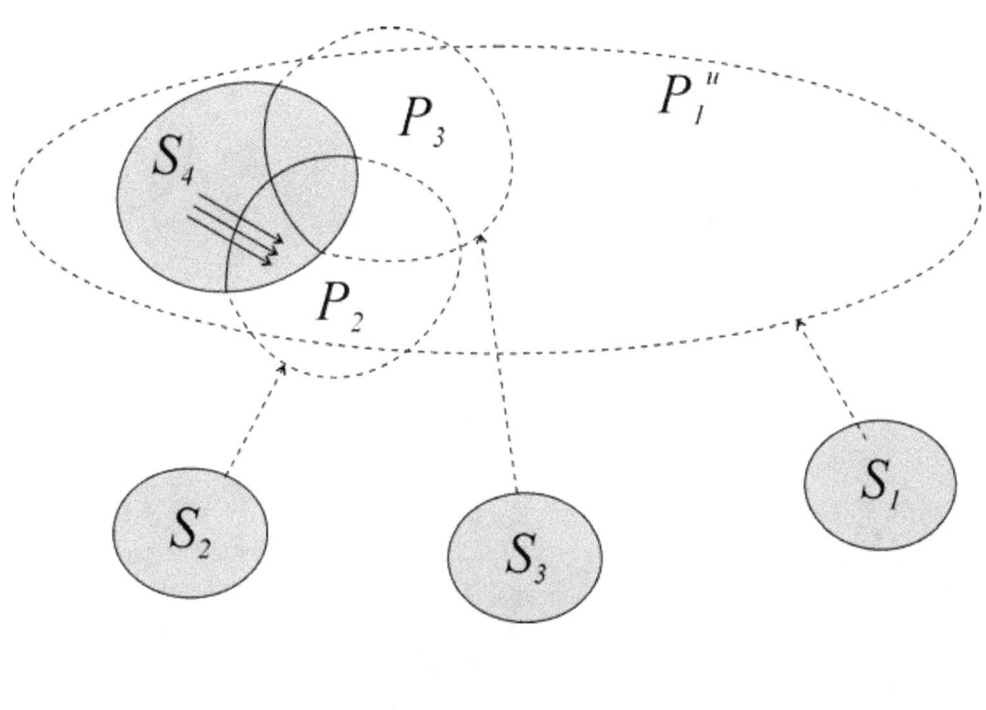

The sequent and definitions are the same as in Figure 5a. The change is between any two of E_1, E_2, E_3 or E_4. In Figure 5b, the change E_1 *Izhen* E_4 is shown.

For this change, the preceding state is defined as follows: S_1, S_2, S_3 and E_1 must be filled, E_2, E_3 and E_4 may be filled. The consequent state is: S_1 and E_4 must be filled, S_2, S_3, E_2 and E_3 may be filled, E_1 may be (remain) filled if transition is not complete, or must be (become) empty otherwise.

If the transition takes place between events E_4 and E_2 (E_4 *Izhen* E_2) in the same soprobytie (Figure 5c; the sequent and definitions as before), the preceding state is: S_1 and E_4 must be filled, S_2, S_3, E_1, E_2, E_3 may be filled or not, and the consequent is: S_1, S_2, and E_2 must be filled, S_3, E_1, E_3 may be filled or not, and E_4 may be filled if transition is not complete, or must be empty otherwise.

- Modus 3 requires more than one multitude in one universum (Figure 5d).

For the case depicted in Figure 5d, another sequent should be formulated:

$$S_1 \rightarrow P^u{}_1;\ S_2 \rightarrow P_2;\ S_3 \rightarrow P_3;\ S_4 - P^u{}_1,\ P_2,\ P_3;\ S_4 - P^u{}_1,\ P_2,\ {\sim}P_3;\ S_4 - P^u{}_1,\ {\sim}P_2,\ P_3;\ S_4 - P^u{}_1,\ {\sim}P_2,\ {\sim}P_3.\ S_5 - P^u{}_1.$$

The events of S_4 are the same as before:

$$E_1 = S_4 - P^u{}_1,\ P_2,\ P_3;\ E_2 = S_4 - P^u{}_1,\ P_2,\ {\sim}P_3;\ E_3 = S_4 - P^u{}_1,\ {\sim}P_2,\ P_3;\ E_4 = S_4 - P^u{}_1,\ {\sim}P_2,\ {\sim}P_3.$$

Figure 5d. Examples of modi of the izhen connective, modus 3 – between meanings of things (one event, one predmet) belonging to the same universum (mind that predmet in this example is a universum event).

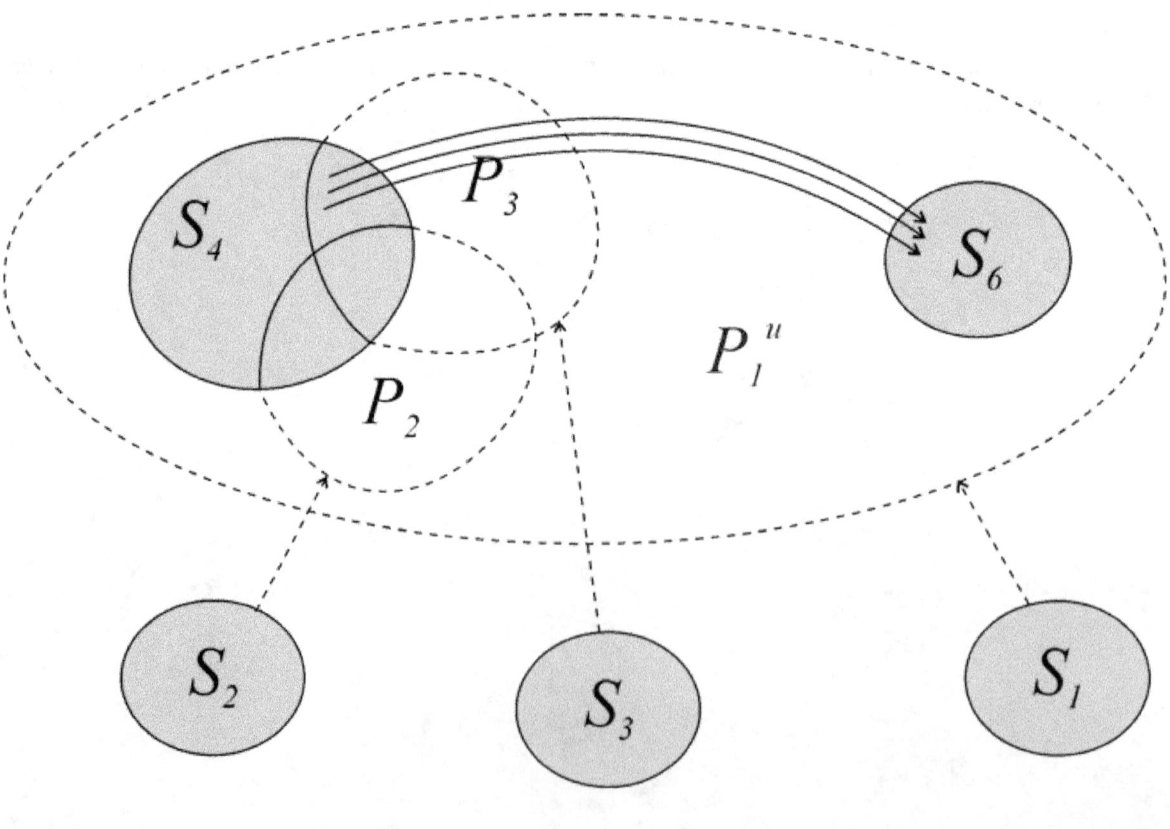

Figure 5e. Examples of modi of the izhen connective, modus 3 – between events of two different predmets; mind that the primeta that serves as a universum for these events (P_3) is not necessarily universum for their predmets (exactly this is the case here).

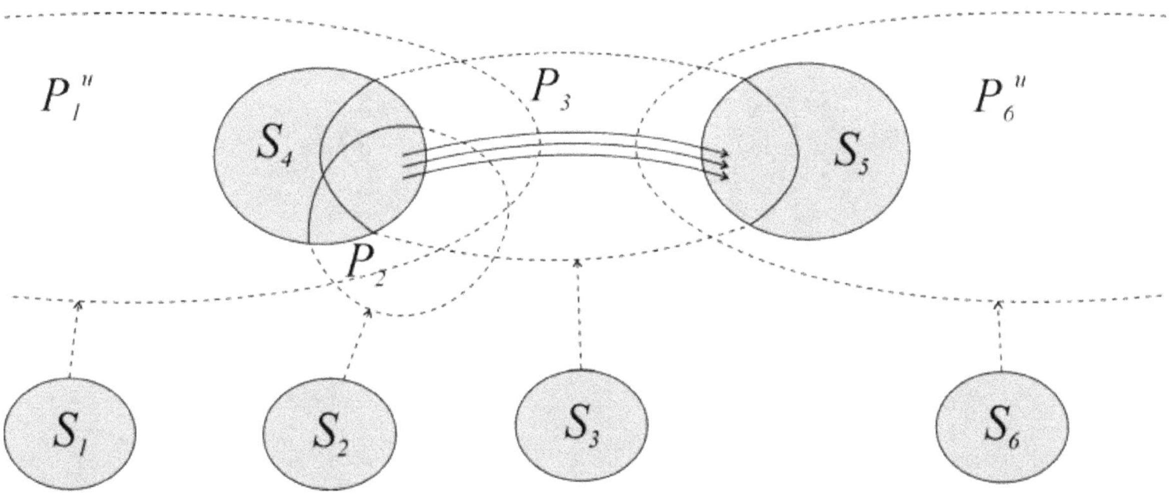

The change E_3 *Izhen* S_5 implies the following preceding state: S_1, S_3 and E_3 must be filled, S_2, S_5, E_1, E_2, E_4, E_5 may be filled. The consequent state should be: S_1, S_5 must be filled, S_2, S_3, E_1, E_2, E_4 may be filled, E_3 may be filled if transition is not complete, or must be empty otherwise.

Another illustration of this modus is presented in Figure 5e. Here the sequent is:

$S_1 \rightarrow P^u_1$; $S_2 \rightarrow P_2$; $S_3 \rightarrow P_3$; $S_6 \rightarrow P^u_6$; $S_4 - P^u_1, P_2, P_3$; $S_4 - P^u_1, P_2, \sim P_3$; $S_4 - P^u_1, \sim P_2, P_3$; $S_4 - P^u_1, \sim P_2, \sim P_3$. $S_5 - P^u_6, \sim P_3$; $S_5 - P^u_6, P_3$.

In addition to the events of S_4, those of S_5 are defined:

$E_1 = S_4 - P^u_1, P_2, P_3$; $E_2 = S_4 - P^u_1, P_2, \sim P_3$; $E_3 = S_4 - P^u_1, \sim P_2, P_3$; $E_4 = S_4 - P^u_1, \sim P_2, \sim P_3$; $E_5 = S_5 - P^u_6, \sim P_3$; $E_6 = S_5 - P^u_6, P_3$.

Importantly, the primeta P_3 here is a wife to predmets (S_4 and S_5) but is universum (mother) to some of their events. Exactly this fact allows us to set another change: E_1 *Izhen* E_6. The preceding state is: S_1, S_2, S_3 and E_1 must be filled, S_5, S_6, E_2, E_3, E_4, E_5, E_6 may be filled. The consequent state is: S_3, S_6, E_6 must be filled, S_1, S_2, E_2, E_3, E_4, E_5 may be filled, E_1 may be filled if transition is not complete, or must be empty otherwise.

Modus 4 applies when transition occurs between multitudes in different universa. Two examples of this are shown in Figures 5f, 5g.

The sequent that describes the soprobytie in Figure 5f is the same as that for Figs. 5a, b, c:

$S_1 \rightarrow P^u_1$; $S_2 \rightarrow P_2$; $S_3 \rightarrow P_3$; $S_4 - P^u_1, P_2, P_3$; $S_4 - P^u_1, P_2, \sim P_3$; $S_4 - P^u_1, \sim P_2, P_3$; $S_4 - P^u_1, \sim P_2, \sim P_3$,

with the same designation of events of S_4:

Figure 5f. Examples of modi of the izhen connective, modus 3 – between meanings of things: from event of one predmet to another predmet that is not in the same universum.

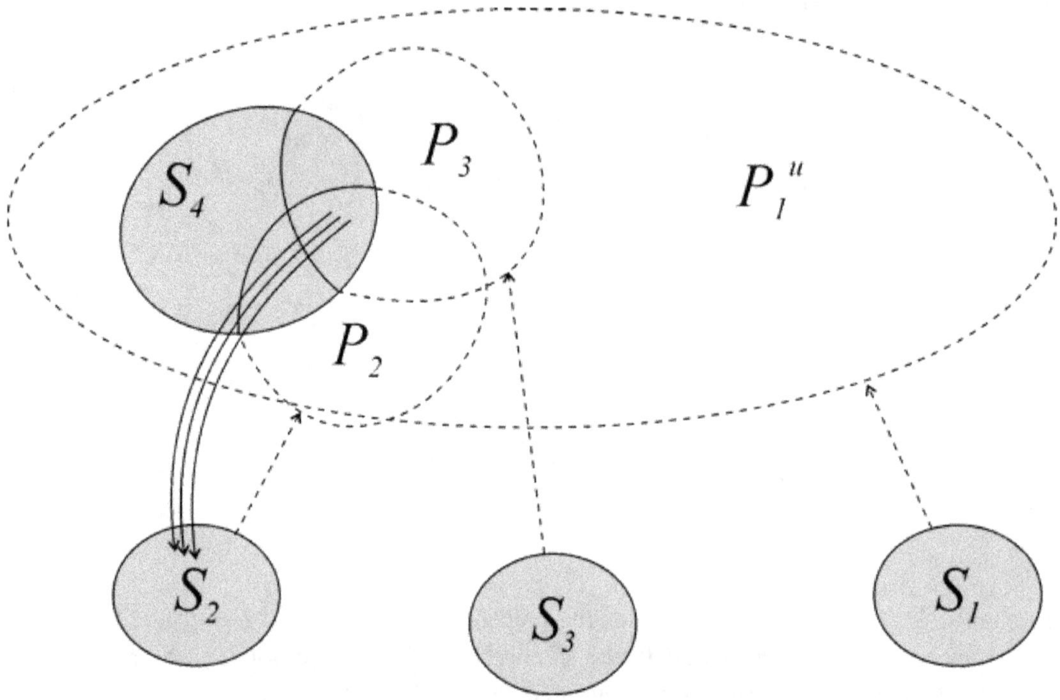

Figure 5g. Examples of modi of the izhen connective, modus 3 – backward.

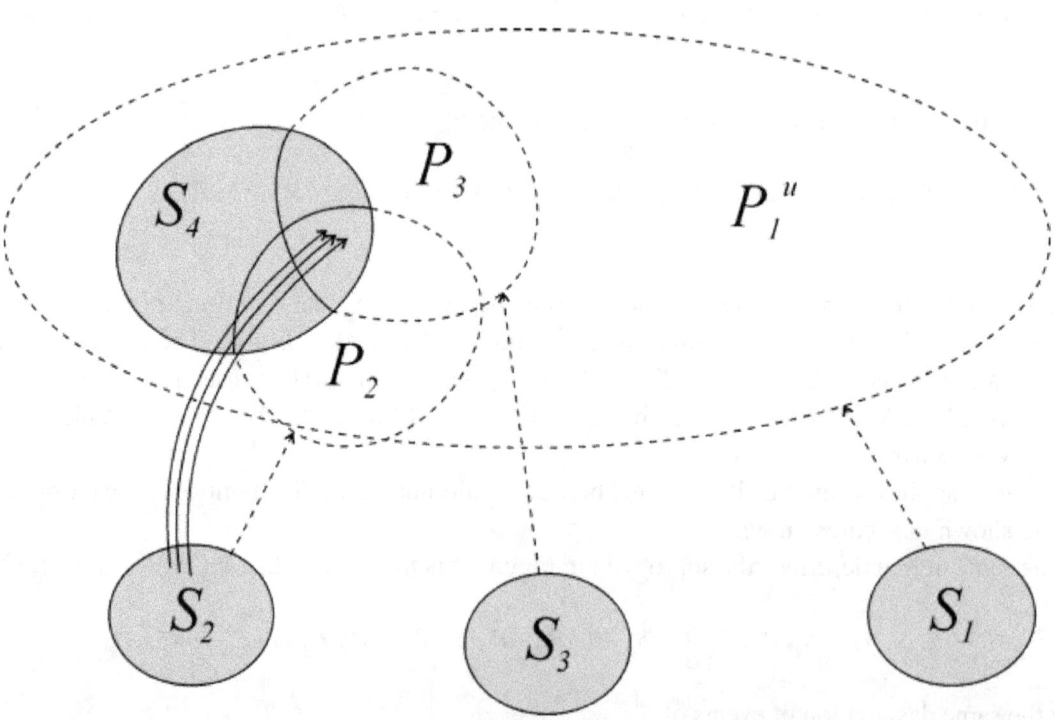

$E_1=S_4-P^u{}_1, P_2, P_3; E_2=S_4-P^u{}_1, P_2, \sim P_3; E_3=S_4-P^u{}_1, \sim P_2, P_3; E_4=S_4-P^u{}_1, \sim P_2, \sim P_3.$

Nonetheless, the change occurs between an event (E_1 in this figure) and a predmet outside of the universum $P^u{}_1$ (S_2 in this figure): E_1 *Izhen* S_2. Its preceding state should be described as: S_1, S_2, E_1 must be filled, S_3, E_2, E_3, E_4 may be filled, and the consequent – S_2 must be filled, S_1, S_3, E_2, E_3, E_4 may be filled, and E_1 may be filled if transition is not complete, or must be empty otherwise.

The example depicted in Figure 5g represents the reverse change in the same soprobytie, S_2 *Izhen* E_1. However, the preceding and consequent states do not mirror those of the previous example because the source ("outgoing") multitude here is predmet whose daughter is the universum for the other predmet, and therefore it may not be empty – in other words, complete transition is not possible here. The preceding state is: S_2 must be filled, S_1, S_3, E_1, E_2, E_3, E_4 may be filled. The consequent state is S_1, S_2, S_3, E_1 must be filled, E_2, E_3, E_4 may be filled.

The next connective, *yon*, is exemplified in Figure 6. For simplicity, it is assumed that transition is directed from one only to two multitudes. Modus 1 is absent in all connectives except the *izhen*. For the modus 2, the same soprobytie as in Figs. 5a, b, c, and the corresponding sequent are used (Figure 6a):

$S_1 \rightarrow P^u{}_1; S_2 \rightarrow P_2; S_3 \rightarrow P_3; S_4-P^u{}_1, P_2, P_3; S_4-P^u{}_1, P_2, \sim P_3; S_4-P^u{}_1, \sim P_2, P_3; S_4-P^u{}_1, \sim P_2, \sim P_3.$

The events of S_4 are:

$E_1=S_4-P^u{}_1, P_2, P_3; E_2=S_4-P^u{}_1, P_2, \sim P_3; E_3=S_4-P^u{}_1, \sim P_2, P_3; E_4=S_4-P^u{}_1, \sim P_2, \sim P_3.$

Figure 6a. Examples of modi of the yon connective – modus 2, within a predmet: change of fill from one event to n events (in the considered case, n=2).

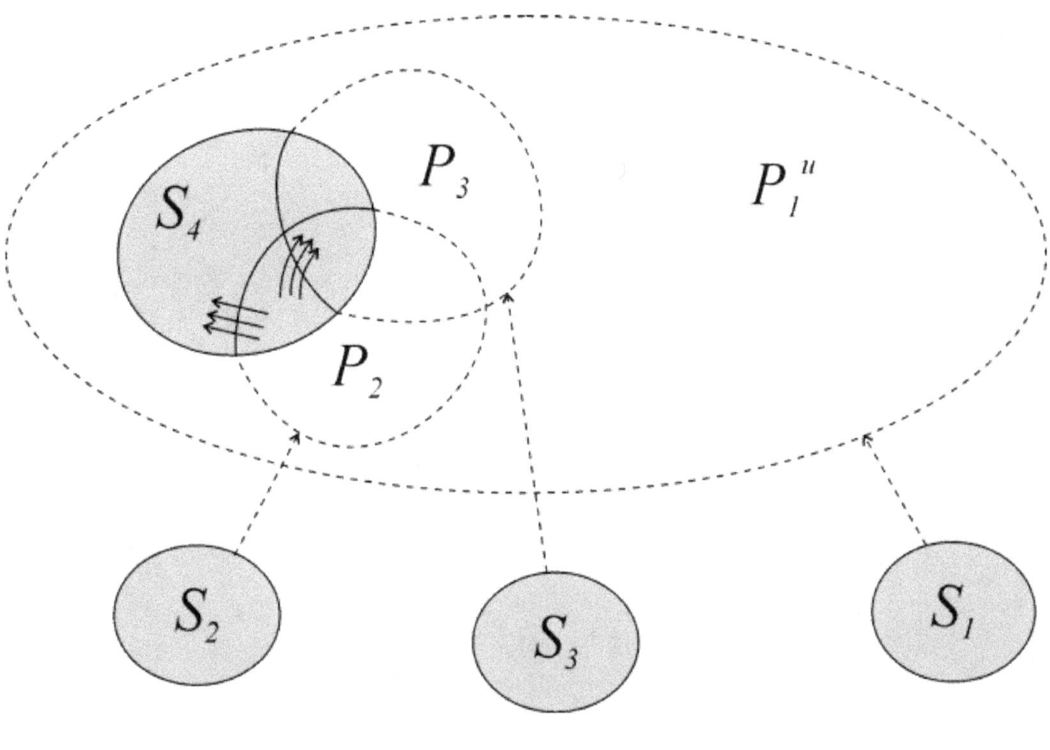

Figure 6b. Examples of modi of the yon connective – modus 3, from one meaning of thing to n meanings of things (in this case: n=2, all meanings of things are predmets).

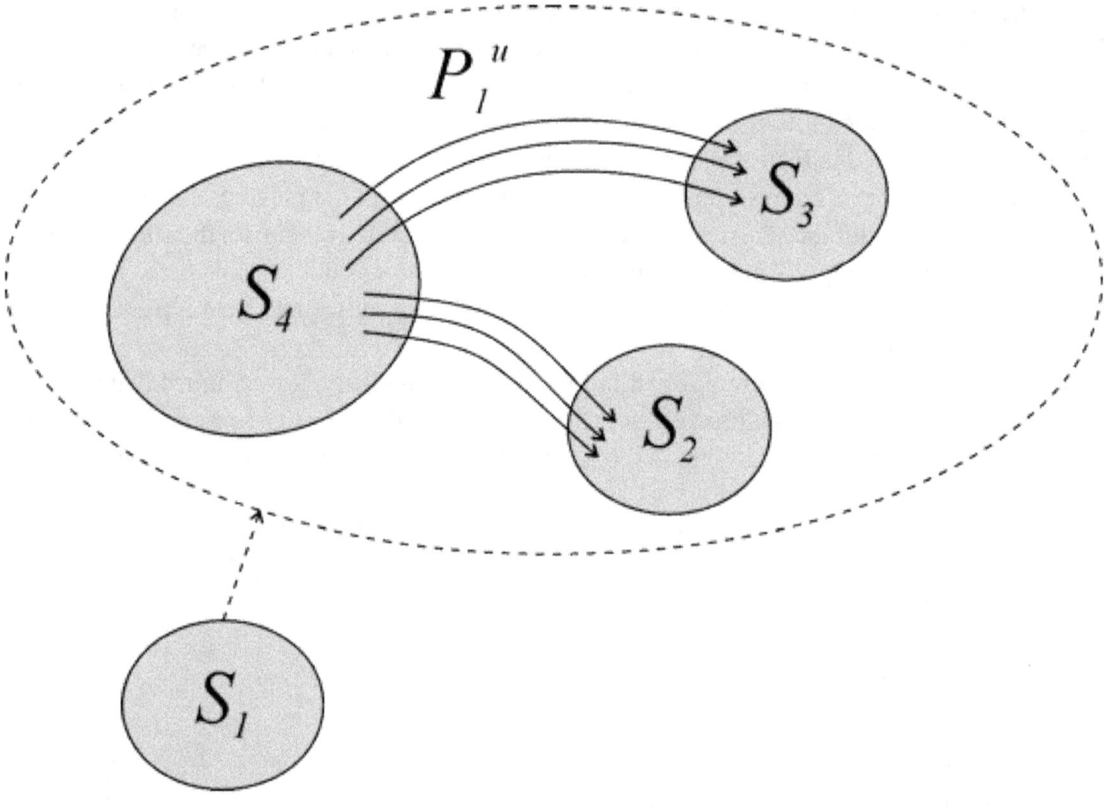

Figure 6c. Examples of modi of the yon connective – modus 4, from one predmet to another predmet and event of yet another predmet.

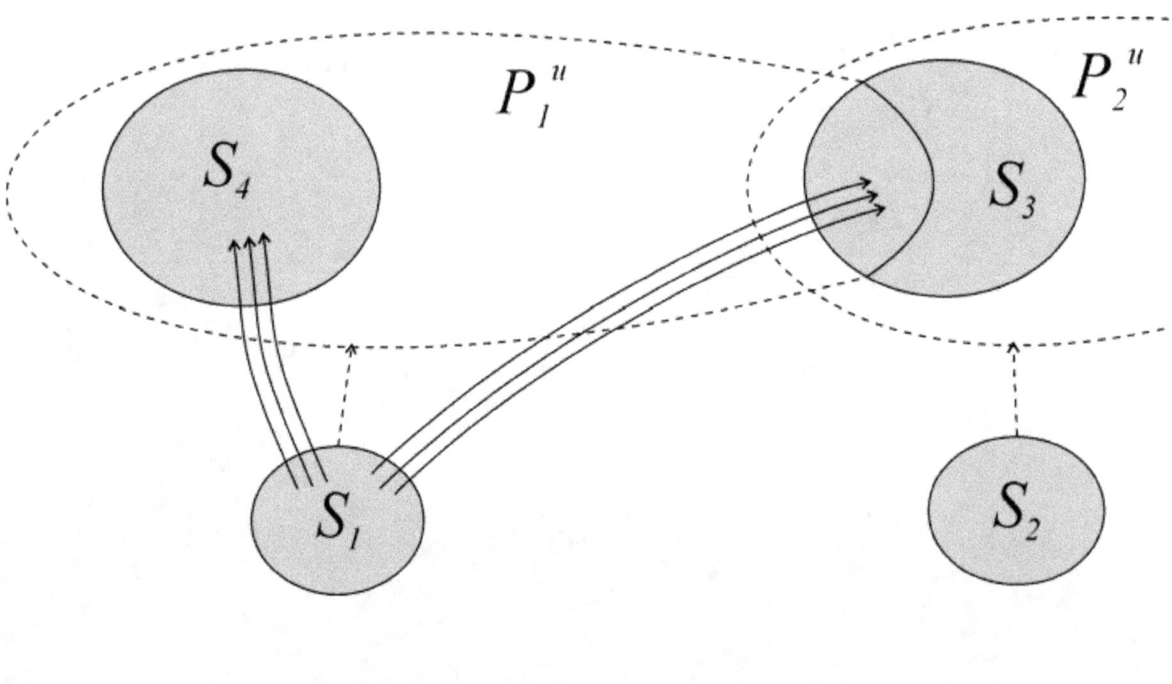

The change in question would be from any of these to any two others, e.g., E_2 *Yon* E_1, E_4. The preceding state for this change is: S_1, S_2, E_2 must be filled, E_1, E_3, E_4 may be filled. The subsequent state: S_1, S_2, S_3, E_1, E_4 must be filled, E_3 may be filled, E_2 may be filled if transition is not complete, or must be empty otherwise.

- Modus 3 of the *yon* connective is illustrated by the example in Figure 6b.

The sequent describing this soprobytie is: $S_1 \rightarrow P^u{}_1$; $S_2 - P^u{}_1$; $S_3 - P^u{}_1$; $S_4 - P^u{}_1$. Mind that if primeta intersects with three or more predmets, the relationship between the events she forms with each of these predmets would be exactly the same and the following change would be applicable as well: S_4 *Yon* S_2, S_3.

The preceding state is S_1, S_4 must be filled, S_2, S_3 may be filled; the subsequent state: S_1, S_2, S_3 must, and S_4 may be filled if transition is not complete, or must be empty otherwise. S_2, S_3, S_4 may well be events of predmets intersecting with $P_1{}^u$.

- Modus 4 is illustrated by four examples (Figure 6c, d, e, f).

In Figure 6c, the transition occurs into the universum from outside. The soprobytie is different, and the sequent is: $S_1 \rightarrow P_1$; $S_2 \rightarrow P^u{}_2$; $S_4 - P_1$; $S_3 - P^u{}_2$, P_1; $S_3 - P^u{}_2$, $\sim P_1$. The primeta is a mother (universum) for S_4 and a wife for S_3.

The events of S_3 are: $E_1 = S_3 - P^u{}_2$, P_1; $E_2 = S_3 - P^u{}_2$, $\sim P_1$.

The change is: S_1 *Yon* S_4, E_1.

The preceding state: S_1 must be filled, S_2, S_4, E_1, E_2 may be filled; the subsequent one: S_1, S_2, S_4, E_1 must be filled, E_2 may be filled, full transition not allowed.

In Figure 6d, the soprobytie and sequent are the same but the change is different: S_4 *Yon* S_1, E_1.

Figure 6d. Examples of modi of the yon connective – modus 4, from one predmet or event to (i) predmet/an event of another predmet and (ii) yet another predmet.

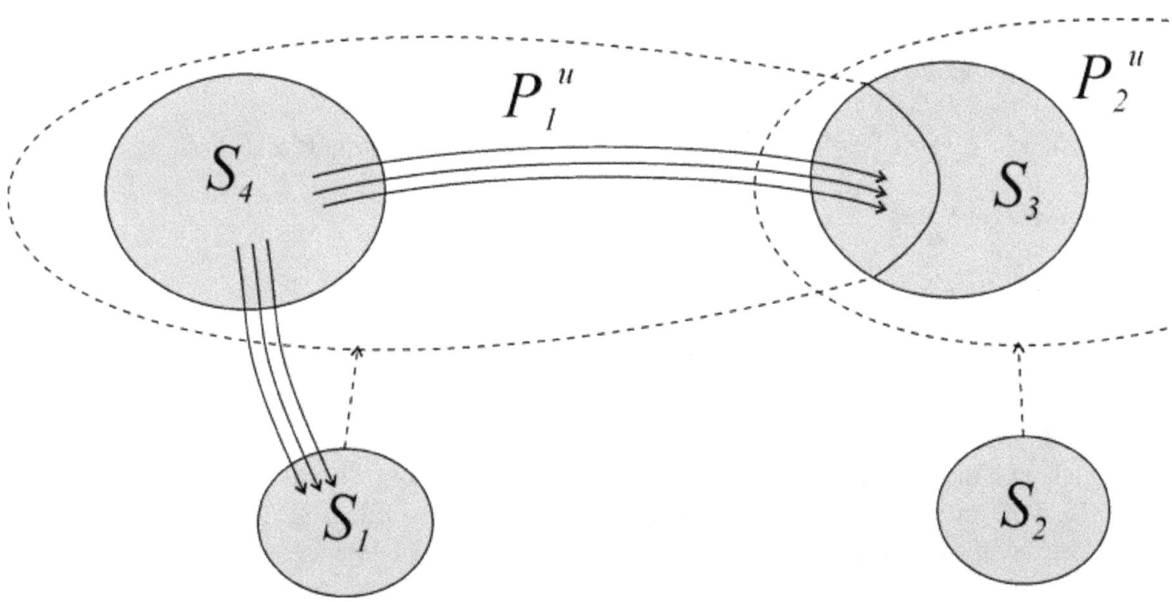

Figure 6e. Examples of modi of the yon connective – modus 4, from one predmet to n events of another predmet (n=2).

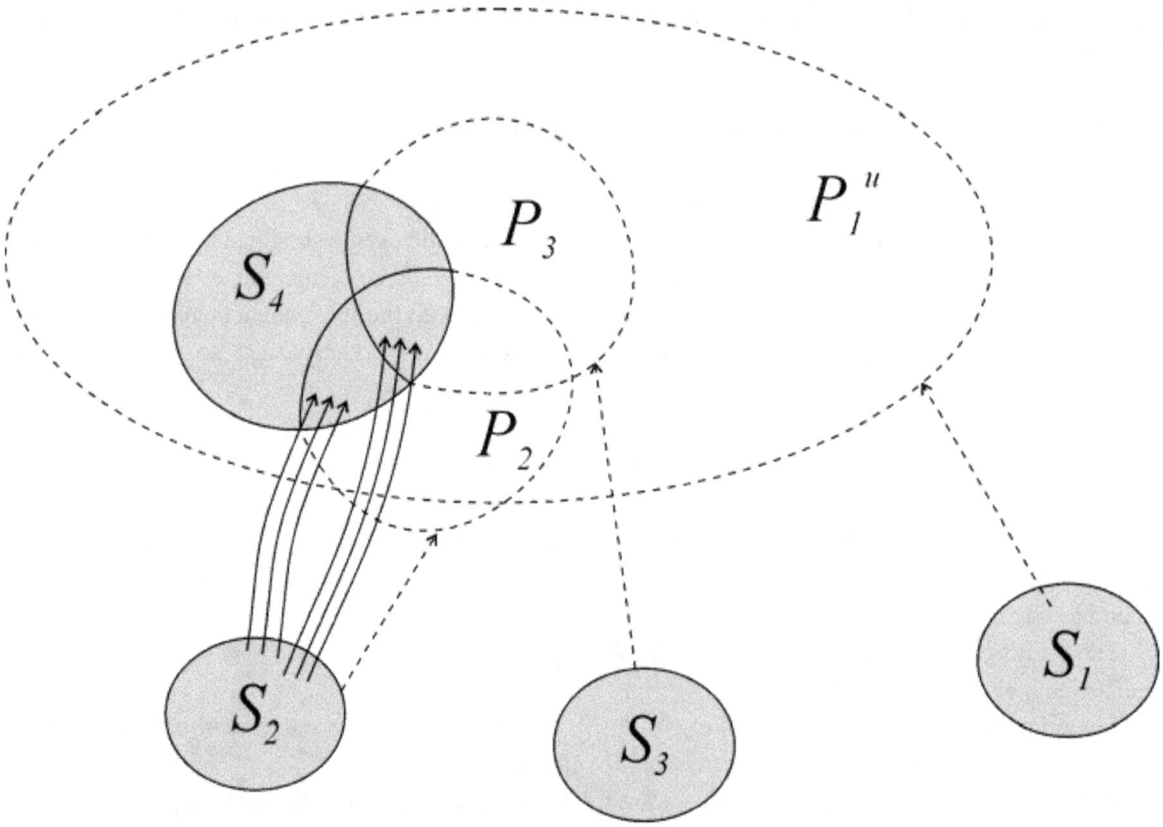

Correspondingly, the preceding state is: S_1, S_4 must be filled, S_2, E_1, E_2 may be filled; in the subsequent state, S_1, S_2, E_1 must be filled, E_2 may be filled, and S_4 may be filled if transition is not complete, or must be empty otherwise.

Two more cases are demonstrated by the same soprobytie as in Figures 5a, 5b, 5c:

$S_1 \rightarrow P^u_1;\ S_2 \rightarrow P_2;\ S_3 \rightarrow P_3;\ S_4 - P^u_1, P_2, P_3;\ S_4 - P^u_1, P_2, \sim P_3;\ S_4 - P^u_1, \sim P_2, P_3;\ S_4 - P^u_1, \sim P_2, \sim P_3,$

with the events of S_4:

$E_1 = S_4 - P^u_1, P_2, P_3;\ E_2 = S_4 - P^u_1, P_2, \sim P_3;\ E_3 = S_4 - P^u_1, \sim P_2, P_3;\ E_4 = S_4 - P^u_1, \sim P_2, \sim P_3.$

In Figure 6e, the change is: S_2 *Yon* E_1, E_2.

Its preceding state is: S_2 must be filled, S_1, S_3, E_1, E_2, E_3, E_4 may be filled, and subsequent state: S_1, S_2, S_3, E_1, E_2 must be filled, E_3, E_4 may be filled. No complete transition allowed.

In Figure 6f, the change is: E_1 *Yon* S_2, S_3. The preceding state for this case is: S_1, S_2, S_3, E_1 must be filled, E_2, E_3, E_4 may be filled. The subsequent state: S_2, S_3 must be filled, S_1, E_2, E_3, E_4 may be filled, and E_1 may be filled if transition is not complete, or must be empty otherwise.

Figure 6f. Examples of modi of the yon connective – modus 4, from event(s) of the same predmet to n other predmets (n=2).

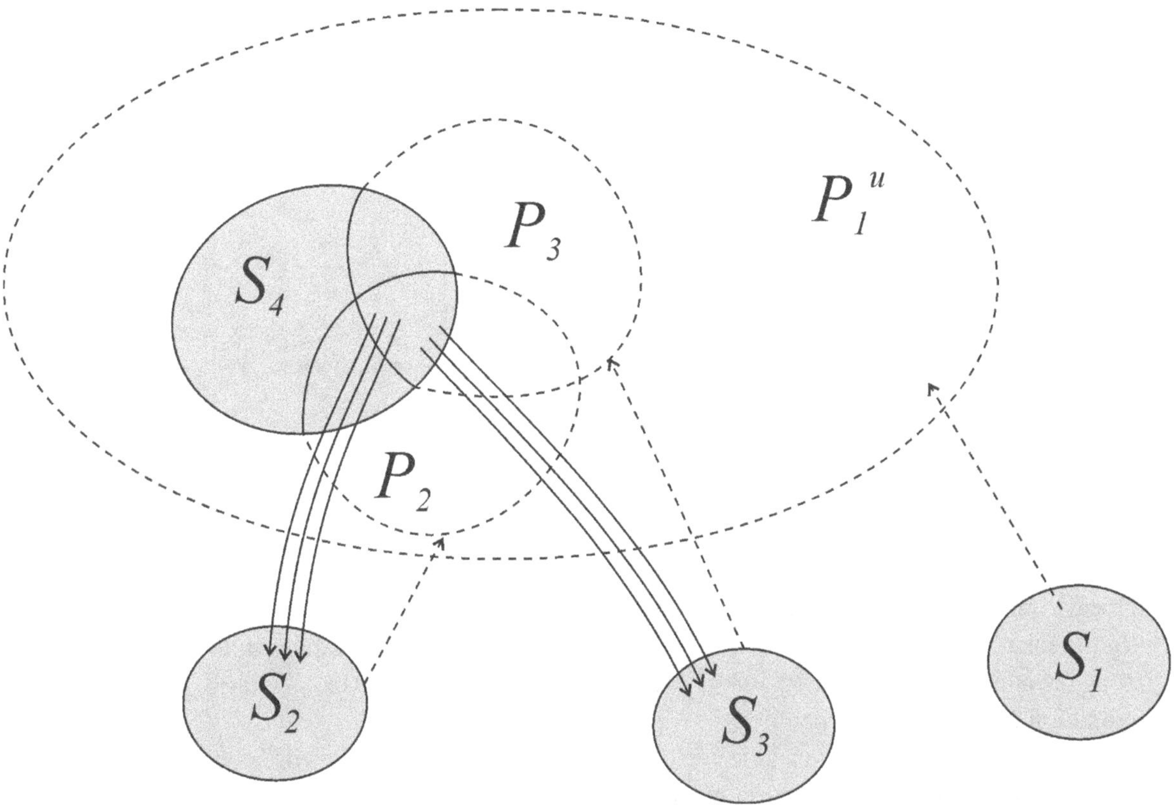

The connective of *shta* is structurally the same as *yon* but the direction of change is the opposite. It is not considered here in detail. Also, omitted is the connective of *zhivot* because of its complexity and due to the fact that it is a combination of *yon* and *shta*. Complete formal definition and description of modi all four types of connective is still ahead. Still, for the purposes of knowledge representation the formalization suggested above looks quite sufficient at the moment.

Further classification of transitions described by each connective should be based on the type of fill, whether the transition is immediate and whether the entire fill moves or not. Detailed classification should be a task for a special study.

Having introduced connectives, one may proceed from the properties of multitudes to the properties of environments.

Behavior of Environments

Two changes in an environment are linked, if at least some of the multitudes of the right list of one change join the left list of the other. Succession of linked changes is termed *scenario*. Two options are possible in an environment, either all its multitudes join scenarios or some multitudes do not. If the latter, this may be because either there are probyties in the environment or it consists of soprobyties only

Figure 7. Classification of environments. See comments in the text.

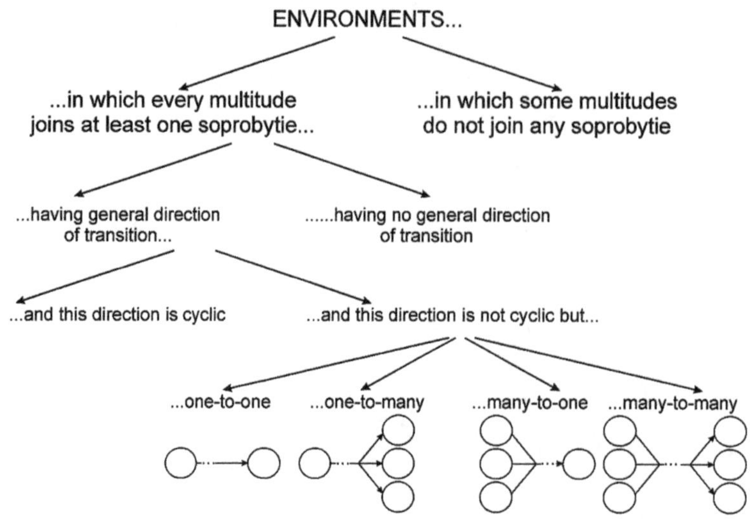

but primetas of at least some of these soprobyties have no relation to predmets of any other soprobytie. Such environment will be referred to as *intermittent*.

In the former case, henceforth denoted as *through* environment, all changes in the environment are directly or indirectly linked. Direct link is enforced by transition of fill (i.e., scenario), and indirect, by interrelation of primetas with predmets.

Through environments may include cycles or not. Cyclic environments will not be addressed in this work. General classification of environments see in Figure 7.

A division of modeling environments, though quite intuitive and incomplete, was suggested by Pshenichny and Kanzheleva (2011) for the sake of the Earth sciences. These authors recognized (i) "no-change" environments that depict only bodies and their properties and (ii) changing environments that describe how bodies and properties change with time, in space or under other conditions. The latter were divided into environments of non-directed change, invariant directed change and alternative directed change.

The no-change environments of Pshenichny and Kanzheleva (2011) include all environments that consist of predmets or probyties only, all intermittent environments, because at least some of the soprobyties are unrelated in them both directly and indirectly, and also some through environments, because in a through environment the transition of fill is possible but not mandatory.

The rest three types of environment, non-directed change, invariant directed change and alternative directed change, may be well combined in through environment and therefore should be considered its extreme cases.

In a through environment, the following observations can be made. Due to the structure of individual soprobytie, (i) any change in the environment, including even the "nothing changes" case (see Figure 5), requires at least two meanings of things be filled either in the left, or in the right, or in both lists and (ii) filling of father predmet activates daughter primeta, which implies not less opportunities of change (and, hence, various continuations of scenario) than the number of predmets (husbands and/or sons) this primeta relates to. Then it looks straightforward to combine scenarios in the environment by merging or ramifying them via common primetas.

For these two reasons, the structure of environment totally based on the soprobytie will be henceforth termed *multifluent*, meaning by "fluent", first, actual transition of fill and, also, influence that father predmets or events always have on this flow via the daughter primetas. Even if the change is that of Modus 1 (transition is actually absent), this influence is always here, and the change similarly requires minimum two meanings be filled.

This shows that the multifluence has two aspects. The transition of fill per se expressed as changes is its "material", or syntactic, aspect, and the relations between meanings that determine this transition make up the semantic aspect of multifluence. The mentioned classification of Pshenichny and Kanzheleva (2011) did not draw distinction between these and is therefore is not free from ambiguity.

Still, if to focus solely on the syntactic aspect of multifluence, i.e., the transition of fill, the category theory (Barr and Wells, 2012) in conjunction with the directed graph theory (Ehrig et al., 2006) seem to offer an efficient tool to formally handle the diversity of contexts by structure. Wolter et al. (2017, see Chapter 3 of this book) have proposed this approach to provide formal grounds for the event bush method. Generalizing their approach, one may suppose that the general case of dynamic environment can be thought of as a directed graph with vertices that are either events or connectives. The graph is simple in the sense that there is at most one edge between two vertices in the graph. There must be no vertices (events or connectives) unconnected by edges. Edges go only from an event vertex to a connective vertex or the other way around. There are no edges between event vertices and no edges between connective vertices, respectively. This is the type graph for all the cases and subcases of environment considered above. The morphism relations should be accurately traced down the hierarchy to achieve the definitions of particular methods, which can be thus used equally well for knowledge engineering/representation and for modeling the phenomena this knowledge is about. Some of these methods, like UML behavior diagrams or Petri nets, are well known as syntactic constructs but the universal semantics offered by the Novorossian logic brings them all in a unite framework and urges to reconsider their definitions. Other methods, like the event bush, are quite new and have not been formalized enough. Building the theory of event bush may appear the core of the methodology of dynamic knowledge representation and modeling the dynamic environments. A pending question herewith is whether the type graph formalization starts at the level of dynamic environment or may begin one level up, at the environment per se, or even two levels up, at the predmetka chart (representing an uninhabited or, better say, not necessarily inhabited, context). Perhaps this should be the first issue to address in the research on categorization of the environments. Nonetheless, this approach, being good in modeling the transition of fill, is unable to express the crucial relations between meanings these transitions are based on and determined by.

Considering the semantic and syntactic aspects jointly, one may formulate theoretical foundations of different methods used, or just designed, for representation and engineering of dynamic knowledge.

It should be noted herewith that even the *izhen*-type one-to-one transition of fill requires at least one more predmet be full (have bytie) both before and after the change. Therefore, a single-root tree-like shape of representation of knowledge adopted, e.g., in the event/probability trees, may be considered adequate for changing environments only if accompanied by additional knowledge at each fork, or be replaced by "polytree" network patterns (Pearl, 1988).

Different types of polytree graphs can be specified for through environments. One possible type, the activity diagram, which is a kind of UML behavior diagram, is shown by Naumov et al. (2018, Chapter 12 of this book). Another type of polytree for through environments called event bush is defined by the following constraints.

Figure 8. Interrelation of the types of event in the event bush (after Pshenichny et al., 2009)

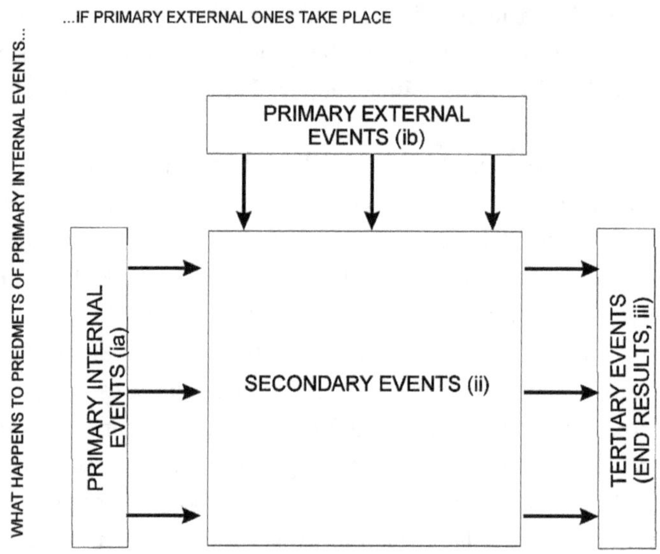

1. Each event belongs to one of four types: (*ia*) primary internal, (*ib*) primary external, (*ii*) secondary and (*iii*) tertiary. In addition to predmet and primetas, it has position: *iaE, ibE, iiE* or *iiiE.*
2. Father predmets for primetas of the (*ia*) and (*ib*) events are not included in the environment.
3. The following changes are to be set between events:
   - Every *(ia)* event alone must lead to one *(ii)* or one *(iii)*;
   - Every *(ib)* event must lead to one or more *(ii)* and do so only together with *(ia)* or another *(ii)* events;
   - Every pair of *(ia)* and *(ib)* events may lead, if ever, only to one event (of the *(ii)* type);
   - Every *(ii)* event must be led to either by one *(ia)*, or one *(ia)* and one *(ib)* or one or more other *(ii)* events;
   - Every *(iii)* event must be led to by a *(ia)* or *(ii)* event;
   - Not a single *(iii)* event may lead to any other event;
   - Not a single event may lead to *(ia)* or *(ib)*;

   This implies some corollaries:

   Event *(ia)* without an event *(ib)* or *(ii)* may lead only to *(ii)* or one event *(iii)*;
   - Event *(ia)* may not lead to any event together with another *(ia)*;
   - Event *(ib)* may lead only to an event *(ii)* and only together with *(ia)* or another *(ii)*; it leads to nothing alone.

   Based on the allowed and forbidden changes, the general interrelation of the four types of event can be expressed graphically as shown in Figure 8.
4. These relations are enforced by the connectives of the event bush, which are specifications of the connectives of the theory of multitudes. Herewith, the following is postulated or defined.
   - The *Flux* connective of the event bush:

   E_i *Flux* E_j

Figure 9. Graphic notation for the flux connective of the event bush (modified from Pshenichny et al., 2009)

(Figure 9) or

S_1–P^u_1, …, P^u_m, P_{m+1}, P_{m+2},…, ~P_{n-1}, ~P_n *Flux* S_2–P^u_k, …, P^u_r, P_{r+1}, P_{r+2},…, ~P_{t-1}, ~P_t

is an *izhen* connective (any modus) under assumption that at least one father predmet/event of any asserted primeta of either event is not included in the event bush.

In case of Modus 1, the position of event changes. Three options are possible (Figure 10):

(ia)S_1–P^u_1, …, P^u_m, P_{m+1}, P_{m+2},…, ~P_{n-1}, ~P_n *Flux Modus 1 (ii)*S_1–P^u_1, …, P^u_m, P_{m+1}, P_{m+2},…, ~P_{n-1}, ~P_n;

(ia)S_1–P^u_1, …, P^u_m, P_{m+1}, P_{m+2},…, ~P_{n-1}, ~P_n *Flux Modus 1 (iii)*S_1–P^u_1, …, P^u_m, P_{m+1}, P_{m+2},…, ~P_{n-1}, ~P_n;

and

(ii)S_1–P^u_1, …, P^u_m, P_{m+1}, P_{m+2},…, ~P_{n-1}, ~P_n *Flux Modus 1 (iii)*S_1–P^u_1, …, P^u_m, P_{m+1}, P_{m+2},…, ~P_{n-1}, ~P_n.

Figure 10. Three options of the izhen modus 1 in the event bush (modified from Pshenichny and Kanzheleva, 2011): (a) change from primary internal event to secondary, (b) from primary internal to tertiary, (iii) from secondary to tertiary.

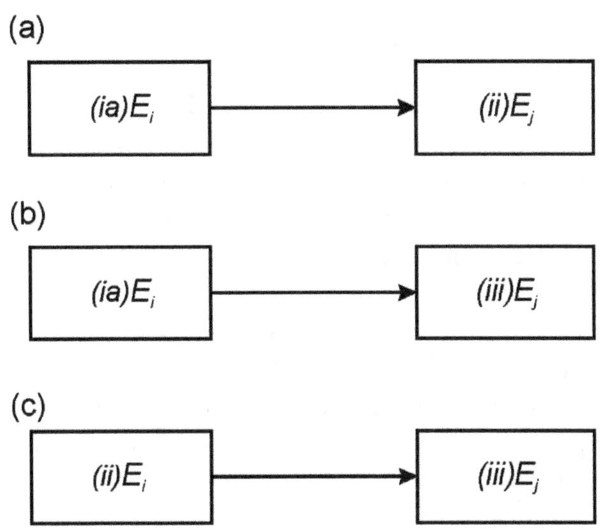

Figure 11. Graphic notation for the influx connective of the event bush (modified from Pshenichny and Kanzheleva, 2011)

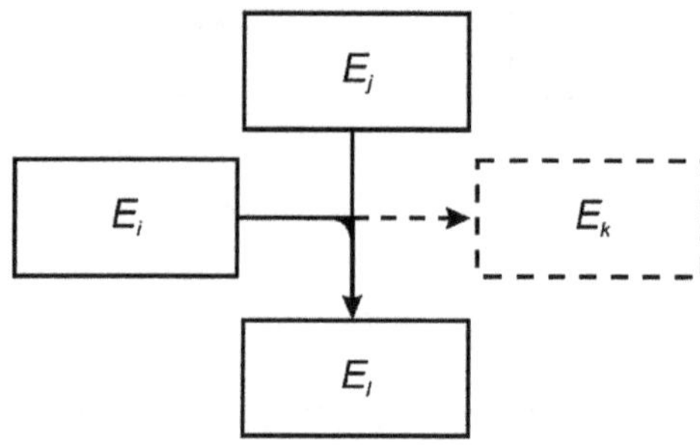

If predmet is the same on the left and on the right, not more than one primeta may be asserted or negated by one *Flux* connective.

- The *Influx* connective of the event bush

$E_i\,^{Ej}\ \mathit{Influx}\ E_l$

(Figure 11) or

$S_1{-}P^u{}_1, \ldots, P^u{}_m, P_{m+1}, P_{m+2}, \ldots, {\sim}P_{n-1}, {\sim}P_n\ ^{E2}\ \mathit{Influx}\ S_1{-}P^u{}_1, \ldots, P^u{}_m, P_{m+1}, P_{m+2}, \ldots, P_{n-1}, {\sim}P_n$

is an *izhen* connective (any modus) implying that the father predmet/event of the primeta P_{n-1} (negated on the left and asserted on the right) is included in the event bush. This may take place in one of two ways: either P_{n-1} has father predmet S_2, and then E_2 is a universum event: $E_2{=}S_2{-}P^u{}_k, \ldots, P^u{}_r$, or P_{n-1} has father event E_2: $E_2\ S_2{-}P^u{}_k, \ldots, P^u{}_r, P_{r+1}, P_{r+2}, \ldots, {\sim}P_{t-1}, {\sim}P_t$, and then the primeta P_{n-1} is produced not just by S_2 but by the entire combination of the S_2 predmet, asserted and even negated primetas, which jointly make up the event.

- *Split* connective of the event bush

$E_i\ \mathit{Split}\ E_{i+1}, \ldots, E_n$

(Figure 12) is a specification of the *yon* modus 2. It applies if secondary event *(ii)*$S_1{-}P^u{}_1, \ldots, P^u{}_m, P_{m+1}, P_{m+2}, \ldots, {\sim}P_{n-1}, {\sim}P_n$ leads to several secondary events, each differing from the causal one only by one negation sign (i.e., one primeta asserted in the causal one is negated here or vice versa). Theoretically, there may be as many such options as many combinations of non-universum primetas differing by one negation sign from the causal event can be (accounting for exclusion where relevant).

- With the same (*ia*), (*ib*), (*ii*) and (*iii*) events, there may be more than one event bush that differ in the changes that link the events. Such event bushes are called the *family* of event bushes. If there is observed situation that E_i *Flux* E_{i+1} is present in one event bush and only there, E_i *Flux* E_{i+2} in another and only there and so forth, up to E_i *Flux* E_n in the last one, this is called *furcation*:

Figure 12. Graphic notation for the split connective of the event bush

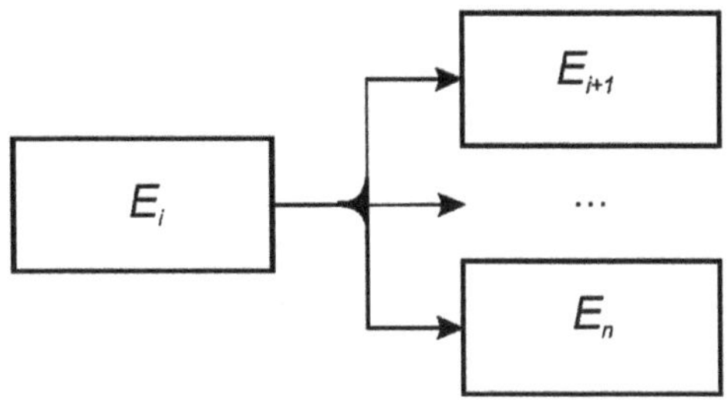

E_i *Furcation* $E_{i+1}, \ldots, E_n$

and marked by a black circle. Though resembling the *yon*, *furcation* is neither split nor a special connective. It is nothing else but sign of several *flux* and/or *split* modus 2 connectives in different, though related, event bushes.

- The *Conflux* connective of the event bush

  $E_1, E_2, \ldots, E_{n-1}$ *Conflux* E_n

  (Figure 13) is a *shta* connective of the theory of multitudes without any specificity.
- The *zhivot* connective is not used in the event bush.
- The event bush must include the connectives of flux and influx and may include the connectives of conflux and split.
- *Flow* is a scenario in the event bush such that the left part of the first change is an *(ia)* event, the right part of the last change is a *(iii)* event, and the right part of every preceding change is the left part of the following change.
  Flow is always followed from *(ia)* to *(iii)*. If there is a change with influx in it, the modifier of this influx (*(ib)* or *(ii)*) is included, but not the events that caused it, unless there are successive influx changes. There may be no events in a bush not included into any flow.

5. Cycles are not admitted in the event bush. (Otherwise there perhaps may be another tool not considered here).

Figure 13. Graphic notation for the influx connective in the event bush (modified from Pshenichny and Kanzheleva, 2011)

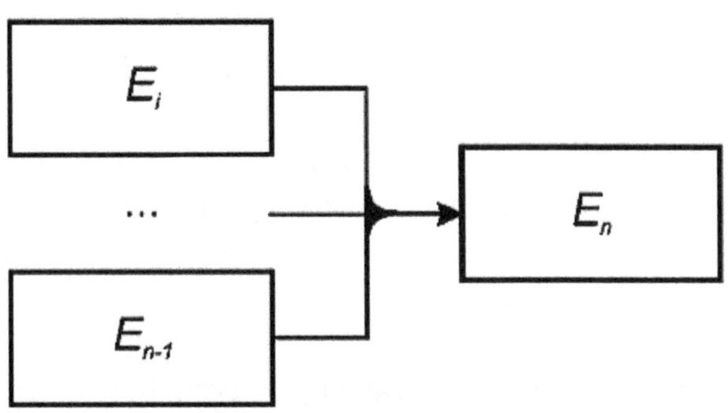

This is the formal description of the event bush. In practice of event bush-based modeling, though, some more rules are informally introduced – e.g., claiming some primetas irrelevant, uniting several events in one and so on, and their formalization is still ahead, but these are local peculiarities of the event bush approach that seem not to deny the above statements.

Introducing constraints on the fill – e.g., putting the requirement of grainy fill – one may formulate the semantic grounds for Petri nets, thus making them better focused on representation of knowledge the transition of fill is determined by, or suggest new formalisms.

APPLICATION

Example of Proof

The reported theory aims at representation of knowledge about changing world in a computer-tractable form. Historical knowledge is one of the main targets of this theory. Below we shall examine by a particular, though imaginary, case, whether, and how, this can be done.

Suppose a knowledge engineer or journalist interviews an expert – a veteran of Anti-Hitler Coalition troops fighting in West Europe in 1944-1945, and particularly suffering from the German counter-raid in the Ardennes brilliantly described in the Kurt Vonnegut's "Slaughterhouse Five, or The Children's Crusade". We can imagine nearly the following narration expressing one of many subjective standpoints.

"Our guys were heavily bombed by the Nazis; many were wounded or killed. But much worse it was for those who faced them and capitulated. The civilians from wherever who were captured by Germans became prisoners and were immediately subject to military crimes. Captured military men at first were treated more or less tolerably. But one should not be misled – they were already prisoners. And sooner or later any prisoner of Hitler concentration camps became subject to a military crime."

In practice of thinking, speaking, and analyzing the speech we rarely separate concerns, distinguishing the meaning and its fill (if any). Conversely, we normally see by our mental eye the filled meanings (i.e., the multitudes) only, and among these, mainly the explicitly given ones. This unavoidably directs the mind of a knowledge engineer and, in fact, misleads it. The approach developed below based on the theory of multitudes would hopefully overcome this serious shortcoming.

In doing so, as it used to be in the knowledge engineering, the expert's knowledge will be both recorded and plotted aiming to understand its internal structure. The process of parsing the expert's knowledge is the process of building a graphic construct. To plot it, one of the tools of dynamic knowledge engineering discussed in Section 4 should be used with account of the predmet-primeta-event grammar, which actually gives any of these methods the power of visual language. Among these, the event bush seems to be the first candidate to consider. The process of eventual building the event bush in process of parsing the knowledge was demonstrated by Mouromtsev et al. (2014). Here this process will be presented in Figure 14a-h.

Parsing the above monolog by meaning, knowledge engineer may do this in a number of ways, but, armed with the Novorossian logic, most likely, one would pin down the following:

Our guys – S_1–...

Simultaneously, this can be expressed graphically by an event bush (Figure 14a).

Figure 14a. The construction of event bush as a result of progressive understanding of the expert's knowledge: The first step.

Figure 14b. The construction of event bush as a result of progressive understanding of the expert's knowledge: the second step.

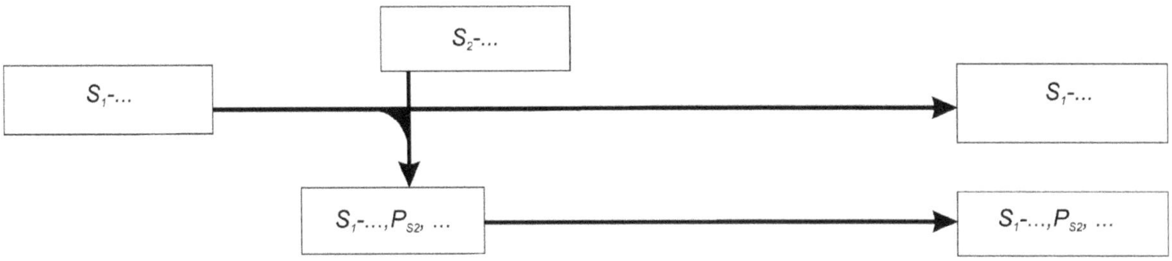

...by the Nazis – S_2–... (importantly, what is meant here by Nazis is not military men but their weapons – this phrase says that Nazi weapons do represent a predmet in this context);

...bombed by the Nazis; many were wounded or killed – $S_2 \rightarrow P_{S2}$ (be bombed, be wounded and be killed are considered as essentially the same meaning in this context – be somehow affected by weapons);

Our guys were bombed by the Nazis; many were wounded or killed – S_1–..., P_{S2} (Figure 14b);

...faced them (Germans) – i.e., Wehrmacht appears on scene – S_3–...;

...faced them (Germans) and capitulated – i.e., Wehrmacht is capturing – $S_3 \rightarrow P^u_{S3}$;

...those (military men) who faced them (Germans) and capitulated – S_1–..., P_{S3} (Figure 14c);

Passing to the next fragments, about civilians who were captured by Germans, like in the first example, one should first introduce a new predmet: S_4–..., (Figure 14d), then consider another predmet that is father to one of the primetas, and only then, finally, put the whole phrase: "civilians who were captured by Germans" – S_4–..., P_{S3} (Figure 14e). Although, such pathway of thought looks not quite natural for humans.

Carrying on, "...civilians who were captured...became prisoners" – S_5–P^u_{S3}, ... (new predmet, prisoners, appears and its universum, "captured", is clearly indicated; the fact that another predmet, civilians, "becomes" this predmet by moving the fill, is best expressed graphically in Figure 14f).

Captured military men at first were treated more or less tolerably – S_5–..., ~P_5 (Figure 14g). Finally, the fact that sooner or later any prisoner of Hitler concentration camps became subject to a military crime is expressed by the last step of building a preliminary event bush (Figure 14h); here we only note that some of the prisoners were not tortured from the very beginning.

Figure 14c. The construction of event bush as a result of progressive understanding of the expert's knowledge: (c) the third step

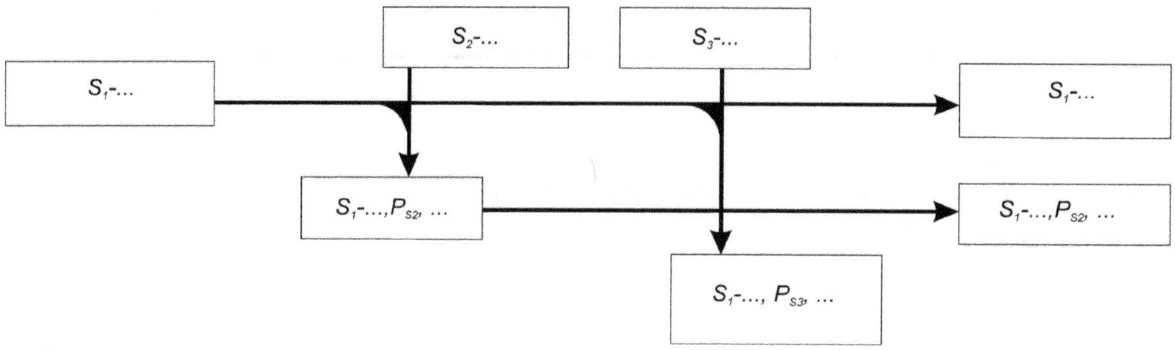

Figure 14d. The construction of event bush as a result of progressive understanding of the expert's knowledge: the fourth step.

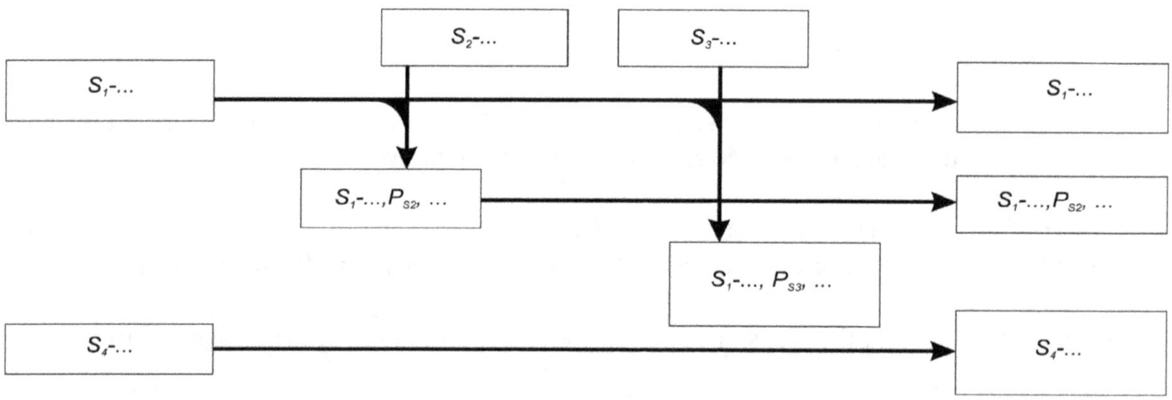

Figure 14e. The construction of event bush as a result of progressive understanding of the expert's knowledge: the fifth step.

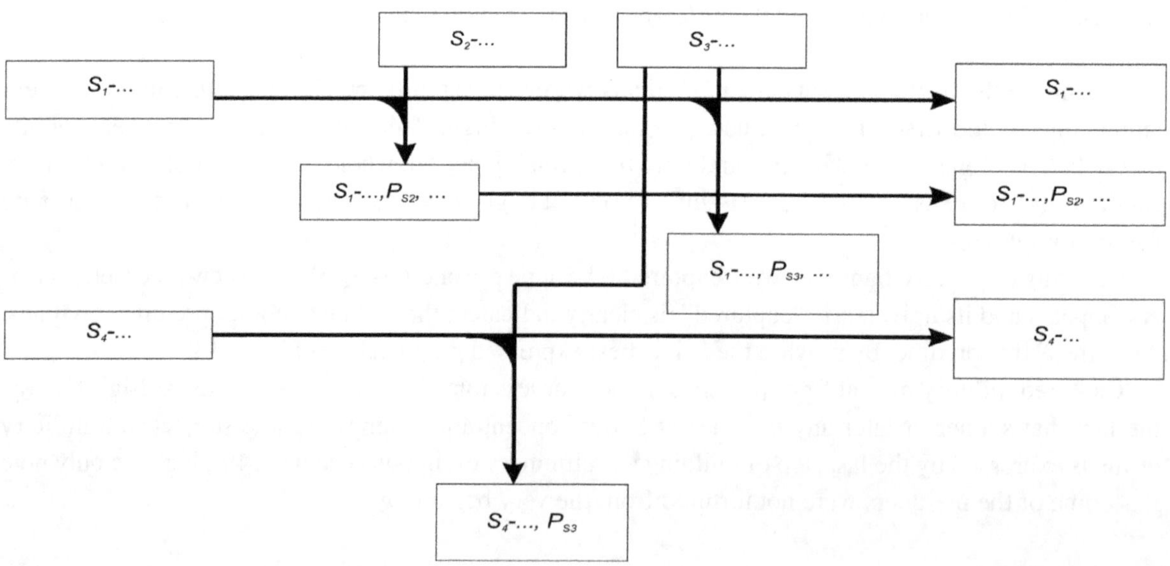

Figure 14f. The construction of event bush as a result of progressive understanding of the expert's knowledge: the sixth step.

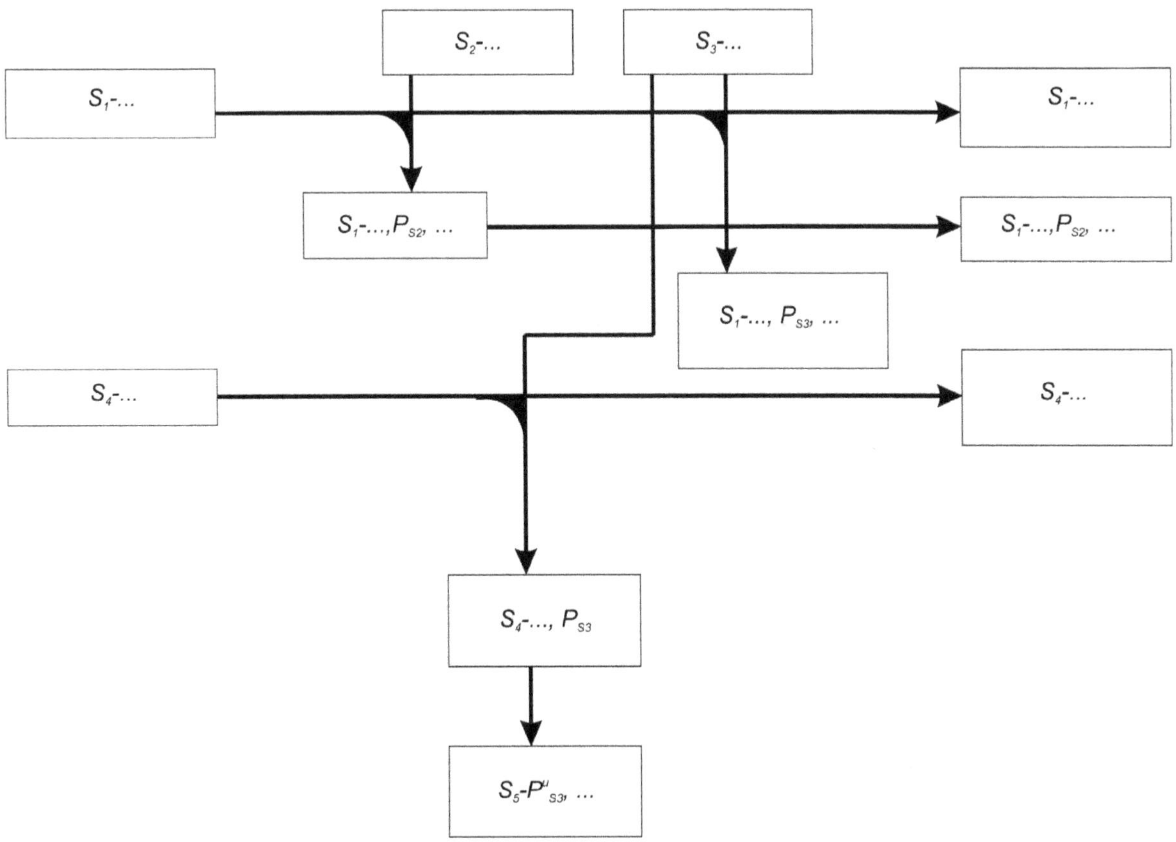

Formulations of those meanings required by the rules of inference but absent in the raw sequent (predmets or primetas, e.g., P^u_1 – occur at the front; P^u_2 – occur in occupation) may be picked up by asking additional questions to expert(s) or from other sources as is being normally done in the knowledge engineering.

From the communication with the expert, assisting ourselves by building an event bush (which is still incomplete, as incomplete are formulations of its events), the following raw sequent appears:

S_1–...; S_1–..., P_{S2}; S_2–...; $S_2 \rightarrow P_{S2}$, ...; S_3–...; $S_3 \rightarrow P^u_{S3}$; S_4–..., P_{S3}; S_1–..., P_{S3}, ...; S_5–P^u_{S3}, ...; S_5–..., $\sim P_5$.

The thread of inference is as follows. At each step, the formula(e) to be transformed and the result of transformation are given in **Underlined Bold**.

Figure 14g. The construction of event bush as a result of progressive understanding of the expert's knowledge: the seventh step.

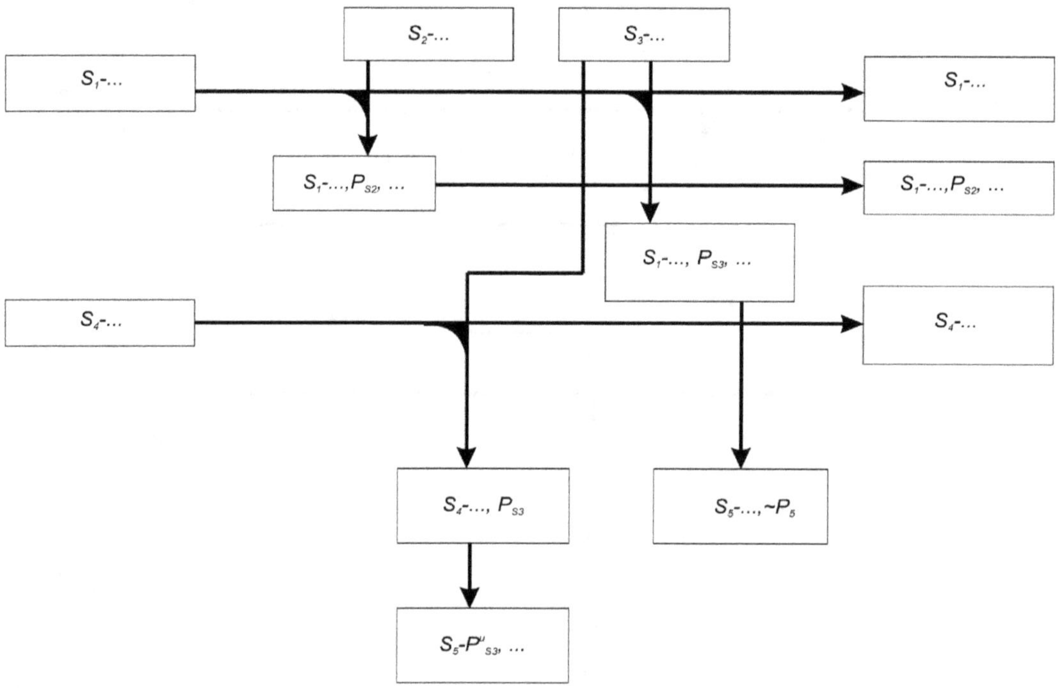

Figure 14h. The construction of event bush as a result of progressive understanding of the expert's knowledge: the eighth step.

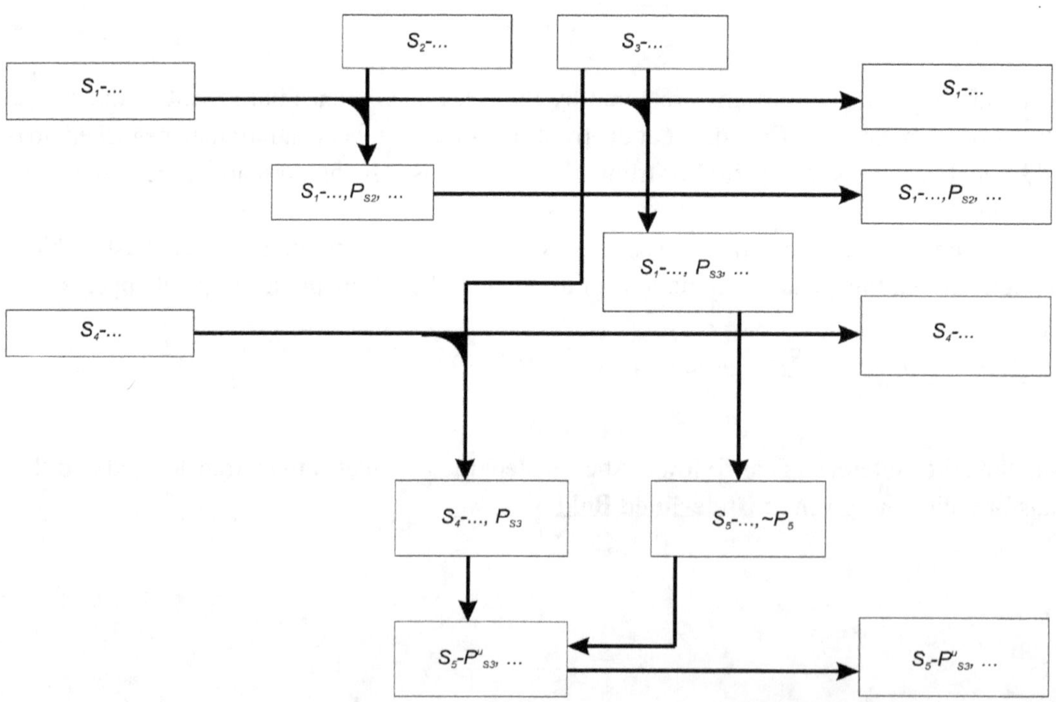

Step 1 (rule SM1)

$\underline{\mathbf{S_1–\ldots}}; S_1–\ldots, P_{S2}, \ldots; S_2–\ldots; S_2\rightarrow P_{S2}; S_3–\ldots; S_3\rightarrow P^u_{S3}; S_4–\ldots, P_{S3}, \ldots; S_1–\ldots, P_{S3}, \ldots; S_5–P^u_{S3}, \ldots; S_5–\ldots, \sim P_5, \ldots\Rightarrow$

$\underline{\mathbf{\Rightarrow S_1–P^{u}{'}, \ldots}}; S_1–\ldots, P_{S2}, \ldots; S_2–\ldots; S_2\rightarrow P_{S2}; S_3–\ldots; S_3\rightarrow P^u_{S3}; S_4–\ldots, P_{S3}, \ldots; S_1–\ldots, P_{S3}, \ldots; S_5–P^u_{S3}, \ldots; S_5–\ldots, \sim P_5, \ldots$

Step 2 (rule ST2: $P'=P^u_1$)

$\underline{\mathbf{S_1–P^{u}{'}, \ldots}}; S_1–\ldots, P_{S2}, \ldots; S_2–\ldots; S_2\rightarrow P_{S2}; S_3–\ldots; S_3\rightarrow P^u_{S3}; S_4–\ldots, P_{S3}, \ldots; S_1–\ldots, P_{S3}, \ldots; S_5–P^u_{S3}, \ldots; S_5–\ldots, \sim P_5, \ldots\Rightarrow$

$\underline{\mathbf{\Rightarrow S_1–P^u_1, \ldots}}; S_1–\ldots, P_{S2}, \ldots; S_2–\ldots; S_2\rightarrow P_{S2}; S_3–\ldots; S_3\rightarrow P^u_{S3}; S_4–\ldots, P_{S3}, \ldots; S_1–\ldots, P_{S3}, \ldots; S_5–P^u_{S3}, \ldots; S_5–\ldots, \sim P_5, \ldots$

Step 3 (rule SM8a)

$\underline{\mathbf{S_1–P^u_1, \ldots; S_1–\ldots, P_{S2}, \ldots}}; S_2–\ldots; S_2\rightarrow P_{S2}; S_3–\ldots; S_3\rightarrow P^u_{S3}; S_4–\ldots, P_{S3}; S_1–\ldots, P_{S3}, \ldots; S_5–P^u_{S3}, \ldots; S_5–\ldots, \sim P_5, \ldots\Rightarrow$

$\underline{\mathbf{\Rightarrow S_1–P^u_1, P_{S2}, \ldots; S_1–P^u_1, \sim P_{S2}, \ldots}}; S_2–\ldots; S_2\rightarrow P_{S2}; S_3–\ldots; S_3\rightarrow P^u_{S3}; S_4–\ldots, P_{S3}, \ldots; S_1–\ldots, P_{S3}, \ldots; S_5–P^u_{S3}, \ldots; S_5–\ldots, \sim P_5, \ldots$

Step 4 (rule SM8a)

$\underline{\mathbf{S_1–P^u_1, P_{S2}, \ldots; S_1–P^u_1, \sim P_{S2}, \ldots}}; S_2–\ldots; S_2\rightarrow P_{S2}; S_3–\ldots; S_3\rightarrow P^u_{S3}; S_4–\ldots, P_{S3}, \ldots; S_1–\ldots, P_{S3}, \ldots; S_5–P^u_{S3}, \ldots; S_5–\ldots, \sim P_5, \ldots\Rightarrow$

$\underline{\mathbf{\Rightarrow S_1–P^u_1, P_{S2}, P_{S3}, \ldots; S_1–P^u_1, \sim P_{S2}, P_{S3}, \ldots; S_1–P^u_1, P_{S2}, \sim P_{S3}}}, \ldots; S_1–P^u_1, \sim P_{S2}, \sim P_{S3}, \ldots; S_2–\ldots; S_2\rightarrow P_{S2}; S_3–\ldots; S_3\rightarrow P^u_{S3}; S_4–\ldots, P_{S3}, \ldots; S_5–P^u_{S3}, \ldots; S_5–\ldots, \sim P_5, \ldots$

Step 5 (rule ST3)

$\underline{\mathbf{S_1–P^u_1, P_{S2}, P_{S3}, \ldots; S_1–P^u_1, \sim P_{S2}, P_{S3}, \ldots; S_1–P^u_1, P_{S2}, \sim P_{S3}, \ldots}}; S_1–P^u_1, \sim P_{S2}, \sim P_{S3}, \ldots; S_2–\ldots; S_2\rightarrow P_{S2}; S_3–\ldots; S_3\rightarrow P^u_{S3}; S_4–\ldots, P_{S3}, \ldots; S_5–P^u_{S3}, \ldots; S_5–\ldots, \sim P_5, \ldots\Rightarrow$

$\underline{\mathbf{\Rightarrow S_1–P^u_1, P_{S2}, P_{S3}; S_1–P^u_1, \sim P_{S2}, P_{S3}; S_1–P^u_1, P_{S2}, \sim P_{S3}; S_1–P^u_1, \sim P_{S2}, \sim P_{S3}}}; S_2–\ldots; S_2\rightarrow P_{S2}; S_3–\ldots; S_3\rightarrow P^u_{S3}; S_4–\ldots, P_{S3}, \ldots; S_5–P^u_{S3}, \ldots; S_5–\ldots, \sim P_5, \ldots$

Step 6 (rule SM10)

Herewith, let us accept an axiom, $(S_1–P^u{}_{1'}, P_{S2'}, P_{S3})=\varnothing$.

$\underline{\mathbf{S_1–P^u{}_{1'} P_{S2'} P_{S3}; S_1–P^u{}_{1'} \sim P_{S2'} P_{S3}; S_1–P^u{}_{1'} P_{S2'} \sim P_{S3}; S_1–P^u{}_{1'} \sim P_{S2'} \sim P_{S3}}}; S_2–\ldots; S_2\to P_{S2}; S_3–\ldots; S_3\to P^u{}_{S3}; S_4–\ldots, P_{S3}, \ldots; S_5–P^u{}_{S3}, \ldots; S_5–\ldots, \sim P_5, \ldots\Rightarrow$

$\Rightarrow\underline{\mathbf{S_1–P^u{}_{1'} \sim P_{S2'} P_{S3}; S_1–P^u{}_{1'} P_{S2'} \sim P_{S3}; S_1–P^u{}_{1'} \sim P_{S2'} \sim P_{S3}}}; S_2–\ldots; S_2\to P_{S2}; S_3–\ldots; S_3\to P^u{}_{S3}; S_4–\ldots, P_{S3}, \ldots; S_5–P^u{}_{S3}, \ldots; S_5–\ldots, \sim P_5, \ldots$

Step 7 (rule SM2a)

$S_1–P^u{}_{1'}, \sim P_{S2'}, P_{S3}; S_1–P^u{}_{1'}, P_{S2'}, \sim P_{S3}; S_1–P^u{}_{1'}, \sim P_{S2'}, \sim P_{S3}; S_2–\ldots; S_2\to P_{S2}; S_3–\ldots; S_3\to P^u{}_{S3}; \underline{\mathbf{S_4–\ldots, P_{S3}, \ldots}}; S_5–P^u{}_{S3}, \ldots; S_5–\ldots, \sim P_5, \ldots\Rightarrow$

$\Rightarrow S_1–P^u{}_{1'}, \sim P_{S2'}, P_{S3}; S_1–P^u{}_{1'}, P_{S2'}, \sim P_{S3}; S_1–P^u{}_{1'}, \sim P_{S2'}, \sim P_{S3}; S_2–\ldots; S_2\to P_{S2}; S_3–\ldots; S_3\to P^u{}_{S3}; \underline{\mathbf{S_4–P'', P_{S3}, \ldots; S_4–P'', \sim P_{S3}, \ldots}}; S_5–P^u{}_{S3}, \ldots; S_5–\ldots, \sim P_5, \ldots$

Step 8 (rule ST2: $P^{u''}=P^u{}_2$)

$S_1–P^u{}_{1'}, \sim P_{S2'}, P_{S3}; S_1–P^u{}_{1'}, P_{S2'}, \sim P_{S3}; S_1–P^u{}_{1'}, \sim P_{S2'}, \sim P_{S3}; S_2–\ldots; S_2\to P_{S2}; S_3–\ldots; S_3\to P^u{}_{S3}; \underline{\mathbf{S_4–P'', P_{S3}, \ldots; S_4–P'', \sim P_{S3}, \ldots}}; S_5–P^u{}_{S3}, \ldots; S_5–\ldots, \sim P_5, \ldots\Rightarrow$

$\Rightarrow S_1–P^u{}_{1'}, \sim P_{S2'}, P_{S3}; S_1–P^u{}_{1'}, P_{S2'}, \sim P_{S3}; S_1–P^u{}_{1'}, \sim P_{S2'}, \sim P_{S3}; S_2–\ldots; S_2\to P_{S2}; S_3–\ldots; S_3\to P^u{}_{S3}; \underline{\mathbf{S_4–P^u{}_2, P_{S3}, \ldots; S_4–P^u{}_2, \sim P_{S3}, \ldots}}; S_5–P^u{}_{S3}, \ldots; S_5–\ldots, \sim P_5, \ldots$

Step 9 (rule SM1):

$S_1–P^u{}_{1'}, \sim P_{S2'}, P_{S3}; S_1–P^u{}_{1'}, P_{S2'}, \sim P_{S3}; S_1–P^u{}_{1'}, \sim P_{S2'}, \sim P_{S3}; \underline{\mathbf{S_2–\ldots}}; S_2\to P_{S2}; S_3–\ldots; S_3\to P^u{}_{S3}; S_4–P^u{}_2, P_{S3}, \ldots; S_4–P^u{}_2, \sim P_{S3}, \ldots; S_5–P^u{}_{S3}, \ldots; S_5–\ldots, \sim P_5, \ldots\Rightarrow$

$\Rightarrow S_1–P^u{}_{1'}, \sim P_{S2'}, P_{S3}; S_1–P^u{}_{1'}, P_{S2'}, \sim P_{S3}; S_1–P^u{}_{1'}, \sim P_{S2'}, \sim P_{S3}; \underline{\mathbf{S_2–P^{u'''}, \ldots}}; S_2\to P_{S2}; S_3–\ldots; S_3\to P^u{}_{S3}; S_4–P^u{}_2, P_{S3}, \ldots; S_4–P^u{}_2, \sim P_{S3}, \ldots; S_5–P^u{}_{S3}, \ldots; S_5–\ldots, \sim P_5, \ldots$

Step 9 (rule ST2: $P^{u'''}=P^u{}_3$)

$S_1–P^u{}_{1'}, \sim P_{S2'}, P_{S3}; S_1–P^u{}_{1'}, P_{S2'}, \sim P_{S3}; S_1–P^u{}_{1'}, \sim P_{S2'}, \sim P_{S3}; \underline{\mathbf{S_2–P^{u'''}, \ldots}}; S_2\to P_{S2}; S_3–\ldots; S_3\to P^u{}_{S3}; S_4–P^u{}_2, P_{S3}; S_4–P^u{}_2, \sim P_{S3}, \ldots; S_5–P^u{}_{S3}, \ldots; S_5–\ldots, \sim P_5, \ldots\Rightarrow$

$\Rightarrow S_1–P^u{}_{1'}, \sim P_{S2'}, P_{S3}; S_1–P^u{}_{1'}, P_{S2'}, \sim P_{S3}; S_1–P^u{}_{1'}, \sim P_{S2'}, \sim P_{S3}; \underline{\mathbf{S_2–P^u{}_3, \ldots}}; S_2\to P_{S2}; S_3–\ldots; S_3\to P^u{}_{S3}; S_4–P^u{}_2, P_{S3}, \ldots; S_4–P^u{}_2, \sim P_{S3}, \ldots; S_5–P^u{}_{S3}, \ldots; S_5–\ldots, \sim P_5, \ldots$

Step 10 (rule SM1):

$S_1 - P^u_{1'}, \sim P_{S2'}, P_{S3'}; S_1 - P^u_{1'}, P_{S2'}, \sim P_{S3'}; S_1 - P^u_{1'}, \sim P_{S2'}, \sim P_{S3'}; S_2 - P^u_{3'}, \ldots; S_2 \to P_{S2'}; \underline{\mathbf{S_3 - \ldots}}; S_3 \to P^u_{S3'}; S_4 - P^u_{2'}, P_{S3'}, \ldots; S_4 - P^u_{2'}, \sim P_{S3'}, \ldots; S_5 - P^u_{S3'}, \ldots; S_5 - \ldots, \sim P_{5'}, \ldots \Rightarrow$

$\Rightarrow S_1 - P^u_{1'}, \sim P_{S2'}, P_{S3'}; S_1 - P^u_{1'}, P_{S2'}, \sim P_{S3'}; S_1 - P^u_{1'}, \sim P_{S2'}, \sim P_{S3'}; S_2 - P^u_{3'}, \ldots; S_2 \to P_{S2'}; \underline{\mathbf{S_3 - P^{u''''}, \ldots}}; S_3 \to P^u_{S3'}; S_4 - P^u_{2'}, P_{S3'}, \ldots; S_4 - P^u_{2'}, \sim P_{S3'}, \ldots; S_5 - P^u_{S3'}, \ldots; S_5 - \ldots, \sim P_{5'}, \ldots$

Step 11 (rule ST2: $P^{u''''} = P^u_4$)

$S_1 - P^u_{1'}, \sim P_{S2'}, P_{S3'}; S_1 - P^u_{1'}, P_{S2'}, \sim P_{S3'}; S_1 - P^u_{1'}, \sim P_{S2'}, \sim P_{S3'}; S_2 - P^u_{3'}, \ldots; S_2 \to P_{S2'}; \underline{\mathbf{S_3 - P^{u''''}, \ldots}}; S_3 \to P^u_{S3'}; S_4 - P^u_{2'}, P_{S3'}, \ldots; S_4 - P^u_{2'}, \sim P_{S3'}, \ldots; S_5 - P^u_{S3'}, \ldots; S_5 - \ldots, \sim P_{5'}, \ldots \Rightarrow$

$\Rightarrow S_1 - P^u_{1'}, \sim P_{S2'}, P_{S3'}; S_1 - P^u_{1'}, P_{S2'}, \sim P_{S3'}; S_1 - P^u_{1'}, \sim P_{S2'}, \sim P_{S3'}; S_2 - P^u_{3'}; S_2 \to P_{S2'}; \underline{\mathbf{S_3 - P^u_{4'}, \ldots}}; S_3 \to P^u_{S3'}; S_4 - P^u_{2'}, P_{S3'}, \ldots; S_4 - P^u_{2'}, \sim P_{S3'}, \ldots; S_5 - P^u_{S3'}, \ldots; S_5 - \ldots, \sim P_{5'}, \ldots$

Step 12 (rule ST3)

$S_1 - P^u_{1'}, \sim P_{S2'}, P_{S3'}; S_1 - P^u_{1'}, P_{S2'}, \sim P_{S3'}; S_1 - P^u_{1'}, \sim P_{S2'}, \sim P_{S3'}; S_2 - P^u_{3'}; S_2 \to P_{S2'}; \underline{\mathbf{S_3 - P^u_{4'}, \ldots}}; S_3 \to P^u_{S3'}; S_4 - P^u_{2'}, P_{S3'}, \ldots; S_4 - P^u_{2'}, \sim P_{S3'}, \ldots; S_5 - P^u_{S3'}, \ldots; S_5 - \ldots, \sim P_{5'}, \ldots \Rightarrow$

$\Rightarrow S_1 - P^u_{1'}, \sim P_{S2'}, P_{S3'}; S_1 - P^u_{1'}, P_{S2'}, \sim P_{S3'}; S_1 - P^u_{1'}, \sim P_{S2'}, \sim P_{S3'}; S_2 - P^u_{3'}, \ldots; S_2 \to P_{S2'}; \underline{\mathbf{S_3 - P^u_{4'}, \ldots}}; S_3 \to P^u_{S3'}; S_4 - P^u_{2'}, P_{S3'}, \ldots; S_4 - P^u_{2'}, \sim P_{S3'}, \ldots; S_5 - P^u_{S3'}, \ldots; S_5 - \ldots, \sim P_{5'}, \ldots$

Step 13 (rule SM9b)

$S_1 - P^u_{1'}, \sim P_{S2'}, P_{S3'}; S_1 - P^u_{1'}, P_{S2'}, \sim P_{S3'}; S_1 - P^u_{1'}, \sim P_{S2'}, \sim P_{S3'}; S_2 - P^u_{3'}, \ldots; S_2 \to P_{S2'}; S_3 - P^u_{4'}, \ldots; S_3 \to P^u_{S3'}; S_4 - P^u_{2'}, P_{S3'}, \ldots; \underline{\mathbf{S_5 - P^u_{S3'}, \ldots; S_5 - \ldots, \sim P_{5'}, \ldots \Rightarrow}}$

$\Rightarrow S_1 - P^u_{1'}, \sim P_{S2'}, P_{S3'}; S_1 - P^u_{1'}, P_{S2'}, \sim P_{S3'}; S_1 - P^u_{1'}, \sim P_{S2'}, \sim P_{S3'}; S_2 - P^u_{3'}, \ldots; S_2 \to P_{S2'}; S_3 - P^u_{4'}, \ldots; S_3 \to P^u_{S3'}; S_4 - P^u_{2'}, P_{S3'}, \ldots; S_4 - P^u_{2'}, \sim P_{S3'}, \ldots; \underline{\mathbf{S_5 - P^u_{S3'}, P_{5'}, \ldots; S_5 - P^u_{S3'}, \sim P_{5'}, \ldots}}$

Step 14 (rule ST3)

$S_1 - P^u_{1'}, \sim P_{S2'}, P_{S3'}; S_1 - P^u_{1'}, P_{S2'}, \sim P_{S3'}; S_1 - P^u_{1'}, \sim P_{S2'}, \sim P_{S3'}; S_2 - P^u_{3'}, \ldots; S_2 \to P_{S2'}; S_3 - P^u_{4'}, \ldots; S_3 \to P^u_{S3'}; S_4 - P^u_{2'}, P_{S3'}, \ldots; \underline{\mathbf{S_5 - P^u_{S3'}, P_{5'}, \ldots; S5 - P^u_{S3'}, \sim P_{5'}, \ldots \Rightarrow}}$

$\Rightarrow S_1 - P^u_{1'}, \sim P_{S2'}, P_{S3'}; S_1 - P^u_{1'}, P_{S2'}, \sim P_{S3'}; S_1 - P^u_{1'}, \sim P_{S2'}, \sim P_{S3'}; S_2 - P^u_{3'}, \ldots; S_2 \to P_{S2'}; S_3 - P^u_{4'}, \ldots; S_3 \to P^u_{S3'}; S_4 - P^u_{2'}, P_{S3'}, \ldots; S_4 - P^u_{2'}, \sim P_{S3'}, \ldots; \underline{\mathbf{S_5 - P^u_{S3'}, P_{5'}; S_5 - P^u_{S3'}, \sim P_5}}$

Step 15 (rule ST3)

$S_1 - P^u_{1'}, \sim P_{S2'}, P_{S3'}; S_1 - P^u_{1'}, P_{S2'}, \sim P_{S3'}; S_1 - P^u_{1'}, \sim P_{S2'}, \sim P_{S3'}; \underline{\mathbf{S_2 - P^u_{3'}, \ldots}}; S_2 \to P_{S2'}; S_3 - P^u_{4'}, \ldots; S_3 \to P^u_{S3'}; S_4 - P^u_{2'}, P_{S3'}, \ldots; S_5 - P^u_{S3'}, P_{5'}; S_5 - P^u_{S3'}, \sim P_5 \Rightarrow$

$\Rightarrow S_{1}-P^{u}_{1'}, \sim P_{S2'}, P_{S3'}; S_{1}-P^{u}_{1'}, P_{S2'}, \sim P_{S3'}; S_{1}-P^{u}_{1'}, \sim P_{S2'}, \sim P_{S3'}; \underline{\mathbf{S_{2}-P^{u}_{3'}}}; S_{2}\rightarrow P_{S2'}; S_{3}-P^{u}_{4'}, \ldots; S_{3}\rightarrow P^{u}_{S3'}; S_{4}-P^{u}_{2'}, P_{S3'}, \ldots; S_{4}-P^{u}_{2'}, \sim P_{S3'}, \ldots; S_{5}-P^{u}_{S3'}, P_{5'}; S_{5}-P^{u}_{S3'}, \sim P_{5}$

Step 16 (rule ST3)

$S_{1}-P^{u}_{1'}, \sim P_{S2'}, P_{S3'}; S_{1}-P^{u}_{1'}, P_{S2'}, \sim P_{S3'}; S_{1}-P^{u}_{1'}, \sim P_{S2'}, \sim P_{S3'}; S_{2}-P^{u}_{3'}; S_{2}\rightarrow P_{S2'}; \underline{\mathbf{S_{3}-P^{u}_{4}, \ldots}}; S_{3}\rightarrow P^{u}_{S3'}; S_{4}-P^{u}_{2'}, P_{S3'}, \ldots; S_{5}-P^{u}_{S3'}, P_{5'}; S_{5}-P^{u}_{S3'}, \sim P_{5}\Rightarrow$

$\Rightarrow S_{1}-P^{u}_{1'}, \sim P_{S2'}, P_{S3'}; S_{1}-P^{u}_{1'}, P_{S2'}, \sim P_{S3'}; S_{1}-P^{u}_{1'}, \sim P_{S2'}, \sim P_{S3'}; S_{2}-P^{u}_{3'}; S_{2}\rightarrow P_{S2'}; \underline{\mathbf{S_{3}-P^{u}_{4'}}}; S_{3}\rightarrow P^{u}_{S3'}; S_{4}-P^{u}_{2'}, P_{S3'}, \ldots; S_{4}-P^{u}_{2'}, \sim P_{S3'}, \ldots; S_{5}-P^{u}_{S3'}, P_{5'}; S_{5}-P^{u}_{S3'}, \sim P_{5}$

Step 17 (rule ST3)

$S_{1}-P^{u}_{1'}, \sim P_{S2'}, P_{S3'}; S_{1}-P^{u}_{1'}, P_{S2'}, \sim P_{S3'}; S_{1}-P^{u}_{1'}, \sim P_{S2'}, \sim P_{S3'}; S_{2}-P^{u}_{3'}; S_{2}\rightarrow P_{S2'}; S_{3}-P^{u}_{4'}; S_{3}\rightarrow P^{u}_{S3'}; \underline{\mathbf{S_{4}-P^{u}_{2'}, P_{S3'}, \ldots}}; \underline{\mathbf{S_{4}-P^{u}_{2'}, \sim P_{S3'}, \ldots}}; S_{5}-P^{u}_{S3'}, P_{5'}; S_{5}-P^{u}_{S3'}, \sim P_{5}\Rightarrow$

$\Rightarrow S_{1}-P^{u}_{1'}, \sim P_{S2'}, P_{S3'}; S_{1}-P^{u}_{1'}, P_{S2'}, \sim P_{S3'}; S_{1}-P^{u}_{1'}, \sim P_{S2'}, \sim P_{S3'}; S_{2}-P^{u}_{3'}; S_{2}\rightarrow P_{S2'}; S_{3}-P^{u}_{4'}; S_{3}\rightarrow P^{u}_{S3'}; \underline{\mathbf{S_{4}-P^{u}_{2'}, P_{S3'}}}; \underline{\mathbf{S_{4}-P^{u}_{2'}, \sim P_{S3'}}}; S_{5}-P^{u}_{S3'}, P_{5'}; S_{5}-P^{u}_{S3'}, \sim P_{5}$

Metasymbols are not present in the sequent right of the turnstile. Verifying the disclosure of meanings in this sequent, it is readily seen that the following incomplete formulae should be added:

$\ldots\rightarrow P^{u}_{1'};$

$\ldots\rightarrow P^{u}_{3'};$

$\ldots\rightarrow P^{u}_{4'};$

$\ldots\rightarrow P^{u}_{2'};$

$\ldots\rightarrow P_{5'}.$

Applying the rules SM7a, b, we obtain

$V'\rightarrow P^{u}_{1'};$

$V''\rightarrow P^{u}_{3'};$

$V'''\rightarrow P^{u}_{4'};$

$V''''\rightarrow P^{u}_{2'};$

$V'''''\rightarrow P_{5'}.$

Now it is on our discretion what relation axiom to choose to each *V'*, *V=S* or *V=E*. Once we see no need to widen our consideration, we may choose the former option to all:

$S_6 \to P^u_1$;

$S_7 \to P^u_3$;

$S_8 \to P^u_4$;

$S_9 \to P^u_2$;

$S_{10} \to P_5$

and thus finish the story. Then the sequent in question will take the form:

$S_1{*}{-}P^u_1{*}, \sim P_{S2}{*}, P_{S3}{*}; S_1{*}{-}P^u_1{*}, P_{S2}{*}, \sim P_{S3}{*}; S_1{*}{-}P^u_1{*}, \sim P_{S2}{*}, \sim P_{S3}{*}; S_2{*}{-}P^u_3{*}; S_2{*}{\to}P_{S2}{*}; S_3{*}{-}P^u_4{*}; S_3{*}{\to}P^u_{S3}{*}; S_4{*}{-}P^u_2{*}, P_{S3}{*}; S_4{*}{-}P^u_2{*}, \sim P_{S3}{*}; S_5{*}{-}P^u_{S3}{*}, P_5{*}; S_5{*}{-}P^u_{S3}{*}, \sim P_5{*};$ $\underline{\mathbf{S_6{*}{\to}P^u_1{*}; S_7{*}{\to}P^u_3{*}; S_8{*}{\to}P^u_4{*}; S_9{*}{\to}P^u_2{*}; S_{10}{*}{\to}P_5{*}}}$.

This sequent is mature. The proof is over. Steps 2, 6, 8, 9 and 11 show that initial sequent was not enough for it, and if no information is added, mature sequent would not be inferred. We expect, though cannot prove this at present, that not more than one mature sequent may follow from a raw one. The mature sequent discloses all provable relations between the multitudes that join its formulae. The procedure of Novorossian inference seems to be readily algorithmized and automated.

The sequent proven above can be expressed graphically as shown in Figure 15. Based on this, final event bush can be obtained (Figure 16).

It would be pertinent now to list all the meanings that have finally occurred in the considered context:

- S_1 – draftees of anti-Hitler coalition countries;
- S_2 – Wehrmacht weapons;
- S_3 – Wehrmacht;
- S_4 – civilians of anti-Hitler coalition countries;
- S_5 – prisoners of Hitler concentration camps;
- S_6 – front;
- S_7 – fire;
- S_8 – capture;
- S_9 – occupation;
- S_{10} – military crime;
- P^u_1 – to occur at the front;
- P^u_2 – to occur in occupation;
- P^u_3 – to deliver fire;
- P^u_4 – to capture;

Figure 15. Predmetka showing the relations between the composite meanings in the proven sequent – see comments in the text

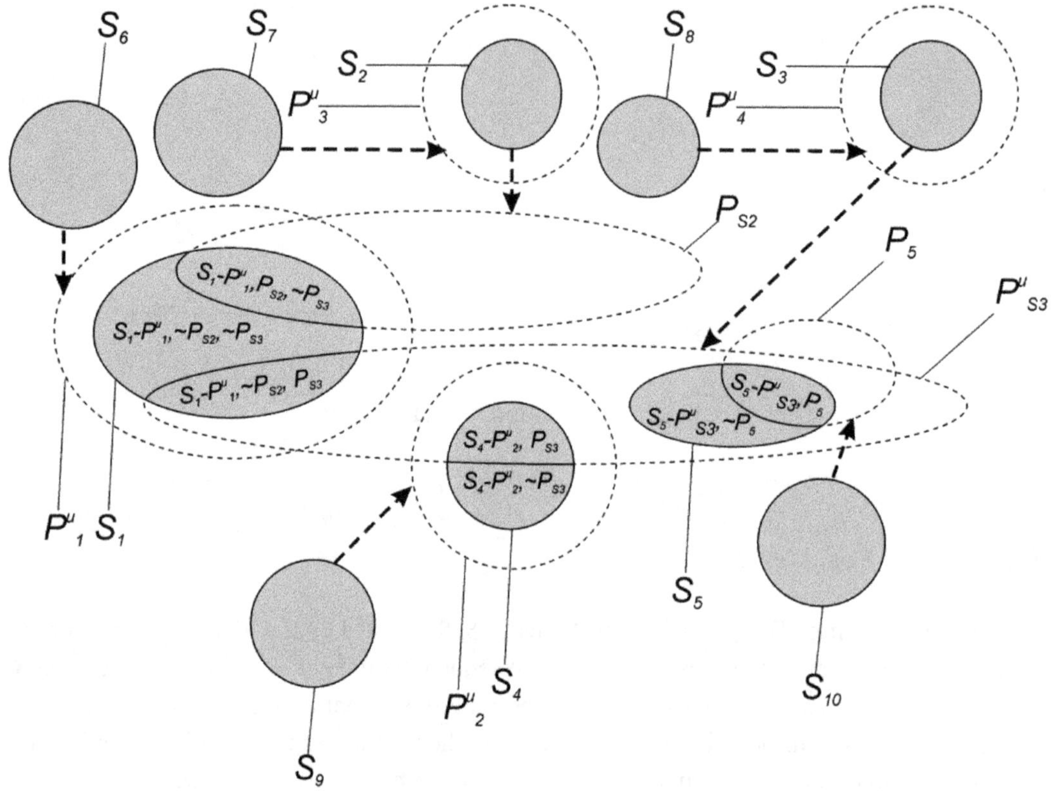

Figure 16. The event bush in final form based on the predmetka

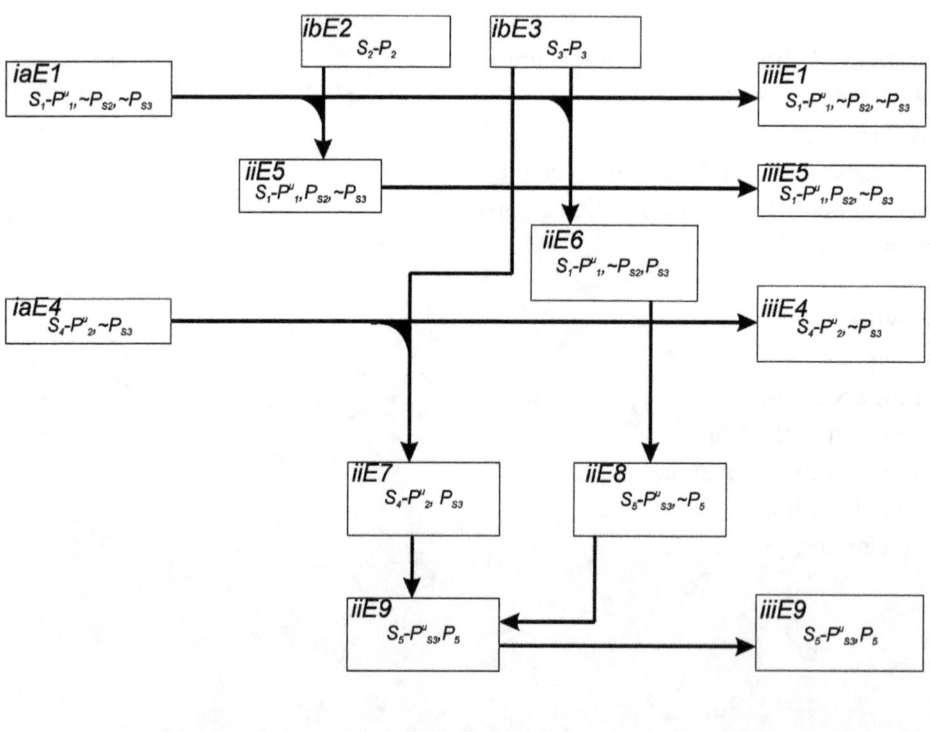

- P_5 – to be victim of military crime;
- P_{S2} – to be affected by a weapon;
- P_{S3} – to be captured.

The fill of the predmets is different. S_1, S_3, S_4, S_5 have grainy fill, and S_2, S_6, S_7, S_8, S_9, S_{10}, lump fill. Importantly, S_2 and S_{10} have lump fill too because are considered as one thing, without specification.

S_1 (draftees of anti-Hitler coalition countries) has indiscernible atoms implying "draftees" as a virtually countless mass of people, and so do other predmets. Then, aiming to consider a destiny of particular draftee Billy Piligrim (or his prototype Kurt Vonnegut) from the US who was captured, one should introduce a predmet "Billy Piligrim" ("Kurt Vonnegut") with lump fill, form a primeta "drafted by an anti-Hitler coalition country" by father-daughter relation from the predmet S_1 and consider this primeta as a universum for the predmet "Kurt Vonnegut". Then in the same way primeta should be formed from the predmet S_5 (prisoners of Hitler concentration camps) and applied as another universum for "Billy Piligrim" ("Kurt Vonnegut").

Another Interpretation

The mature sequent and predmetka in Figure 15, as well as the event bush based on it, represent a formal model that may have various interpretations. Thus, it may appear relevant in a few domains of knowledge, e.g., in geology. Suppose the following:

- S_1 – sandstones;
- S_2 – tectonic uplift;
- S_3 – tectonic downsagging;
- S_4 – granites;
- S_5 – silicic crustal magmatic melt;
- S_6 – formation by cementation of sand;
- S_7 – earthcrust block uplift;
- S_8 – earthcrust block drown;
- S_9 – formation by crystallization of silicic magmatic melt;
- S_{10} – crystallization;
- P^u_1 – to be formed by cementation of sand;
- P^u_2 – to be formed by crystallization of silicic magmatic melt;
- P^u_3 – to lift up blocks of the earthcrust;
- P^u_4 – to drop down blocks of the earthcrust;
- P_5 – to crystallize;
- P_{S2} – to disintegrate;
- P_{S3} – to be melted.

The expert's knowledge in this field may be expressed as follows.

Sandstones may equally well disintegrate in case of tectonic uplift, or, on the contrary, if downsagging occurs, be fused at depth and become crustal magmatic melt of silicic composition. Similar stuff may originate from melting of granites if they are driven down by tectonic downsagging. The silicic melt, even in crustal conditions, once appeared, unlikely will immediately crystallize back.

What a knowledge engineer may do with this knowledge is presented below.
Sandstones – S_1–... (the first predmet has appeared);

... in case of tectonic uplift – S_2–... (another predmet);

... may ... disintegrate in case of tectonic uplift – $S_2 \rightarrow P_{S2}$;

Sandstones may ... disintegrate – S_1–..., P_{S2};

...if downsagging occurs... – i.e., tectonic downsagging appears on scene – S_3–...;

...if downsagging occurs, be fused at depth – i.e., downsagging, similar to uplift, gives birth to a primeta of another predmet (sandstone) – $S_3 \rightarrow P_{S3}$;

Sandstones may... be fused at depth – S_1–..., P_{S3};

... melting of granites – S_4–..., P_{S3};

Similar stuff (i.e., silicic melt in the earthcrust) may originate – S_5–P^u_{S3}, ... (the fact that new predmet, "similar stuff", comes from another predmet, granites, given it has primeta "be melted" born by the predmet "downsagging", is clearly seen in the event bush structure);

The silicic melt, even in crustal conditions, once appeared, unlikely will immediately crystallize back – S_5–..., $\sim P_5$.

The rest of the pathway is the same as in the case of historical knowledge, and the fill of the corresponding predmets is similar.

FUTURE WORK

Further development of quantitative modeling methodology based on the multitude theory may proceed in two interrelated dimensions:

- Elaboration of structures in an environment and
- Linking with existing mathematical theories.

Further research into the regularities of context structure is an intriguing endeavor that could, the same time, be helpful in a vast diversity of fields, from conceptual modeling to building ideological doctrines. This must be the topmost part of the Novorossian logic.

Different mathematical theories are compatible with the reported formalism contingent to the type of fill of the events – singular element (or none), multiple elements, or flow. In case of singular element, transition from event to event may be crisp or fuzzy as prescribed by the fuzzy sets theory (Zadeh, 1965).

Multiple elements may move all or not – then the probabilistic approach including Bayesian computation (Jensen, 2001) is possible in regard of one element or *n* elements. For singular and multiple elements, postbinary computing may appear promising for more accurate description of transition of elements (Anoprienko and Ivanitsa, 2001). For a flow, the whole diversity of models describing the flow and field are applicable. In all cases, the belief theory of Dempster and Shafer (Shafer, 1976) can be used to assess the situation externally, from the observer's viewpoint. Also, traditional deterministic mathematical and physical models can be well parsed at least by the event bush, as demonstrated by Diviacco et al. (2014).

The whole kit of mathematical formalisms that can be associated with the modeling of environments may be termed, by analogy with the Novorossian logic, the Novorossian mathematics. The opportunities the Novorossian mathematics can open to modeling the world are yet hard to encompass but will hopefully be inspirational for the scientific community.

CONCLUSION

Based on the properties of meaning, a formal theory called Novorossian logic is developed. It suggests its own calculus that includes both structural and semantic rules of inference. This calculus allows to identify whether the pieces of information available are enough to build an integer, finite, complete and self-consistent context.

Disclosure of all provable relations in a context should be considered an ideal result of a knowledge engineering work. Various methods of elicitation of knowledge (sequence diagrams, event bush and others) used or usable in the knowledge engineering lead to results that should be regarded as preliminary to, or a part of, an integer mature sequent. Theoretical grounds for these methods can be formulated in terms of the Novorossian logic.

Inhabiting some of the multitudes in this context with various kinds of fill (atoms, lump or field), a powerful tool for modeling is obtained, well compatible with various mathematical theories and modeling methodologies. This tool, the Novorossian mathematics, may give a universal framework for various and diverse mathematical and computational methods and lead from representation and engineering of dynamic knowledge about the world to modeling and understanding the world itself.

ACKNOWLEDGMENT

The author is deeply obliged to Uwe Wolter for his constructive criticism and valuable insights partly developed here, and partly still awaiting their turn.

REFERENCES

Anoprienko, A. Ya., & Ivanitsa, S. V. (2001). Postbinarnyi kompiuting i interval'nye vychisleniya v kontekste kodo-logicheskoi evolutsii [Post-Binary Computing and Interval Computation in the Context of Code-Logical Evolution]. Donetsk National Technical University – UNITEX Publishers.

Barr, M., & Wells, C. (2012). Category Theory for Computing Science. *Reprints in Theory and Applications of Categories*, *22*, 1–538.

Diviacco, P., Pshenichny, C., Carniel, R., Khrabrykh, Z., Shterkhun, V., Mouromtsev, D., & Pascolo, P. et al. (2014). Organization of a geophysical information space by using an event-bush-based collaborative tool. *Earth Science Informatics*, *4*(3), 139–160. doi:10.1007/s12145-014-0182-2

Ehrig, H., Ehrig, K., Prange, U., & Taentzer, G. (2006). *Fundamentals of Algebraic Graph Transformations*. Berlin: Springer.

Gentzen, G. (1934). Untersuchungen über das logische Schliessen [Investigations into logical deduction]. In The Collected Papers of Gerhard Gentzen (pp. 68-128). Amsterdam: North-Holland.

Gruber, T. R. (1993). Toward Principles for the Design of Ontologies Used for Knowledge Sharing. *International Journal Human-Computer Studies, 43,* 907-928.

Jensen, F. V. (2001). *Bayesian Networks and Decision Graphs*. New York: UCL Press. doi:10.1007/978-1-4757-3502-4

Naumov, A., Popov, I., Bondarenko, I., Krylov, B., Timonin, R., & Ofitserov, I. (2018). Modeling of Transaction Scripts Using Dynamic Knowledge Representation. In C. A. Pshenichny, P. Diviacco, & D. I. Mouromtsev (Eds.), *Representation of Dynamic Knowledge in Scientific Domains*. IGI Global.

Pearl, J. (1988). *Probabilistic Reasoning in Intelligent Systems: Networks of Plausible Inference*. San Mateo, CA: Morgan Kaufmann.

Petri, C. A. (1973). Concepts of Net Theory. Mathematical Foundations of Computer Science. *Proc. of Symposium and Summer School*, 137-146.

Pshenichny, C. A. (2018). Theory of Multitudes as an Alternative to the Set Theory. In C. A. Pshenichny, P. Diviacco, & D. I. Mouromtsev (Eds.), *Representation of Dynamic Knowledge in Scientific Domains*. IGI Global.

Pshenichny, C. A., & Kanzheleva, O. M. (2011). Theoretical Foundations of the Event Bush. In K. Sinha, L. Gundersen, J. Jackson, & D. Arctur (Eds.), Geological Society of America Special Paper: Vol. 482. Societal Challenges and Geoinformatics (pp. 139–165). Academic Press. doi:10.1130/2011.2482(12)

Pshenichny, C. A., & Mouromtsev, D. I. (2015). Grammar of Dynamic Knowledge for Collaborative Knowledge Engineering and Representation. In Collaborative Knowledge in Scientific Research Networks. IGI Global. doi:10.4018/978-1-4666-6567-5.ch016

Pshenichny, C.A., Nikolenko, S.I., Carniel, R., Vaganov, P.A., Khrabrykh, Z.V., Moukhachov, V.P., … Rezyapkin, A.A. (2009). The Event Bush as a Semantic-based Numerical Approach to Natural Hazard Assessment (Exemplified by Volcanology). *Computers and Geosciences, 35*(5), 1017-1034. doi: 10.1016/j.cageo.2008.01.009

Shafer, G. (1976). *A Mathematical Theory of Evidence*. Princeton, NJ: Princeton University Press.

Wolter, U., Korableva, O., & Solovyov, N. (2018). The Event Bush Method in the Light of Typed Graphs. In C. A. Pshenichny, P. Diviacco, & D. I. Mouromtsev (Eds.), *Representation of Dynamic Knowledge in Scientific Domains*. IGI Global.

Zadeh, L. (1965). Fuzzy sets. *Information and Control*, *8*(3), 338–353. doi:10.1016/S0019-9958(65)90241-X

Chapter 3
Scenarios in Event Bushes:
A Formal Proposal

Uwe Wolter
University of Bergen, Norway

ABSTRACT

This chapter presents a further step of the formalization of the event bush method by means of graphs and graph homomorphisms. First, the authors revise an analysis of the structure of event bushes by means of of bipartite structure graphs that are equipped with a multi-flow structure. They consider event bushes as a compact description of a set of alternative scenarios of directed changes of environments. They give a first formalization of the concept of scenario. A snapshot describes the state of an environment, which evolves in accordance with a given scenario at a certain moment in time. A sequence of snapshots describes how an environment may evolve over time. Cartoons are maximal sequences of snapshots. The authors discuss, define, and exemplify the concepts scenario, snapshot, and cartoon.

INTRODUCTION

Event bushes constitute a visual and semi-formal method to represent knowledge about real world phenomena. The event bush method has been developing over the years by applying it to a variety of areas besides the original motivating area of geoscience. For the original description of event bushes and examples, we refer the reader to (Pshenichny & Kanzheleva 2011, Wolter et al. 2018) and the literature referenced in these papers. Despite its many applications, the event bush method is still lacking a proper mathematical foundation, which also establishes a proper basis for designing and implementing computer-based tools. The development of such a proper mathematical foundation is the objective of our ongoing research. In (Wolter et al. 2018) we formalized the structure of event bushes by means of directed graphs and homomorphisms between directed graphs. The present paper addresses the dynamic aspects of the event bush method.

More precisely, we are going to discuss and to outline a formalization of the purpose of event bushes as proclaimed in (Pshenichny & Kanzheleva 2011):

DOI: 10.4018/978-1-5225-5261-1.ch003

The event bush method addresses a particular yet very wide type of (geo) environment, that of directed alternative changes, which is likely to occur in many information domains.

By environment, we mean a part of the real world we are interested in to describe. This may be a volcano, a lake, a transport system in a town, a segment of market of pharmaceutical technologies, a computer game, or others. To answer the question "What is changing?" we have to introduce a concept of state of an environment and discuss in what way an event bush describes, at the same time, as well the possible states as the possible changes of the states of an environment. We have to identify the sources of alternatives of changes and we have to formulate the properties of an event bush ensuring directedness of changes.

The paper is organized as follows. In section "Structure of event bushes", we give a revised description of the structure of event bushes by means of bipartite structure graphs that are equipped with a multiflow structure. We consider event bushes as a compact description of a set of alternative scenarios of directed changes of an environment. In section "Scenarios", we discuss this viewpoint and outline a formalization of the concept of scenario. In section "Snapshots and cartoons" we discuss snapshots that describe the state of an environment, that evolves in accordance with a given scenario, at a certain moment in time. A sequence of snapshots describes how an environment evolves over time in accordance with a given

Figure 1. A sample event bush "Blocked transport"

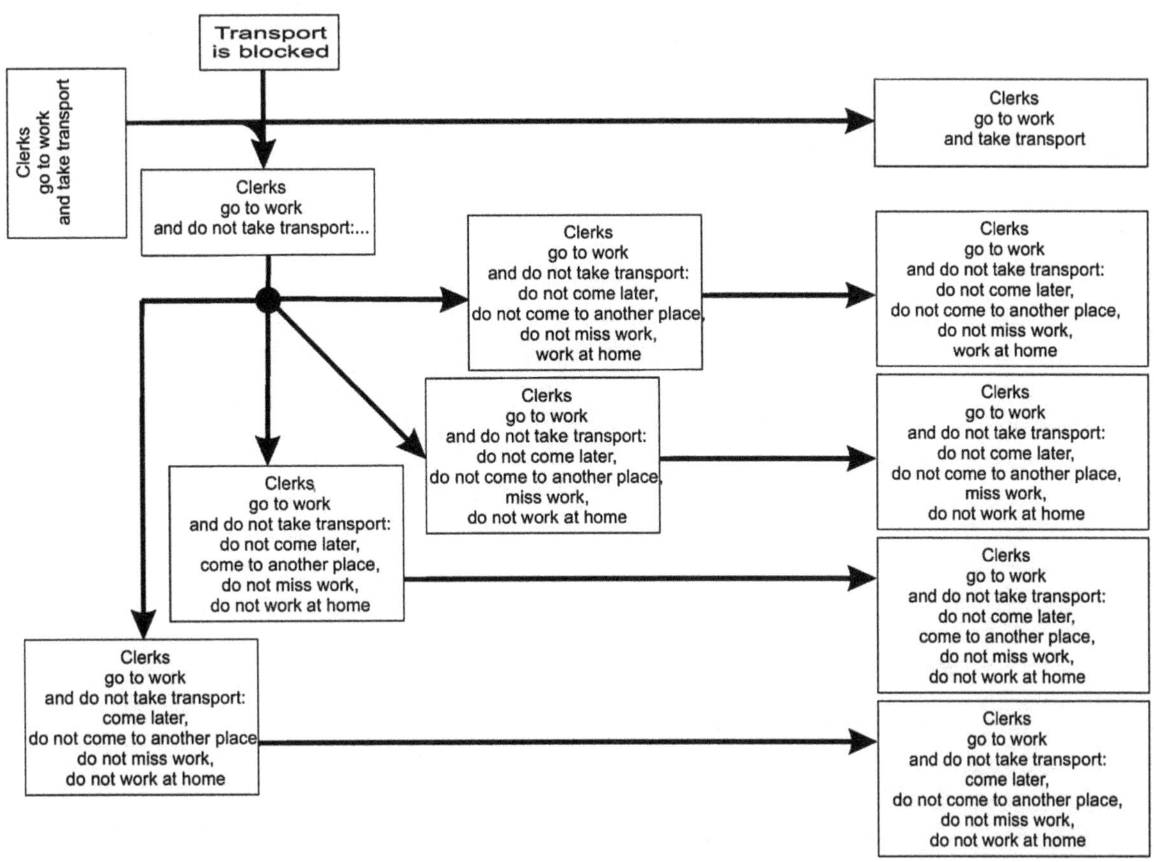

scenario. Cartoons are defined as maximal sequences of snapshots. We discuss, define, and exemplify the concepts snapshot and cartoon. We conclude with some remarks concerning the continuation of the research presented in the paper.

STRUCTURE OF EVENT BUSHES

An event bush consists of events, visualized as rectangles, and connectives connecting the events. Figure 1 presents a simple sample event bush, taken from (Wolter et al. 2018), to give an impression of event bushes. We refer the reader to (Pshenichny & Kanzheleva 2011, Wolter et al. 2018) for a more comprehensive introduction into the event bush method and more complex examples.

The Theory of Multitudes, presented in a first version in (Pshenichny 2018), serves as a mathematical foundation of events. For this paper, it is sufficient to consider an event to be given by a subject (predmet), as "clerk" for example, with a finite set of positive or negated properties (primeta), as "go to work", "come late", and "do not miss work" for example. The dynamics of an environment is reflected as well by changes of the properties of subjects as by changes of subjects.

Event bushes are build up by means of four connectives: The *flux* connective (Figure 2, a) describes one event E_1 producing another event E_2. The *influx* connective (Figure 2, b) describes that an event E_1 produces another event E_3, if an additional "external" event E_2 is present. If E_2 is not present, nothing changes and event E_1 evolves to event E_4, with the same subject and the same properties as E_1 but at a later moment in time. The *furcation* connective (Figure 2, c) describes that an event E_1 produces exactly one of the incompatible alternative events $E_2, \ldots, E_n$. The *conflux* connective (Figure 2, d) describes the production of an event E_n by the simultaneous presence of events $E_1, \ldots, E_{n-1}$.

Figure 2. Connectives and their representations as graphs

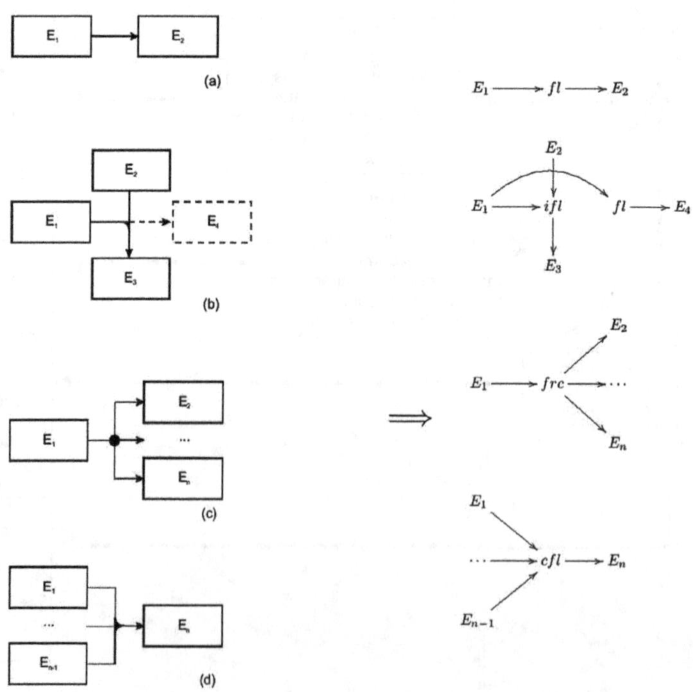

Event Bushes as Graphs

To define the structure of event bushes we use the concept of directed graph. A *directed graph* $\boldsymbol{G} = (\boldsymbol{G}_V, \boldsymbol{G}_E)$ is given by a set $\boldsymbol{G}_V$ of vertices (nodes) and a set $\boldsymbol{G}_E \subseteq \boldsymbol{G}_V \times \boldsymbol{G}_V$ of edges where each edge is an ordered pair of vertices $(x,y) \in \boldsymbol{G}_V \times \boldsymbol{G}_V$. An edge (x,y) is an outgoing edge for x and an incoming edge for y, respectively.

To obtain an appropriate formalization of event bushes we separate the structure of an event bush from the content of its events. For each "event box" in an event bush, we introduce a corresponding distinct event identifier and we denote the set of all event identifiers for a given event bush by $\mathcal{E}_{Event}$. The content of the "boxes" at the different positions can be described then by a function from $\mathcal{E}_{Event}$ into a corresponding "universe of events" (Pshenichny, 2018). We will not discuss "content functions" in the present paper. We concentrate, instead, on the structure of event bushes. Note, that the separation of structure and content allows us to have distinguished "boxes" with the same content as it happens, for example, in the event bush in Figure 1. For each of the five boxes in the rightmost column in Figure 1 there is another box in the event bush with the same content, where a flux connective connects any two boxes with the same content.

Besides "event boxes" we find another kind of vertex in event bushes, namely, special symbols to indicate connectives: A "right turn" for influx, a "double right turn" for conflux and a circle for furcation connectives (Pshenichny & Kanzheleva 2011). To treat connectives in a uniform way, we introduce abstract identifiers for all occurrences of connectives and we take into account that any influx connective is always accompanied by a flux connective (Pshenichny & Kanzheleva 2011, p. 152). Compare the description of the influx connective above. This gives us a translation of the original description of connectives into directed graphs as described in the right column in Figure 2.

We use indices to indicate the different occurrences of connectives in an event bush and denote by $\mathcal{E}_{Conn}$ the set of all those "connective vertices" in an event bush. According to the definition of event bushes, we do have either edges from events to connectives or the other way around. In summary, the structure of an event bush can be described by a *structure graph* $\mathcal{E} = (\mathcal{E}_V, \mathcal{E}_E)$ with a set $\mathcal{E}_V = \mathcal{E}_{Event} \cup \mathcal{E}_{Conn}$ of vertices and a set $\mathcal{E}_E \subseteq \mathcal{E}_{Event} \times \mathcal{E}_{Conn} \cup \mathcal{E}_{Conn} \times \mathcal{E}_{Event}$ of edges. Figure 3 gives an example of a simple event bush and its structure graph.

Figure 3. A simple event bush and its structure graph

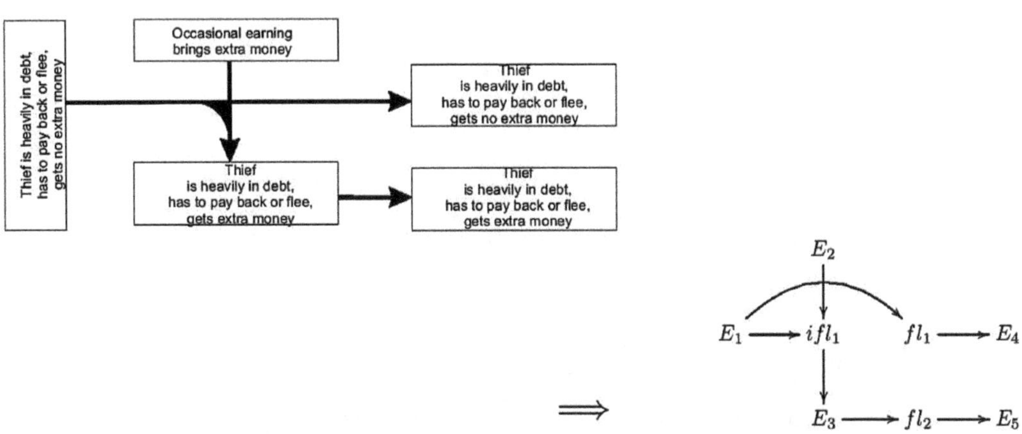

Event Bushes as Bipartite Graphs

A bipartite graph, is a set of graph vertices decomposed into two disjoint sets such that no two graph vertices within the same set are adjacent. We defined the structure graph $\mathcal{E} = (\mathcal{E}_V, \mathcal{E}_E)$ of an event bush as a bipartite graph: $\mathcal{E}_V$ is divided into two disjoint sets – a set $\mathcal{E}_{Event}$ of "event(vertice)s" and a set $\mathcal{E}_{Conn}$ of "connective(vertices)s". Edges go either from "event vertices" to "connective vertices" or vice versa. There are no edges between events and no edges between connectives, respectively!

We draw attention to the possibility to represent the concept "bipartite" by a graph $\boldsymbol{B} = (\boldsymbol{B}_V, \boldsymbol{B}_E)$ with $\boldsymbol{B}_V = \{Event, Conn\}$ and $\boldsymbol{B}_E = \{(Event, Conn), (Conn, Event)\}$

$\boldsymbol{B}$: $Event \rightleftarrows Conn$

The requirement that the structure graph $\mathcal{E} = (\mathcal{E}_V, \mathcal{E}_E)$ of an event bush is bipartite can be elegantly expressed by requiring a graph homomorphism $\tau_{\mathcal{E}}: \mathcal{E} \to \boldsymbol{B}$, i.e., a function $\tau_{\mathcal{E}}: \mathcal{E}_V \to \boldsymbol{B}_V$ such that each edge $(x, y) \in \mathcal{E}_E$ entails an edge $(\tau_{\mathcal{E}(x)}, \tau_{\mathcal{E}(y)}) \in \boldsymbol{B}_E$ in $\boldsymbol{B}$. Since $\boldsymbol{B}_V$ has two elements, a graph homomorphism $\tau_{\mathcal{E}}: \mathcal{E} \to \boldsymbol{B}$ generates a partition of $\mathcal{E}_V$ into two disjoint sets, namely the pre-images $\mathcal{E}_{Event} := \tau_{\mathcal{E}}^{-1}(Event) = \{x \in \mathcal{E}_V \mid \tau_{\mathcal{E}}(x) = Event\}$ and $\mathcal{E}_{Conn} := \tau_{\mathcal{E}}^{-1}(Conn) = \{x \in \mathcal{E}_V \mid \tau_{\mathcal{E}(x)} = Conn\}$. Moreover, there cannot be edges (e, e') in $\mathcal{E}_E$ with $\tau_{\mathcal{E}(e)} = \tau_{\mathcal{E}(e')} = Event$ or $\tau_{\mathcal{E}(e)} = \tau_{\mathcal{E}(e')} = Conn$, respectively, since there are no loops *(Event, Event)* or *(Conn, Conn)* in $\boldsymbol{B}_E$. This validates that graph homomorphisms into $\boldsymbol{B}$ formalize indeed the concept "bipartite graph".

Example 1. For our example in Figure 3 the graph homomorphism $\tau_{\mathcal{E}}: \mathcal{E} \to \boldsymbol{B}$, that describes the structure graph $\mathcal{E}$ as a bipartite graph, is given by the assignments $\tau_{\mathcal{E}}(E_1) = \tau_{\mathcal{E}}(E_2) = \tau_{\mathcal{E}}(E_3) = \tau_{\mathcal{E}}(E_4) = \tau_{\mathcal{E}}(E_5) = Event$ and $\tau_{\mathcal{E}}(ifl_1) = \tau_{\mathcal{E}}(fl_1) = \tau_{\mathcal{E}}(fl_2) = Conn$ (see Figure 4).

Multiflow Structure

An essential characteristic feature of even bushes is the so-called multiflow structure which classifies the events by four different types (Pshenichny & Kanzheleva, 2011; see Figure 5).

Figure 4. A sample bipartite graph

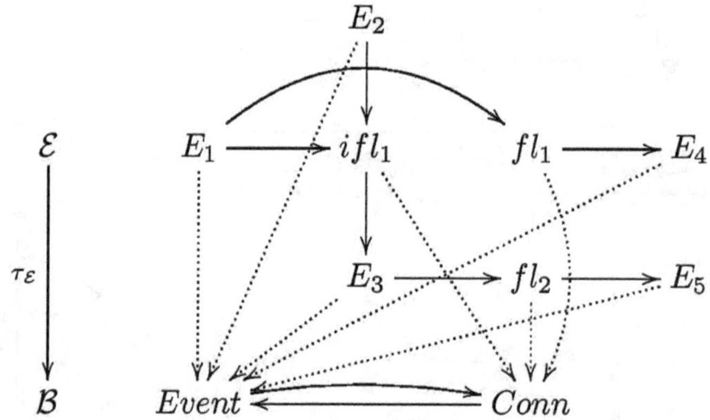

Figure 5. Multiflow structure

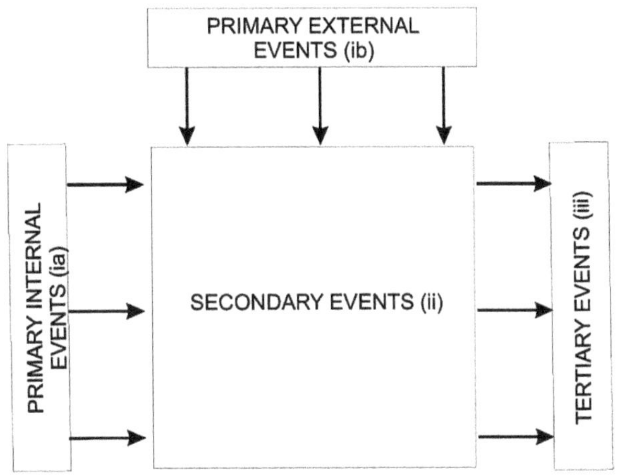

1. Primary internal events. These are primary, non-unique inputs representing the "passive conditions". These would determine any further course of events ("happenings").
2. Primary external events meaning the "invading agents". They may affect basic inputs or influence their further, indirect manifestations, thus "shaping up" different "happenings".
3. Secondary events that result from primary internal inputs with or without the contribution of primary external ones – the "happenings" proper formulated in a strict concise way indicating their core features determined by the causes, following the principle "one more cause – one change of event".
4. Tertiary events that denote end results, generated either by primary internal or by secondary events, with or without primary external ones. Tertiary events document the completed "happenings".

This multiflow structure together with the classification of connective vertices as either flux, influx, furcation, or conflux connectives can be represented formally by the graph ***ST*** depicted in Figure 6, declaring the types for events and connectives.

Figure 6. Graph ST declaring the types for events and connectives

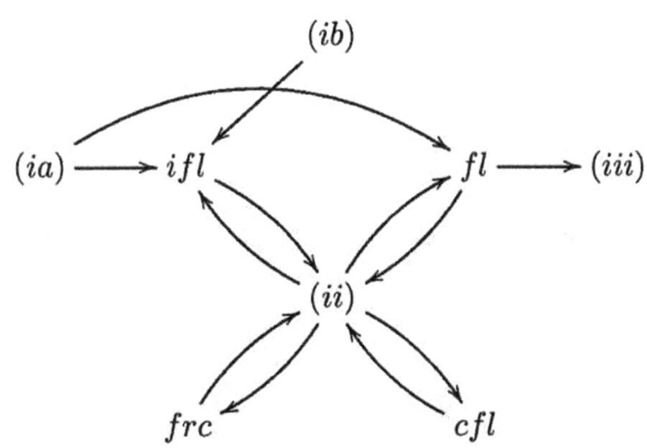

The graph ***ST*** represents also some of the structural restrictions for event bushes formulated in (Pshenichny & Kanzheleva 2011) like

1. Not a single event may lead to *(ia)* or *(ib)*.
2. Event *(ia)* without an event *(ib)* or *(ii)* may lead only to *(ii)* or an event *(iii)*.
3. Event *(ib)* may lead only to an event *(ii)*.
4. Event *(iii)* may not lead to any other event.
5. Only the flux connective may produce tertiary events *(iii)*.
6. All events united by conflux are secondary.

Restriction (1) is ensured since neither the vertex *(ia)* nor the vertex *(ib)* in ***ST*** do have an incoming edge. The vertex (ia) in ***ST*** has only the two outgoing edges *(ia,ifl)* and *(ia,fl)*, respectively, while *(fl)*, in turn, has only the two outgoing edges *(fl,ii)* and *(fl,iii)*. This reflects restriction (2). Restriction (3) is ensured since *(ib)* has the only outgoing edge *(ib,ifl)* while *(ifl,ii)* is the only outgoing edges for *(ifl)*. Restriction (4) means that *(iii)* has no outgoing edge at all while restriction (5) means that *(fl,iii)* is the only incoming edge for *(iii)*. Finally, restriction (6) is ensured since *(ii,cfl)* is the only incoming edge for *(cfl)*.

There are, however, restrictions that cannot be expressed by just graphs and graph homomorphisms, respectively. This concerns, especially, multiplicity restrictions, like "furcation has exactly one incoming event", "conflux has exactly one out coming event" and so on. To express those more involved restrictions we have, in the future, to develop and to use appropriate diagram predicates along the methodology proposed by the Diagram Predicate Framework (DPF) (Diskin & Wolter 2008, Rutle et al. 2012, Rossini et al. 2014).

We consider ***ST*** as a bipartite graph with the obvious graph homomorphism τ_{ST}: ***ST*** $\rightarrow$ ***B*** sending the vertices *(ia)*, *(ib)*, *(ii)*, *(iii)* to the vertex *Event* in ***B*** and the vertices *ifl*, *fl*, *cfl*, *frc* to the vertex *Conn*.

We can summarize our revised analysis of the structure of event bushes as follows: The structure of an event bush is given by a directed graph $\mathcal{E}$, called the **structure graph** of the event bush, together with

- A graph homomorphism $\tau_{\mathcal{E}}$: $\mathcal{E} \rightarrow$ ***B***, that declares the vertices in $\mathcal{E}$ either as event vertices or as connective vertices, respectively, and
- A graph homomorphism $\mu_{\mathcal{E}}$: $\mathcal{E} \rightarrow$ ***ST***, that declares as well the multiflow structure of the event bush as the types of the connective vertices.
- Moreover, the obvious compatibility, that $\mu_{\mathcal{E}}$ composed with τ_{ST} is equal to $\tau_{\mathcal{E}}$, holds.

SCENARIOS

In the last section, we have introduced two very basic formal concepts – directed graphs and homomorphisms between them. By means of these two concepts, we have been able to formalize the basic structural features of event bushes. For directed graphs and graph homomorphisms we do have many concepts, constructions and results available (compare, for example, Ehrig et al. 2006). Our overall research interest is to exploit this "toolbox" to formalize and to deal with other features, relations and desired constructions in the area of event bushes.

In the remaining sections, we exploit directed graphs and graph homomorphisms to discuss and to formalize the purpose of event bushes to describe environments of *directed alternative changes.*

First, we point out that "directed" can be ensured by the requirement that the structure graph $\mathcal{E}$ of an event bush is *acyclic*, i.e., has no cycles. *In the rest of the paper, we consider only event bushes with acyclic structure graphs.* Second, let us discuss the aspect "alternative": There are essentially two sources for "alternatives" in event bushes (compare Pshenichny & Kanzheleva 2011): A furcation connective results in "mutually incompatible events" and an influx connective is "always accompanied by a flux connective" meaning "that the initial event would not change unless affected by another event". We base our discussion and our first attempt to formalize the idea of "directed alternative changes" on the following methodological hypothesis:

An event bush describes, in a compact way, a set of alternative "scenarios" of directed changes of an environment.

We follow the heuristics that all the behavioural aspects of an event bush are reflected in its structure graph independent of the content of the events. We do have to answer therefore the question what a "scenario" described by the structure graph of an event bush should be: In our case of an acyclic event bush, a *scenario* ***SC*** is a subgraph of the corresponding structure graph $\mathcal{E}$ that does not include any alternatives:

- For any furcation chose exactly one of the incompatible alternatives.
- For any influx we have either influx, if external event is present, or accompanied flux, if external event is not present.

SC, as a subgraph of $\mathcal{E}$, inherits from $\mathcal{E}$ as well the bipartition $\tau_\mathcal{E}$ as the multiflow structure $\mu_\mathcal{E}$. More formally, the absence of alternatives is expressed by the following two requirements:

- Any connective vertex C in ***SC*** with $\mu_\mathcal{E}(C) = frc$ has exactly one outgoing edge in ***SC.***
- For any event vertex E in ***SC*** there is an edge (E,C) in ***SC*** with $\mu_\mathcal{E}(C) = fl$, only if, there is no edge (E,C') in ***SC*** with $\mu_\mathcal{E}(C') = ifl$.

In addition, ***SC*** has to satisfy the following connectivity requirements:

- Any connective vertex C in ***SC*** has as well an incoming as an outgoing edge.
- Any event vertex E in ***SC*** with $\mu_\mathcal{E}(E) = ii$ has as well an incoming as an outgoing edge.
- Any event vertex E in ***SC*** with $\mu_\mathcal{E}(E) = ia$ or with $\mu_\mathcal{E}(E) = ib$ has an outgoing edge.
- Any event vertex E in ***SC*** with $\mu_\mathcal{E}(E) = iii$ has an incoming edge.

Finally, we require completeness for conflux connectives:

- For any connective vertex C in ***SC*** with $\mu_\mathcal{E}(C) = cfl$ we require that (E,C) is also an edge in ***SC*** for all edges (E,C) in $\mathcal{E}$.

As an example, we consider the structure graph in Figure 7 derived from the sample event bush in Figure 1 by dropping two of the alternatives of the furcation. Dependent on the presence or non-presence of the primary external event E_2 and due to the two remaining alternatives of the furcation we obtain three alternative scenarios as visualized in Figure 8.

Figure 7. Another sample structure graph

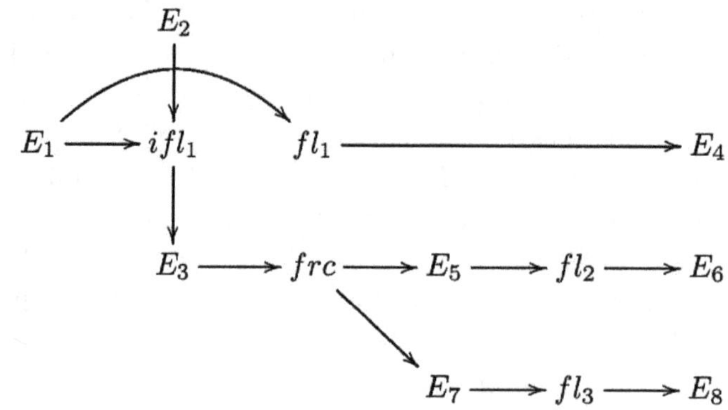

Figure 8. Three alternative scenarios

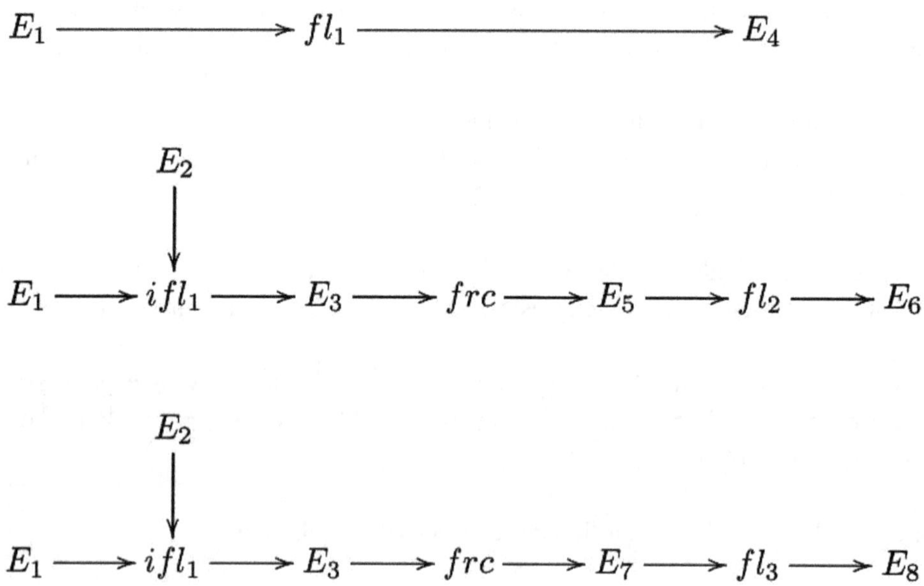

SNAPSHOTS AND CARTOONS

After we outlined that, an event bush describes a set of alternative scenarios, we want to discuss now what it means that an environment evolves in accordance with a given scenario. First, we have to remind that scenarios, as subgraphs of acyclic graphs, are acyclic as well.

Now we consider a given scenario ***SC***. An environment, which evolves in accordance with this scenario, is in a certain state at each “moment in time”. Those states do have the following characteristic properties:

- At each state the set of event vertices in ***SC*** is divided into present events, past events and future events, respectively.

- Correspondingly, the set of connective vertices is divided into connectives that have been "firing" once in the past and connectives that will "fire" in the future.
- The past influences the future only via the present events.
- All present events are independent from each other.

We use the term snapshot for those states. Formally, a *snapshot* of a scenario ***SC*** is a graph homomorphism σ: ***SC*** → ***P***, where the (bipartite) graph ***P*** is given by

past pconn present fconn future

Figure 9 visualizes all four possible snapshots of the second sample scenario in Figure 8.

Figure 9. All snapshots of the second sample scenario in Figure 8

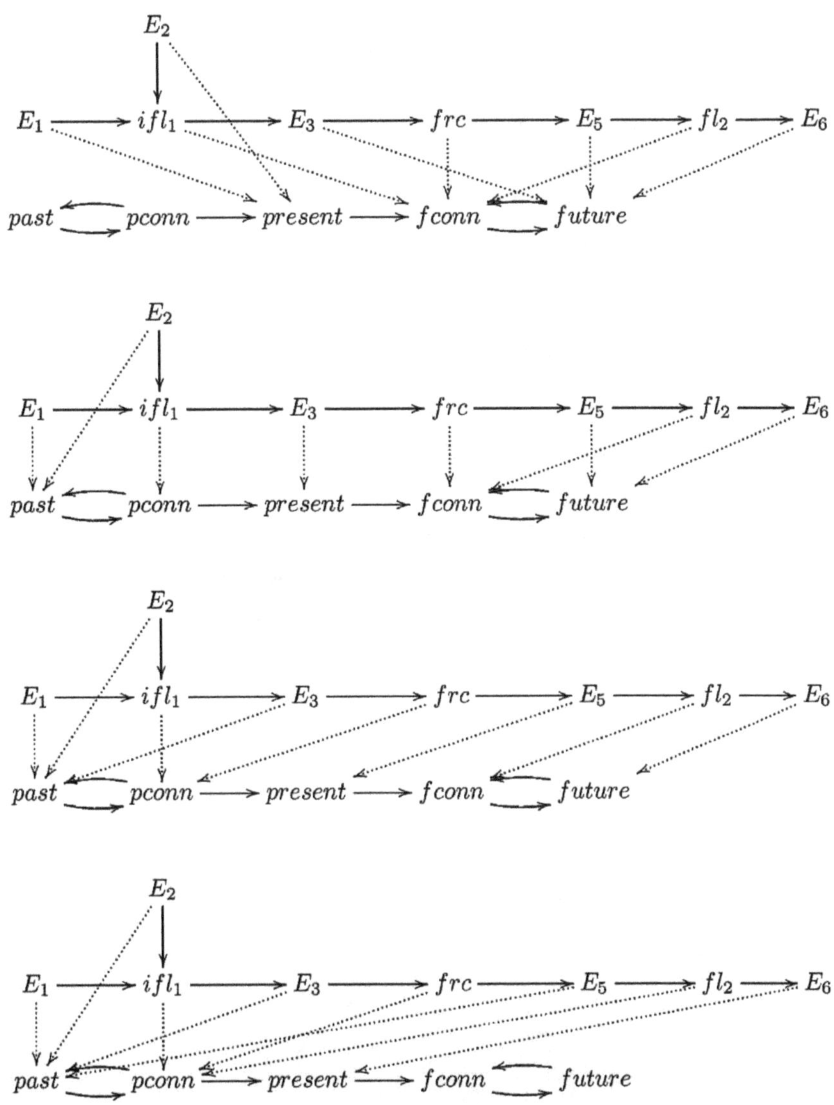

The "present events" in the first snapshot are exactly all the primary events in the scenario, while the "present events" in the last snapshot are exactly all the tertiary events in the scenario! That a scenario ***SC*** is acyclic ensures, especially, that there exists for any event E in ***SC*** a snapshot σ: $\boldsymbol{SC} \rightarrow \boldsymbol{P}$, with $\sigma(E) =$ *present*. Moreover, the property "the past influences the future only via the present events" is reflected by the fact that any snapshot is uniquely determined by the set of "present events"!

An environment will change from one state to another state. Based on our outlined formalization this means that we move step by step from one snapshot to other snapshots. Moving from one snapshot to a *consecutive snapshot* is not deterministic and can be defined analogously to Petri nets (Petri 1973):

- Given a snapshot σ: $\boldsymbol{SC} \rightarrow \boldsymbol{P}$ a connective $C \in \boldsymbol{SC}_V$ is "enabled" iff $\sigma(C) =$ *fconn* and for all events $E \in \boldsymbol{SC}_V$ we have that $(E, C) \in \boldsymbol{SC}_E$ implies $\sigma(E) =$ *present*.
- We can choose any non-empty set of "enabled" connectives and "fire" them in parallel.
- This results in a new consecutive snapshot such that:
  - All the events, that are inputs of "fired" connectives, are moved from *present* to *past*.
  - The "fired" connectives are moved from *fconn* to *pconn*.
- All the events, that are outcomes of "fired" connectives, are moved from *future* to *present*.

The evolution of an environment appears as a sequence of consecutive snapshots. A *cartoon* is a sequence of consecutive snapshots that is "maximal" in the following sense:

- Each event in the scenario has the status "present" in at least one of the snapshots of the cartoon.
- The "present events" in the first snapshot of the "cartoon" are exactly all the primary events in the scenario.
- The "present events" in the last snapshot of the "cartoon" are exactly all the tertiary events in the scenario!

A scenario is *deterministic* if there is only one way to realize it as a "cartoon", i.e., if for any snapshot there is exactly one connective enabled. All the sample scenarios in Figure 8 are deterministic and the four consecutive snapshots in Figure 9 give the unique cartoon for the second scenario in Figure 8.

A scenario is *non-deterministic* if it can be scheduled in different ways as a cartoon, i.e., if there is at least one snapshot where we have different choices for a non-empty set of "enabled" connectives".

Figure 10. A non-deterministic scenario

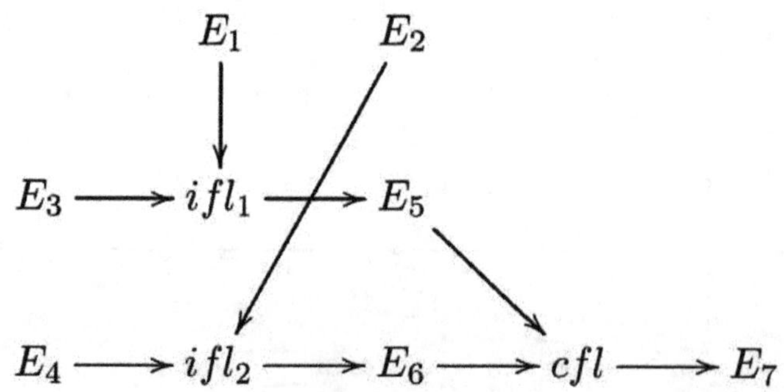

Figure 11. The acyclic graph of snapshots of the non-deterministic scenario in Figure 10

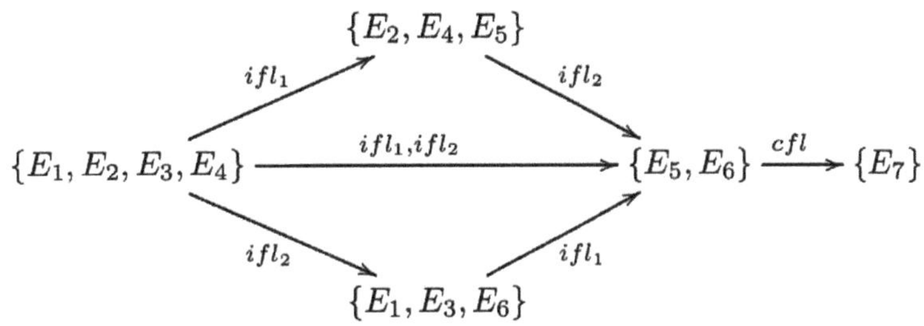

Figure 10 presents an example of a non-deterministic scenario, with two influx and one conflux connective, together with the corresponding acyclic graph of consecutive snapshots drawn in Figure 11. Note, that the snapshots are uniquely represented by their corresponding sets of “present events”. The three possible cartoons for the non-deterministic scenario correspond to the three different paths in this acyclic graph from the first snapshot, where all primary events are “present events”, to the last snapshot, where all tertiary events are the “present events”.

CONCLUSION

We used the very basic formal concepts – directed graphs and homomorphisms between directed graphs – to formalize static and dynamic aspects of the event bush method:

1. The structure of an event bush is defined by a bipartite structure graph that is equipped with a multiflow structure.
2. An event bush describes a set of alternative scenarios while a scenario is defined as a subgraph of the corresponding structure graph satisfying certain requirements.
3. An environment, which evolves in accordance with a given scenario, is in a certain state at each “moment in time”. Those states are formalized as snapshots, i.e., certain graph homomorphisms.
4. In analogy to Petri nets, we define how a certain snapshot could give rise to other consecutive snapshots.
5. Cartoons are introduced as maximal sequence of consecutive snapshots. They describe how a certain environment may evolve, in possibly different ways, in accordance with a given scenario.

We are convinced that the proposed formalization is an adequate basis for the development of computer tools supporting the analysis of real world scenarios as in the context of hazard assessment, for example. One task will be to find all possible scenarios for a given set of primary events in an event bush.

At the moment, we can identify, at least, two extensions of the present formalization:

1. In some of the sample event bushes in the literature, we find other sources for alternatives, like an event with two outgoing edges to two different flux connectives. The open question is, if this a general feature of event bushes or just a design flaw. There are no principal problems to extend our definition of scenarios to include other sources of alternatives.

2. In the present paper, we considered only acyclic event bushes. We can extend, however, the definition of scenarios to cyclic event bushes as well. Scenarios will be no longer subgraphs of the corresponding structure graph, but acyclic graphs together with a graph homomorphism into the structure graph.

REFERENCES

Diskin, Z., & Wolter, U. (2008). A Diagrammatic Logic for Object-Oriented Visual Modeling. *ENTCS, 203*(6), 19–41.

Ehrig, H., Ehrig, K., Prange, U., & Taentzer, G. (2006). *Fundamentals of Algebraic Graph Transformations. EATCS Monographs on Theoretical Computer Science.* Berlin: Springer.

Petri, C. A. (1973). Concepts of Net Theory. *Mathematical Foundations of Computer Science. Proc. of Symposium and Summer School*, 137-146.

Pshenichny, C. A. (2018a). Theory of Multitudes as an Alternative to the Set Theory. In C. A. Pshenichny, P. Diviacco, & D. I. Mouromtsev (Eds.), *Representation of Dynamic Knowledge in Scientific Domains.* IGI Global.

Pshenichny, C. A., Carniel, R., & Diviacco, P. (2013). Engineering of Dynamic Knowledge inExact Sciences: First Results of Application of the Event Bush Method in Physics. MSEPS 2013 - Modeling States, Events, Processes and Scenarios, Proceedings, 60–73.

Pshenichny, C. A., & Kanzheleva, O. M. (2011). Theoretical Foundations of the Event Bush. In K. Sinha, L. Gundersen, J. Jackson, & D. Arctur (Eds.), Geological Society of America Special Paper: Vol. 482. Societal Challenges and Geoinformatics (pp. 139–165). Academic Press. doi:10.1130/2011.2482(12)

Rossini, A., de Lara, J., Guerra, E., Rutle, A., & Wolter, U. (2014). A formalisation of deep metamodelling. *Formal Aspects of Computing*, 1–38.

Rutle, A., Rossini, A., Lamo, Y., & Wolter, U. (2012). A formal approach to the specification and transformation of constraints in MDE. *Journal of Logic and Algebraic Programming, 81*(4), 422–457. doi:10.1016/j.jlap.2012.03.006

Wolter, U., Korableva, O. N., & Solovyov, N. A. (2018). The event bush method in the light of typed graphs illustrated by common sense reasoning. In C. A. Pshenichny, P. Diviacco, & D. I. Mouromtsev (Eds.), Representation of Dynamic Knowledge in Scientific Domains. IGI Global.

Chapter 4
On Formalization and Representation in Collaborative Research

Paolo Diviacco
Istituto Nazionale di Oceanografia e di Geofisica Sperimentale (OGS), Italy

ABSTRACT

The aim of this chapter is to delve into the issues related to the formalization and transmission of knowledge within the scope of collaborative scientific research and to propose a new approach to address such difficulties. Analyzing methods and practices of collaborative research, the authors highlight that observation and reasoning are systematically prone to flaws, so that theorization is made highly conjectural. To gain reliability, points of views and visions need then a support from a community; in other words, they become public. To allow convergence to take place, conceptualizations need to be understood by people with possibly different cognitive models. Therefore, the authors propose using artifacts that can be strongly structured in individual use while weakly structured in common use. These artifacts take place generally as graphic representations, and as in the case of the arts, they can be realistic or abstract, they can emphasize, hide, or allow different, contrasting and concurrent interpretations of the exposed knowledge.

Keywords: Collaborative Research, Scientific reasoning, Knowledge Representation, Knowledge Formalization, Boundary objects.

INTRODUCTION

This should start asking why do people collaborate?

Most of the scientific research conventional iconography depicts scientists working in isolation in their laboratory. Can it still be like this?

The answer is of course not. As a general trend, scientific topics moved from narrow domain specific problems to systemic issues, where by systems we mean complex sets of interacting and cross dependent mechanisms resulting in complex phenomena.

DOI: 10.4018/978-1-5225-5261-1.ch004

At the same time, technologies available in scientific research made a huge leap forward triggering high specialization so that no scientist, anymore, can cover all the needed topics and activities.

Full encyclopedic knowledge coverage by a single person is not possible anymore.

Collaborative research is invoked then to overcome such limitations. Researches get together, bringing their contributions to understand and solve complex issues and problems that they would not be able to address alone.

In addition to this, a peculiar loop exists between collaborative science and large scientific facilities such as, for example: satellites, particle colliders, telescopes and the like. If on one hand scientists need such big facilities for their work, on the other hand, the need to share the costs of such infrastructures triggers collaborative science (Georghiou, 1998).

Collaborative research brings in the picture big promises but also problems. High specialization of researchers can result, during teamwork, in possible lacks of understanding and difficulties in handling data in a consistent way.

This is only seldom addressed.

Current trends in collaborative science support mostly focus on the technical issues related to the consolidation of data to a central facility, where it can be processed using high quality resources and from where data can be retrieved remotely, many times and by multiple researchers.

This, of course, is a very respectable goal, but considering the above mentioned issues is unlikely to be enough.

On the contrary, we think necessary to pave the road to a new vision of collaborative science that aims to bridge the gaps between scientific domains and ways of thinking. But first we should start from a strict definition of what we mean by collaboration.

The terms collaboration, cooperation and coordination are often used interchangeably. Actually, this is not correct.

Cooperation means working together, using possibly the same resources, but separating goals in order that each partner achieves its own. The main motivation in participating in a cooperation is that partners gain something that would not be available should they go alone.

Collaboration has the fundamental difference that all the parties work towards the construction of a common goal.

In the first case the cognitive overlap among partners is limited, being each partner activities linked to their "separated" goal. In the second case a common cognitive space must be created, maintained and referred to by all partners. Such as in the old medieval maps, where dangerous areas had to be highlighted, we can say "Hic sunt leones" (here are the lions).

To understand how to address such problems we need to start from the very beginning.

The Scientific Method

As everyone knows, the difference between science and superstition is the scientific method. The Oxford Dictionary (2016) defines the scientific method as "a method or procedure that has characterized natural science since the 17th century, consisting in systematic observation, measurement, and experiment, and the formulation, testing, and modification of hypotheses". Essentially: a series of observations (such as for example *"I saw"*, lets say, *"10 black crows"*) result in a generalization *("I saw only black crows")* on top of which a theory is built *("Only black crows exist")*. The loop is closed by the experiment, where a scientist plans to check whether the theory is correct ("Let's see whether the next 10 crows will be black") and if observations are consistent with what expected, then the theory is likely to be true.

Of course this has been largely demonstrated to be mostly a myth and that in reality science takes rather different paths.

One of the key points to explain this is to consider the set of issues related to the process of observation. Scientific laws should be founded on observations that should be reproducible experimentally at any time.

If in some domains, this could be true, although someone doubts even of this, in most domains, and especially in the natural, social, economic or human sciences, observations are very difficult to break down into elementary processes.

Phenomena involve complex interactions over a wide range of spatial and temporal scales. This is complicated even more by the fact that, often, only a limited number of cases are available, so that eventually analogies could be only hardly recognized (Engelhardt and Zimmermann 1982). This results in a situation where correlation does not necessarily implies causation and where relevant parameters can often be underestimated (Lawlord et al. 2004)

If all this limits heavily the phase of generalization, it complicates even more the phase of verification. Considering the difficulties in choosing what to monitor during the phase of verification/falsification the overall possibility to state the likeliness of a theory is put at stake. In addition, following Popper (1963), positive outcomes of experiments cannot confirm a scientific theory, because a single counter-example can demolish it.

Scientific Reasoning

Following the schema proposed above, it is possible to identify 2 main areas in scientific explanations. Induction, that is the area which leads from observation to generalization, and deduction, which leads from theory to prediction. Only the process of deduction yields a strict logic inference, while, as it works within a closed system, it is basically a tautology. Meaning that it simply changes the configuration of knowledge, and therefore it cannot discover anything new (Wittgenstein, 1921).

On the other hand following a red line that connects authors such as Hume [1748] (1910) or Popper (1979), the very logical foundation of induction has been questioned by the fact that there is no way to guarantee the truth of the conclusions of induction even if the premises are true. Inductive inference proceeds from statements about a known phenomenon to amplified statements about phenomena that resemble the known (Peirce, 1958).

Now, if deduction cannot build new ideas and induction is prone to flaws, considering the above mentioned issues with verification/falsification and the limits of observation, how much can we consider scientific reasoning reliable? Probably the answer is: not as much as normally is thought. Seemingly scientific theories are more conjectural than most of the people claim them to be.

In fact there is another mode of reasoning that is often underestimated and neglected, but is very important in scientist's everyday life, which is abduction. This is a way of reasoning that allow to build new ideas from seemingly disparate and unconnected facts and observations (Eco, 1981). It is a form of inference that relies on the intuition of scientists. Quoting Peirce (1958):

Abduction merely suggests that something may be.

The main consequence of abductive reasoning is that multiple or even contrasting views can have the same authoritative voice on the explanation of a phenomena. At the same time each vision will try to preserve its existence, building a set of peripheral and auxiliary hypotheses (the protective belt) that when under attack can be changed in order to preserve the kernel of the theory (Lakatos, 1970)

Following Kuhn (1962) researchers gather around paradigms. These are the environments, the backgrounds and the way of thinking defined by each researcher's personal history and which influences his/her position in the scientific community. Becher and Trowler (2001) suggest that paradigms result in the creation of sub-communities that resemble "tribes" that tend to evolve separately developing separate cognitive models

Status and identity of researchers and tribes can be defined by the possibility to control facilities and resources of research (Whitley, 2000). At the same time, the lives of scientists and experts are conditioned by the network of allies and by the strategic position each researcher is able to maintain in the community. Following Latour and Woolgar (1979) "Science is a social construct", which extends the scope of what we are discussing here from the realm of logics to that of socio-logy and communication.

Communication

Communication can be layered into syntax, semantics and pragmatics (Morris, 1971, Diviacco, 2013a). Briefly; syntax is concerned with the transmission of meaning, semantics enter meaning, while pragmatics consider meaning within the context in which it is used. Difficulties can emerge in all those levels. In Diviacco (2016) we proposed to classify these issues in two groups:

- **Epistemic and Cognitive**: That are constructed on the structural, intrinsic, often innate and unconscious limits of communication, such as tacit knowledge (Polanyi, 1966) and theory ladenness (Hanson, 1958)
- **Socio-Economic**: That are conscious and voluntary. In these cases, parties aim to exert an influence on a process, project or activity, in order to obtain a goal or a position (David and Bunn, 1988; Latour & Woolgar, 1979).

So far, most of the work done to overcome problems in communication, ever since the seminal work of Shannon and Weaver (1949) was mainly focused on the syntactic level and was more related to the technical implementation of communication. Here, although often exceptions apply, agreeing on standards is an easier goal to achieve. At higher abstraction levels of semantics and pragmatics, both cognitive and socio-economic epistemic issues irrupt into the picture. Here a convergence of different communities living in different paradigms can become then difficult to obtain.

Summing up, a new vision emerges that questions science as being related to real and measurable facts, an that suggests, instead, a more reasonable and vague conception that demands judgments to be done by a group of sane, rational and sufficiently informed people. This should not be performed by the single person but need confrontation with others.

Knowledge

Knowledge is associated with the acquisition and processing of information to retain meaning that under certain conditions can be used, expressed and transmitted.

Often knowledge is associated with science because this latter is supposed to be able to warrant the truth. And as we have seen this assumption can be questioned.

Knowledge can be systematized using multiple means, and language is only one of them. The "deliteralization" of Knowledge opens the door to inclusion of multiple forms of human expression that can be gathered around the concept of Culture. Here Science and Technology are only a part of a whole picture that is composed also of values (ideologies), social structures and conventions (norms and taboos), Religion and aesthetics (Art, Music, Literature, Fashion, Architecture, Design) (Diviacco 2015).

Remaining within the domain of collaborative science, when we need to confront our knowledge with other people we need methods to pack, transmit, and decode it. First of all, we need to describe what we retain in our minds. In doing this two paths can be followed:

- Knowledge formalization, which is based on the identification and definition of the concepts that constitute that particular abstraction, point of view, or vision of the world and of the relations that link those concepts.
- Non formalized knowledge, which is based on the introduction of artifacts that "resonate" knowledge into someone else's mind.

In the first case, concepts and relations must be defined as comprehensively as possible, in the second case, the evocative power of the artifact does not require concepts and link to be explicitly defined.

Now, the main problem with knowledge formalization emerges right here.

To be able to formalize an abstraction we need to be able to identify and properly describe all the pieces of the puzzle. This is almost never possible, because a portion of what we know cannot be made explicit as it is obscured in socio-political phenomena or embedded in practices or even tools. This is called tacit knowledge. Following Polanyi (1966) "we know more than we can tell" and therefore we know more than we can formalize, which means of course that the whole process of formalization can be then considered questionable.

Representation

A pragmatic vision that considers all limitations in scientific reasoning, would be based on the fact that the explanation of a phenomenon would be grounded in sets of claims or judgments which truth values are not determined privately. Here knowledge would be based on a form of intentional portray of things, such as, for example, beliefs, percepts, sentences and pictures, that can be defined representation. Representations can be theories and models, linguistic and mathematical entities, computer simulations, concrete objects and so on and so forth (Diviacco, 2015; Diviacco & Busato, 2015). Representation can be based on natural language, software, algorithms, or graphs such as diagrams or maps. What is important here, is that, when necessary, the message can be kept implicit, that it can accommodate different (possibly contrasting) points of view and that it can be biased by cognitive and epistemic issues, by ideology or politics; or concurrently by all these factors. Any representation can be considered then generally false (Chakravartty, 1995). Following Callender and Cohen (2006), representation has features reality does and does not have. It is characterized by omissions and commissions, where omission is the act of neglecting some possible causes among those that can explain a phenomenon; commission is the act of deliberately changing the network of possible causes and explanations of a phenomenon. (Diviacco, 2015). Representation can highlight, change or even hide some pieces of the puzzle in order to make the conceptualization clearer, such as, for example, in the case of physics, where some models concentrate all the mass of a body in a point: this of course is unrealistic but ease the understanding of the problem. As much as the representation deviates from common sense or from observations available, so much it becomes un-realistic or abstract. When simple objects, phenomena or concepts are to be described, the iconic use of an artifact is often easy to grasp by anyone. When conventions need to be introduced, the situation becomes complex because it requires a previous convergence of a community which is not a given.

Another interesting point in the comparison between formalization and representation is the need of maintenance. Updates in conceptualization must be mirrored in the relative knowledge formalization. This can be time and resources consuming. Representation moves such problems to the user so that maintenance is a less relevant problem

During the nineties Susan Leigh Star and James R. Griesemer (1989) found that different communities, characterized by different languages, cultures and traditions can collaborate effectively notwithstanding the fact that they could have rather different cognitive models. They realized that the process of integration of the two cultures was facilitated by the presence of artifacts that gather users in the same cognitive space bridging concurrent cognitive models through abstraction from each partner's domain. In this, artifacts must be weakly structured in common use, but strongly structured in individual use (Diviacco, 2012; Star, & Griesemer, 1989).

A boundary object can take place in many ways. Experiences in non-computational knowledge formalization applied to several disciplines and study cases have been reported by Diviacco (2005, 2013, 2014; Diviacco & Busato, 2013) and Diviacco et al (2015, 2014, 2011, 2010), Pshenichny and Mouromtsev (2014) and Rogalciuk and Solomin (2014). Here, knowledge formalization takes place as a graphic map where nodes are essentially labels of concepts/entities. At the same time, these nodes can be considered repositories for data and information. In this perspective the system with its map and data, information and knowledge that populate each node becomes the status of the discourse and the public knowledge of the designated community, integrating what is already in each partner's mind.

So far the solutions proposed (Diviacco 2012,2013,2015) can suffer some limitations due to the fact that currently data management and association with nodes is decided during upload only. We are currently working on an extension of this method to incorporate semantics. First results seem encouraging

CONCLUSION

This work aims at highlighting the often underestimated issues related to knowledge formalization in scientific collaborative work. These can be traced back to the very basic moment of reasoning and theorization. We showed how much scientific reasoning is actually based on conjectures, and how much this creates problems in the definition of authoritative voices in the scientific community. This eventually results in multiple and concurrent "tribes" of scientists that develop their own personal way of thinking and cognitive modes. To let them collaborate it is necessary to support their activities using tools that can be used flexibly to harmonize their contribution. Knowledge formalization have limits in this because not all knowledge can be made explicit. Representation instead, similarly to what happens in the Arts, allow interpretation of artifacts to take place in order that each partner is enabled to find in it what it needs, while at the same time share his asset to the other partners.

REFERENCES

Becher, T., & Trowler, P. R. (2001). *Academic tribes and territories intellectual enquiry and culture of disciplines*. Buckingham, UK: The Society for Research into Higher Education and Open University Press.

Bishop, M. A. (1992). Theory-Ladenness of Perception Arguments. In *Proceedings of the Biennial Meeting of the Philosophy of Science Association* (pp. 287-299). Academic Press.

Callender, C., & Cohen, J. (2005). There Is No Special Problem About Scientific Representation. *Theoria*, *55*, 67–81.

Chakravartty, A. (1995). Truth and Representa- tion in Science: Two Inspirations from Art. In R. Frigg & M. Hunter (Eds.), *Beyond Mimesis and Convention: Representation in Art and Science* (pp. 33-50). Berlin: Springer.

David, P. A., & Bunn, J. A. (1988). The economics of gateway technologies and network evolution. *Information Economics and Policy*, *3*(2), 165–202. doi:10.1016/0167-6245(88)90024-8

Diviacco, P. (2005). An open source, web based, simple solution for seismic data dissemination and collaborative research. *Computers & Geosciences*, *31*(5), 599–605. doi:10.1016/j.cageo.2004.11.008

Diviacco, P. (2013). Addressing Conflicting Cognitive Models in Collaborative e-Research: A Case Study in Exploration Geophysics. In *Collaborative and Distributed e-Research: Innovations in Technologies, Strategies and Applications*. Hershey, PA: IGI Global.

Diviacco, P. (2015) Reconciling Knowledge and Collaborative e-Research. In Collaborative Knowledge in Scientific Research Networks. Hershey, PA: IGI Global. doi:10.4018/978-1-4666-6567-5.ch001

Diviacco, P., & Busato, A. (2013). The Geo-Seas Seismic data viewer: A tool to facilitate control of data access. *Bollettino di Geofisica Teorica ed Applicata*, *54*(3), 257–270.

Diviacco, P., & Busato, A. (2015). Maps, graphs and annotations as Boundary Objects in Knowledge Networks, Distributed Cognition and Collaborative e-Research. In P. Diviacco, P. Fox, C. Pshenichny, & A. Leadbetter (Eds.), *Collaborative Knowledge in Scientific Research Networks*. Hershey, PA: IGI Global. doi:10.4018/978-1-4666-6567-5.ch019

Diviacco, P., DeCawer, K., Leadbetter, A., Sorribas, J., Stojanov, Y., Busato, A., & Cova, A. (2014). *Bridging semantically different paradigms in the field of marine acquisition event logging, Earth Science Informatics*. Springer. doi:10.1007/s12145-014-0192-0

Diviacco, P., & Pshenichny, C. A. (2010). Concept-referenced spaces in computer-supported collaborative work. *Geophysical Research Abstracts*, (12).

Diviacco, P., & Pshenichny, C. A. (2011). *A case study on the use of event bushes as a formal representation for Computer Supported Collaborative Work in the GeoSciences. European Geosciences Union*. EGU.

Diviacco, P., Pshenichny, C. A., Carniel, R., Khrabrykh, Z., Shterkhun, V., Mouromtsev, D., & Pascolo, P. et al. (2014). Organization of a geophysical information space by using an event-bush-based collaborative tool. *Earth Science Informatics*, *10*. doi:10.1007/s12145-014-0182-2

Eco, U. (1981). Guessing: From Aristotle to Sherlock Holmes. *Versus, 30*.

Engelhardt, W., & Zimmerman, J. (1982). *Theory of earth science*. Cambridge, UK: Cambridge University Press.

Geourghiou, L. (1998, September). 1998, Global cooperation in research. *Research Policy*, *27*(6), 611–626. doi:10.1016/S0048-7333(98)00054-7

Hanson, N. R. (1958). *Patterns of Discovery*. Cambridge, UK: Cambridge University Press.

Hume, D. (1910). *[1748]. An enquiry concerning human understanding*. New York, NY: P. F. Collier & Son.

Kuhn, T. S. (1962). *The structure of scientific revolutions*. Chicago, IL: University of Chicago Press.

Lakatos, I. (1970). Falsification and the methodology of scientific research programmes. In Criticism and the Growth of Knowledge. Cambridge, UK: Cambridge University Press. doi:10.1017/CBO9781139171434.009

Latour, B., & Woolgar, S. (1979). *Laboratory life: The construction of scientific facts*. Princeton, NJ: Princeton University Press.

Lawlor, D. A., Davey Smith, G., & Ebrahim, S. (2004). Commentary: The hormone replace- ment-coronary heart disease conundrum: Is this the death of observational epidemiology? *International Journal of Epidemiology*, *33*(3), 464–467. doi:10.1093/ije/dyh124 PMID:15166201

Morris, C. (1971). *Foundations of the theory of signs*. Chicago, IL: University of Chicago Press.

Peirce, C. S. (1958). *Collected Writings* (Vols. 1-8; C. Hartshorne, P. Weiss, & A. W. Burks, Eds.). Cambridge, MA: Harvard University Press.

Polanyi, M. (1966). *The tacit dimension*. New York, NY: Anchor Day Books.

Popper, K. (1963). *Conjectures and refutations*. London, UK: Routledge.

Pschenichny, C., & Mouromtsev, D. (n.d.). Grammar of Dynamic Knowledge for Collaborative Knowledge Engineering and Representation. In Collaborative Knowledge in Scientific Research Networks. Hershey, PA: IGI Global.

Pshenichny, C. A. (2004). Classical logic and the problem of uncertainty. Geological Society, 239, 111-126.

Scardamalia, M., & Bereiter, C. (2006). Knowledge building: Theory, pedagogy, and technology. In R. K. Sawyer (Ed.), *The Cambridge handbook of the learning sciences* (pp. 97–115). New York, NY: Cambridge University Press.

Scientific method. (n.d.). In *Oxford dictionary*. Retrieved from https://en.oxforddictionaries.com/definition/scientific_method

Shannon, C. E., & Weaver, W. (1949). *A mathematical model of communication*. Chicago, IL: University of Illinois Press.

Star, S. L., & Griesemer, J. R. (1989). Institutional ecology, translations and boundary objects: Amateurs and professionals in Berkeley's museum of vertebrate zoology. *Social Studies of Science*, *19*(4), 387–420. doi:10.1177/030631289019003001

Whitley, R. (2000). *The intellectual and social organization of the sciences*. Oxford, UK: Clarendon Press.

Wittgenstein, L. (1921). Logisch-philosophiche Abhandlung. Annalen der Naturphilosophie, 14, 185–262.

KEY TERMS AND DEFINITIONS

Boundary Object: An artifact that allows two different and incommensurable communities to nevertheless collaborate. They are weakly structured in common use and become strongly structured in individual use.

Tacit Knowledge: Knowledge that cannot be made explicit because it is obscured in socio-political phenomena or embedded in practices or even tools.

Theory Ladenness: Refers to the fact that experience is categorized and preconditioned by prior belief, and that the process of building knowledge always involves the use of concepts based on some previous view. This holds both from a logical and a perceptive point of view.

Section 2

Dynamic Knowledge Representation in Geosciences

Chapter 5
Qualitative Modeling of Ice Sheet Accumulation for Identification of Mass Balance Variation

Kirill Khvorostovsky
Nansen Environmental and Remote Sensing Center, Norway

Pavel Lunev
Polar Marine Geosurvey Expedition, Russia

Victoria Shterkhun
Karpinsky All-Russian Research Geological Institute (VSEGEI), Russia

ABSTRACT

Formation and evolution of ice sheets is one of the "hot" problems of modern geosciences, as it has direct implication on the issues of climate change and sea level rise. Different methods of measurement or computing the mass balance of modern ice sheets based on various physical models sometimes give conflicting results. To understand them, one should first reconcile the models they are based on. This, in turn, requires one to decipher the vision different researchers have on the generation and evolution of ice sheets. This vision is initially qualitative. Hence, a qualitative model is desired that would reconcile various, and sometimes conflicting, physical models. This chapter proposes this model.

INTRODUCTION

In the studies of present climate changes the mass balance of the Greenland ice sheets (GrIS) is of particular interest because its potential melting is critical for sea-level change and freshwater impact on ocean circulation. Wide range of satellite measurements during the last two and a half decades allowed significant improving of our knowledge about such extreme polar environments as ice sheets. Various

DOI: 10.4018/978-1-5225-5261-1.ch005

Figure 1. The surface of the ice sheet in the Central Antarctica

estimates show that the GrIS mass balance over this period is negative and mass loss is accelerating in time (AMAP, 2011).

Previously it was considered that mass changes of the ice sheets on the decadal time scale are primarily caused by surface mass balance, i.e. difference between snow accumulation and water runoff due to summer surface melting. Area extent of the surface melting varies significantly from year to year, but water runoff occurs only in the narrow zone over GrIS margins, while above the equilibrium line most water from surface melting refreezes in the firn (Figure 1). In this case variability of snow accumulation is the main parameter of the mass balance over most area of the ice sheet. At the same time water runoff from melting accounts for about half of the mass loss from the ice sheet. Another source of mass loss is the ice discharge through iceberg calving (Figure 2a, b). Rate of the ice discharge depends on the ice flow velocity, which was considered to vary very slowly as a response to mass balance changes hundreds and thousands years ago and therefore may be assumed as constant (Figure 4).

However recent studies have shown that ice flow velocity may vary considerably on inter-decadal and even inter-annual time scales (e.g., Rignot et al., 2011; Moon et al., 2012). Significant ice loss over the GrIS low-elevation areas occurred during the last one and a half decades is explained partly by the intense surface melting due to increased temperature and partly by the acceleration of the ice flow. This acceleration is observed not only over the ice streams and outlet glaciers, characterized by the high flow velocities of up to several kilometers per year, but also propagates inland of the ice sheet. These rapid changes in ice flow velocities were not predicted by the ice flow models, which were not able to resolve short time scale variations of the ice dynamics and did not account for the effect of all external forcing components.

Currently two mechanisms responsible for the increased ice flow velocities are considered. One of them is related to changes in basal lubrication and sliding of the ice following increased melting on the surface and drainage through moulins (Zwally et al., 2002; Joughin et al., 2008; Van de Wal et al., 2008; Bartholomew et al., 2010). The other mechanism is connected with changes in buttressing of outlet glaciers caused by removal of their floating ice tongues due to warmer waters and allowing ice-flow speed-up (e.g. Thomas et al., 2004; Nick et al., 2009). However although these processes may explain recent ice flow acceleration it is not obvious that they will remain dominant over longer-term scale. For example, it was shown (van de Wal et al., 2008) that rapid drainage of melt water significantly speeds-up ice flow over the period of only several days implying that the englacial hydraulic system adjusts constantly to the meltwater input. As for another mechanism, high sensitivity of tidewater outlet glaciers to changes

Figure 2. The front edges of ice sheets in the Antarctic: (a) in the Antarctic Peninsula area, (b) near the Erebus volcano.

Figure 3. View on the world's largest Lambert Glacier, Antarctica

in their terminus boundary conditions may reflect only short-term (of several years) dynamical adjustments but cannot be maintained in the long term (Nick et al., 2009).

However not only understanding of the current GrIS mass changes, but also assessment of the mass balance still remains uncertain because various studies give different estimates. Three different methods are used for estimation of the GrIS (as well as the Antarctic ice sheet) mass balance (Alley et al., 2007; AMAP, 2011). One is based on using radar or laser altimeter measurements from satellite or aircraft (e.g. Krabil et al., 2004; Johannessen et al., 2005; Zwally et al., 2005; Thomas et al., 2006; Zwally et al., 2011; Sørensen et al., 2011; Khvorostovsky, 2012). This method gives direct assessment of the mass balance through estimating of ice volume changes from measured changes in the elevation. The mass budget method calculates input and output ice fluxes separately. Surface mass balance is determined from models and reanalysis, while survey of the ice discharge can be made from measurements of ice flow velocities along the coastline using satellite images (e.g. Rignot and Kanagnaratnam 2006; van den Broeke et al., 2009). The third method is the use of satellite gravimetric measurements such as the Gravity Recovery and Climate Experiment (GRACE) that measure directly the difference between mass input and output (e.g. Luthke et al. 2006; Chen et al., 2006; Velicogna, 2009; Ramilen et al., 2006; Wouters et al., 2008, Wu et al., 2010). The results obtained by different methods are in general agreement showing that the GrIS was near in balanced in 1990th, while from 2000th significant mass loss has began, and this loss accelerates gradually. As it was noted above, current negative mass balance is primarily caused by the mass loss over the GrIS low-elevation areas due to increased surface melting and accelerating of the ice flow. In the interior regions the observed changes are comparatively small, although in some of these areas ice loss propagates inland that may be caused by the decrease of accumulation as well as by the effects of ice flow speed-up over margins. At the same time, although reconciled estimate of ice-sheet mass balance has been reported in (Shepherd et al., 2012) different mass balance estimates vary in a wide range even when obtained by the same method (Hurkmans et al., 2014), and further studies are needed to come to an agreement.

To do this, in our opinion, the glacier dynamics should be better conceptualized, including its mass gain and mass loss by different mechanisms and their possible interplay. This would allow us to fix the list of "players" (i.e., portions of water coming to the glacier, portions of ice and firn and others), the external conditions and the rules of the "game" based on general glaciological, meteorological, hydrological and other related knowledge, trace the scenarios of glacier evolution under various conditions and their change, consider all possible relations between the real-world entities and physical parameters and, finally, link physical models and assumptions to the qualitatively outlined scenarios more or less in the way it was done in seismology for the site effects by the method of event bush (Carniel et al., 2011). The obtained conceptual model is expected to put reasonable qualitative constraints on the physical and numerical modeling and decrease the uncertainty in interpretation of the observation data.

This claim, being quite challenging, requires an extensive and continuous effort and unlikely can be entirely fulfilled in one publication. However, in this pioneering work we may give an outline of the study and make the first step to ensure its feasibility. In particular, we will

1. Suggest the general approach and architecture of the conceptual model;
2. Fill the first block of the model and
3. Consider the issues of quantification of this model to proceed, in the future, to physical models and quantitative assessments.

GENERAL APPROACH AND ARCHITECTURE OF THE MODEL

Keeping in mind the ultimate goal – the reconciliation of physical models – the methodology of the study should allow (i) strict management of qualitative terms and (ii) description of changing states of objects, as well as appearance and disappearance of objects. The theory of multitudes pretends to offer a tool, the Novorossian logic, to meet both these claims. Among the methods based on this methodology, the event bush (Pshenichny et al., 2009) seems to be best developed semantically. Besides, it was designed specially for the geoscientific tasks.

General approach to the modeling is determined by the properties of the modeled environment. The environment is the ice sheet and the landscape in which it originates, evolves, and comes to end. At the beginning, it looks reasonable to focus not on the specificity of Greenland ice sheets, but on the common features of ice sheets throughout the world. Therefore, the modeling presented in this study would bear little to no regional specificity.

Traditionally, glaciers are divided into the zones of dominant mass gain (accumulation) and dominant mass loss (ablation). The mass loss may proceed in a number of ways, all of them being directly or indirectly related to heat, and in some glaciers, e.g., in the Antarctic, ablation per se is minor.

The architecture of the conceptual model should mimic this bi-to-tripartite structure of the glacier and surrounding landscape. The conditions favoring the generation of glacier in high mountains and polar regions, both mountainous and flat, qualitatively are similar. They differ only in numbers (size, height, thickness and so forth), but not conceptually. Importantly, in both cases the surface at the initial stage of glacier evolution must have no outer slopes to ensure accumulation of ice. Therefore, the first block of the model should be equally applicable to ice sheets and mountainous glaciers.

FIRST BLOCK OF THE MODEL: FORMATION OF GLACIER

The mechanism of formation of a glacier from snow through the stages of neve and firn is well known and described in numerous publications. Based on these, it can be generally recited as follows.

In the region of permanently negative air temperature (i.e., in polar regions and/or high enough in the mountains) the snow falls down on the surface of the solid Earth that is either ground or the surface of existing glacier. If not blown off by wind, the snow accumulates, survives summer, its lower part becomes compact (neve) and finally recrystallizes into another substance called firn, which, in turn, being further compacted and recrystallized, turns into glacier ice. When the batch of ice becomes thick enough, it begins to flow, and the batches of firn and snow bedded above are consequently mobilized and flow too. Importantly, once the snow has survived on the solid surface, the slope angle of this surface, being certainly influential, is not critical anymore for further accumulation, transformation and even flowage of ice, firn and snow. As for their mechanical properties, they are irrelevant too because they are virtually the same everywhere in the world where glaciers form. Only the thickness of ice matters, and the flow at this earliest stage occurs radially off the thickest zone, be it a vast flatland in Greenland in Late Pleistocene or a small cirque in a mountain ridge. At the next stage, which is supposed to be covered by the subsequent model, the topography underlying the glacier base becomes critical, and the difference between mountainous glacier and ice sheet is obvious. Also, any kind of melting due to external heat (solar radiation, friction at the glacier base, penetration of water and others) occurs further, below the snowline, and will not be addressed here.

Finally, at this initial step of modeling we refrain from considering the recrystallization process in detail and avoid the issues of internal melting, air trapping and others, not to make the model worryingly complex from the very beginning. Nevertheless, we reserve the "slot" to introduce all these.

At present, full model includes:

Predmets:

S_1 – Name;
S_2 – Role;
S_3 – Subject;
S_4 – Object;
S_5 – Earth;
S_6 – Solidity;
S_7 – Surface;
S_8 – Thickness;
S_9 – Coalescence;
S_{10} – Particle;
S_{11} – Batch;
S_{12} – Rock;
S_{13} – Snow;
S_{14} – Firn;
S_{15} – Ice;
S_{16} – Top;
S_{17} – Wind;
S_{18} – Flowage;
P_1 – To be named;
P_2 – To have role;
P_3 – To be of the Earth;
P_4 – To be solid;
P_5 – To be thick;
P_6 – To coalesce;
P_7 – To be of rock;
P_8 – To be of snow;
P_9 – To be of firn;
P_{10} – To be of ice;
P_{11} – To be blown off;
P_{12} – To flow;

P_{13} – To be in relation to top;
P_{14} – (For "someone") to top (something);
P_{15} – (For "someone") to be topped (by something);
P_{16} – To top the batch of rocks;
P_{17} – To top thick batch of snow;
P_{18} – To top not a thick batch of snow;
P_{19} – To top thick batch of firn;
P_{20} – To top not a thick batch of firn;
P_{21} – To top thick batch of ice;
P_{22} – To top not a thick batch of ice;
P_{23} – To be topped;
P_{24} – To be topped by particles of snow;
P_{25} – To be topped by thick batch of snow;
P_{26} – To be topped by not a thick batch of snow;
P_{27} – To be topped by thick batch of firn;
P_{28} – To be topped by not a thick batch of firn;
P_{29} – To be topped by thick batch of ice;
P_{30} – To be topped by not a thick batch of ice.

Exclusions (expressions making no sense in the considered context):

$S_3 – P_1, \sim P_2, \ldots = \varnothing$; $S_4 – P_1, \sim P_2, \ldots = \varnothing$ (subject or object does not have role);

$S_{11} – P_{u1}, P_{u4}, \sim P_3, \ldots = \varnothing$ (Batch of something is not related to the Earth);

$S_{11} – P_{u1}, P_{u4}, P_3, P_7, P_{12}, \ldots = \varnothing$ (Batch of rocks is solid);

$S_{11} – P_{u1}, P_{u4}, P_3, \sim P_5, P_8, P_{12}, \ldots = \varnothing$; $S_{11} – P_{u1}, P_{u4}, P_3, \sim P_5, P_9, P_{12}, \ldots = \varnothing$; $S_{11} – P_{u1}, P_{u4}, P_3, \sim P_5, P_{10}, P_{12}, \ldots = \varnothing$

(Thin batch of snow, firn or ice may not flow – note that the model relates only to the accumulation zone of a glacier. The reasons for these are different: thin ice may not flow because it does not form unstable slopes, and thin firn and snow do not flow in the accumulation zone simply because only thick enough firn and snow are assumed to cover the thick batch of ice. For the same reason, if a batch of ice or firn are flowing, they may not be overlain by thin batches of firn or snow, correspondingly).

$S_{10} – P_{u1}, P_{u4}, P_3, P_6, P_8, P_{11}, \ldots = \varnothing$ (Snow flakes may not coalesce when blown off by wind);

$S_{10} – P_{u1}, P_{u4}, P_3, P_8, P_{11}, P_{14}, \ldots = \varnothing$ (Snow flakes may not top anything when blown off by wind);

S_{10}–P_{u1}, P_{u4}, P_3, P_8, ~P_{11}, ~P_{14}, … =∅ (Snow flakes must either top a batch of something or be blown off by wind – one option must be necessarily taken);

S_{11}–P_{u1}, P_{u4}, ~P_5, P_7, …=∅ (Batch of rock may not be thin);

S_{11}–P_{u1}, P_{u4}, P_3, P_7, P_{16}, … =∅ ; S_{11}–P_{u1}, P_{u4}, P_3, P_8, P_{17}, … =∅ ; S_{11}–P_{u1}, P_{u4}, P_3, P_9, P_{16}, … =∅ ; S_{11}–$Pu1$, P_{u4}, P_3, P_{10}, P_{16}, P_{21} или P_{22} =∅ (Batch of rock, of snow, of firn or ice may not be topped by itself.)

S_{11}–P_{u1}, P_{u4}, P_3, P_{10}, P_{25}, … =∅; S_{11}–P_{u1}, P_{u4}, P_3, P_{10}, P_{26}, … =∅; … (Batch of ice may not be covered by batch of snow without firn; firn is always below snow, and so forth.)

S_{11}–P_{u1}, P_{u4}, P_9, P_{25}, P_{26}, … … =∅ Batch cannot be covered by another batch that is thick and thin simultaneously in one place;

S_{11}–P_{u1}, P_{u4}, P_{10}, P_{27}, P_{28}, … … =∅ The same.

With these limitations in mind, the full sequent should be as follows. To save time and space, we will not present the entire pathway of inference. Rather, below we give the final result of this inference.

$S_1 \rightarrow P_1$;
$S_2 \rightarrow P_2$;
$S_6 \rightarrow P_4$;
S_5–P_{u1}, P_4; S_5–P_1, ~P_4;
(S_5–P_{u1}, P_4)$\rightarrow P_3$;
S_{16}–P_{u3};
$S_{16} \rightarrow P_{13}$;
S_3–P_{u1}, P_2, P_{13}; S_3–P_{u1}, P_2, ~P_{13}; S_3–P_{u1}, ~P_2, P_{13}; S_3–P_{u1}, ~P_2, ~P_{13};
(S_3–P_{u1}, P_2, P_{13})$\rightarrow P_{14}$;
S_4–P_{u1}, P_2, P_{13}; S_4–P_{u1}, P_2, ~P_{13}; S_4–P_{u1}, ~P_2, P_{13}; S_4–P_{u1}, ~P_2, ~P_{13};
(S_4–P_{u1}, P_2, P_{13})$\rightarrow P_{15}$;
S_{12}–P_{u1}, P_3; S_{12}–P_{u1}, ~P_3;
(S_{12}–P_{u1}, P_3)$\rightarrow P_7$;
S_{13}–P_{u1}, P_3; S_{13}–P_{u1}, ~P_3;
(S_{13}–P_{u1}, P_3)$\rightarrow P_8$;
S_{14}–P_{u1}, P_3; S_{14}–P_{u1}, ~P_3;
(S_{14}–P_{u1}, P_3)$\rightarrow P_9$;
S_{15}–P_{u1}, P_3; S_{15}–P_{u1}, ~P_3;
(S_{15}–P_{u1}, P_3)$\rightarrow P_{10}$;

$S_{17}–P_{u1}, P_{3}; S_{15}–P_{u1}, \sim P_{3};$

$(S_{17}–P_{u1}, P_{3})\rightarrow P_{11};$

$S_{18}–P_{u1}, P_{3}; S_{15}–P_{u1}, \sim P_{3};$

$(S_{18}–P_{u1}, P_{3})\rightarrow P_{12};$

$S_{10}–P_{u1}, P_{u4}, P_{3}, \sim P_{6}, P_{8}, \sim P_{11}, P_{14};$

$S_{10}–P_{u1}, P_{u4}, P_{3}, \sim P_{6}, P_{8}, P_{11}, \sim P_{14};$

$S_{10}–P_{u1}, P_{u4}, P_{3}, P_{6}, P_{8}, \sim P_{11}, P_{14}$

$(S_{10}–P_{u1}, P_{u4}, P_{3}, \sim P_{6}, P_{8}, \sim P_{11}, P_{14})\rightarrow P_{24}$

$S_{11}–P_{u1}, P_{u4}, P_{3}, P_{5}, \sim P_{7}, P_{8}, \sim P_{9}, \sim P_{10}, P_{12}, P_{24};$

$S_{11}–P_{u1}, P_{u4}, P_{3}, P_{5}, \sim P_{7}, P_{8}, \sim P_{9}, \sim P_{10}, \sim P_{12}, \sim P_{24};$

$S_{11}–P_{u1}, P_{u4}, P_{3}, \sim P_{5}, \sim P_{7}, P_{8}, \sim P_{9}, \sim P_{10}, \sim P_{12}, P_{24};$

$S_{11}–P_{u1}, P_{u4}, P_{3}, \sim P_{5}, \sim P_{7}, P_{8}, \sim P_{9}, \sim P_{10}, \sim P_{12}, \sim P_{24};$

$(S_{11}–P_{u1}, P_{u4}, P_{3}, P_{5}, \sim P_{7}, P_{8}, \sim P_{9}, \sim P_{10}, P_{12}, P_{24})\rightarrow P_{25};$

$(S_{11}–P_{u1}, P_{u4}, P_{3}, P_{5}, \sim P_{7}, P_{8}, \sim P_{9}, \sim P_{10} \sim P_{12}, \sim P_{24})\rightarrow P_{25};$

$(S_{11}–P_{u1}, P_{u4}, P_{3}, \sim P_{5}, \sim P_{7}, P_{8}, \sim P_{9}, \sim P_{10}, \sim P_{12}, P_{24})\rightarrow P_{26};$

$(S_{11}–P_{u1}, P_{u4}, P_{3}, \sim P_{5}, \sim P_{7}, P_{8}, \sim P_{9}, \sim P_{10}, \sim P_{12}, \sim P_{24})\rightarrow P_{26};$

$S_{11}–P_{u1}, P_{u4}, P_{3}, P_{5}, \sim P_{7}, \sim P_{8}, P_{9}, \sim P_{10}, P_{12}, P_{25};$

$S_{11}–P_{u1}, P_{u4}, P_{3}, P_{5}, \sim P_{7}, \sim P_{8}, P_{9}, \sim P_{10}, \sim P_{12}, P_{25};$

$S_{11}–P_{u1}, P_{u4}, P_{3}, P_{5}, \sim P_{7}, \sim P_{8}, P_{9}, \sim P_{10} \sim P_{12}, P_{26};$

$S_{11}–P_{u1}, P_{u4}, P_{3}, \sim P_{5}, \sim P_{7}, \sim P_{8}, P_{9}, \sim P_{10}, \sim P_{12}, P_{25};$

$S_{11}–P_{u1}, P_{u4}, P_{3}, \sim P_{5}, \sim P_{7}, \sim P_{8}, P_{9}, \sim P_{10}, \sim P_{12}, P_{26};$

$(S_{11}–P_{u1}, P_{u4}, P_{3}, P_{5}, \sim P_{7}, \sim P_{8}, P_{9}, \sim P_{10}, P_{12}, P_{25})\rightarrow P_{27};$

$(S_{11}–P_{u1}, P_{u4}, P_{3}, P_{5}, \sim P_{7}, \sim P_{8}, P_{9}, \sim P_{10} \sim P_{12}, P_{25})\rightarrow P_{27};$

$(S_{11}–P_{u1}, P_{u4}, P_{3}, P_{5}, \sim P_{7}, \sim P_{8}, P_{9}, \sim P_{10} \sim P_{12}, P_{26})\rightarrow P_{27};$

$(S_{11}–P_{u1}, P_{u4}, P_{3}, \sim P_{5}, \sim P_{7}, \sim P_{8}, P_{9}, \sim P_{10}, \sim P_{12}, P_{25})\rightarrow P_{28};$

$(S_{11}–P_{u1}, P_{u4}, P_{3}, \sim P_{5}, \sim P_{7}, \sim P_{8}, P_{9}, \sim P_{10}, \sim P_{12}, P_{26})\rightarrow P_{28};$

$S_{11}–P_{u1}, P_{u4}, P_{3}, P_{5}, \sim P_{7}, \sim P_{8}, \sim P_{9}, P_{10}, P_{12}, P_{27};$

$S_{11}–P_{u1}, P_{u4}, P_{3}, P_{5}, \sim P_{7},$ ~P8, ~P9, P10, ~P12, P27;

$S_{11}–P_{u1}, P_{u4}, P_{3}, P_{5}, \sim P_{7}, \sim P_{8}, \sim P_{9}, P_{10}, \sim P_{12}, P_{28};$

$S_{11}–P_{u1}, P_{u4}, P_{3}, \sim P_{5}, \sim P_{7}, \sim P_{8}, \sim P_{9}, P_{10}, \sim P_{12}, P_{27};$

$S_{11}–P_{u1}, P_{u4}, P_{3}, \sim P_{5}, \sim P_{7}, \sim P_{8}, \sim P_{9}, P_{10}, \sim P_{12}, P_{28};$

$S_{11}–P_{u1}, P_{u4}, P_{3}, P_{5}, P_{7}, \sim P_{8}, \sim P_{9}, \sim P_{10}, P_{24}, \sim P_{27}, \sim P_{28};$

$S_{11}–P_{u1}, P_{u4}, P_{3}, P_{5}, P_{7}, \sim P_{8}, \sim P_{9}, \sim P_{10}, \sim P_{24}, P_{27}, \sim P_{28};$

$S_{11}–P_{u1}, P_{u4}, P_{3}, P_{5}, P_{7}, \sim P_{8}, \sim P_{9}, \sim P_{10}, \sim P_{24}, \sim P_{27}, P_{28}.$

Table 1. Formulations of events in Figure 4

Event	Description
iaE_1	Particles of snow do not top the solid surface, have not been blown off the solid surface, do not coalesce
ibE_2	Solid surface is of the batch of rock
ibE_3	Wind is strong enough
iiE_1	Particles of snow do not top the solid surface, have not been blown off the solid surface, do not coalesce
iiE_4	Particles of snow top the solid surface, have not been blown off the solid surface, do not coalesce
iiE_5	Particles of snow top the solid surface, have not been blown off the solid surface, coalesce
iiE_6	Particles of snow top the solid surface, have been blown off the solid surface, do not coalesce
iiE_7	Batch of snow is not thick, tops the batch of rocks, does not top not a thick batch of firn, does not top thick batch of firn, does not flow
iiE_8	Particles of snow do not top the solid surface, have been blown off the solid surface, do not coalesce
iiE_9	Solid surface is of the batch of snow
iiE_{10}	Batch of snow is thick, tops the batch of rocks, does not top not a thick batch of firn, does not top thick batch of firn, does not flow
iiE_{11}	Batch of snow is not thick does not top the batch of rocks, tops not a thick batch of firn, does not top thick batch of firn, does not flow
iiE_{12}	Batch of firn is not thick, tops the batch of rocks, does not top not a thick batch of ice, does not top thick batch of ice, does not flow
iiE_{13}	Batch of snow is thick, does not top the batch of rocks, tops not a thick batch of firn, does not top thick batch of firn, does not flow
iiE_{14}	Batch of snow is thick, does not top the batch of rocks, does not top not a thick batch of firn, tops thick batch of firn does not flow
iiE_{15}	Batch of firn is thick, tops the batch of rocks, does not top not a thick batch of ice, does not top thick batch of ice, does not flow
iiE_{16}	Batch of firn is not thick, does not top the batch of rocks, tops not a thick batch of ice, does not top thick batch of ice, does not flow
iiE_{17}	Batch of ice is not thick, tops the batch of rocks, does not flow
iiE_{18}	Batch of firn is thick, does not top the batch of rocks, does not top not a thick batch of ice, tops thick batch of ice, does not flow
iiE_{19}	Batch of ice is thick, tops the batch of rocks, does not flow
iiE_{20}	Batch of ice is thick, tops the batch of rocks, flows
iiE_{21}	Flowage of ice occurs
iiE_{22}	Batch of firn is thick, does not top the batch of rocks, does not top not a thick batch of ice, tops thick batch of ice, flows
iiE_{23}	Flowage of firn occurs
iiE_{24}	Batch of snow is thick, does not top the batch of rocks, does not top not a thick batch of firn, tops thick batch of firn, flows
$iiiE_1$	Particles of snow do not top the solid surface, have not been blown off the solid surface, do not coalesce
$iiiE_7$	Batch of snow is not thick, tops the batch of rocks, does not top not a thick batch of firn, does not top thick batch of firn, does not flow
$iiiE_{10}$	Batch of snow is thick, tops the batch of rocks, does not top not a thick batch of firn, does not top thick batch of firn, does not flow
$iiiE_{11}$	Batch of snow is not thick does not top the batch of rocks, tops not a thick batch of firn, does not top thick batch of firn, does not flow
$iiiE_{12}$	Batch of firn is not thick, tops the batch of rocks, does not top not a thick batch of ice, does not top thick batch of ice, does not flow

continued on following page

Table 1. Continued

Event	Description
$iiiE_{13}$	Batch of snow is thick, does not top the batch of rocks, tops not a thick batch of firn, does not top thick batch of firn, does not flow
$iiiE_{14}$	Batch of snow is thick, does not top the batch of rocks, does not top not a thick batch of firn, tops thick batch of firn does not flow
$iiiE_{15}$	Batch of firn is thick, tops the batch of rocks, does not top not a thick batch of ice, does not top thick batch of ice, does not flow
$iiiE_{16}$	Batch of firn is not thick, does not top the batch of rocks, tops not a thick batch of ice, does not top thick batch of ice, does not flow
$iiiE_{17}$	Batch of ice is not thick, tops the batch of rocks, does not flow
$iiiE_{18}$	Batch of firn is thick, does not top the batch of rocks, does not top not a thick batch of ice, tops thick batch of ice, does not flow
$iiiE_{19}$	Batch of ice is thick, tops the batch of rocks, does not flow
$iiiE_{20}$	Batch of ice is thick, tops the batch of rocks, flows
$iiiE_{22}$	Batch of firn is thick, does not top the batch of rocks, does not top not a thick batch of ice, tops thick batch of ice, flows
$iiiE_{24}$	Batch of snow is thick, does not top the batch of rocks, does not top not a thick batch of firn, tops thick batch of firn, flows

From the events of this sequent, an event bush could be built as described by Pshenichny (2018b, see Chapter 2 of this book). Not all events are included in the bush but only those whom it seems reasonable to us to populate with fill. These are the events, which, in our opinion, will best correspond to the physical models of the processes in the considered environment. Moreover, to avoid unnecessary duplication of connectives in the event bush, we considered it possible to take two different primetas (to land on the surface of rock batch and to land on the surface of snow batch) as one (to land on a solid surface) if all the consequences considered further are the same. In the formulation of events we omit most general universa ("to be named", "to be related to the Earth" and the like). The resulting events, grouped in accordance with the basic syntax of the event bush (see, e.g., Pshenichny and Kanzheleva, 2011), are listed in Table 1.

These events are grouped into an event bush as follows (see Figure 4).

The event bush, reflecting the quantitative transitions in an environment qualitatively modeled by a Novorossian sequent, is a model that can be "run" (on paper or, in the future, on computer) to simulate the behavior of natural objects. We imply that it describes possible scenarios in given time, in given place. The virtue of the event bush in Figure 1 is that it can equally well start from "no glacier" (iaE_1, ibE_2 *Influx* iiE_4) and from "existing glacier and new snowfall" (iaE_1, ibE_9 *Influx* iiE_4) situations. The above implies that *either* we have bare (or almost bare) rock or a glacier in the location in question, but not both.

The slots for detailed description of recrystallization and melting are in the events $iiE_4 - iiE_{24}$. Primetas "responsible" for physical properties may be added to these, and connectives of the event bush may be attributed the sense of mathematical relations between these properties or assumptions about them adopted in the physical models.

The fill of the events is as follows: iaE_1, $iiE_1 - iiE_6$, iiE_8, $iiiE_1$ grainy, all the rest, lump fill. Grainy fill, by Pshenichny (2018b; Chapter 2 of this volume) means that there are "atoms" that bear the said meaning (in the considered case, each "particle of snow", i.e., snowflake). Lump fill, by the same author, implies that there is only one thing (material or ideal) under consideration that satisfies this meaning, i.e., "batch of snow" or "flowage".

Figure 4. The event bush of formation of glacier above snowline

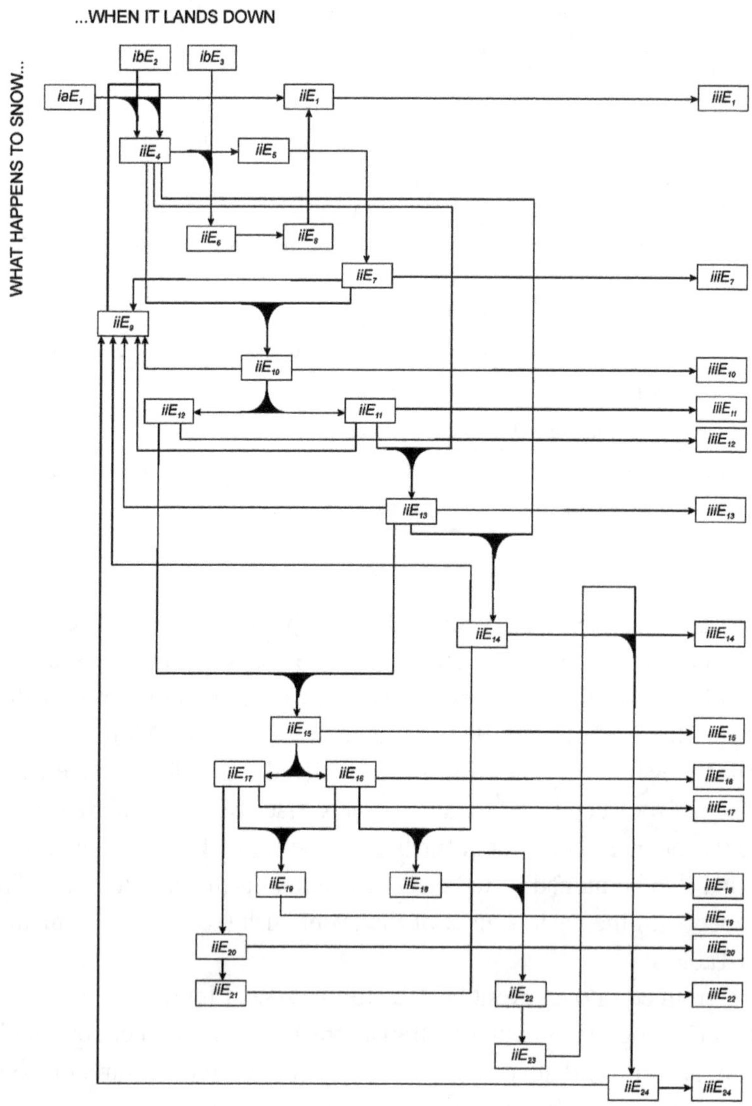

Semantically, the distinction between types of fill allows us to operate with such meanings as, say, "particles of snow" and "batch of snow" simultaneously, even if we consider the batch of snow consisting of particles of snow, both entities can be present in the bush without a fear to "count twice" those particles that make up the batch. This opens an opportunity of adding a "parallel line" of events in the discussed event bush starting from the change iiE_5 *Flux* iiE_7 down to iiE_{24}. In the mentioned change, a passage from micro-level to macro-level occurs, i.e., from particles to a body that consists of them. Still, one may continue considering the behavior of particles (their compaction, warming up, melting at the edge, production of interstitial water, possible migration of this water, refreezing it and cementation of particles by refrozen water and so forth) *along with* the behavior of the body (i.e., the batch of snow/firn/ice). This is a possible and favorable development of the event bush model of glacier generation.

However, in any case three groups of scenarios are recognized in this model, be it developed further or not. The first group includes the behavior of snow that is not captured at the surface:

1.1 iaE_1 *Flux* iiE_1, iiE_1 *Flux* $iiiE_1$,
1.2 iaE_1, ibE_2 *Influx* iiE_4, iiE_4, ibE_3 *Influx* iiE_6, iiE_6 *Flux* iiE_1, iiE_1 *Flux* $iiiE_1$ and
1.3 iaE_1, iiE_9 *Influx* iiE_4, iiE_4, ibE_3 *Influx* iiE_6, iiE_6 *Flux* iiE_1, iiE_1 *Flux* $iiiE_1$.

Of these three, scenarios 1.2 and 1.3 are totally incompatible with each other because either there is solid surface of rock or of a preexisting glacier. Though, both 1.1 and 1.2 are incompatible with 1.1 only if we assume that all particles behave uniformly (either all settle down or all are in air) that is an apparent idealization. Otherwise one may expect that some particles behave in the way 1.1, and some, either in 1.2 or in 1.3.

The second and third groups include scenarios of formation of a static (second group) or flowing (third group) body of snow, firn and snow, or ice, firn, and snow, with a prerequisite that only a body that includes all three components may flow. These scenarios are quite numerous, and unlikely it is reasonable to cite them all. Rather, we will give their general characterization and examples.

All scenarios of the second and third groups begin like 1.2 or 1.3, that is, either with iaE_1, ibE_2 *Influx* iiE_4 or with iaE_1, iiE_9 *Influx* iiE_4. The distinction starts at iiE_4, where another change occurs that is common for all the scenarios of the second and third groups, iiE_4 *Flux* iiE_5, that means, snowflakes are not blown off the surface and "freely" coalesce with each other. As in the cases of 1.2 vs. 1.1 and 1.3 vs. 1.1 scenarios, this change may be considered incompatible only under assumption that all snowflakes behave uniformly.

After the change iiE_4 *Flux* iiE_5, the transition occurs that binds events with two different types of fill, iiE_5 *Flux* iiE_7. This is transition from granular to lump fill, in other words, transition from individual snowflakes to a body, or batch, of snow, which may evolve as a singular whole. Simultaneously, as was noted above, we may continue parallel reasoning in terms of particles. The simplest scenario of the second group ends here by formation of a small batch of snow: iiE_7 *Flux* $iiiE_7$.

However, if the snow continues to come and the snow batch grows, these two events together would give another event (iiE_{10}), which also may mark the end of the story: iiE_{10} *Flux* $iiiE_{10}$ – or be continued, if the snow continues to come, and so on, up to the event iiE_{19}. This event means formation of a glacier with thick ice layer, but still static. At each stage the flux leading to a tertiary event is incompatible with a conflux leading to continuation of the story (e.g., iiE_7 *Flux* $iiiE_7$ is incompatible with iiE_7, iiE_4 *Conflux* iiE_{10}, and so forth).

Scenarios of the second group end by the secondary event iiE_{19} and the flux iiE_{19} *Flux* $iiiE_{19}$. The third group events include the next event, iiE_{20}, denoting the flowing batch of ice. This event, leading to one end result per se (iiE_{20} *Flux* $iiiE_{20}$), also leads to a phenomenon of flowage of ice, which, in turn, entails flowing batches of firn (iiE_{22}) and snow (iiE_{24}). Thus there are two scenarios in the third group, starting either from iaE_1, ibE_2 *Influx* iiE_4 or from iaE_1, iiE_9 *Influx* iiE_4 and having, in both cases, three end results: flowing ice ($iiiE_{20}$), flowing firn ($iiiE_{22}$) and flowing snow ($iiiE_{24}$) batches. The corresponding tertiary events appear, therefore, always compatible with each other.

Only the scenarios of the third group may be continued in the second block of the model describing the flowage of a glacier in an accumulation zone or in a zone transitional from accumulation to ablation (or another type of discharge). The three end results of these scenarios will become the primary internal events of the corresponding event bushes of the second block. However, in contrast to the first block represented by only one event bush, these bushes are expected to be different not only for the mountainous glaciers and ice sheets, but perhaps also for different types of ice sheet, e.g., GrIS and Antarctic ones.

DISCUSSION

The conceptual model of the accumulation areas of ice sheets reported here is the first step in the conceptual modeling of the phenomenon of glaciers by means of the dynamic knowledge engineering. From this point, even omitting the mountainous glaciers, this model can be developed in a number of interrelated directions. First, the ablation (or, better say, ice loss) zone should be covered, and desirably a through event bush should be constructed to qualitatively follow the behavior of an ice sheet in the ice loss zone depending on its "biography" in the accumulation zone. Then, the model should be specified to qualitatively account for the contribution of atmospheric CO_2 to the mass balance variation (Foster et al., 2017), the response of the earthcrust (isostasy) and gravity field to increasing and decreasing ice load (which, in the latter case, may lead to partial isolation of ice sheets from the sea – Alley et al., 2005; Ridley et al., 2010; Ritz et al., 2015; DeConto & Pollard, 2016). Another important effect to be addressed at least in Greenland is that as the ice sheet melts, the elevations become lower, and the temperature increases above its surface, which, again, amplifies melting (Alley et al., 2005; Ridley et al., 2010). With these peculiarities, we believe that conceptual model of ice sheet evolution may achieve the state, in which it becomes readily coincided with the physical and numerical models of mass balance variation, which was the main motivation of this research.

Still, this model will be able to address much wider range of issues of physical modeling than solely the mass balance variation. It should be wholly coincided with the fundamentals and state-of-the-art of physical modeling of ice sheets. Some key points have been summarized by Cuffey & Paterson (2010), but dozens of papers appear every year. For instance, the regularities were discovered in the behavior of the Antarctic and Greenland ice sheets over centennial and millennial timescales (Clark et al., 2016). Another important issue is that the Greenland ice sheet has 3 stable modes, which can be reached after a threshold is crossed. For example, it is suspected that Greenland ice sheet will escape the current stable mode if the global temperature exceeds 1.5°C. Yet greater challenge is the long-term prediction of glaciation behavior. None of the CMIP-5 model experiments have been able to model glaciation over the next 70000 years (IPCC, 2013), and application of dynamic knowledge engineering approach here may become the only solution. Therefore, the general question is, how to relate the event bush model to any kind of physical model of an ice sheet?

This question, once put, immediately invokes generalization. Indeed, in the light of dynamic knowledge representation, it bears little specificity to glaciers per se. Exactly the same can be asked about, say, volcanoes, slope processes or interaction of celestial bodies, and general solution, to our knowledge, has not been suggested.

Developing this issue in seismology, Carniel et al. (2011) proposed that each connective in an event bush may be specified as one step of computation or one assumption, explicit or hidden. The same approach was used by Pshenichny et al. (2013) for the general case of physical knowledge, yet by a very simple example of Mendeleev-Clapeyron Law.

Still, this approach, informally entitled as "parsing physical models" by the event bush (Pshenichny et al., 2013), remains highly intuitive, as intuitive have been the definitions of connectives of the event bush (Pshenichny & Kanzheleva, 2011). With the latest developments of the theory of multitudes (Pshenichny,

Figure 5. One-to-one transition between two multitudes, A and Б, modeled by physical model B as displayed in the predmetka chart: (a) – the transition per se; (b) – the transition interpreted by a physical model; (c) – the transition interpreted by a physical model as a part of the context depicted by predmetka.

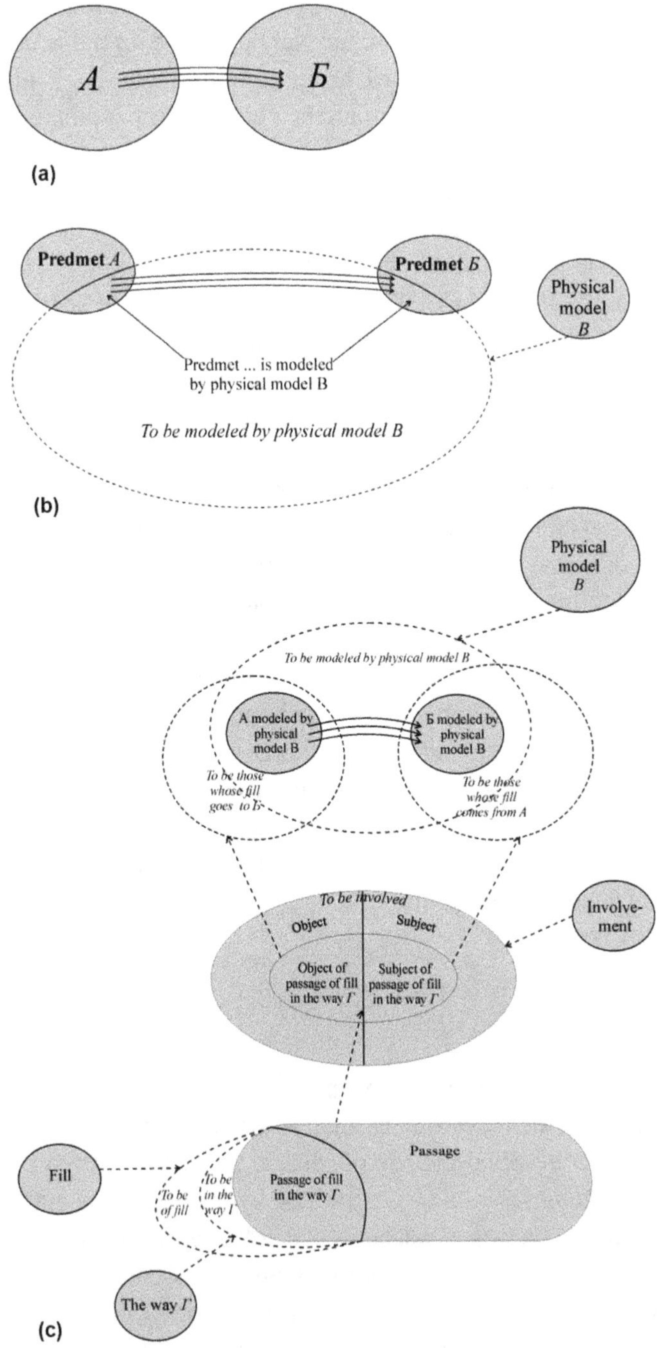

2018a, b – see chapters 1 and 2 of this book), a part of which is the event bush method, one may reasonably query, how the suggested parsing of physical models is expressed in the predmet-primeta-event language of this theory. This query, in fact, is how the physical model per se is expressed in this language.

Prior to answering it, let us put forth that there is crucial difference between bare mathematical expression and physical formula, even if they look formally the same. In mathematical sense, stating that $A=B$ we just note that the values of *A* and *B* have happened to be the same. In physics, saying that $A=B$, we imply that both *A* and *B* necessarily change, and their equality is maintained at any change they may undergo. In terms of the theory of mulitudes, this means that any physical model should relate to an environment that allows transitions (Pshenichny, 2018b). Transitions are effected by connectives of the theory of multitudes. Connectives of the event bush are interpretation of the said connectives.

Consider the simplest, one-to-one transition of multitude *A* to multitude *Б*. In the reported event bush, it can be exemplified, e.g., by the iiE_6 *FLUX* iiE_8 change. Let us try to see how this transition would be displayed in predmetka – the chart disclosing the relations between meanings and possibly showing transitions between them (Pshenichny, 2018b). This is demonstrated in Figure 5.

Note that both *A* and *Б* may be predmets or events, or even events of one predmet. For simplicity, we will consider them predmets. When subject to modeling, they are intersected by primeta "To be modeled by physical model *B*" and form events, "*A/Б* modeled by physical model *B*". Then, the fact of being modeled should be expressed as narrowing the transition between *A* and *Б* to the transition between these very events (Figure 5b). Now the question is, how to express the fact that model *B* is a kind of model formally able to reflect such transitions. Based on templates for displaying relations between meaning given by Pshenichny (2018a), this can be done as shown in Figure 5c. This plot shows that the transition between *A* and *Б* modeled by the model *B* is a transition, or passage, made in the way *Г*, and this very *Г* is also present in the context. Another, although important, issue is whether "to be in the way *Г*" applies only to *A* and *Б* modeled by the model *B* or to *A* and *Б* per se. Perhaps this could be a criterion of adequacy of the model (the model is adequate in the latter case and dubious in the formed). *Г* may include both qualitative and quantitative properties that we consider relevant. These, in turn, may (and should) be expressed in terms of the theory of multitudes, but this is the next part of research that should be considered an issue of separate publication.

CONCLUSION

Rounding up this initial stage of research on conceptualization of ice sheet evolution, we should admit that the methodology chosen – the Novorossian logic – and the particular method, the event bush, are efficient in development of conceptual models of glaciations at least at the first approximation. The general approach to this modeling should be determined by the properties and variability of this environment observed in the nature, such as: (i) clear division of all glaciers into predominant accumulation and predominant ablation (discharge) zones, possibly with a transition zone starting directly below the snowline; (ii) qualitative similarity of processes operating in the accumulation zones of mountainous glaciers and ice sheets and increasing diversity downwards. Likewise, the model should consist of three blocks corresponding to the three zones of glaciers, representing interrelated event bushes, such that some of the tertiary events of the previous bush become primary internals of the next one. Herewith, the first block may be modeled by single event bush, and starting from the second block, specific bushes should be developed for mountainous glaciers and ice sheets of different types (including, likely, different event bushes for the Greenland and Antarctic ice sheets).

The event bush describing the first block of the model, the zone of predominant accumulation of ice sheet material, should include three groups of scenarios: (i) those describing the behavior of snow that is not captured at the surface, (ii) those describing the formation of a static body of snow, firn and snow, or ice, firn, and snow, and (iii) those describing the the formation of a flowing body of ice, firn and snow. Only the third group scenarios are translated further into the second block models. Some of the scenarios of this event bush also account for natural cyclicity of snow accumulation on the surface of an already existing glacier.

The events of the event bush are filled either with "atoms" (particles of snow, i.e., snowflakes) or with a "lump" (bunch of snow, or firn, or ice, or just the fact of flowage of these materials). For the sake of physical modeling, both visions may be needed. Moreover, there exists an opportunity in the event bush to treat natural processes in two ways simultaneously, at the particle level (e.g., compaction of snowflakes, melting of their edges, production of meltwater and its refreezing, cementation of particles by the newly formed interstitial ice) and at the level of the whole body (degree of compaction and densification, load on the underlying batch and so forth). This option has not been used in the current version of the event bush model but can be readily applied if required to incorporate corresponding physical models.

REFERENCES

Alley, R. B., Clark, P. U., Huybrechts, P., & Joughin, I. (2005). Ice-sheet and sea-level changes. *Science, 310*(5747), 456–460. doi:10.1126/science.1114613 PMID:16239468

Alley, R. B., Spencer, M. K., & Anandakrishnan, S. (2007, October). Ice-sheet mass balance: Assessment, attribution and prognosis. *Annals of Glaciology*, *46*(1), 1–7. doi:10.3189/172756407782871738

AMAP. (2011). Snow, Water, Ice and Permafrost in the Arctic (SWIPA): Climate Change and the Cryosphere. Arctic Monitoring and Assessment Programme (AMAP).

Bartholomew, I., Nienow, P., Mair, D., Hubbard, A., King, M. A., & Sole, A. (2010). Seasonal evolution of subglacial drainage and acceleration in a Greenland outlet glacier. *Nature Geoscience*, *3*(6), 408–411. doi:10.1038/ngeo863

Carniel, R., Pshenichny, C., Khrabrykh, Z., Shterkhun, V., & Pascolo, P. (2011). Modeling Models: Understanding of Structure of Geophysical Knowledge by Means of the Event Bush Method. *Mathematical geosciences at the crossroads of theory and practice, IAMG Proceedings*, 1336-1350. doi doi:10.5242/iamg.2011.0074

Chen, J. L., Wilson, C. R., & Tapley, B. D. (2006). Satellite gravity measurements confirm accelerated melting of Greenland ice sheet. *Science*, *313*(5795), 1958–1960. doi:10.1126/science.1129007 PMID:16902089

Clark, P. U., Shakun, J. D., Marcott, S. A., Mix, A. C., Eby, M., Kulp, S., & Plattner, G.-K. et al. (2016). Consequences of twenty-first-century policy for multi-millennial climate and sea-level change. *Nature Climate Change*, *6*(4), 360–369. doi:10.1038/nclimate2923

Cuffey, K. M., & Paterson, W. S. B. (2010). *The physics of glaciers*. Burlington, MA: Butterworth-Heinemann/Elsevier.

DeConto, R. M., & Pollard, D. (2016). Contribution of Antarctica to past and future sea-level rise. *Nature, 531*(7596), 591–597. doi:10.1038/nature17145 PMID:27029274

Foster, G. L., Royer, D. L., & Lunt, D. J. (2017). Future climate forcing potentially without precedent in the last 420 million years. *Nature Communications*, 8. PMID:28375201

Hurkmans, R. T. W. L., Bamber, J. L., Davis, C. H., Joughin, I. R., Khvorostovsky, K. S., Smith, B. S., & Schoen, N. (2014). Time-evolving mass loss of the Greenland Ice Sheet from satellite altimetry. *The Cryosphere*, *8*(5), 1725–1740. doi:10.5194/tc-8-1725-2014

IPCC. (2013). Retrieved from http://www.ipcc.ch/report/ar5/wg1/

Johannessen, O. M., Khvorostovsky, K., Miles, M. W., & Bobylev, L. P. (2005). Recent ice-sheet growth in the interior of Greenland. *Science*, *310*(5750), 1013–1016. doi:10.1126/science.1115356 PMID:16239440

Joughin, I., Das, S. B., King, M. A., Smith, B. E., Howat, I. M., & Moon, T. (2008). Seasonal speedup along the western flank of the Greenland Ice Sheet. *Science*, *320*(5877), 781–783. doi:10.1126/science.1153288 PMID:18420901

Khvorostovsky, K. (2012). Merging and analysis of elevation time series over Greenland ice sheet from satellite radar altimetry. *IEEE Transactions on Geoscience and Remote Sensing*, *50*(1), 1. doi:10.1109/TGRS.2011.2160071

Krabill, W. & 12 others (2004). Greenland Ice Sheet: increased coastal thinning. *Geophysical Research Letters, 31*(2), L24402. 10.1029/2004GL021533

Luthcke, S. B., Zwally, H. J., Abdalati, W., Rowlands, D. D., Ray, R. D., Nerem, R. S., & Chinn, D. S. et al. (2006). Recent Greenland ice mass loss by drainage system from satellite gravity observations. *Science*, *314*(5803), 1286–1289. doi:10.1126/science.1130776 PMID:17053112

Moon, T., Joughin, I., Smith, B., & Howat, I. (2012). 21st-Century Evolution of Greenland Outlet Glacier Velocities. *Science*, *336*(6081), 576–578. doi:10.1126/science.1219985 PMID:22556249

Nick, F. M., Vieli, A., Howat, I. M., & Joughin, I. (2009). Large-scale changes in Greenland outlet glacier dynamics triggered at the terminus. *Nature Geoscience*, *2*(2), 110–114. doi:10.1038/ngeo394

Pshenichny, C. A. (2018a). Theory of Multitudes as an Alternative to the Set Theory. In C. A. Pshenichny, P. Diviacco, & D. I. Mouromtsev (Eds.), *Representation of Dynamic Knowledge in Scientific Domains*. IGI Global.

Pshenichny, C. A. (2018b). Qualitative and Quantitative Calculi in the Theory of Multitudes. In C. A. Pshenichny, P. Diviacco, & D. I. Mouromtsev (Eds.), *Representation of Dynamic Knowledge in Scientific Domains*. IGI Global.

Pshenichny, C., Carniel, R., & Diviacco, P. (2013). Engineering of Dynamic Knowledge in Exact Sciences: First Results of Application of the Event Bush Method in Physics. *MSEPS 2013 – Modeling States, Events, Processes and Scenarios: workshop proceedings*, 60-73. Retrieved from http://iccs2013.hbcse.tifr.res.in/workshops/copy2_of_text.pdf, http://ceur-ws.org/Vol-1223

Pshenichny, C. A., & Kanzheleva, O. M. (2011). Theoretical foundations of the event bush method. In K. Sinha, L. Gundersen, J. Jackson, & D. Arctur (Eds.), Geological Society of America Special Paper: Vol. 482. *Societal Challenges and Geoinformatics* (pp. 139–165). doi:10.1130/2011.2482(12)

Pshenichny, C. A., Nikolenko, S. I., Carniel, R., Vaganov, P. A., Khrabrykh, Z. V., Moukhachov, V. P., & Rezyapkin, A. A. et al. (2009). The event bush as a semantic-based numerical approach to natural hazard assessment (exemplified by volcanology). *Computers & Geosciences*, *35*(5), 1017–1034. doi:10.1016/j.cageo.2008.01.009

Ramillien, G., Lombart, A., Cazenave, A., Ivins, E. R., Llubes, M., Remy, F., & Biancale, R. (2006). Interannual variations of the mass balance of the Antarctica and Greenland ice sheets from GRACE. *Global and Planetary Change*, *53*(3), 198–208. doi:10.1016/j.gloplacha.2006.06.003

Ridley, J., Gregory, J. M., Huybrechts, P., & Lowe, J. (2010). Thresholds for irreversible decline of the Greenland ice sheet. *Climate Dynamics*, *35*(6), 1049–1057. doi:10.1007/s00382-009-0646-0

Rignot, E., & Kanagaratnam, P. (2006). Changes in the velocity structure of the Greenland Ice Sheet. *Science*, *311*(5673), 986–990. doi:10.1126/science.1121381 PMID:16484490

Rignot, E., Velicogna, I., van den Broeke, M. R., Monaghan, A., & Lenaerts, J. T. M. (2011). Acceleration of the contribution of the Greenland and Antarctic ice sheets to sea level rise. *Geophysical Research Letters*, *38*(5), L05503. doi:10.1029/2011GL046583

Ritz, C., Edwards, T. L., Durand, G., Payne, A. J., Peyaud, V., & Hindmarsh, R. C. (2015). Potential sea-level rise from Antarctic ice-sheet instability constrained by observations. *Nature*, *528*(7580), 115. PMID:26580020

Shepherd, A., Ivins, E. R., A, G., Barletta, V. R., Bentley, M. J., Bettadpur, S., & Zwally, H. J. et al. (2012). A Reconciled Estimate of Ice-Sheet Mass Balance. *Science*, *338*(6111), 1183–1189. doi:10.1126/science.1228102 PMID:23197528

Sørensen, L. S., Simonsen, S. B., Nielsen, K., Lucas-Picher, P., Spada, G., Adalgeirsdottir, G., & Hvidberg, C. S. et al. (2011). Mass balance of the Greenland ice sheet (2003–2008) from ICESat data – the impact of interpolation, sampling and firn density. *The Cryosphere*, *5*(1), 173–186. doi:10.5194/tc-5-173-2011

Thomas, R., Frederick, E., Krabill, W., Manizade, S., & Martin, C. (2006). Progressive increase in ice loss from Greenland. *Geophysical Research Letters, 33*(10), L10503. 10.1029/2006GL026075

Thomas, R. H., Rignot, E. J., Kanagaratnam, K., Krabill, W. B., & Casassa, G. (2004). Force-perturbation analysis of Pine Glacier, Antarctica, suggests cause for recent acceleration. *Annals of Glaciology*, *39*, 133–138. doi:10.3189/172756404781814429

Van de Wal, R. S. W., Boot, W., van den Broeke, M. R., Smeets, C. J. P. P., Reijmer, C. H., Donker, J. J. A., & Oerlemans, J. (2008). Large and rapid meltinduced velocity changes in the ablation zone of the Greenland Ice Sheet. *Science*, *321*(5885), 111–113. doi:10.1126/science.1158540 PMID:18599784

Van den Broeke, M. R., Bamber, J., Ettema, J., Rignot, E., Schrama, E., van de Berg, W. J., & Wouters, B. et al. (2009). Partitioning recent Greenland mass loss. *Science*, *2434*(326), 984–986. doi:10.1126/science.1178176 PMID:19965509

Velicogna, I. (2009). Increasing rates of ice mass loss from the Greenland and Antarctic ice sheets revealed by GRACE. *Geophysical Research Letters*, *36*(19), L19503. doi:10.1029/2009GL040222

Wouters, B., Chambers, D., & Schrama, E. J. O. (2008). GRACE observes small-scale mass loss in Greenland. *Geophysical Research Letters*, *35*(20), L20501. doi:10.1029/2008GL034816

Wu, X., Heflin, M. B., Schotman, Y., Vermeersen, B. L. A., Dong, D., Gross, R. S., & Owen, S. E. et al. (2010). Simultaneous estimation of global present-day water transport and glacial isostatic adjustment. *Nature Geoscience*, *3*(9), 642–646. doi:10.1038/ngeo938

Zwally, H. J., Abdalati, W., Herring, T., Larson, K., Saba, J., & Steffen, K. (2002). Surface melt-induced acceleration of Greenland ice-sheet flow. *Science*, *297*(5579), 218–222. doi:10.1126/science.1072708 PMID:12052902

Zwally, H. J., Giovinetto, M. B., Li, J., Cornejo, H. G., Beckley, M. A., Brenner, A. C., & Yi, D. et al. (2005). Mass changes of the Greenland and Antarctic ice sheets and shelves ... to sea level rise: 1992-2002. *Journal of Glaciology*, *51*(175), 509–527. doi:10.3189/172756505781829007

Zwally, H. J., Li, J., Brenner, A. C., Beckley, M., Cornejo, H. G., DiMarzio, J., & Wang, W. et al. (2011). Greenland ice sheet mass balance: Distribution of increased mass loss with climate warming; 2003-07 versus 1992–2002. *Journal of Glaciology*, *57*(201), 88–102. doi:10.3189/002214311795306682

Chapter 6
Conceptualization of the Dike Distribution Analysis Aiming at Identification of Eruptive Centers

Sophie Koneva
Karpinsky All-Russian Research Geological Institute (VSEGEI), Russia

Cyril Pshenichny
ITMO University, Russia

ABSTRACT

Formation of dikes is typical for most of the volcanic centers. Different types of eruptive centers tend to create different patterns of dike distribution by strike. Looking at this distribution, one may suggest, by purely geometrical reasons, where the vent should be (if buried) or should have been (if removed by denudation or new eruptions). However, wide application of this technique raises the issue of interpretation of the results in the context of the geological history of a particular volcano. Impartial and universal interpretation may be possible in a framework of knowledge of eruptive center evolution that can be built by means of the event bush method. Such a framework, however, may have wide application in volcanology extending far from the initial task it has been developed for.

INTRODUCTION

Volcanic eruptions are accompanied by intrusion of dikes that often indicate the location of past eruptive centers. This has led Carniel et al. (2017) to develop an algorithm aiming to identify the location of buried or denudated eruptive centers based on the documented dikes where their density allows to do so. The algorithm was implemented as an *R* opensource script entitled FIERCE (FInding volcanic ERuptive CEnters) and tried at a number of volcanoes and volcanic areas, both modern and ancient, with encouraging results. Position of the vents computed by this algorithm coincided with the observed (for modern centers) and was in concordance with the distribution of volcanic products and other evidence of

DOI: 10.4018/978-1-5225-5261-1.ch006

their location. Naturally, in every case the reliability of the computed position of eruptive center should be supported by the interpretation of volcanological and geological history of the volcano.

Every time this interpretation was made ad hoc, based on the geological data available, and it seems to be a good luck that so far we have been finding a self-consistent and palpable explanation for the results of computation. Still, this good luck is more or less explained by that (i) we have addressed a very small number of volcanoes and that (ii) almost all of these volcanoes have had a long record of geologic research. Query, what happens if we interpret dozens or hundreds of datasets? Wouldn't there be some typical interpretations? Wouldn't different scenarios of geological evolution be potentially responsible for the same dike pattern? Is the diversity of these patterns infinite or it can be somehow constrained?

To answer these questions, one should systematize the existing interpretations and put them in a coherent framework of volcano-tectonic knowledge. To build such framework is the purpose of this work.

Below in this chapter we will briefly overview the FIERCE methodology and the results achieved, suggest a knowledge framework to host these results and discuss its use in volcanology.

PREVIOUS WORK

FIERCE Methodology

Dikes in volcanic areas usually have radial, tangential (circumferential), or regional patterns (e.g., Chadwick & Howard, 1991; Takada, 1997; Acocella & Neri, 2009). The radial patterns suggest control by a local stress field caused by the load of the edifice, an effect that increases with edifice height, while tangential dikes are often related to caldera collapses (McGuire & Pullen, 1989; Corbi et al., 2015). Distinguishing with certainty between radial and tangential dikes is generally possible only if the vent position is known. Conversely, when denudated volcanoes preserve dike segments, their orientations may prove highly useful for identifying the position of its central or peripheral vents. Once we identify the vent's location, we can then determine which dikes are radial or tangential and hence understand the morphological and structural evolution of a volcanic edifice, i.e., what caused its shape (Fiske & Jackson 1972).We can also identify evidence for features like sector collapses, because the orientation of the dikes may be controlled by the collapse scarps (e.g., McGuire & Pullen1989; Acocella et al. 2006a, 2013; Neri et al. 2008; Battaglia et al. 2011).

By Carniel et al. (2017), one should first consider a very simple case and assume that all dikes are radial with respect to their unknown eruptive center. In such an ideal case, the directions of all the dikes should intersect exactly in the sought eruptive center. To find such an intersection, one needs to mathematically define the direction to which each of the dike segments belongs.

In reality, some dikes may belong to the classes of tangential and regional stress dikes. If they are erroneously assumed as radial with respect to the eruptive center, they will actually do no harm in the computation, as their directions basically become random, i.e., will not lead to a wrong solution. However, tangential dikes also preserve useful information that is: if a dike is tangential, any section of it observed in the field can be approximated by the line joining its ends. This line must be tangential to an imaginary circle surrounding the eruptive center. Then a perpendicular line that passes through the midpoint of this line would act in the same way as a radial direction.

Finally, if the dike direction is determined mainly by the regional stress distribution, these dikes cannot provide direct information about the location of the eruptive center, but nonetheless they will not bring about any radial convergence to a wrong solution, therefore remaining completely neutral in the analysis. Therefore, they need not to be removed from the dataset.

The algorithm starts with a polyline shapefile, exported from a GIS, that contains a set of dikes. It is essential to know their strike, which can also be derived directly by their graphic representation. There are no limitations on the minimum set of dikes or dike segments, but obviously the more dikes we have, the more reliable the results. It is worth noting that it is not necessary to know the age or composition of the different dike segments, although we can use this information to build more homogenous datasets to be processed separately. Details of its implementation in the open-source *R* script named FIERCE are given in the supplementary material by Carniel et al. (2017).

Validation

The algorithm was tested by five volcanic objects of various age, structure and preservation.

1. Summer Coon is an eroded Oligocene, basaltic andesite to rhyolite, stratovolcano in southern CO, USA (Lipman, 1968). The cone must had a basal diameter of 14 km, a summit elevation of 4200 m and an average flank slope of 17° (Moats, 1990). Poland et al. (2008) mapped 20 silicic dikes that reach up to about 7 km in length and are formed by various offset segments. The dikes of this volcano are almost perfectly radial, and they thicken with increasing radial distance. Offset segments and fingers form along the upper peripheries of the intrusions. These dikes are interpreted as feeding flank eruptions, but their radial distribution provides information about the location of the main vent of the volcano (Poland et al. 2008).
2. Etna stratovolcano, situated on the east coast of Sicily, is Europe's largest and most active volcano. It has been active since the Middle Pleistocene and covers an area of ~1250 km2. Massive debris flows, associated with the formation of a large (4 × 6 km) and deep (up to 1 km) denudational depression, the Valle del Bove, occurred between ~15 and 4 ka (Coltelli et al. 1994). This morphological depression is situated high on the eastern slope of the volcano and its scarps show the relicts of ancient (up to ~110 ka) volcanic rocks. The present volcanic edifice, reaching a height of ~3330 m in 2010 (Behncke et al. 2016), has arisen from a succession of central apparatuses and associated flank eruptions from parasitic cones roughly during the last 200 ka (Corsaro et al. 2002; Branca et al. 2015).

The products and, in some cases, the general shape and the presumed position of the central vent of the succession of these ancient volcanoes can be recognized along the Valle del Bove sides (Coltelli et al. 1994; Calvari et al. 1994, 2004; Branca et al. 2015). From east to west, they are as follows: the Rocche Centers, which reached a maximum height of ~1800 m, 110–101 ka, the Trifoglietto Center (2400 m, 107–99 ka), the Salifizio-Serra Giannicola Grande-Cuvigghiuni Centers (2700 m, 85–65 ka), the Ellittico Center (3800 m, 65–15 ka), and, finally, the present Mongibello (3330 m, <15 ka). The Ellittico and Mongibello volcanoes have the same central eruptive conduit.

A significant characteristic of the current magmatism is the prevalent lateral movement of magma radially from the central conduit zone of the volcano along a distinctive series of rifts, aligned broadly towards the NE, S, and W away from the summit area (Garduño et al. 1997; Neri et al. 2011; Cappello et al. 2012; Giammanco et al. 2016). They are fed by steeply dipping dike systems. Exposures of these dikes, particularly in the walls of the Valle del Bove, display subhorizontal flow direction indicators consistent with radial movement of magma (Garduño et al. 1997; Acocella et al. 2009; Geshi and Neri 2014; Falsaperla and Neri 2015); in a few cases, the dikes are originated by plugs, corresponding to the conduits of the ancient eruptive centers (Ferrari et al. 1991; Calvari et al. 1994).

3. Stromboli island volcano is one of the six islands of the Aeolian Volcanic Arc (which includes also several seamounts) in the southern Tyrrhenian Basin, in Italy. The islands of Panarea and Stromboli are located at the eastern margin of the arc, which is aligned NE-SW. At Stromboli, the regional NNW-SSE extension direction is dominant and is manifested by the NE-SW and E-W alignments of several morpho-structural elements, i.e., summit vents, dikes, faults, and eruptive fractures (Pasquarè et al. 1993; Tibaldi 2001; Acocella et al. 2006a; Tibaldi et al. 2009). Today, Stromboli is characterized by a continuous strombolian activity, occasionally interrupted by paroxysmal events (Jaquet and Carniel 2003). Morphologically, the most visible present-day feature is the Sciara del Fuoco (~5 ka old; Tibaldi 2001), a deep depression opened in the NW flank, both above and below the sea level. But the history of the volcano is characterized by previous several collapses that occurred in the last 13 ka (Tibaldi 2001; Tibaldi et al. 2009), often concurrent with changes in the magma composition from calc-alkaline to potassic series, through high-K calc-alkaline and shoshonitic ones (Francalanci et al. 1993).

On the basis of stratigraphic and morpho-structural unconformities and rock composition, the volcanic succession of Stromboli has been divided into six major periods of activity, represented by their lithosomes: Paleostromboli I (85–67 ka), Paleostromboli II (67–56 ka), Paleostromboli III (56–26 ka), Vancori (16–13 ka), Neostromboli (13–5 ka), and Recent Stromboli (Pizzo lithosome and present-day activity; <5 ka) (Corazzato et al. 2008; Lucchi et al. 2013, and references therein).

4. The Somma-Vesuvius volcanic complex, in Italy, is an active stratovolcano ~1250 m high, lying between NE-SW and NW-SE regional fault systems (Bruno and Rapolla 1999 and references therein). During the last ~30 ka, it erupted silica-undersaturated and potassic basic products (Scandone et al. 1993; Santacroce and Sbrana 2003). Vesuvius is the most recent volcano, active during the last 2 ka. To the north, Vesuvius is bordered by the remnants of the Mt. Somma edifice (~30–18 ka), which forms a scarp derived by a multistage caldera developed between ~18 ka and 79 AD; the scarp exposes a hundred dikes that are coeval with the Somma edifice and consequently these dikes may not be representative of the present-day feeding system of the Vesuvius cone (Cioni et al. 1999; Acocella et al. 2006b; Porreca et al. 2006). The most recent activity occurred in 1944 (Cioni et al. 1999).

The dikes are all concentrated in the actual collapse scarp of Mt. Somma, to the north of the Vesuvius active crater (Porreca et al. 2006). The absence of other dikes is probably the result of their disappearance due to the collapse and/or to the burial under younger deposits.

5. Vicuña Pampa Volcanic Complex (VPVC) is located at the SE margin of the Altiplano-Puna plateau, in Argentina. The Vicuña Pampa was recently interpreted as a complex massif that underwent intense degradational processes giving rise to a large (18 × 13 km) morphological depression

(Guzmán et al. 2017a). VPVC is a Middle Miocene volcano, formed by plugs and necks, lava flows, block and ash flow deposits, and ballistic fall-out tephra, whose composition mainly ranges from basaltic andesite to andesite. Guzmán et al. (2017) interpreted that VPVC had at least two stages of evolution separated by a considerable time (enough for the erosion of the central depression). During the construction of VPVC, there were also multiple episodes of dike intrusions that today are mostly exposed at the central and western parts of the eroded volcano.

Vertical and inclined sheets are seen all over the complex, but the densest concentration is recognized in the western portion of the depression. These are mostly vertical dikes, while moderate to low angle (<60°) inclined sheets are less common. Dikes intrude different volcanic sequences of the Vicuña Pampa Volcanic Complex, some being feeder dikes of small lava flows. Several intrusion episodes occurred during the construction of the main volcanic edifice and after the formation of the central depression (Guzmán et al. 2017b).

EVENT-BUSH-BASED FRAMEWORK OF VOLCANO-TECTONIC KNOWLEDGE

Interaction between volcanism and tectonism is, broadly speaking, one of the central issues of the science of geology. Even if to omit the aspects of global tectonics and narrow this problem to the interrelation between rising magma and opening faults in, below and directly around the volcanic apparatus, this topic would include magma chamber evolution, volcano-plutonic associations, seismic signatures of magma intrusion and eruption, formation of large collapse calderas and volcano-tectonic depressions and many others. Still, for the purposes of subvolcanic dike strike analysis, we may try to narrow it yet further and focus it as follows: how a primitive scoria cone or lava dome may evolve and what types of dike would develop therein, based on the general volcanological knowledge (see, e.g., Cas & Wright, 1987; Branney & Kokelaar, 2002; Parfitt & Wilson, 2008, and many others) and our own field experience. These knowledge and experience form our vision of the field of knowledge that will be converted into a formalism. Of course, this vision is subjective and obviously far from the only possible, perhaps even misleading, and definitely not fixed forever. In process of its formalization, the options not covered by it will be elucidated and briefly discussed.

This vision will be formulated in descriptive terms, which will be chosen to be as connotation-free as possible. Nonetheless, even descriptive terms in volcanology often bear genetic or other connotation, more or less shared by the research community and sometimes even not felt by the researchers. Herewith, the consensus on use of a term may arise not only from shared understanding of the natural phenomenon this term relates to, but also from:

- Compromise (which is always achieved at the expense of ignoring some evidence);
- Common bias or
- Common lack of knowledge.

In many cases, it looks more efficient to introduce "fresh", connotation-free terms directly relating to meanings we have in mind; however, these terms may be met unfriendly by volcanologists, so we will try to use existing terms and explicate their connotation whenever possible.

This said, we proceed to its narration first in the conventional form of a text.

Basic Vision

By eruptive center we mean any geological structure expressed as a landform (i.e., morphostructure), through which magma or other material related to magma (including boiling groundwater) is being erupted. Eruptive center may be a positive landform or negative one. Positive landform will be called edifice. In case of aerial volcanism, by eruptive center we mean each conduit and the landform it creates (positive or negative). Primitive eruptive center is the one that has erupted only once. Scoria cones and lava domes (both of effusive or extrusive origin) are considered primitive positive eruptive centers, or primitive eruptive edifices. Volcano henceforth is referred to as an eruptive edifice that has erupted more than once. Primitive negative eruptive center may be maar that erupted once, or any other kind of a pit, with negligibly small or without walls, through which something (even steam) was erupted. Summit crater on a volcano is not primitive eruptive center, because it is not eruptive center at all, but only a part of eruptive center (i.e., volcano). Caldera is a negative eruptive center that erupted once or more, which is the same size as volcano or bigger. We axiomatically accept that primitive negative eruptive center may not be the same size as volcano or bigger. This is consistent with the traditional understanding of caldera, because an eruptive center must be really big enough to develop the typical features of caldera, such as high caldera rim, postcaldera faults, differentiated vertical displacement of the floor, extrusive domes, intracaldera tuff and others, up to new primitive eruptive centers (positive or negative) and even volcanoes growing up inside it.

Formation of eruptive center may be accompanied by formation of dikes. Carniel et al. (2017) say that formation of dikes is controlled by the local stress field caused by the load of the edifice, an effect that increases with edifice height. Other mechanisms of dike intrusion in and around the eruptive centers may also be thought of (see, e.g., Acocella & Neri, 2009) but, based solely on observations reported by these researchers, we may say that various types of eruptive center may be accompanied, or documented, by various dike patterns.

Now, suppose there is a primitive eruptive center.

Unlikely it would be accompanied a series of radial dikes but definitely has one that feeds it with magma, and formally speaking, this dike should be regarded as radial, for geometrically it passes through the point of the conduit. Still, it may be well created by the regional stress or be a radial or tangential dike of a larger eruptive center. Moreover, magma has much better chance to reach the Earth surface at intersection of faults, and hence, it is possible that two intersecting dikes feed this eruptive center, and both, being geometrically radial, are likely determined by the regional stress.

If magma delivery continues, this primitive eruptive edifice grows and eventually turns either into a central type volcano, or, in case of drastic gas release in the conduit, a primitive crater. Gas may be juvenile, and then the crater would be more or less like maar, which has commonly basic to ultrabasic magma but may well be of trachytic or andesitic composition (Leyrit et al., 2016), or just water vapor from the host rock, and the crater then would be either maar, or phreatic or phreatomagmatic crater.

Craters are not accompanied by dikes and unlikely may evolve further. Therefore, this scenario is of no interest for us. Still, sometimes new lava or scoria cones grow in such craters, but, as neither cones nor craters create their own dikes, we may disregard this circumstance and consider such cones growing from “bare nothing”.

The other scenario, transformation of primitive eruptive edifice into a central type volcano, has a few options.

Given the continued effusive activity, the volcano would be built dominantly of lave, with slope angle determined by magma composition and, hence, viscosity – from quite gentle (for basalts, then it would be termed shield volcano) to rather steep (for andesites; however, purely effusive andesitic volcanoes are not common). If the magma is yet more silicic, then a big extrusion dome would grow – e.g., Diky Greben Volcano in Kamchatka, Russia, or Soufrierre Hills in Montserrat, the Caribbean. If, in addition, gas release is high enough during eruptions, explosions occur, pyroclasts are being produced, and a stratovolcano is formed or dome collapses happen. In any case, however, the edifice that is being formed may be termed central type volcano, and in every case this volcano would be accompanied by a series of radial dikes pointing to its vent.

Then, two principal opportunities arise: dissection of an edifice by fault(s) and partial emptying of magma chamber below. They may not occur together but the former may occur any number of times, and the latter may follow the former or take place immediately.

Faults originally are a consequence of regional stress. Volcanoes may be subject to faulting along the regional directions or orthogonally, or at some regular angle to these, and faulting may be accompanied or not by displacement of blocks of rocks, vertical or horizontal. Though, strictly speaking, any fault implies displacement, but this may be a displacement of integer blocks of the earthcrust or just compaction of rock at the opening of the fault. The latter case will be regarded as "no displacement".

Abundant literature on structural geology discusses faulting and related rock deformation in much detail (see, e.g., Moores, & Twiss, 1995; Van der Pluijm & Marshak, 1997; Holdsworth, Strachan & Dewey, 1998; Hancock, 1994), but in concern of volcanic edifices the following should be highlighted. Faults without displacement, in our opinion, indicate bare strain and are likely more or less vertical. Perhaps other understanding of faulting without displacement of rock blocks is possible, but what was said above is our vision.

If faults are not deep enough, they are soon filled by volcanic debris from the top and are not of interest for dike analysis. However, if deep enough, they may open new pathways to magma from the chamber and thus form new vents instead of older ones, which in most cases tend to solidify. Thus, in our vision, regional faulting without displacement is responsible for opening new vents in environments like Stromboli or Etna, while displacement is always accompanied by compression and, hence, impedes magma ascent. There are some examples of occurrence of volcanoes under compression, in which the overpressure of magma takes place. However, this case looks rather exotic and is not accounted for here.

One may argue that at least horizontal displacement may also favor magma ascent, but this happens exactly due to the local strain caused by such displacement, and to give the way to magma, a secondary fault must appear along the maximum strain, so this fault would be without displacement. We do admit that vertical dislocation may release pressure and give rise to magma, but, to our mind, this magma would intrude the host rock and stop at some level if its rise is not supported by the strain until the surface, i.e., if there is no fault without displacement stretching from the magma intrusion depth on to the surface. New vents may be singular or multiple, clearly parasitic or compete with the main one, or even become the main vent. Some volcanoes have several summit craters (e.g., Etna, minimum four at present, Elbrus in the Caucasus, Russia, has two, and Zhupanovsky in Kamchatka, Russia, five).

In our view, magma rise to the surface may also result in a fissure flank eruption, when no secondary edifice is built.

In both cases, the volcano would be dissected by faults without displacement, but in the former case it would also become complex, or yet more complex, by morphology. Still, if new edifice tends to create its own series of radial dikes different from the previous ones, fissure eruption, being fed by a segment of dike, leaves no new dikes except its feeder. The dikes feeding secondary cones or fissure eruptions may be radial (or older edifices) or regional by origin.

Faults with horizontal displacement of blocks just dissect volcanic edifices, create weakened zones at the surface further developed by erosion and/or used by lava, pyroclastic flows or lahars, but do not add much to the morphostructure of volcanoes. At some depth, they may be used by intruding magma and give some dikes irrelevant for identification of vents. On the contrary, they may spoil the ornament of older relevant dikes tearing them into parts, some of which may become buried or denudated. Also, cureferencestting dikes into shorter segments may make them shorter than the sensitivity threshold and thus exclude them from consideration.

Unlike the horizontal, vertical (or dominantly vertical) displacements in volcanic edifices are not only often pronounced in volcanic topography as scarps but may create catastrophic collapses of parts of edifice, e.g., Sciara del Fuoco in Stromboli. Moreover, cutting the potentially active stratovolcanoes or lava domes, vertical displacements expose their hot interior that may produce eruption, e.g., of a pyroclastic density current or hot block avalanche, immediately. Nonetheless, this would not imply a creation of another eruptive center – rather, an unexpected "reuse" of the existing one. We do not believe that magma can intrude these faults event at some depth, but their contribution to the dike ornament may be controversial. From one end, normal faults may push downward and obscure forever any portion of older dikes, from another, they may also expose previously obscured dikes in the hanging wall. Finally, Acocella et al. (2006a) suggest, based on the work of predecessors, that younger dikes may deviate from radiality when reach very shallow depths (0–200 m) being blocked by the underside of the normal fault that may be the case at Sciara del Fuoco.

Another principal option pertains to volcanoes that exhibit notable explosive activity. Of course, some explosivity is present even in the Hawaiian type eruptions, but what is meant here is production of considerable amounts of pyroclasts expressed as Volcanic Explosivity Index (VEI) values of 3-4 or more.

Explosion is the only way to transport huge amounts of magma almost instantaneously, given the rapid exsolution of gas from the magmatic melt. Once this happened, voluminous pyroclastic density current deposits (ignimbrites) are formed on the earth surface, and magma chamber is notably exhausted. Its replenishment with magma takes time, and a void may form in its upper part that would lead to gravitational subsidence or collapse of its roof. This is the way to form a caldera. It may top a large volcano, being more or less the same size as it, or totally engross it and take even larger area, depending on the size of magma chamber and the void in it.

Still, many scientists state that there are other ways of caldera formation, e.g., with subsidence of volcano at the onset of eruption or even caldera-forming eruption without any preceding volcano at all (see Aguirre-Díaz, 2008; Cas et al., 2011; Petrinovich et al., 2010, and references therein). Such scenarios are outside of the conceptualization presented here but will be hopefully included in it later.

Calderas develop a pattern of tangential faults and dikes. However, this pattern is clear dominantly at the periphery of caldera, more or less near the caldera rim, while inside the rim, at the caldera floor, the pattern is much more complicated and even chaotic due to numerous vertical faults, along which blocks of caldera floor may go up and down. Moreover, in many calderas the feeding system (that possibly preexisted the caldera collapse) may be not from the caldera rims, but may start from a central vent or from cones sheets with other distribution, both radial and tangential, depending on the preceding history of this eruptive center.

Along with this, continuous ascent of viscous magma may continue, and some of these faults, instead of displacing blocks, may conduct magma and feed small extrusion domes at the caldera floor. Further on, new volcanic edifices may develop there, and even become polygenetic and complex, with their own kit of radial dikes.

We presume that calderas, unlike volcanoes, unless entirely fossilized, may not be roughly split into two parts by a fault, the more so to be displaced one part against the other (horizontally or vertically), because represent a complex ensemble of blocks on a vivid basement of magma chamber. In the available literature we found an evidence of highly complex response of alive calderas to tectonic movements along regional faults (see, e.g., Bursik, 2009).

However, the largest known calderas (such as Toba, Taupo, Yellowstone or La Garita) seem to be formed by a number of reasons, which are not fully understood but likely include, inter alia, purely tectonic contribution. However, these great calderas represent a separate issue in the geosciences, and the dike strike analysis is unlikely an adequate tool to locate their main vents because of complexity of their structure and, arguably, multiplicity of such vents and probably even presence of fissures erupting pyroclastic density currents.

With time, any volcano, if not transformed into caldera, becomes extinct, but once magma chamber remains alive for quite longer time, it may be reactivated with change of some tectonic or magmatic conditions and the vent becomes active again even though the chemical composition of magma and consequently, the mode of eruption may well change. This does not apply, however, to lava or scoria cones or small extrusions because if they have not evolved into a mature volcano, they have poor chance to revive after extinction via the same conduit. Even if this happens, what forms then may be considered a brand new volcanic apparatus. For calderas, such reactivation is normal modus vivendi, and the larger the caldera, the longer the return period of its reactivation, up to dozens of thousands of years. Though some researchers oppose this viewpoint and tend to consider each collapse as a birth of a new caldera, this is not crucial for the conceptualization presented here, because if the magma chamber is in the same place and is active again, the scenario of magmatic activity continues, call it the same or new caldera, and only if this period is well beyond and no activity is observed, the caldera may be confidently considered fossil.

Formalization of the Vision

Our reasoning will be based on the formalism of event bush (Pshenichny et al., 2009; Pshenichny and Kanzheleva, 2011), which is the key method of dynamic knowledge representation (Pshenichny and Mouromtsev, 2015; Pshenichny, 2017 – see Chapter 2 of this book). Its basic syntax includes four types of event: primary internal, primary external and tertiary, all consisting of one predmet or predmet event (more or less equivalent to subject in the modern logic) and one or more primetas (the properties). The kit of predmets for the primary events is determined by the title of the bush. Let us put it as follows: "How primitive eruptive edifice evolves with changing magmatic and tectonic conditions" – keeping in mind that each state of the evolution of the edifice likely produces subvolcanic dikes. By primitive eruptive edifice we mean a monogenetic apparatus that is either a cone (lava or scoria) or a lava dome (extrusion) formed in one, however continuous, eruption. Of course, neither there are strict criteria for "one eruption", nor for "continuous", nor for "monogenetic". Still, abstracting from real objects, whose behavior is always fuzzy and who are "hanging between" various concepts, we will reason about the concepts per se, which are quite crisp.

The following event predmets will be introduced:

S_1 – Primitive eruptive edifice (may produce radial dikes if is big enough or at least has one that feeds it);

S_2 – Primitive crater (is right on the ground, has no to small walls ("tuff ring"), may be of magmatic or phreatic origin; resembles and probably includes maar but, unlike the latter, may have whatever chemical composition of products; not necessarily forms in place of a cone/dome and may appear just by itself, but this is outside of the scope of this study; produces no dikes; may host primitive eruptive edifices on its floor; still, as it produces no dikes itself, this scenario does not differ from growth of primitive eruptive edifices from "bare nothing");

S_3 – Central type volcano (stratovolcano, shield or big extrusion dome; produces radial dikes);

S_4 – Caldera (produces tangential dikes);

S_5 – Magma (the substance that includes melt, dissolved gas, crystals and xenoliths);

S_6 – Gas (in free form, either magmatic or external; dominantly, but not necessarily, water vapor);

S_7 – Void (that may form in magma chamber);

S_8 – Fault (with displacement or not);

S_9 – Chamber (that hosts magma);

S_{10} – Bursting (of craters);

S_{11} – Growth (of eruptive edifices – primitive ones or volcanoes).

The following primetas will relate to these predmet events:

P_1 – to be on land (universum);

P_2 – to be alive (i.e., erupting or dormant but not extinct or fossil);

P_3 – to be continuously delivered (for magma; to the surface; in excessive amount and/or duration);

P_4 – to release in the conduit (for gas);

P_5 – to increase drastically (for gas);

P_6 – to form in upper part of magma chamber (for the void);

P_7 – to form below the eruptive center (for the void);

P_8 – to dissect the eruptive center (for the faults);

P_9 – to displace horizontally a part of eruptive center (for the faults);

P_{10} – to displace vertically a part of eruptive center (for the faults);

P_{11} – to reactivate (for the chamber);

P_{12} – to be affected by continued magma delivery (for eruptive edifices – primitive edifices or central type volcanoes);

P_{13} – to be affected by drastic gas release in the conduit (i.e., to be explosive – for eruptive edifices – primitive edifices or central type volcanoes);

P_{14} – to be affected by void underneath (i.e., to undergo gravitational collapse – for central type volcanoes);

P_{15} – to be affected by faults without displacement (for eruptive edifices – primitive edifices or central type volcanoes);

P_{16} – to be affected by horizontal displacement along faults (for eruptive edifices – primitive edifices or central type volcanoes);

P_{17} – to be affected by vertical displacement along faults (for eruptive edifices – primitive edifices or central type volcanoes);

P_{18} – to be complex (i.e., include several eruptive edifices – for volcanoes);

P_{19} – to be nested (for calderas);

P_{20} – to have voluminous extracaldera ignimbrite (for calderas);

P_{21} – to exhibit intense internal faulting and subsidence (for calderas);

P_{22} – to have primitive eruptive edifices inside (for calderas);

P_{23} – to have volcanoes inside (for calderas);

P_{24} – to host magma (for the chamber);

P_{25} – to relate to craters (be "...of craters");

P_{26} – to relate to primitive eruptive edifices (be "...of primitive eruptive edifice");

P_{27} – to relate to central type volcanoes (be "...of central type volcano");

P_{28} – to have craters inside (for calderas).

The following primary events will be specified:

Internal (iaE_1):

Primitive eruptive edifice (S_1)

is on land (P_1),

is alive (P_2),

unaffected by continued magma delivery ($\sim P_{12}$),

unaffected by drastic gas release in the conduit ($\sim P_{13}$),

unaffected by void underneath ($\sim P_{14}$),

unaffected by reactivating faults without displacement ($\sim P_{15}$),

unaffected by horizontal displacement along fault ($\sim P_{16}$),

unaffected by vertical displacement along fault ($\sim P_{17}$);

in the accepted event bush notation, $iaE_1 = S_1 – P_1, P_2, \sim P_{12}, \sim P_{13}, \sim P_{14}, \sim P_{15}, \sim P_{16}, \sim P_{17}$; ~ is negation.

External:

Magma

is continuously delivered

($ibE_2 = S_5 – P_1, P_3$)

(that implies, given no other constraints, transition to a larger eruptive edifice);

Gas release

in the conduit

drastically increases

($ibE_3 = S_6 – P_1, P_5$)

(that implies voluminous pyroclastic eruption);

Fault

dissects the eruptive center

without horizontal displacement,

without vertical displacement

($ibE_4=S_8-P_1, P_8, \sim P_9, \sim P_{10}$)

Fault

dissects the eruptive center and

horizontally displaces a block of the earthcrust

including part of eruptive center

($ibE_5=S_8-P_1, P_8, P_9, \sim P_{10}$)

Fault

dissects the eruptive center and

vertically displaces a block of the earthcrust

including part of eruptive center

($ibE_6=S_8-P_1, P_8, \sim P_9, P_{10}$).

Void

forms in upper part of magma chamber

below the eruptive center

($ibE_7=S_7-P_1, P_6, P_7$).

Magma chamber

is reactivated

($ibE_8=S_9-P_1, P_{24}, P_{11}$).

These events, in terms of the Novorossian logic (Pshenichny, 2017, see Chapter 2 of this book), represent a raw sequent and may entail the secondary and tertiary events of an event bush. Raw sequent includes all events of the event bush and may include some others that make the sequent formally complete. However, inference of a mature sequent is quite a laborious task that well deserves a separate work, in which the constraints on the predmets discussed and indicated above will be transformed into exclusion axioms. Still, even this raw sequent allows us to build an event bush that would serve as a solicited framework of volcano-tectonic knowledge to implement the FIERCE methodology and be not more intuitive than other knowledge frameworks in the geosciences known to the authors used, e.g., to organize concepts in volcanology to optimize hazard assessment (Newhall & Hoblitt, 2002; Neri, 2008; Marzocchi et al., 2008; Marti et al., 2008; Tadini et al., 2017, and many others).

This event bush appears too large to be published on paper in graphic form. Instead, it is given in the appendices in the end of the chapter. Appendix 1 presents definitions of events, and Appendix 2, the changes (events united by connectives), from which relevant scenarios can be constructed. To save space, the model is given only in symbolic form in Appendix 2, and still is not presented entirely.

In verbal form, only a small part of the model may be shown on a page, as is done in Fig. 1. Still, even there some formulations were forcedly truncated (i.e., most of negations rejected) and the boxes indicating events put as tight as possible.

One peculiarity of this event bush is that it has "double finish" to better fit the nature of the phenomenon (eruptive centers) described by the modeled knowledge. Natural state of eruptive center is to erupt, exhibit unrest or remain in quiescence (dormancy), but be not extinct. The duration of quiescence varies in very wide range, and this issue is not under discussion in the present work. We denote this state (eruption/unrest/dormancy given hot magma in the chamber) as "alive". With time, naturally, every alive eruptive center becomes "not alive" (i.e., extinct), when the magma solidifies in the chamber, but, as this process is quite long, reactivation of the chamber and revival of eruptive centers others than those termed here as "primitive" is possible. Therefore, every event meaning an alive eruptive center must lead via simple Flux connective (i.e., requiring no other causes) to the event describing the same eruptive center being not alive. The tertiary events denoting the end of scenario should refer to extinct eruptive centers with no hot magma underneath. Still, there must be an option to "restart" the story right before complete solidification, and this is done by the influx of the secondary events describing "not alive" eruptive centers with the primary external event "Magma chamber is reactivated" ($ibE_8 = S_9 – P_1, P_{24}, P_{11}$).

Representation of Test Objects

This event bush is used to "locate" the concepts corresponding to the test objects of the FIERCE technique. Fig. 1 represents a part of the bush that corresponds to one of the objects that is considered the simplest, the Summer Coon volcano.

Eruptive centers discussed above are related to some scenarios tracked in the event bush. Any scenario in the event bush starts from a primary internal event. In this bush, there is only one such event, iaE_1, hence, this is the starting point of all scenarios in the bush. The scenario is followed from this event to the event at which the name of the eruptive center is put. Importantly, not necessarily these scenarios are finished, i.e., end up with a tertiary event. This is true only for the eruptive centers that are surely extinct – in our case, these are Summer Coon and Vicuña Pampa. Also, perhaps the Somma volcano of the Somma-Vesuvius complex may be regarded as such if taken apart from Vesuvius. In all other cases, the scenarios that correspond to the volcano in question would "stop" at some secondary event, possibly having passed through "reactivation of the magma chamber" (i.e., "restarting the game"). These are Etna, Stromboli and Vesuvius (though, strictly speaking, Vesuvius itself has not been studied by the FIERCE technique).

All test objects except Stromboli appear to be linked not to one, but to a number of scenarios. This is because of that the information our interpretations are based on (see Carniel et al., 2017) is rather brief and not always complete. This leads to uncertainty in choice of scenario. Another source of the uncertainty is the issue of scale of various volcanic events, e.g., numerous faults with horizontal displacement are observed on Etna, but how relevant they are? In such cases, it would be safer to admit several opportunities and track several scenarios, keeping in mind one that seems to be the best match so far. More detailed geological data would shed light and help choose only one of them.

The queried volcanic centers are interpreted as follows.

1. **Summer Coon:** As stratovolcano, it is definitely central type. Denudated Oligocene apparatus should be definitely regarded as not alive. Its supposedly regular conic shape reported by the researchers suggests that it likely was not complex, i.e. had only one vent. Exactly this is confirmed by the FIERCE-based analysis. Nonetheless, some dikes are interpreted as feeders of flank eruptions. Then we should assume that these eruptions were from fissures without notable edifices, but anyway faulting without displacement did take pace. Also, various offset segments seen in dikes imply that the volcano was dissected with displacement as well, and probably many times. We have not enough data to assert whether these were vertical or horizontal displacements, so, to keep on the safe side, unless new data are found, we admit both. Finally, the composition of products (basaltic andesite to rhyolite) and stratovolcanic shape favors that Summer Coon must have experienced explosive eruption, which could be strong (while strong an voluminous basaltic andesitic basaltic explosive eruptions are uniquely rare). Therefore it looks reasonable to attribute the primeta "to be affected by drastic gas release in the conduit". The same time, no evidence of central collapse (subsidence) has been reported, so the explosive eruptions must not have led to formation of a void under the volcano.

Thereafter, at present state of knowledge, we tend to think that Summer Coon Volcano is best described by the scenario leading to the event $iiiE_{36}$ of the event bush model:

Central type volcano (S_3)
is on land (P_1),
is not alive ($\sim P_2$),
is not complex ($\sim P_{18}$),
affected by drastic gas release in the conduit (P_{13}),
unaffected by void underneath ($\sim P_{14}$),
affected by faults without displacement (P_{15}),
affected by horizontal displacement along faults (P_{16}),
affected by vertical displacement along faults (P_{17}):
$S_3 – P_1, \sim P_2, \sim P_{18}, P_{13}, \sim P_{14}, P_{15}, P_{16}, P_{17}$.

Figure 1. Fragment of event bush model of volcano-tectonic knowledge about possible evolution of a primitive eruptive edifice. See comments in the text.

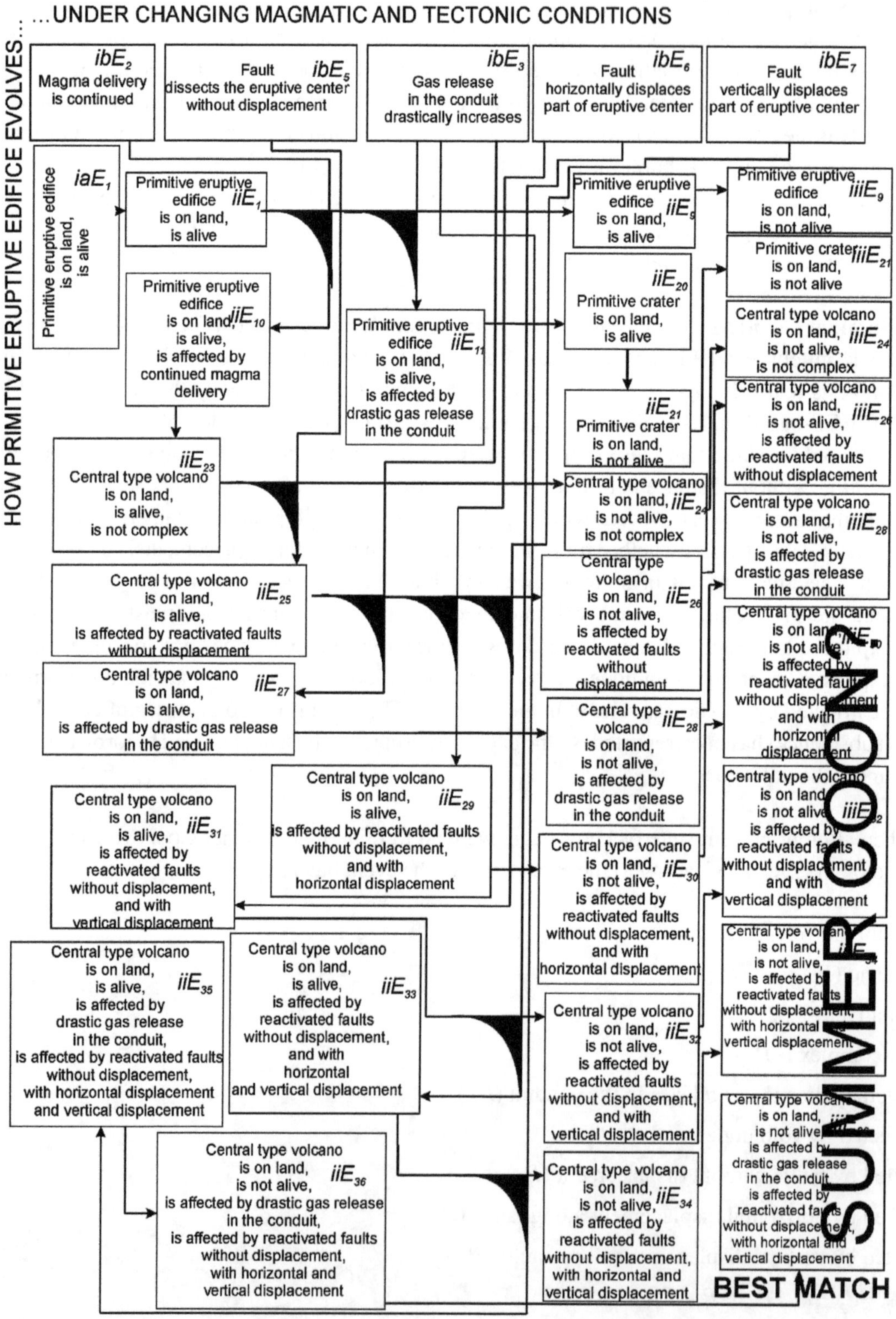

Other possible options include scenarios leading to the events that either admit no drastic gas release in the conduit, or not all kinds of faulting, but not "neither", as this option looks low-probable to us.

2. **Etna:** Stratovolcano means central type. Etna started as a submarine volcano but now all its active vents are definitely on land. It is surely active, but formation such a large denudational depression as Valle del Bove points to at least one break in magmatic activity and, consequenty, reactivation of the magma chamber. Once the edifice has arisen from a succession of central apparatuses and associated flank eruptions from parasitic cones, the volcano is complex. The composition of products (predominantly basalt) and observed eruptive styles do not suppose notable explosive events (excluding the rare cases of phreatomagmatic eruptions reported by Behncke et al. (2003) and Behncke (2009) and modeled by Behncke and Pshenichny (2009) by means of the event bush), nor collapses of the volcanic edifice due to exhausting the magma chamber. Meanwhile, lateral magma movement points, in our opinion, to abundant faults without displacement. Still, horizontal displacement of some parasitic cones on Etna was observed by us in the field (although, the matter of scale can be debated herewith – how relevant these minor displacements are). The same time, considering Valle del Bove denudational feature, we observe no cases of formation of scarps or other evidence of vigorous vertical displacement of rocks on the volcano. Therefore, we interpret Etna as follows:

Central type volcano (S_3)

is on land (P_1),

is alive (P_2),

is complex (P_{18}),

unaffected by drastic gas release in the conduit ($\sim P_{13}$),

unaffected by void underneath ($\sim P_{14}$),

affected by faults without displacement (P_{15}),

affected by horizontal displacement along fault (P_{16}),

unaffected by vertical displacement along fault ($\sim P_{17}$):

$$S_3 - P_1, P_2, P_{18}, \sim P_{13}, \sim P_{14}, P_{15}, P_{16}, \sim P_{17}.$$

3. **Stromboli:** This is also stratovolcano, and therefore also belongs to the central type. It is well known as one of the most active volcanoes of the world, hence, should be regarded as alive. Though it is an island in the Aeolian Volcanic Arc, the eruptions take place in the subaerial environment, so the volcano that exists since the Paleostromboli I phase (85–67 ka) should be considered as located on land (made by the volcano per se). Identification of several distinct and separated phases of its eruptive history with corresponding lithosomes, Paleostromboli I (85–67 ka), Paleostromboli II (67–56 ka), Paleostromboli III (56–26 ka), Vancori (16–13 ka), Neostromboli (13–5 ka), and Recent Stromboli (Pizzo lithosome and present-day activity; <5 ka), implies that the volcano

has undergone several reactivations of the magma chamber. Different researchers pointed that at different phases the central vents were also different, but at present the vent is one that makes us consider the present-day Stromboli not complex. We assume that it was the same in the past, and the data of the FIERCE-based dike strike analysis support the point that Stromboli had one vent in previous phases of its history. Nonetheless, the same data suggest that this vent migrated and generally followed the regional NE-SW trend. This is interpreted as action of a regional fault without displacement of blocks of the volcano. The displacement, however, occurred in the Holocene at formation of Sciara del Fuoco, which we refer to vertical displacement along fault. Finally, the eruptive style of Stromboli – continuous strombolian activity, occasionally interrupted by paroxysmal events – represents mainly the lava fountaining with minor explosive events, i.e., absence of drastic gas release in the conduit. Naturally, no evidence for subsidence and caldera formation is found on Stromboli, that means that no void underneath the volcano is being formed.

Then, the event that most likely describes the present state of Stromboli is:

Central type volcano (S_3)

is on land (P_1),

is alive (P_2),

is not complex ($\sim P_{18}$),

unaffected by drastic gas release in the conduit ($\sim P_{13}$),

unaffected by void underneath ($\sim P_{14}$),

affected by faults without displacement (P_{15}),

unaffected by horizontal displacement along fault ($\sim P_{16}$),

affected by vertical displacement along fault (P_{17}):

$S_3 – P_1, P_2, \sim P_{18}, \sim P_{13}, \sim P_{14}, P_{15}, \sim P_{16}, P_{17}.$

4. **Somma-Vesuvius:** The volcanic complex of Somma-Vesuvius including the multistage caldera developed between ~18 ka and 79 AD must be described as caldera located on land, alive (both because of its own historic eruption and activity of Vesuvius), nested, with voluminous extracaldera ignimbrite, with intense internal faulting and subsidence, with primitive eruptive edifices inside (extrusion domes), without primitive craters inside and with volcanic edifices (modern Vesuvius). However, exact position of the corresponding event and, hence, the scenario that leads to it are not shown in the model, for they fall into one of its blocks, which, due to overwhelming complexity, have just externally defined but not elaborated.
5. 5. **Vicuña Pampa:** The Vicuña Pampa was recently interpreted as a complex massif that gives us ground to define it as a central type volcano located on land and complex. As it was active in Middle Miocene, it is not alive now. Its two stages of evolution separated by a considerable time (enough for the denudation of the central depression) clearly indicate reactivation of the magma

chamber. The rest of the detail may not be confidently stated, but, based also on Aramayo et al. (2017), we select several incompatible tertiary events, one of which must fit the case of Vicuña Pampa. The one that looks the most probable to us is:

Central type volcano (S_3)

is on land (P_1),

is not alive ($\sim P_2$),

is complex (P_{18}),

affected by drastic gas release in the conduit (P_{13}),

unaffected by void underneath ($\sim P_{14}$),

affected by faults without displacement (P_{15}),

affected by horizontal displacement along fault (P_{16}),

affected by vertical displacement along fault (P_{17}):

S_3–P_1, $\sim P_2$, P_{18}, P_{13}, $\sim P_{14}$, P_{15}, P_{16}, P_{17}.

Final choice for this and other test objects will be made with more data available.

POSSIBLE APPLICATIONS

The framework of volcano-tectonic knowledge presented in this chapter represents a full group of formalized interpretations (hypothesized scenarios) of evolution of an eruptive center based on our subjective vision. Even accepting this vision, some of the interpretations may be thought unrealistic by some volcanologists. Still, each of the interpretations considered realistic may be instantiated by particular volcanic centers, alive or not, for which the dike distribution analysis can also be performed. The ultimate goal of this work is to become able to relate the results of this analysis directly to particular scenario of evolution of a volcanic center. Here it is done via the name of the center in question that serves as a "tag" both for a scenario or scenarios in the bush and for a result or results of FIERCE-based computation. In Fig. 2 the output of a computation run for the Summer Coon volcano is linked to the event that marks the scenario believed most probable for the Summer Coon formation.

From this point, a number of applications can be developed. The most straightforward is to put the event-bush-based framework in a collaborative research Internet-based environment like COLLA (Diviacco, 2012). The event bushes have been imported into the COLLA and served as knowledge frameworks (or knowledge models) for virtual collaborative projects in seismology (Diviacco et al., 2014a) and volcanology (Diviacco et al., 2014b). Hoping that the method becomes widely used, it looks natural to unite the efforts of FIERCE practitioners around the world in a virtual community around an integer knowledge framework, which itself could be corrected by each of the users in any way not violating the syntactic and semantic rules of the event bush. Then, each user could upload his computation results in graphic or any other form linking them to particular scenarios, query and retrieve other scenarios and

Figure 2. Linking an interpretation of the Summer Coon Volcano history to the results of dike distribution computation. See comments in the text.

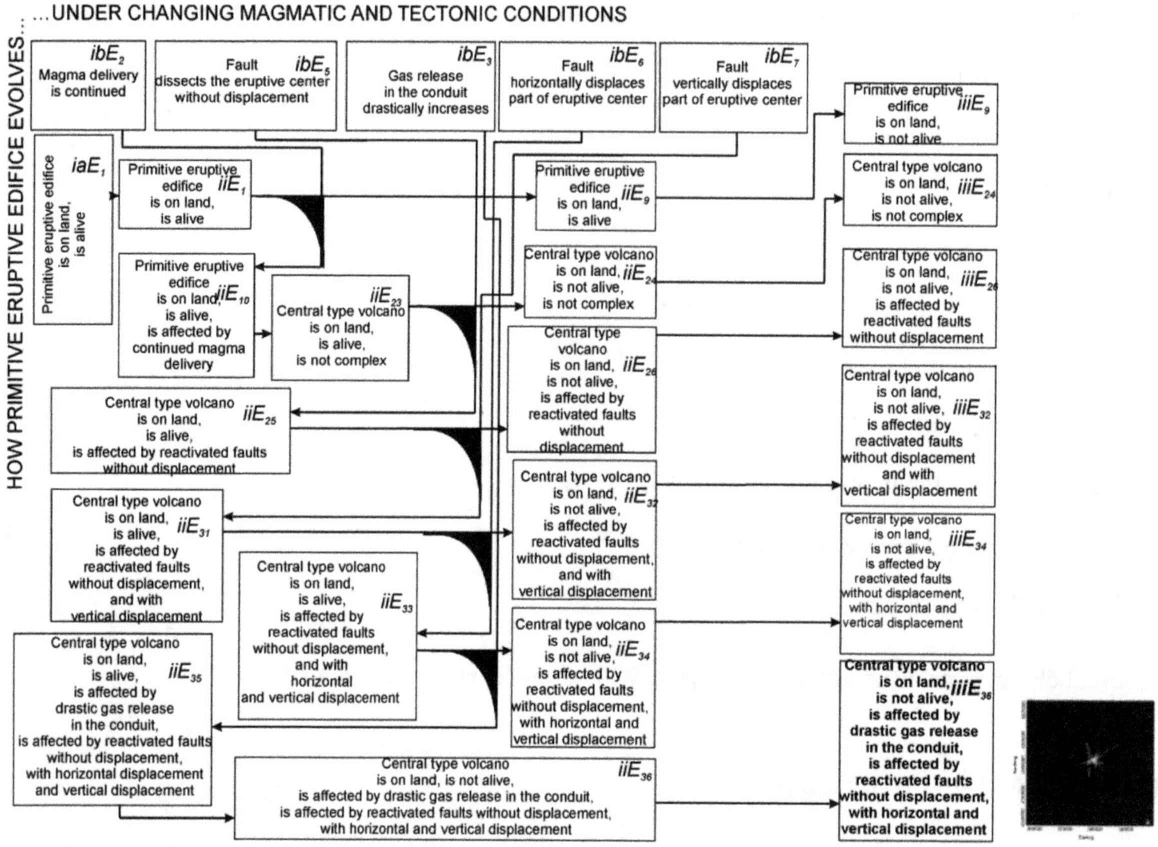

results of other researchers. The interface of COLLA allows users to do all the abovementioned, and also to launch an open discussion of the uploaded material.

Another useful application could be an expert/decision support system, whose knowledge base is filled either by expert judgment in the form of production rules (how dike patterns *should* look for different scenarios) or by supervised, or even unsupervised, learning. In the latter case, for any eruptive center in question the dike ornament is studied and interpretation of geological history is independently made and related to the event bush as done above in this work, so that the system records and remembers both and, with growing amount of studies, looks for similarities in the dike ornaments relating to the same or similar scenarios. In addition to the formal framework presented above, such systems, especially those using human expert judgment, may use supporting information, such as composition of magma, geologic structure of the region, duration of volcanic activity and magma chamber solidification, and even accessibility of sites and reliability of data.

An alternative approach could be a "strong artificial intelligence" (sensu McCarthy, see Giarratano & Riley, 1988) in a form of computer simulation model building dike patterns for each scenario based on the simple rules defining the formation of radial, tangential and regional dikes by different types of volcanic apparatuses and regional stress. Artificially generated dike patterns could be used as reference for the results of computation over actual datasets, and comparison can be made by means of pattern recognition.

In addition to the above, the framework of volcano-tectonic knowledge developed for the purpose of identification of eruptive centers may be used for other purposes – e.g., modeling the formation of typical sequences of volcanic deposits (erupted units), specification of eruptive regimes and many others, including education.

Similar frameworks could be built for other aspects of volcanology and the branch of petrology informally entitled plutonology (e.g., functionation of magma chambers, ascent and intrusion or eruption of magma, timing, seismic signatures and geophysical constraints of magma chamber processes). Interrelation of these frameworks would make the field of volcanology largely covered by the artificial intelligence and automated reasoning.

However, in this case, the existing framework should be elaborated to the mature sequent by the rules of inference of the Novorossian logic (see Pshenichny, 2017, Chapter 2 of this book), and the same should be done with other knowledge frameworks developed in volcanology.

CONCLUSION

The study presented in this chapter shows that various scenarios of evolution of eruptive centers can be brought into one conceptual framework built by means of the event bush method of dynamic knowledge engineering. This would make it easier and less intuitive to interpret the results of dike distribution computation aiming at identification of position of modern and past eruptive centers.

The framework of volcano-tectonic knowledge obtained for this purpose provides a compelling interpretation of the volcanic centers used to validate the method of computation. Linking the results of the computation to the scenarios of the event bush in a collaborative research network, decision support system or a computer simulation model may help to identify the position of hidden or ruined eruptive vents worldwide. Still, this framework of knowledge may be used separately for purposes other than the dike distribution analysis, and, if brought to formally perfect form by means of the Novorossian inference, serve as one of the key components of a hypothetic artificial intelligence system covering the entire field of volcanology.

REFERENCES

Acocella, V., & Neri, M. (2009). Dike propagation in volcanic edifices: Overview and possible developments. *Tectonophysics*, *471*(1-2), 67–77. doi:10.1016/j.tecto.2008.10.002

Acocella, V., Neri, M., & Norini, G. (2013). An overview of analogue models to understand a complex volcanic instability: Application to Etna, Italy. *Journal of Volcanology and Geothermal Research*, *251*, 98–111. doi:10.1016/j.jvolgeores.2012.06.003

Acocella, V., Neri, M., & Scarlato, P. (2006a). Understanding shallow magma emplacement at volcanoes: Orthogonal feeder dikes during the 2002–2003 Stromboli (Italy) eruption. *Geophysical Research Letters*, *33*(17), L17310. doi:10.1029/2006GL026862

Acocella, V., Neri, M., & Sulpizio, R. (2009). Dike propagation within active central volcanic edifices: Constraints from Somma-Vesuvius, Etna and analogue models. *Bulletin of Volcanology*, *71*(2), 219–223. doi:10.1007/s00445-008-0258-2

Acocella, V., Porreca, M., Neri, M., Massimi, E., & Mattei, M. (2006b). Propagation of dikes at Vesuvio (Italy) and the effect of Mt. Somma. *Geophysical Research Letters*, *33*(8), L08301. doi:10.1029/2005GL025590

Aguirre-Díaz, G. (2008). Types of collapse calderas. *IOP Conference Series: Earth and Environmental Science*, 3. doi:10.1088/1755-1307/3/1/012021

Aramayo, A., Guzmán, S., Hongn, F., del Papa, C., Montero López, C., & Sudo, M. (2017). A Middle Miocene (13.5–12 Ma) deformational event constrained by volcanism along the Puna – Eastern Cordillera border, NW Argentina. *Tectonophysics*, *703-704*, 9–22. doi:10.1016/j.tecto.2017.02.018

Battaglia, M., Di Bari, M., Acocella, V., & Neri, M. (2011). Dike emplacement and flank instability at Mount Etna: Constraints from a poro-elastic model of flank collapse. *Journal of Volcanology and Geothermal Research*, *199*(1-2), 153–164. doi:10.1016/j.jvolgeores.2010.11.005

Behncke, B. (2009). Hazards from pyroclastic density currents at Mt. Etna (Italy). *Journal of Volcanology and Geothermal Research*, *180*(2), 148–160. doi:10.1016/j.jvolgeores.2008.09.021

Behncke, B., Fornaciai, A., Neri, M., Favalli, M., Ganci, G., & Mazzarini, F. (2016). LiDAR surveys reveal eruptive volumes and rates at Etna, 2007-2010. *Geophysical Research Letters*, *43*(9), 4270–4278. doi:10.1002/2016GL068495

Behncke, B., Neri, M., & Carniel, R. (2003). An exceptional case of lava dome growth spawning pyroclastic avalanches at Mt. Etna (Italy): The 1999 Bocca Nuova eruption. *Journal of Volcanology and Geothermal Research*, *124*, 15–128. doi:10.1016/S0377-0273(03)00072-6

Behncke, B., & Pshenichny, C. A. (2009). Modeling unusual eruptive behavior of Mt. Etna, Italy, by means of event bush. *Journal of Volcanology and Geothermal Research*, *185*(3), 157–171. doi:10.1016/j.jvolgeores.2009.04.020

Branca, S., Coltelli, M., & Groppelli, G. (2015). Carta Geologica del Vulcano Etna. *Mem Descr Carta Geol D'Italia, 98.*

Branney, M. J., & Kokelaar, P. (2002). Pyroclastic Density Currents and the Sedimentation of Ignimbrites. Geological Society Memoir, 27.

Bruno, P. P., & Rapolla, A. (1999). Study of the sub-surface structure of Somma-Vesuvius (Italy) by seismic reflection data. *Journal of Volcanology and Geothermal Research*, *92*(3-4), 373–387. doi:10.1016/S0377-0273(99)00093-1

Bursik, M. (2009). A general model for tectonic control of magmatism: Examples from Long Valley Caldera (USA) and El Chichón (México). *Geofísica Internacional*, *48*(1), 171–183.

Calvari, S., Tanner, L. H., Groppelli, G., & Norini, G. (2004). A comprehensive model for the opening of the Valle del Bove depression and hazard evaluation for the eastern flank of Etna volcano. AGU (Geophysical Monograph), 143, 65–75.

Cappello, A., Neri, M., Acocella, V., Gallo, G., Vicari, A., & Del Negro, C. (2012). Spatial vent opening probability map of Mt. Etna volcano (Sicily, Italy). *Bulletin of Volcanology*, *74*(9), 2083–2094. doi:10.1007/s00445-012-0647-4

Carniel, R., Guzmán, S., & Neri, M. (2017). FIERCE: FInding volcanic ERuptive CEnters by a grid-searching algorithm in R. *Bulletin of Volcanology*, *79*(2), 19. doi:10.1007/s00445-017-1102-3

Cas, R. A. F., Wright, H. M. N., Folkes, C. B., Lesti, C., Porreca, M., Giordano, G., & Viramonte, J. G. (2011). The flow dynamics of an extremely large volume pyroclastic flow, the 2.08-Ma Cerro Galán Ignimbrite, NW Argentina, and comparison with other flow types. *Bulletin of Volcanology*, *73*(10), 1583–1609. doi:10.1007/s00445-011-0564-y

Cas, R. A. F., & Wright, J. V. (Eds.). (1987). Volcanic Successions. Modern and Ancient. *Geological Magazine*, *125*(2), 189. doi:10.1017/S0016756800009602

Chadwick, W. W. Jr, & Howard, K. A. (1991). The pattern of circumferential and radial eruptive fissures on the volcanoes of Fernandina and Isabela islands, Galapagos. *Bulletin of Volcanology*, *53*(4), 259–275. doi:10.1007/BF00414523

Cioni, R., Santacroce, R., & Sbrana, A. (1999). Pyroclastic deposits as a guide for reconstructing the multi-stage evolution of the Somma-Vesuvius caldera. *Bulletin of Volcanology*, *60*(4), 207–222. doi:10.1007/s004450050272

Coltelli, M., Garduño, V. H., Neri, M., Pasquarè, G., & Pompilio, M. (1994). Geology of northern wall of Valle del Bove, Etna (Sicily). *Acta Vulcanologica*, *5*, 55–68.

Corazzato, C., Francalanci, L., Menna, M., Petrone, C., Renzulli, A., Tibaldi, A., & Vezzoli, L. (2008). What controls sheet intrusion in volcanoes? Petrological and structural characters of the Stromboli sheet complex, Italy. *Journal of Volcanology and Geothermal Research*, *173*, 26–54. doi:10.1016/j.jvolgeores.2008.01.006

Corbi, F., Rivalta, E., Pinel, V., Maccaferri, F., Bagnardi, M., & Acocella, V. (2015). How caldera collapse shapes the shallow emplacement and transfer of magma in active volcanoes. *Earth and Planetary Science Letters*, *431*, 287–293. doi:10.1016/j.epsl.2015.09.028

Corsaro, R. A., Neri, M., & Pompilio, M. (2002). Paleo-environmental and volcano-tectonic evolution of the south-eastern flank of Mt. Etna during the last 225 ka inferred from volcanic succession of the «Timpe», Acireale, Sicily. *Journal of Volcanology and Geothermal Research*, *113*(1-2), 289–306. doi:10.1016/S0377-0273(01)00262-1

Diviacco, P. (2012). Addressing Conflicting Cognitive Models in Collaborative E-Research: A Case Study in Exploration Geophysics in Collaborative and Distributed E-Research. In Collaborative and Distributed E-Research: Innovations in Technologies, Strategies and Applications. IGI Global. doi:10.4018/978-1-4666-0125-3.ch012

Diviacco, P., Pshenichny, C., Carniel, R., Khrabrykh, Z., Shterkhun, V., Mouromtsev, D., & Pascolo, P. et al. (2014). Organization of a geophysical information space by using an event-bush-based collaborative tool. *Earth Science Informatics*, *4*(3), 139–160. doi:10.1007/s12145-014-0182-2

Diviacco, P., Pshenichny, C. A., Behncke, B., & Kanzheleva, O. M. (2014b). Sozdanie sistem podderzhki sovmestnykh nauchnykh issledovanii v naukah o Zemle v srede COLLA na osnove metoda kusta sobytii [in Russian]. *Vestnik ITMO*, *5*(93), 83–91.

Falsaperla, S., & Neri, M. (2015). Seismic footprints of shallow dyke propagation at Etna, Italy. *Scientific Reports*, *5*(1), 11908. doi:10.1038/srep11908 PMID:26173557

Ferrari, L., Garduño, V. H., & Neri, M. (1991). I dicchi dellaValle del Bove, Etna: Un metodo per stimare le dilatazioni di un apparato vulcanico [in Italian]. *Memorie della Societa Geologica Italiana*, *47*, 495–508.

Fiske, R. S., & Jackson, E. D. (1972). Orientation and growth of Hawaiian volcanic rifts: The effect of regional structure and gravitational stresses. *Proceedings of the Royal Society of London*, *329*(1578), 299–326. doi:10.1098/rspa.1972.0115

Francalanci, L., Manetti, P., Peccerillo, A., & Keller, J. (1993). Magmatological evolution of the Stromboli volcano (Aeolian Arc, Italy): Inferences from major and trace element and Sr-isotopic composition of lavas and pyroclastic rocks. *Acta Vulcanologica*, *3*, 127–151.

Garduño, V. H., Neri, M., Pasquarè, G., Borgia, A., & Tibaldi, A. (1997). Geology of NE rift of Mount Etna, Sicily (Italy). *Acta Vulcanologica*, *9*(1/2), 91–100.

Geshi, N., & Neri, M. (2014). Dynamic feeder dyke systems in basaltic volcanoes: The exceptional example of the 1809 Etna eruption (Italy). *Frontiers of Earth Science*, *2*, 13. doi:10.3389/feart.2014.00013

Giammanco, S., Melián, G., Neri, M., Hernández, P. A., Sortino, F., Barrancos, J., & Perez, N. M. et al. (2016). Active tectonic features and structural dynamics of the summit area of Mt. Etna (Italy) revealed by soil CO2 and soil temperature surveying. *Journal of Volcanology and Geothermal Research*, *311*, 79–98. doi:10.1016/j.jvolgeores.2016.01.004

Giarratano, J. C., & Riley, G. D. (1998). *Expert Systems: Principles and Programming*. Thomson Course Technology.

Guzmán, S., Strecker, M. R., Martí, J., Petrinovic, I., Schildgen, T. F., Grosse, P., & Sudo, M. et al. (2017). Construction and degradation of a broad volcanic massif: The Vicuña Pampa Volcanic Complex, southern Central Andes, NW Argentina. *Geological Society of America Bulletin*, *129*(5-6), 750–766. doi:10.1130/B31631.1

Hancock, P. L. (1994). *Continental Deformation*. Pergamon Press.

Holdsworth, R. E., Strachan, R. A., & Dewey, J. F. (1998). Continental Transpressional and Transtensional Tectonics. *Geological Society of London, Special Publications*, 135.

Jaquet, O., & Carniel, R. (2003). Multivariate stochastic modelling: Towards forecasts of paroxysmal phases at Stromboli. *Journal of Volcanology and Geothermal Research*, *311*(1-3), 261–271. doi:10.1016/S0377-0273(03)00259-2

Leyrit, H., Zylberman, W., Lutz, P., Jaillard, A., & Lavina, P. (2016). Characterization of Phreatomagmatic Deposits from the Eruption of the Pavin Maar (France). In Lake Pavin: History, geology, biogeochemistry, and sedimentology of a deep meromictic maar lake. Springer.

Lipman, P. W. (1968), Geology of the Summer Coon volcanic center, eastern San Juan Mountains, Colorado. Colorado School of Mines Quarterly, 63, 211–236.

Lucchi, F., Keller, J., De Astis, G., Francalanci, L., & Tranne, C. A. (2013). *Geological Map of Stromboli (Aeolian Archipelago)*. Università di Bologna, Universitat Freiburg, Istituto Nazionale di Geofisica e Vulcanologia, Università di Firenze.

Martí, J., Aspinall, W. P., Sobradelo, R., Felpeto, A., Geyer, A., Ortiz, R., & Lopez, C. et al. (2008). A long-term volcanic hazard event tree for Teide-Pico Viejo stratovolcanoes (Tenerife, Canary Islands). *Journal of Volcanology and Geothermal Research*, *178*(3), 543–552. doi:10.1016/j.jvolgeores.2008.09.023

Marzocchi, W., Sandri, L., & Selva, J. (2008). BET_EF: A probabilistic tool for long- and short-term eruption forecasting. *Bulletin of Volcanology*, *70*(5), 623–632. doi:10.1007/s00445-007-0157-y

McGuire, W. J., & Pullen, A. D. (1989). Location and orientation of eruptive fissures and feeder-dykes at Mount Etna: Influence of gravitational and regional stress regimes. *Journal of Volcanology and Geothermal Research*, *38*(3-4), 325–344. doi:10.1016/0377-0273(89)90046-2

Moats, W. P. (1990). *Geometry and emplacement of silicic dikes at Summer Coon Volcano, Rio Grande and Saguache counties, Colorado* (MS Thesis). Arizona State University, Tempe, AZ.

Moores, E. M., & Twiss, R. J. (1995). *Tectonics*. Freeman.

Neri, A., Aspinall, W. P., Cioni, R., Bertagnini, A., Baxter, P. J., Zuccaro, G., & Woo, G. et al. (2008). Developing an Event Tree for probabilistic hazard and risk assessment at Vesuvius. *Journal of Volcanology and Geothermal Research*, *178*(3), 397–415. doi:10.1016/j.jvolgeores.2008.05.014

Neri, M., Acocella, V., Behncke, B., Giammanco, S., Mazzarini, F., & Rust, D. (2011). Structural analysis of the eruptive fissures at Mount Etna (Italy). *Annales Geophysicae*, *54*(5), 464–479. doi:10.4401/ag-5332

Neri, M., Lanzafame, G., & Acocella, V. (2008). Dike emplacement and related hazard in volcanoes with sector collapse: The 2007 Stromboli eruption. *Journal of the Geological Society*, *165*(5), 883–886. doi:10.1144/0016-76492008-002

Newhall, C. G., & Hoblitt, R. P. (2002). Constructing event trees for volcanic crises. *Bulletin of Volcanology*, *64*(1), 3–20. doi:10.1007/s004450100173

Parfitt, E., & Wilson, L. (2008). *Fundamentals of Physical Volcanology*. Wiley-Blackwell.

Pasquarè, G., Francalanci, L., Garduño, V. H., & Tibaldi, A. (1993). Structure and geologic evolution of the Stromboli volcano, Aeolian islands, Italy. *Acta Vulcanologica*, *3*, 79–89.

Petrinovich, I. A., Marti, J., Aguirre-Diaz, G. J., Guzman, S., Geyer, A., & Salado Paz, N. (2010). The Cerro Aguas Calientes caldera, NW Argentina: An example of a tectonically controlled polygenetic collapse caldera, and its regional significance. *Journal of Volcanology and Geothermal Research*, *194*(1-3), 15–26. doi:10.1016/j.jvolgeores.2010.04.012

Poland, M. P., Moats, M. P., & Fink, J. H. (2008). A model for radial dike emplacement in composite cones based on observations from Summer Coon volcano, Colorado, USA. *Bulletin of Volcanology*, *70*(7), 861–875. doi:10.1007/s00445-007-0175-9

Porreca, M., Acocella, V., Massimi, E., Mattei, M., Funiciello, R., & De Benedetti, A. A. (2006). Geometric and kinematic features of the dike complex at Mt. Somma, Vesuvio (Italy). *Earth and Planetary Science Letters*, *245*(1-2), 389–407. doi:10.1016/j.epsl.2006.02.027

Pshenichny, C. A., & Kanzheleva, O. M. (2011). Theoretical foundations of the event bush method. In K. Sinha, L. Gundersen, J. Jackson, & D. Arctur (Eds.), Societal Challenges and Geoinformatics (pp. 139-165). Boulder, CO: GSA Special Paper 482. doi:10.1130/2011.2482(12)

Pshenichny, C. A., & Mouromtsev, D. I. (2015). Grammar of Dynamic Knowledge for Collaborative Knowledge Engineering and Representation. In P. Diviacco, P. Fox, C. Pshenichny, & A. Leadbetter (Eds.), *Collaborative Knowledge in Scientific Research Networks* (pp. 323–354). IGI Global. doi:10.4018/978-1-4666-6567-5.ch016

Pshenichny, C. A., Nikolenko, S. I., Carniel, R., Vaganov, P. A., Khrabrykh, Z. V., Moukhachov, V. P., & Rezyapkin, A. A. et al. (2009). The event bush as a semantic-based numerical approach to natural hazard assessment (exemplified by volcanology). *Computers & Geosciences*, *35*(5), 1017–1034. doi:10.1016/j.cageo.2008.01.009

Santacroce, R., & Sbrana, A. (2003). Carta Geologica del Vesuvio. *Scala*, *1*, 15000.

Scandone, R., Giacomelli, L., & Gasparini, P. (1993). Mount Vesuvius: 2000 years of volcanological observations. *Journal of Volcanology and Geothermal Research*, *58*(1-4), 4–25. doi:10.1016/0377-0273(93)90099-D

Tadini, A., Bevilacqua, A., Neri, A., Cioni, R., Aspinall, W. P., Bisson, M., & Pistolesi, M. et al. (2017). Assessing future vent opening locations at the Somma-Vesuvio volcanic complex: 2. Probability maps of the caldera for a future Plinian/sub-Plinian event with uncertainty quantification. *Journal of Geophysical Research. Solid Earth*, *122*(6), 4357–4376. doi:10.1002/2016JB013860

Takada, A. (1997). Cyclic flank-vent and central-vent eruption patterns. *Bulletin of Volcanology*, *58*(7), 539–556. doi:10.1007/s004450050161

Tibaldi, A. (2001). Multiple sector collapses at Stromboli volcano: How they work. *Bulletin of Volcanology*, *63*(2-3), 112–125. doi:10.1007/s004450100129

Tibaldi, A., Corazzato, C., Marani, M., & Gamberi, F. (2009). Subaerial and submarine evidence of structures feeding magma to Stromboli Volcano, Italy, and relations with edifice flank failure and creep. *Tectonophysics*, *469*(1-4), 112–136. doi:10.1016/j.tecto.2009.01.031

Van der Pluijm, B. A., & Marshak, S. (1997). *Earth Structure: An introduction to structural geology and tectonics*. McGraw-Hill.

APPENDIX 1

Definitions of Events

$$iaE_1 = S_1 - P_1, P_2, \sim P_{12}, \sim P_{13}, \sim P_{14}, \sim P_{15}, \sim P_{16}, \sim P_{17}$$

$$ibE_2 = S_5 - P_1, P_3$$

$$ibE_3 = S_6 - P_1, P_5$$

$$ibE_4 = S_7 - P_1, P_6, P_7$$

$$ibE_5 = S_8 - P_1, P_8, \sim P_9, \sim P_{10}$$

$$ibE_6 = S_8 - P_1, P_8, P_9, \sim P_{10}$$

$$ibE_7 = S_8 - P_1, P_8, \sim P_9, P_{10}$$

$$ibE_8 = S_9 - P_1, P_{24}, P_{11}$$

$$iiE_1 = S_1 - P_1, P_2, \sim P_{12}, \sim P_{13}, \sim P_{14}, \sim P_{15}, \sim P_{16}, \sim P_{17}$$

$$iiE_9 = S_1 - P_1, \sim P_2, \sim P_{12}, \sim P_{13}, \sim P_{14}, \sim P_{15}, \sim P_{16}, \sim P_{17}$$

$$iiE_{10} = S_1 - P_1, P_2, P_{12}, \sim P_{13}, \sim P_{14}, \sim P_{15}, \sim P_{16}, \sim P_{17}$$

$$iiE_{11} = S_1 - P_1, P_2, \sim P_{12}, P_{13}, \sim P_{14}, \sim P_{15}, \sim P_{16}, \sim P_{17}$$

$$iiE_{12} = S_1 - P_1, P_2, \sim P_{12}, \sim P_{13}, \sim P_{14}, P_{15}, \sim P_{16}, \sim P_{17}$$

$$iiE_{13} = S_1 - P_1, \sim P_2, \sim P_{12}, \sim P_{13}, \sim P_{14}, P_{15}, \sim P_{16}, \sim P_{17}$$

$$iiE_{14} = S_1 - P_1, P_2, \sim P_{12}, \sim P_{13}, \sim P_{14}, \sim P_{15}, P_{16}, \sim P_{17}$$

$$iiE_{15} = S_1 - P_1, \sim P_2, \sim P_{12}, \sim P_{13}, \sim P_{14}, \sim P_{15}, P_{16}, \sim P_{17}$$

$$iiE_{16} = S_1 - P_1, P_2, \sim P_{12}, \sim P_{13}, \sim P_{14}, \sim P_{15}, \sim P_{16}, P_{17}$$

$$iiE_{17} = S_1 - P_1, \sim P_2, \sim P_{12}, \sim P_{13}, \sim P_{14}, \sim P_{15}, \sim P_{16}, P_{17}$$

$$iiE_{18} = S_1 - P_1, P_2, \sim P_{12}, \sim P_{13}, \sim P_{14}, \sim P_{15}, P_{16}, P_{17}$$

$$iiE_{19} = S_1 - P_1, \sim P_2, \sim P_{12}, \sim P_{13}, \sim P_{14}, \sim P_{15}, P_{16}, P_{17}$$

$$iiE_{20} = S_2 - P_1, P_2, P_{13}, \sim P_{14}, \sim P_{15}, \sim P_{16}, \sim P_{17}$$

$$iiE_{21} = S_2 - P_1, \sim P_2, P_{13}, \sim P_{14}, \sim P_{15}, \sim P_{16}, \sim P_{17}$$

$$iiE_{22} = S_{10} - P_1, P_{25}$$

$$iiE_{23} = S_3 - P_1, P_2, \sim P_{18}, \sim P_{13}, \sim P_{14}, \sim P_{15}, \sim P_{16}, \sim P_{17}$$

$$iiE_{24} = S_3 - P_1, \sim P_2, \sim P_{18}, \sim P_{13}, \sim P_{14}, \sim P_{15}, \sim P_{16}, \sim P_{17}$$

$$iiE_{25} = S_3 - P_1, P_2, \sim P_{18}, \sim P_{13}, \sim P_{14}, P_{15}, \sim P_{16}, \sim P_{17}$$

$$iiE_{26} = S_3 - P_1, \sim P_2, \sim P_{18}, \sim P_{13}, \sim P_{14}, P_{15}, \sim P_{16}, \sim P_{17}$$

$$iiE_{27} = S_3 - P_1, P_2, \sim P_{18}, P_{13}, \sim P_{14}, P_{15}, \sim P_{16}, \sim P_{17}$$

$$iiE_{28} = S_3 - P_1, \sim P_2, \sim P_{18}, P_{13}, \sim P_{14}, P_{15}, \sim P_{16}, \sim P_{17}$$

$$iiE_{29} = S_3 - P_1, P_2, \sim P_{18}, \sim P_{13}, \sim P_{14}, P_{15}, P_{16}, \sim P_{17}$$

$$iiE_{30} = S_3 - P_1, \sim P_2, \sim P_{18}, \sim P_{13}, \sim P_{14}, P_{15}, P_{16}, \sim P_{17}$$

$iiE_{31} = S_3 – P_1, P_2, \sim P_{18}, \sim P_{13}, \sim P_{14}, P_{15}, \sim P_{16}, P_{17}$

$iiE_{32} = S_3 – P_1, \sim P_2, \sim P_{18}, \sim P_{13}, \sim P_{14}, P_{15}, \sim P_{16}, P_{17}$

$iiE_{33} = S_3 – P_1, P_2, \sim P_{18}, \sim P_{13}, \sim P_{14}, P_{15}, P_{16}, P_{17}$

$iiE_{34} = S_3 – P_1, \sim P_2, \sim P_{18}, \sim P_{13}, \sim P_{14}, P_{15}, P_{16}, P_{17}$

$iiE_{35} = S_3 – P_1, P_2, \sim P_{18}, P_{13}, \sim P_{14}, P_{15}, P_{16}, P_{17}$

$iiE_{36} = S_3 – P_1, \sim P_2, \sim P_{18}, P_{13}, \sim P_{14}, P_{15}, P_{16}, P_{17}$

$iiE_{37} = S_3 – P_1, P_2, P_{18}, \sim P_{13}, \sim P_{14}, P_{15}, \sim P_{16}, \sim P_{17}$

$iiE_{38} = S_3 – P_1, \sim P_2, P_{18}, \sim P_{13}, \sim P_{14}, P_{15}, \sim P_{16}, \sim P_{17}$

$iiE_{39} = S_3 – P_1, P_2, \sim P_{18}, P_{13}, \sim P_{14}, \sim P_{15}, \sim P_{16}, \sim P_{17}$

$iiE_{40} = S_3 – P_1, \sim P_2, \sim P_{18}, P_{13}, \sim P_{14}, \sim P_{15}, \sim P_{16}, \sim P_{17}$

$iiE_{41} = S_3 – P_1, P_2, \sim P_{18}, P_{13}, P_{14}, \sim P_{15}, \sim P_{16}, \sim P_{17}$

$iiE_{42} = S_3 – P_1, P_2, P_{18}, P_{13}, \sim P_{14}, P_{15}, \sim P_{16}, \sim P_{17}$

$iiE_{43} = S_3 – P_1, \sim P_2, P_{18}, P_{13}, \sim P_{14}, P_{15}, \sim P_{16}, \sim P_{17}$

$iiE_{44} = S_3 – P_1, P_2, \sim P_{18}, P_{13}, P_{14}, P_{15}, \sim P_{16}, \sim P_{17}$

$iiE_{45} = S_3 – P_1, P_2, P_{18}, P_{13}, \sim P_{14}, P_{15}, P_{16}, \sim P_{17}$

$iiE_{46} = S_3 – P_1, \sim P_2, P_{18}, P_{13}, \sim P_{14}, P_{15}, P_{16}, \sim P_{17}$

$iiE_{47} = S_3 – P_1, P_2, P_{18}, P_{13}, \sim P_{14}, P_{15}, P_{16}, P_{17}$

$iiE_{48} = S_3 – P_1, \sim P_2, P_{18}, P_{13}, \sim P_{14}, P_{15}, P_{16}, P_{17}$

$iiE_{49} = S_3 – P_1, P_2, P_{18}, P_{13}, \sim P_{14}, P_{15}, \sim P_{16}, P_{17}$

$iiE_{50} = S_3 – P_1, \sim P_2, P_{18}, P_{13}, \sim P_{14}, P_{15}, \sim P_{16}, P_{17}$

$iiE_{51} = S_3 – P_1, P_2, \sim P_{18}, \sim P_{13}, \sim P_{14}, \sim P_{15}, P_{16}, \sim P_{17}$

$iiE_{52} = S_3 – P_1, \sim P_2, \sim P_{18}, \sim P_{13}, \sim P_{14}, \sim P_{15}, P_{16}, \sim P_{17}$

$iiE_{53} = S_3 – P_1, P_2, \sim P_{18}, \sim P_{13}, \sim P_{14}, \sim P_{15}, P_{16}, P_{17}$

$iiE_{54} = S_3 – P_1, \sim P_2, \sim P_{18}, \sim P_{13}, \sim P_{14}, \sim P_{15}, P_{16}, P_{17}$

$iiE_{55} = S_3 – P_1, P_2, P_{18}, \sim P_{13}, \sim P_{14}, P_{15}, P_{16}, \sim P_{17}$

$iiE_{56} = S_3 – P_1, \sim P_2, P_{18}, \sim P_{13}, \sim P_{14}, P_{15}, P_{16}, \sim P_{17}$

$iiE_{57} = S_3 – P_1, P_2, P_{18}, \sim P_{13}, \sim P_{14}, P_{15}, P_{16}, P_{17}$

$iiE_{58} = S_3 – P_1, \sim P_2, P_{18}, \sim P_{13}, \sim P_{14}, P_{15}, P_{16}, P_{17}$

$iiE_{59} = S_3 – P_1, P_2, \sim P_{18}, P_{13}, \sim P_{14}, \sim P_{15}, P_{16}, \sim P_{17}$

$iiE_{60} = S_3 – P_1, \sim P_2, \sim P_{18}, P_{13}, \sim P_{14}, \sim P_{15}, P_{16}, \sim P_{17}$

$iiE_{61} = S_3 – P_1, P_2, \sim P_{18}, P_{13}, \sim P_{14}, \sim P_{15}, P_{16}, P_{17}$

$iiE_{62} = S_3 – P_1, \sim P_2, \sim P_{18}, P_{13}, \sim P_{14}, \sim P_{15}, P_{16}, P_{17}$

$iiE_{63} = S_3 – P_1, P_2, \sim P_{18}, \sim P_{13}, \sim P_{14}, \sim P_{15}, \sim P_{16}, P_{17}$

$iiE_{64} = S_3 – P_1, \sim P_2, \sim P_{18}, \sim P_{13}, \sim P_{14}, \sim P_{15}, \sim P_{16}, P_{17}$

$iiE_{65} = S_3 - P_1, P_2, P_{18}, \sim P_{13}, \sim P_{14}, P_{15}, \sim P_{16}, \sim P_{17}$

$iiE_{66} = S_3 - P_1, \sim P_2, P_{18}, \sim P_{13}, \sim P_{14}, P_{15}, \sim P_{16}, \sim P_{17}$

$iiE_{67} = S_3 - P_1, P_2, \sim P_{18}, P_{13}, \sim P_{14}, P_{15}, \sim P_{16}, P_{17}$

$iiE_{68} = S_3 - P_1, \sim P_2, \sim P_{18}, P_{13}, \sim P_{14}, \sim P_{15}, \sim P_{16}, P_{17}$

$iiE_{69} = S_3 - P_1, P_2, P_{18}, P_{13}, \sim P_{14}, P_{15}, \sim P_{16}, \sim P_{17}$

$iiE_{70} = S_3 - P_1, \sim P_2, P_{18}, P_{13}, \sim P_{14}, P_{15}, \sim P_{16}, \sim P_{17}$

$iiE_{71} = S_{11} - P_1, P_{26}$

$iiE_{72} = S_{11} - P_1, P_{27}$

$iiE_{73} = S_4 - P_1, P_2, \sim P_{19}, \sim P_{20}, \sim P_{21}, \sim P_{22}, \sim P_{28}, \sim P_{23}$

$iiE_{74} = S_4 - P_1, \sim P_2, \sim P_{19}, \sim P_{20}, \sim P_{21}, \sim P_{22}, \sim P_{28}, \sim P_{23}$

$iiE_{75} = S_4 - P_1, P_2, \sim P_{19}, P_{20}, \sim P_{21}, \sim P_{22}, \sim P_{28}, \sim P_{23}$

$iiE_{76} = S_4 - P_1, \sim P_2, \sim P_{19}, P_{20}, \sim P_{21}, \sim P_{22}, \sim P_{28}, \sim P_{23}$

$iiE_{77} = S_4 - P_1, P_2, \sim P_{19}, P_{20}, P_{21}, \sim P_{22}, \sim P_{28}, \sim P_{23}$

$iiE_{78} = S_4 - P_1, \sim P_2, \sim P_{19}, P_{20}, P_{21}, \sim P_{22}, \sim P_{28}, \sim P_{23}$

$iiE_{79} = S_4 - P_1, P_2, P_{19}, \sim P_{20}, \sim P_{21}, \sim P_{22}, \sim P_{28}, \sim P_{23}$

$iiE_{80} = S_4 - P_1, \sim P_2, P_{19}, \sim P_{20}, \sim P_{21}, \sim P_{22}, \sim P_{28}, \sim P_{23}$

$iiE_{81} = S_4 - P_1, P_2, P_{19}, P_{20}, \sim P_{21}, \sim P_{22}, \sim P_{28}, \sim P_{23}$

$iiE_{82} = S_4 - P_1, \sim P_2, P_{19}, P_{20}, \sim P_{21}, \sim P_{22}, \sim P_{28}, \sim P_{23}$

$iiE_{83} = S_4 - P_1, P_2, P_{19}, P_{20}, P_{21}, \sim P_{22}, \sim P_{28}, \sim P_{23}$

$iiE_{84} = S_4 - P_1, \sim P_2, P_{19}, P_{20}, P_{21}, \sim P_{22}, \sim P_{28}, \sim P_{23}$

$iiE_{85} = S_4 - P_1, P_2, \sim P_{19}, \sim P_{20}, \sim P_{21}, P_{22}, \sim P_{28}, \sim P_{23}$

$iiE_{86} = S_4 - P_1, \sim P_2, \sim P_{19}, \sim P_{20}, \sim P_{21}, P_{22}, \sim P_{28}, \sim P_{23}$

$iiE_{87} = S_4 - P_1, P_2, \sim P_{19}, P_{20}, \sim P_{21}, P_{22}, \sim P_{28}, \sim P_{23}$

$iiE_{88} = S_4 - P_1, \sim P_2, \sim P_{19}, P_{20}, \sim P_{21}, P_{22}, \sim P_{28}, \sim P_{23}$

$iiE_{89} = S_4 - P_1, P_2, \sim P_{19}, P_{20}, P_{21}, P_{22}, \sim P_{28}, \sim P_{23}$

$iiE_{90} = S_4 - P_1, \sim P_2, \sim P_{19}, P_{20}, \sim P_{21}, P_{22}, \sim P_{28}, \sim P_{23}$

$iiE_{91} = S_4 - P_1, P_2, P_{19}, P_{20}, P_{21}, P_{22}, \sim P_{28}, \sim P_{23}$

$iiE_{92} = S_4 - P_1, \sim P_2, P_{19}, P_{20}, P_{21}, P_{22}, \sim P_{28}, \sim P_{23}$

$iiE_{93} = S_4 - P_1, P_2, P_{19}, P_{20}, \sim P_{21}, \sim P_{22}, \sim P_{28}, \sim P_{23}$

$iiE_{94} = S_4 - P_1, \sim P_2, P_{19}, P_{20}, \sim P_{21}, \sim P_{22}, \sim P_{28}, \sim P_{23}$

$iiE_{95} = S_4 - P_1, P_2, P_{19}, \sim P_{20}, \sim P_{21}, P_{22}, \sim P_{28}, \sim P_{23}$

$iiE_{96} = S_4 - P_1, \sim P_2, P_{19}, \sim P_{20}, \sim P_{21}, P_{22}, \sim P_{28}, \sim P_{23}$

$iiiE_{9} = S_1 - P_1, \sim P_2, \sim P_{12}, \sim P_{13}, \sim P_{14}, \sim P_{15}, \sim P_{16}, \sim P_{17}$

$iiiE_{13} = S_1 - P_1, \sim P_2, \sim P_{12}, \sim P_{13}, \sim P_{14}, P_{15}, \sim P_{16}, \sim P_{17}$

$$iiiE_{15} = S_1 - P_1, \sim P_2, \sim P_{12}, \sim P_{13}, \sim P_{14}, \sim P_{15}, P_{16}, \sim P_{17}$$
$$iiiE_{17} = S_1 - P_1, \sim P_2, \sim P_{12}, \sim P_{13}, \sim P_{14}, \sim P_{15}, \sim P_{16}, P_{17}$$
$$iiiE_{19} = S_1 - P_1, \sim P_2, \sim P_{12}, \sim P_{13}, \sim P_{14}, \sim P_{15}, P_{16}, P_{17}$$
$$iiiE_{21} = S_2 - P_1, \sim P_2, P_{13}, \sim P_{14}, \sim P_{15}, \sim P_{16}, \sim P_{17}$$
$$iiiE_{24} = S_3 - P_1, \sim P_2, \sim P_{18}, \sim P_{13}, \sim P_{14}, \sim P_{15}, \sim P_{16}, \sim P_{17}$$
$$iiiE_{26} = S_3 - P_1, \sim P_2, \sim P_{18}, \sim P_{13}, \sim P_{14}, P_{15}, \sim P_{16}, \sim P_{17}$$
$$iiiE_{28} = S_3 - P_1, \sim P_2, \sim P_{18}, P_{13}, \sim P_{14}, P_{15}, \sim P_{16}, \sim P_{17}$$
$$iiiE_{30} = S_3 - P_1, \sim P_2, \sim P_{18}, \sim P_{13}, \sim P_{14}, P_{15}, P_{16}, \sim P_{17}$$
$$iiiE_{32} = S_3 - P_1, \sim P_2, \sim P_{18}, \sim P_{13}, \sim P_{14}, P_{15}, \sim P_{16}, P_{17}$$
$$iiiE_{34} = S_3 - P_1, \sim P_2, \sim P_{18}, \sim P_{13}, \sim P_{14}, P_{15}, P_{16}, P_{17}$$
$$iiiE_{36} = S_3 - P_1, \sim P_2, \sim P_{18}, P_{13}, \sim P_{14}, P_{15}, P_{16}, P_{17}$$
$$iiiE_{38} = S_3 - P_1, \sim P_2, \sim P_{18}, \sim P_{13}, \sim P_{14}, \sim P_{15}, \sim P_{16}, \sim P_{17}$$
$$iiiE_{40} = S_3 - P_1, \sim P_2, \sim P_{18}, P_{13}, \sim P_{14}, \sim P_{15}, \sim P_{16}, \sim P_{17}$$
$$iiiE_{43} = S_3 - P_1, \sim P_2, P_{18}, P_{13}, \sim P_{14}, P_{15}, \sim P_{16}, \sim P_{17}$$
$$iiiE_{46} = S_3 - P_1, \sim P_2, P_{18}, P_{13}, \sim P_{14}, P_{15}, P_{16}, \sim P_{17}$$
$$iiiE_{48} = S_3 - P_1, \sim P_2, P_{18}, P_{13}, \sim P_{14}, P_{15}, P_{16}, P_{17}$$
$$iiiE_{50} = S_3 - P_1, \sim P_2, P_{18}, P_{13}, \sim P_{14}, P_{15}, \sim P_{16}, P_{17}$$
$$iiiE_{52} = S_3 - P_1, \sim P_2, \sim P_{18}, \sim P_{13}, \sim P_{14}, \sim P_{15}, P_{16}, \sim P_{17}$$
$$iiiE_{54} = S_3 - P_1, \sim P_2, \sim P_{18}, \sim P_{13}, \sim P_{14}, \sim P_{15}, P_{16}, P_{17}$$
$$iiiE_{56} = S_3 - P_1, \sim P_2, P_{18}, \sim P_{13}, \sim P_{14}, P_{15}, P_{16}, \sim P_{17}$$
$$iiiE_{58} = S_3 - P_1, \sim P_2, P_{18}, \sim P_{13}, \sim P_{14}, P_{15}, P_{16}, P_{17}$$
$$iiiE_{60} = S_3 - P_1, \sim P_2, \sim P_{18}, P_{13}, \sim P_{14}, \sim P_{15}, P_{16}, \sim P_{17}$$
$$iiiE_{62} = S_3 - P_1, \sim P_2, \sim P_{18}, P_{13}, \sim P_{14}, \sim P_{15}, P_{16}, P_{17}$$
$$iiiE_{64} = S_3 - P_1, \sim P_2, \sim P_{18}, \sim P_{13}, \sim P_{14}, \sim P_{15}, \sim P_{16}, P_{17}$$
$$iiiE_{66} = S_3 - P_1, \sim P_2, P_{18}, \sim P_{13}, \sim P_{14}, P_{15}, \sim P_{16}, \sim P_{17}$$
$$iiiE_{68} = S_3 - P_1, \sim P_2, \sim P_{18}, P_{13}, \sim P_{14}, \sim P_{15}, \sim P_{16}, P_{17}$$
$$iiiE_{70} = S_3 - P_1, \sim P_2, P_{18}, P_{13}, \sim P_{14}, P_{15}, \sim P_{16}, \sim P_{17}$$
$$iiiE_{74} = S_4 - P_1, \sim P_2, \sim P_{19}, \sim P_{20}, \sim P_{21}, \sim P_{22}, \sim P_{28}, \sim P_{23}$$
$$iiiE_{76} = S_4 - P_1, \sim P_2, \sim P_{19}, P_{20}, \sim P_{21}, \sim P_{22}, \sim P_{28}, \sim P_{23}$$
$$iiiE_{78} = S_4 - P_1, \sim P_2, \sim P_{19}, P_{20}, P_{21}, \sim P_{22}, \sim P_{28}, \sim P_{23}$$
$$iiiE_{80} = S_4 - P_1, \sim P_2, P_{19}, \sim P_{20}, \sim P_{21}, \sim P_{22}, \sim P_{28}, \sim P_{23}$$
$$iiiE_{82} = S_4 - P_1, \sim P_2, P_{19}, P_{20}, \sim P_{21}, \sim P_{22}, \sim P_{28}, \sim P_{23}$$
$$iiiE_{84} = S_4 - P_1, \sim P_2, P_{19}, P_{20}, P_{21}, \sim P_{22}, \sim P_{28}, \sim P_{23}$$
$$iiiE_{86} = S_4 - P_1, \sim P_2, \sim P_{19}, \sim P_{20}, \sim P_{21}, P_{22}, \sim P_{28}, \sim P_{23}$$

$iiiE_{88} = S_4 – P_{1'}\ \sim P_{2'}\ \sim P_{19'}\ P_{20'}\ \sim P_{21'}\ P_{22'}\ \sim P_{28'}\ \sim P_{23}$
$iiiE_{90} = S_4 – P_{1'}\ \sim P_{2'}\ \sim P_{19'}\ P_{20'}\ \sim P_{21'}\ P_{22'}\ \sim P_{28'}\ \sim P_{23}$
$iiiE_{92} = S_4 – P_{1'}\ \sim P_{2'}\ P_{19'}\ P_{20'}\ P_{21'}\ P_{22'}\ \sim P_{28'}\ \sim P_{23}$
$iiiE_{94} = S_4 – P_{1'}\ \sim P_{2'}\ P_{19'}\ P_{20'}\ \sim P_{21'}\ \sim P_{22'}\ \sim P_{28'}\ \sim P_{23}$
$iiiE_{96} = S_4 – P_{1'}\ \sim P_{2'}\ P_{19'}\ \sim P_{20'}\ \sim P_{21'}\ P_{22'}\ \sim P_{28'}\ \sim P_{23}$

APPENDIX 2

Changes of Events in the Event Bush Model

iaE_1 *Flux* iiE_1
$ibE_{2'}\ iiE_1$ *Influx* iiE_{10}
$ibE_{2'}\ iiE_{25}$ *Influx* iiE_{37}
$ibE_{3'}\ iiE_1$ *Influx* iiE_{11}
$ibE_{3'}\ iiE_{23}$ *Influx* iiE_{39}
$ibE_{3'}\ iiE_{25}$ *Influx* iiE_{27}
$ibE_{3'}\ iiE_{33}$ *Influx* iiE_{35}
$ibE_{3'}\ iiE_{37}$ *Influx* iiE_{42}
$ibE_{3'}\ iiE_{73}$ *Influx* iiE_{75}
$ibE_{4'}\ iiE_{39}$ *Influx* iiE_{41}
$ibE_{4'}\ iiE_{42}$ *Influx* iiE_{44}
$ibE_{4'}\ iiE_{47}$ *Influx* iiE_{73}
$ibE_{4'}\ iiE_{73}$ *Influx* iiE_{79}
$ibE_{4'}\ iiE_{75}$ *Influx* iiE_{77}
$ibE_{4'}\ iiE_{75}$ *Influx* iiE_{81}
$ibE_{4'}\ iiE_{77}$ *Influx* iiE_{83}
$ibE_{5'}\ iiE_1$ *Influx* iiE_{12}
$ibE_{5'}\ iiE_{23}$ *Influx* iiE_{25}
$ibE_{6'}\ iiE_1$ *Influx* iiE_{14}
$ibE_{6'}\ iiE_{23}$ *Influx* iiE_{51}
$ibE_{6'}\ iiE_{25}$ *Influx* iiE_{29}
$ibE_{6'}\ iiE_{31}$ *Influx* iiE_{33}
$ibE_{6'}\ iiE_{37}$ *Influx* iiE_{55}
$ibE_{6'}\ iiE_{39}$ *Influx* iiE_{59}

ibE_{6}, iiE_{42} *Influx* iiE_{45}

ibE_{7}, iiE_{1} *Influx* iiE_{16}

ibE_{7}, iiE_{16} *Influx* iiE_{18}

ibE_{7}, iiE_{23} *Influx* iiE_{63}

ibE_{7}, iiE_{25} *Influx* iiE_{31}

ibE_{7}, iiE_{37} *Influx* iiE_{65}

ibE_{7}, iiE_{39} *Influx* iiE_{67}

ibE_{7}, iiE_{42} *Influx* iiE_{49}

ibE_{7}, iiE_{42} *Influx* iiE_{69}

ibE_{7}, iiE_{45} *Influx* iiE_{47}

ibE_{7}, iiE_{51} *Influx* iiE_{53}

ibE_{7}, iiE_{55} *Influx* iiE_{57}

ibE_{7}, iiE_{59} *Influx* iiE_{61}

ibE_{8}, iiE_{24} *Influx* iiE_{23}

ibE_{8}, iiE_{26} *Influx* iiE_{25}

ibE_{8}, iiE_{28} *Influx* iiE_{27}

ibE_{8}, iiE_{30} *Influx* iiE_{29}

ibE_{8}, iiE_{32} *Influx* iiE_{31}

ibE_{8}, iiE_{34} *Influx* iiE_{33}

ibE_{8}, iiE_{36} *Influx* iiE_{35}

ibE_{8}, iiE_{38} *Influx* iiE_{37}

ibE_{8}, iiE_{40} *Influx* iiE_{39}

ibE_{8}, iiE_{43} *Influx* iiE_{42}

ibE_{8}, iiE_{46} *Influx* iiE_{45}

ibE_{8}, iiE_{48} *Influx* iiE_{47}

ibE_{8}, iiE_{50} *Influx* iiE_{49}

ibE_{8}, iiE_{52} *Influx* iiE_{51}

ibE_{8}, iiE_{54} *Influx* iiE_{53}

ibE_{8}, iiE_{56} *Influx* iiE_{55}

ibE_{8}, iiE_{58} *Influx* iiE_{57}

ibE_{8}, iiE_{60} *Influx* iiE_{59}

ibE_{8}, iiE_{62} *Influx* iiE_{61}

ibE_{8}, iiE_{64} *Influx* iiE_{63}

ibE_{8}, iiE_{66} *Influx* iiE_{65}

$ibE_{8'}$ iiE_{68} Influx iiE_{67}
$ibE_{8'}$ iiE_{70} Influx iiE_{69}
$ibE_{8'}$ iiE_{74} Influx iiE_{73}
$ibE_{8'}$ iiE_{76} Influx iiE_{75}
$ibE_{8'}$ iiE_{78} Influx iiE_{77}
$ibE_{8'}$ iiE_{80} Influx iiE_{79}
$ibE_{8'}$ iiE_{82} Influx iiE_{81}
$ibE_{8'}$ iiE_{84} Influx iiE_{83}
$ibE_{8'}$ iiE_{86} Influx iiE_{85}
$ibE_{8'}$ iiE_{88} Influx iiE_{87}
$ibE_{8'}$ iiE_{90} Influx iiE_{89}
$ibE_{8'}$ iiE_{92} Influx iiE_{91}
$ibE_{8'}$ iiE_{94} Influx iiE_{93}
$ibE_{8'}$ iiE_{96} Influx iiE_{95}
iiE_{1} Flux iiE_{9}
iiE_{1} Flux iiE_{71}
iiE_{9} Flux $iiiE_{9}$
iiE_{10} Flux iiE_{23}
iiE_{11} Flux iiE_{20}
iiE_{12} Flux iiE_{13}
iiE_{12} Flux iiE_{71}
iiE_{13} Flux $iiiE_{13}$
iiE_{14} Flux iiE_{15}
iiE_{14} Flux iiE_{71}
iiE_{15} Flux $iiiE_{15}$
iiE_{16} Flux iiE_{17}
iiE_{16} Flux iiE_{71}
iiE_{17} Flux $iiiE_{17}$
iiE_{18} Flux iiE_{19}
iiE_{18} Flux iiE_{71}
iiE_{19} Flux $iiiE_{19}$
iiE_{20} Flux iiE_{21}
iiE_{20} Flux iiE_{22}
iiE_{21} Flux $iiiE_{21}$

iiE_{23} *Flux* iiE_{24}
iiE_{23} *Flux* iiE_{72}
iiE_{24} *Flux* $iiiE_{24}$
iiE_{25} *Flux* iiE_{26}
iiE_{25} *Flux* iiE_{72}
iiE_{26} *Flux* $iiiE_{26}$
iiE_{27} *Flux* iiE_{28}
iiE_{28} *Flux* $iiiE_{28}$
iiE_{29} *Flux* iiE_{30}
iiE_{30} *Flux* $iiiE_{30}$
iiE_{31} *Flux* iiE_{32}
iiE_{32} *Flux* $iiiE_{32}$
iiE_{33} *Flux* iiE_{34}
iiE_{34} *Flux* $iiiE_{34}$
iiE_{35} *Flux* iiE_{36}
iiE_{36} *Flux* $iiiE_{36}$
iiE_{37} *Flux* iiE_{38}
iiE_{37} *Flux* iiE_{72}
iiE_{38} *Flux* $iiiE_{38}$
iiE_{39} *Flux* iiE_{40}
iiE_{39} *Flux* iiE_{72}
iiE_{40} *Flux* $iiiE_{40}$
iiE_{41} *Flux* iiE_{73}
iiE_{42} *Flux* iiE_{43}
iiE_{42} *Flux* iiE_{72}
iiE_{43} *Flux* $iiiE_{43}$
iiE_{44} *Flux* iiE_{73}
iiE_{45} *Flux* iiE_{46}
iiE_{45} *Flux* iiE_{72}
iiE_{46} *Flux* $iiiE_{46}$
iiE_{47} *Flux* iiE_{48}
iiE_{48} *Flux* $iiiE_{48}$
iiE_{49} *Flux* iiE_{50}
iiE_{50} *Flux* $iiiE_{50}$

iiE_{51} *Flux* iiE_{52}

iiE_{52} *Flux* $iiiE_{52}$

iiE_{53} *Flux* iiE_{54}

iiE_{53} *Flux* iiE_{72}

iiE_{54} *Flux* $iiiE_{54}$

iiE_{55} *Flux* iiE_{56}

iiE_{56} *Flux* $iiiE_{56}$

iiE_{57} *Flux* iiE_{58}

iiE_{57} *Flux* iiE_{72}

iiE_{58} *Flux* $iiiE_{58}$

iiE_{59} *Flux* iiE_{60}

iiE_{60} *Flux* $iiiE_{60}$

iiE_{61} *Flux* iiE_{62}

iiE_{61} *Flux* iiE_{72}

iiE_{62} *Flux* $iiiE_{62}$

iiE_{63} *Flux* iiE_{64}

iiE_{63} *Flux* iiE_{72}

iiE_{64} *Flux* $iiiE_{64}$

iiE_{65} *Flux* iiE_{66}

iiE_{65} *Flux* iiE_{72}

iiE_{66} *Flux* $iiiE_{66}$

iiE_{67} *Flux* iiE_{68}

iiE_{67} *Flux* iiE_{72}

iiE_{68} *Flux* $iiiE_{68}$

iiE_{69} *Flux* iiE_{70}

iiE_{69} *Flux* iiE_{72}

iiE_{70} *Flux* $iiiE_{70}$

iiE_{71}, iiE_{73} *Influx* iiE_{85}

iiE_{71}, iiE_{75} *Influx* iiE_{87}

iiE_{71}, iiE_{77} *Influx* iiE_{89}

iiE_{71}, iiE_{79} *Influx* iiE_{95}

iiE_{71}, iiE_{81} *Influx* iiE_{93}

iiE_{71}, iiE_{83} *Influx* iiE_{91}

iiE_{73} *Flux* iiE_{74}

iiE_{74} $Flux$ $iiiE_{74}$
iiE_{75} $Flux$ iiE_{76}
iiE_{76} $Flux$ $iiiE_{76}$
iiE_{77} $Flux$ iiE_{78}
iiE_{78} $Flux$ $iiiE_{78}$
iiE_{79} $Flux$ iiE_{80}
iiE_{80} $Flux$ $iiiE_{80}$
iiE_{81} $Flux$ iiE_{82}
iiE_{82} $Flux$ $iiiE_{82}$
iiE_{83} $Flux$ iiE_{84}
iiE_{84} $Flux$ $iiiE_{84}$
iiE_{85} $Flux$ iiE_{86}
iiE_{86} $Flux$ $iiiE_{86}$
iiE_{87} $Flux$ iiE_{88}
iiE_{88} $Flux$ $iiiE_{88}$
iiE_{89} $Flux$ iiE_{90}
iiE_{90} $Flux$ $iiiE_{90}$
iiE_{91} $Flux$ iiE_{92}
iiE_{92} $Flux$ $iiiE_{92}$
iiE_{93} $Flux$ iiE_{94}
iiE_{94} $Flux$ $iiiE_{94}$
iiE_{95} $Flux$ iiE_{96}
iiE_{96} $Flux$ $iiiE_{96}$

Chapter 7
Formation of Faden Quartz Druses in Mid-Carboniferous Sandstones of the Donetsk Basin

Oleg Krisak
Donetsk National Technical University, Donetsk People's Republic, formerly Ukraine

Vyacheslav Bezrukov
ITMO University, Russia

ABSTRACT

The chapter addresses the mechanism of growth of druses of faden quartz in a tectonically deformed sandstone. A peculiar feature of this type of quartz, the so-called "white tape" represented by fine sub-parallel cracks with fluid inclusions, appears highly informative about the genesis of the mineral and tectonic regime of its growth. Two stages of formation of druses of faden quartz are recognized. The suggested mechanism is checked for contradictions by means of the event bush method. The proposed event bush model appears to describe a wider range of quartz formation environments and therefore may serve as a conceptual framework for various models of quartz growth in sedimentary rocks.

INTRODUCTION

Quartz is one of the most common minerals in the earth's crust. It is abundantly present in sedimentary rocks and occurs, inter alia, in vein mineral assemblages. Its perfectly shaped aggregates in druses of rock crystal, smoky quartz, amethyst and other coloured quartz varieties are desirable specimens for any mineralogical museum. Still, the most valuable at present are considered faden quartz crystals. Quartz with "white tape", or faden quartz, is a quartz crystal with a through white nebulous tape.

The formation of faden quartz has long been debated by mineralogists. Recent studies have brought us close to a solution that seems to stand the test of time. Nonetheless, as is common in geology, this idea, supposedly general in application, was informed mainly by the scientists' intuition, and now we are

DOI: 10.4018/978-1-5225-5261-1.ch007

curious about how complete and self-consistent it is. Also, what would be of interest to us in the near future is to what extent previous ideas are compliant with it. Now a tool for accomplishing these tasks has appeared, the event bush method.

In this chapter, we will give an outline of evolution of ideas about formation of faden quartz, report our results that have brought us to our own solution, briefly describe the event bush method, and then perform and discuss the formalization of this idea by means of the event bush.

EVOLUTION OF IDEAS

The first proper version of faden quartz origin was suggested by Lemmlein (1946). Based on the research results of quartz fields in Polar Urals, he described the formation of quartz with "white tape" in Alpine-type veins of quartz-sericite-chlorite schists. Lemmlein found that the faden quartz formation was somehow related to opening of crack, first slow, then faster. When a growing crack met growing crystal of quartz, it was tearing this crystal in two. Later the break was sealed by pore solutions from both sides. If breakage and sealing repeat, *faden* forms, which is a tape of mineral substance settled from pore solutions. Since the sealing of each break took part mainly in one direction, faden acquired a flattened shape that the growing crystals inherited Lemmlein (1973). Later Rykart, a specialist (1977) Richards (1990) developed a mechanism of discrete formation of faden quartz.

PROPOSED MODEL

A few years ago Krisak and Kupenko (2014) found single crystals of faden quartz in druses in a low-amplitude fault in the Donetsk Basin, Lugansk People's Republic (LPR, formerly part of Ukraine). No earlier evidence, published or unpublished, of occurrence of faden quartz in this basin is known to us. Krisak and Kupenko suggested that the formation of crystals was related to plastic and brittle deformations in the rock.

In 2015 we encountered druses orthogonal to the sedimentary layers, which were completely filled with faden quartz, in the south-western suburb of Zorinsk (Perevalsky District, LPR).

The druses occurred in a mid-Carboniferous sandstone and structurally were confined by the crest of an asymmetric, 3000 m long, 1800 m wide and 300 m high anticline fold with a flat-lying northern limb (10° – 15°) and steep southern limb (50° – 70°). The southern limb was complicated by an upthrust. Fold axis dips 15° westward.

Sandstone containing the faden quartz is medium-grained, with its own quartz grains and aggregates ranging from 0.4 to 1 mm across. All grains in the sandstone are dominantly round- or oval-shaped. Veins that host the druses have thickness from 5 to 30 mm with little variation along strike, the contacts are sharp and clear. Sometimes veins are coated in hematite. Faden quartz in the druses is represented by clusters of crystals orthogonal to the vein contact (1). Crystals are 4 to 15 mm, rarely up to 30 mm long, have flattened or polygonal form, with a predominant development of the two prism faces {1010} and weak development of rhombohedron faces {1011}.

Under the microscope, in transmitted light the white tape appears to consist of thin parallel alternating cracks, which can be traced by chains of single-phase or rarer two-phase fluid inclusions (2).

Figure 1. Faden quartz in druses: (a) photograph; (b) graphic sketch with faden shown in gray.

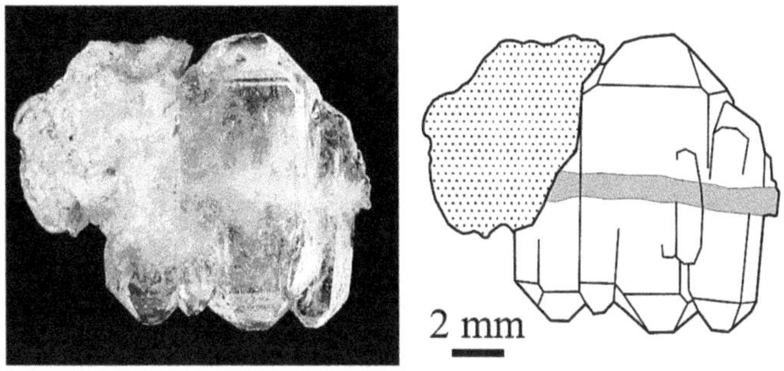

According to Lemmlein's hypothesis and from our own data, we identify two stages of druse formation. The first step is opening of the crack accompanied by displacement of one contact relative to the other. Thus the crack breaks a quartz aggregate enclosed in a sandstone. Hot solutions penetrating into the crack tend to fill the gap between the parts of the aggregate with a new mineral substance. Herewith, small drops of solution may be trapped by solidifying "bridge" (3a). Then the crack may reopen and new portion of solution would seal it (3b). The crack-sealing cycle repeats as long as the crack is widening. As a result, a tape of sealed cracks with trapped fluid inclusions forms, which is used to be termed as faden (3c).

Nearly equal distance between parallel cracks in the faden quartz points at eventual opening of the crack, possible fluctuations not exceeding hundredths and tenths millimetres.

The second stage occurs when the crack opening stops. The bridge that seals the crack is not broken anymore, and faden is the seed that begins to overgrow with crystal bodies (3d). This overgrowth occurs in the same direction, so that they have flattened shape. The parallel microcracks with inclusions are preserved in the crystal in the form of "white tape".

EVENT BUSH FORMALIZATION OF THE MODEL

The event bush method was applied to formalize the model and check it, at intuitive level, for completeness and absence of internal contradictions. Of course, a check using Novorossian inference (Pshenichny, 2017b, Chapter Two of this book) would be strict and formal, but for current purposes of our research it is not necessary, for at present we do not intend to perform mathematical or physical modeling of

Figure 2. The structure of "white tape" in faden quartz under a microscope

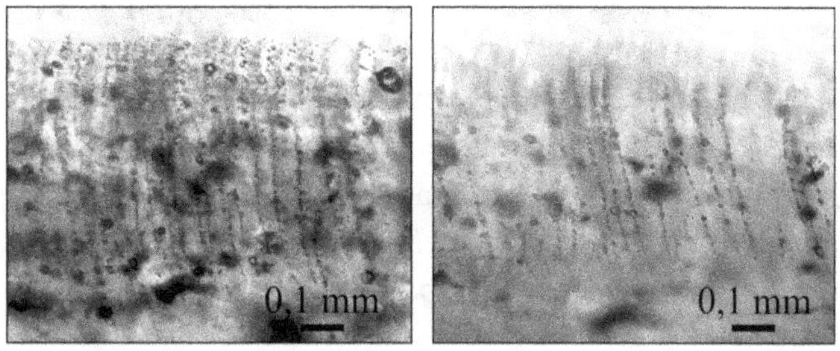

Figure 3. Stages of formation of faden quartz druses. See explanation in the text.

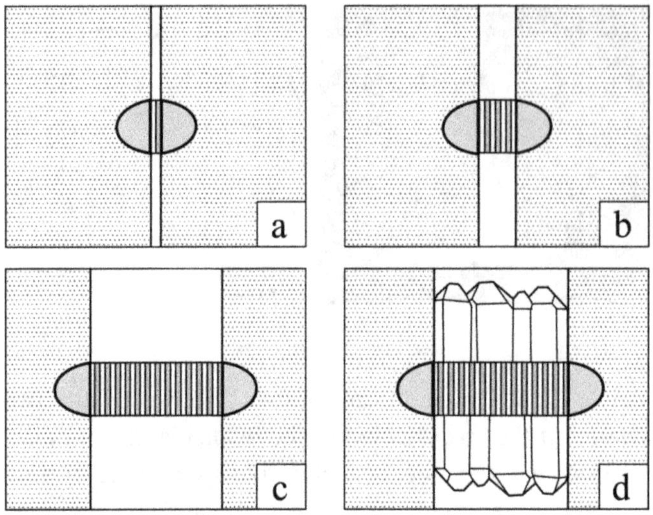

mineral growth or create an extended knowledge base. Rather, we are going just to control ourselves, make sure we have not omitted any relevant details of the faden quartz growth environment, and correct our model if necessary. For this, simple event bush seems to be quite enough.

The event bush is entitled as "What happens to quartz aggregates in a rock of initially sedimentary origin... When it is broken by cracks, in which alpine-type hydrothermal solutions circulate" (see 4). Its composition and formulation of events in the nodes are in line with the rules put forth by Pshenichny et al. (2008; 2009) and Pshenichny and Kanzheleva (2011).

The following events were taken as primary:

Internal: Quartz aggregates grow in rock, are unaffected by cracks, do not respond to cracking, are not deformed: with or without displacement, are not affected by hydrothermal solutions of the Alpine type, nothing grows on them: neither sinters, nor single crystals, nor bridges without single crystals, nor bridges with single crystals;

Externals:

- Cracks open in the rock and
- Hydrothermal solutions of the Alpine type circulate in the rock.

Mind that the latter event makes sense only if the former takes place: in this model, solutions are assumed to circulate only along cracks, not via the interconnected pores even if they are present in the rock. Cracks must be "fresh", newly opening, because older ones would never tear newly growing mineral aggregates apart.

Initially in the event bush model there are three predmets (or, to put it more precisely, predmet events) sensu Pshenichny (2017a; Chapter 1 of this book): quartz aggregates, cracks and hydrothermal solutions. One more predmet, "Quartz bridges", is added in a secondary event to describe the appearance of the "white tape", and one more, "Quartz crystals", the faden quartz proper. Though, of course, much more predmets are implied ("sinter", "breakage" and so forth), and they would appear in predmetka once identified (Pshenichny, 2017a), but the event bush proper directly operates only with primetas formed by these predmets.

By "aggregates" we mean quartz growths potentially of any size and structure that originally were in the rock. A colon followed by a list of options (e.g., "are not deformed: with or without displacement") means subsequent ramification of scenarios by primetas specified after the colon: to be not deformed with displacement and to be not deformed without displacement.

As supposed by the event bush methodology, the model takes the issue in question for one possible scenario and puts it into a wider context that is postulated to include the full group of analogous scenarios starting with the "nothing happens" one (i.e., cracks do not open, solutions do not come). This is clearly seen by the list of tertiary events at the right of 4. Obviously, all of the scenarios are incompatible, but some scenarios (those that include a secondary event with more than one arrow coming out of it) imply more than one tertiary outcome.

After the "nothing happens", there comes the scenario that may be informally defined as "crack only", and if hydrothermal activity is involved, "no breakage" scenario. Then, if breakage does happen but hydrothermal solutions do not penetrate, the "breakage only" scenario is introduced, and if they do, another group of scenarios referred to as "breakage and solution" becomes relevant. Among these, the "crack opens quickly" scenario describes the case of partial sealing of the break by hydrothermal solution, which still is not fast enough to seal the break, and two scenarios of the "crack opens slowly" group relate to the phenomenon in question. Importantly, these scenarios end up not in one but in two tertiary events, one (the same for both scenarios), describing what happens to the broken aggregate, and the other, what happens between its displaced parts. In one case, if the hydrothermal solution stops, this is just formation of a siliceous "bridge" with parallel micro-cracks and trapped drops of solution; in the other, given the solution is virtually unlimited (in terms of the event bush, this means that another input of the corresponding primary external event takes place), the growth of new crystals of quartz begins from this bridge, which represents nothing else but the "white tape", and the new crystals therefore are exactly the faden quartz.

The collection of scenarios presented by the event bush is finite and full. Obviously, one cannot think of any other scenario not included in this model. However, this semantic completeness is achieved at the expense of consideration of details; otherwise this event bush would be much larger and harder to read and to analyze.

DISCUSSION

The model presented above gives a general description of faden quartz formation. It does not go into such detail as size and internal structure of broken quartz aggregate, speed and momentum of crack opening, dynamics and chemistry of hydrothermal solution, nor does it specify the relations between the solution and solidifying "bridge" and mechanism of entrainment of drops of fluid along micro-cracks. Exactly these issues, and especially the quantitative relationships arising herewith, are crucial for understanding of mechanics of formation of faden quartz. Nonetheless, the suggested model (i) outlines the general framework of this phenomenon and sets its relations with similar phenomena in a strict yet intuitively clear "what-if"/"what-if-not" way, and (ii) defines a qualitative but formalized framework to account for all said issues. There has been experience in putting physical and chemical models in the event bush framework, or, following the authors' terminology, parsing these models by the event bush (Carniel et

Figure 4. The event bush model of formation of faden quartz and other options of quartz aggregate deformation in sedimentary rock given the growing cracks and hydrothermal solutions

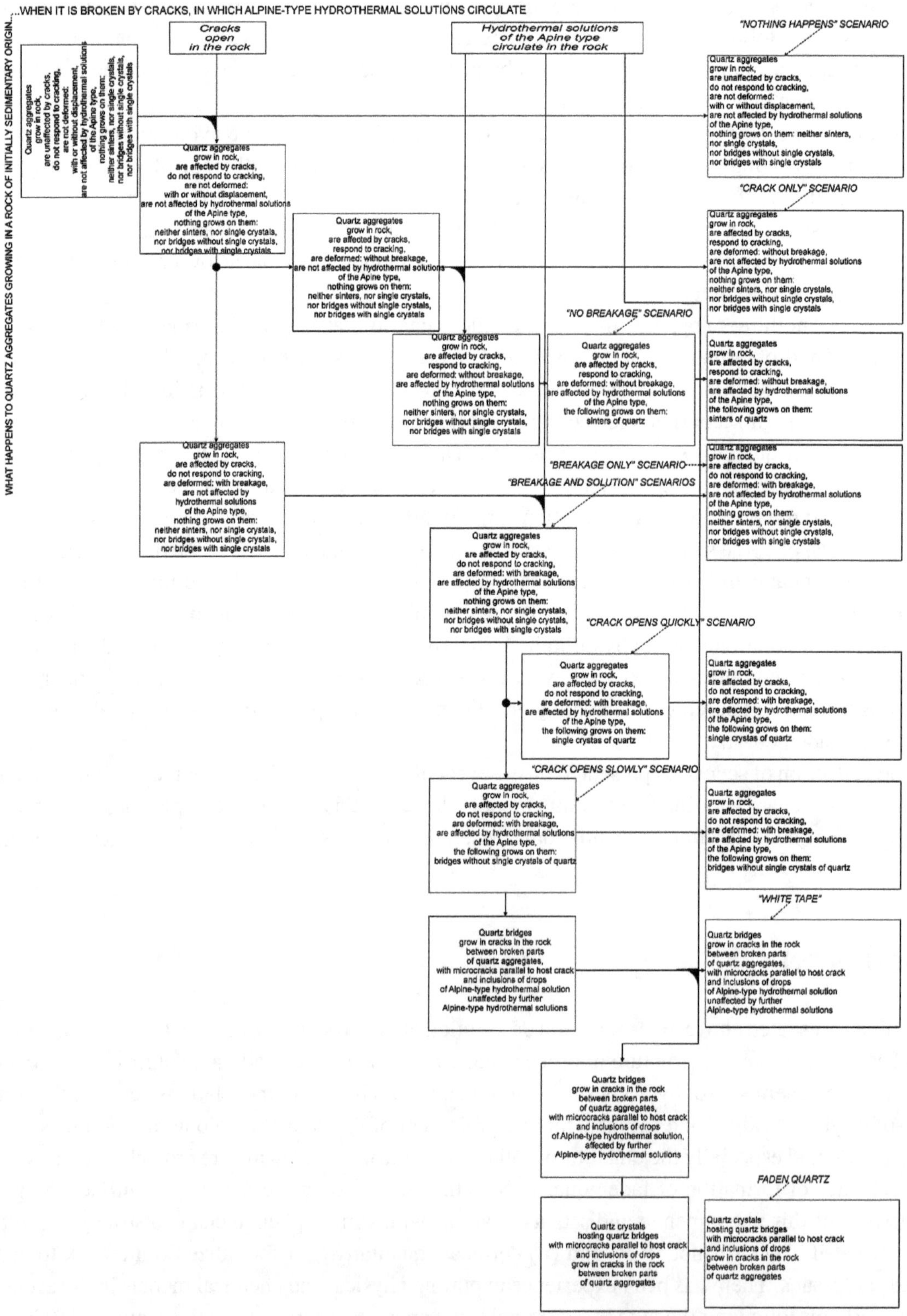

al., 2011; Pshenichny, 2013; Diviacco et al., 2014). Importantly, in all these cases the parsed models related not to one or several particular scenarios but to the entire bush, and if it was needed to narrow the scope, special assumptions would need be considered.

Furthermore, Pshenichny and Kanzheleva (2011) state that an integer event bush corresponds to some more or less "rounded", relatively self-sufficient domain of reality they term environment (which, of course, is always a part of a larger environment, and likewise, any event bush can be thought of as nested within a larger one or chained in a multi-bush structure, as shown by Wolter et al., 2017 – see Chapter 13 of this book).

In our case, an environment would be the sandstone with quartz aggregates in it. In this environment, cracks may develop or not, and if yes, hydrothermal solutions may circulate. Hence, we believe that in the problem of quartz growth in sedimentary rock one should not look for physico-chemical models describing particular phenomena; rather, the aim should be to cover the entire environment – or, at least, to explicitly delineate in the event bush what part of the environment is going to be covered by a model or a family of models.

CONCLUSION

The field work and microscopic studies of faden quartz in druses from a sandstone in Donetsk Basin allow us to conclude that

- Studied veins are confined only to the crest of an anticline fold in the sandstone;
- The "white tape" in faden quartz consists of thin subparallel cracks made visible by chains of fluid inclusions from the hydrothermal solution that had formed this quartz;
- Druses of faden quartz formed in two stages: first, opening cracks, breakage of quartz aggregates and sealing the break by hydrothermal solution under repeated small dislocations along the crack (the result is the "white tape" bridge, and then, growth of quartz crystals filling the gap between the broken parts and inheriting the white tape;
- Faden quartz crystals can be formed not only in metamorphosed shales, but in sedimentary rocks as well, at least in sandstone.

The event bush modelling of faden quartz formation in a crack in sandstone has proven that the model of faden quartz formation as a result of repeated cycle of breakage and sealing followed by crystal growth is complete and has no internal contradictions, at least at semi-intuitive level. The event bush defines a framework for the whole diversity of similar phenomena of quartz growth in sedimentary rock given the cracking and, possibly, circulation of hydrothermal solutions of the Alpine type. Physico-chemical modeling of these phenomena should keep the whole framework in mind. The experience of event bush modeling of faden quartz formation in sandstone could be then extended onto other domains of mineralogy of sedimentary rocks.

REFERENCES

Krisak, O.S., & Kupenko, V.I. (2014). Usloviya formirovania kvartza s beloi polosoi Seleznyovskogo uglenosnogo raiona Donbassa [Conditions of formation of quartz with white tape in Seleznyovsky Coal Basin, Donbass]. In *Problemi Teoretichnoi I Prikladnoi Mineralogii, Geologii, Metallogenii Girnichnodobuval'nikh Regioniv*. Vid. Krivoriz'kogo Nazional'nogo Universitetu, Krivoi Rog. (in Russian)

Lemmlein, G. G. (1946). *O proiskhozhdenii ploskikh kvartzev s beloi polosoi [On the origin of flat quartz with white tape]. Voprosy Mineralogii, Geohimii i Petrografii*. Moscow: Academy of Sciences of the USSR. (in Russian)

Lemmlein, G. G. (1973). *Morfologia i Genesis Kristallov [Morphology and Genesis of Crystals]*. Moscow: Nauka.

Pshenichny, C. A., & Kanzheleva, O. M. (2011). Theoretical foundations of the event bush method. Societal Challenges and Geoinformatics, 482, 139-165. doi:10.1130/2011.2482(12)

Pshenichny, C. A., Nikolenko, S. I., Carniel, R., Sobissevitch, A. L., Vaganov, P. A., Khrabrykh, Z. V., . . . Gusev, E. A. (2008). The event bush as a potential complex methodology of conceptual modelling in the geosciences. *Proceedings, iEMSs – International Congress on Environmental Modelling and Software*, *2*, 900-912.

Pshenichny, C. A., Nikolenko, S. I., Carniel, R., Vaganov, P. A., Khrabrykh, Z. V., Moukhachov, V. P., & Rezyapkin, A. A. et al. (2009). The event bush as a semantic-based numerical approach to natural hazard assessment (exemplified by volcanology). *Computers & Geosciences*, *35*(5), 1017–1034. doi:10.1016/j.cageo.2008.01.009

Richards, R. P. (1990). The origin of faden quartz. The Mineralogical Records, 21.

Rykart, R. (1977). Zum Wachstum plattiger Quarze mit "Faden". Schweizer Strahler, 4.

Chapter 8
An Approach to Improve Bayesian Inference in Assessment of Coast Evolution

Victoria Shterkhun
Karpinsky All-Russian Research Geological Institute (VSEGEI), Russia

Gonéri Le Cozannet
BRGM – French Geological Survey, France

Manuel Garcin
BRGM –French Geological Survey, France

ABSTRACT

The chapter reports the state of the art in the modeling of coastal environments by the method of event bush. This method appears to be an efficient alternative to highly subjective conceptualizations used in the databases that provide data for probabilistic assessment of evolution of the coast. The event bushes themselves may be used for Bayesian computation, but there emerge complications that pose intriguing theoretical tasks. Still, their application requires deep conceptual rethinking of the field of knowledge including both the terminology and concepts traditionally accepted in it.

INTRODUCTION AND MOTIVATION

Shoreline evolution in the coastal zone has received much attention due to the increasing presence of numerous assets of social and economic nature in these areas and because of uncertainties about the actual consequences of future sea level rise. Many factors are involved in causing shoreline changes in the coastal zone: waves, currents, winds, sea level changes. All these may mobilize available sediments leading to morphological changes in the coastal zone. Because of the complex interactions between all factors involved in shoreline mobility at various time and space scales, finding out the relative importance of each process in causing sediment and shoreline mobility is still a challenge.

DOI: 10.4018/978-1-5225-5261-1.ch008

This issue can be approached quantitatively – e.g., by developing indices that aim to aggregate factors such as sea level rise, coastal elevation, historical shoreline movement, geology, geomorphology, wave height, tide range, beach slope, vegetation, and coastal defenses (e.g. Gornitz et al.,1994; Shaw et al., 1998; Coelho et al., 2006). However, evaluation of the relative importance of each factor is still a challenge. To meet it, the Bayesian approach was applied in the coastal studies (Hapke and Plant, 2010; Gutierrez et al.,2011; Plant and Holland, 2011).

In particular, Gutierrez et al. (2011) proposed investigating this issue through the exploration of large coastal databases. They modeled the relations between shoreline mobility and the factors causing it in a Bayesian Belief Network (BBN; Jensen, 2001). In their study, sea level was found the most important cause for shoreline evolution at the eastern coast of the USA. A similar study was undertaken in Europe (Yates and Le Cozannet, 2012), using the Eurosion dataset (Eurosion, 2004). In that study, the most important parameter to explain the shoreline migration was coastal geomorphology. Le Cozannet et al. (2012) investigated the causes of this discrepancy and found that the values of relative importance of the indices largely depend on the conceptualization (data model) accepted in a dataset used to compute the indices.

The problem was aggravated by the fact that these coastal databases had been developed for other purposes. Thus, in Europe, the Eurosion (2004) dataset was collected throughout European Union member state institutions in charge of shoreline surveys in order to provide authorities with homogeneous statistics and GIS data on the coastal zone of Europe. The data model of the GIS dataset was the same as in the previous 'Corine Coastal Erosion' dataset, which was released in 1990. Incidentally, this database became the only European scale-complete dataset available for applying the data exploration method developed by Gutierrez et al. (2011) in Europe.

However, to improve the situation, one should abstract from indices and look at the relationships between real "actors" of the coastal environments – sea, coastal sediments and rocks, coastal landforms, winds, and others. The data model should desirably be a conceptual, or qualitative, model of a natural system (e.g., coast). Then, if at least basic relationships in this system are understood in a more or less uniform way by different scientists building the databases, the data models they use would not be contradicting. Nonetheless, in regard of the coast, this reasonable point immediately brings another hindrance that has got no methodological solution so far, to the authors' knowledge. Existing methods of building data models, essentially the ER (entity-relationship) diagrams of Pin and Chen (1971), express the relationships between static objects, and so does the whole database theory (Codd, 1970). Description of processes can be made only indirectly and implicitly, via an ensemble of properties related to each other and to time, and inevitably leads to ambiguity. Still, coastal systems are highly dynamic, and this urges us to look for another tool to formally and qualitatively express the processes operating in the coastal zone.

This task seems to fall in the scope of a method of knowledge engineering proposed specially to describe changing geoenvironments, the event bush (Pshenichny and Kanzheleva, 2011).

Behncke and Pshenichny (2009) suggested that the event bush could serve as an intermediary method between expert elicitation and event description based on Bayesian networks. Here, we present the first application of the event bush formalism for modeling the coast behavior. Using this preliminary result, we question if the event bush can help improving data model of coastal databases.

PREVIOUS WORK

The EUROSION data were semantically preprocessed by Yates and Le Cozannet (2012) for use in the Bayesian network. The geomorphology was divided into four classes: rocky cliffs and platforms, erodible (e.g. chalk) cliffs, beaches, and wetlands; the geology was divided simplistically into two categories: hard and soft sediments (depending on the erosion potential). The shoreline evolution data were classified as eroding, stable, or accreting.

Based on this, a BBN was created. Its architecture is shown in Fig. 1.

As seen from this figure, the model is built at the highest admittable level of generality. Still, other coastal vulnerability studies use nearly the same set of variables (Thieler & Hammar-Klose, 1999; Coelho et al., 2006; Gutierrez et al., 2011). Also, quite similar set of variables and functions was used by (Karunarathna, & Reeve, 2007; 2008) in their modeling of estuarine environments by means of Boolean algebra.

This model, demonstrating a moderate (>65%) agreement with the EUROSION data, shows that geomorphology was the main forcing factor in determining the coastal evolution. It looks not surprising at all judging by the very architecture of the network (Fig. 1). Geomorphology is partly a response variable, to which all other forcing factors contribute (relative sea-level rise rate, mean significant wave height, mean tidal range and geology). Hence, it acquires importance from each of them. However, the nature and character of links between the factors is not explicated. In a similar study undertaken at the Atlantic coast of the USA, sea level rise was considered the most important. Still, with deep semantic rethinking of the subject area one may never be confident about the numbers returned by the models including the relevant relevance of parameters. Such semantic rethinking, to our knowledge, has never been undertaken.

SEMANTIC ANALYSIS OF THE BBN VARIABLES

The variables of the BBN proposed by Yates and Le Cozannet (2012) were reviewed semantically for the purpose of qualitative explication of cause-effect relations expressed by arrows of the BBN. The results significantly differ from the semantics of the initial BBN.

Figure 1. Bayesian belief network: the black arrows indicate causal relationships linking the forcing factors (gray) and the response variable (coastal evolution, black) – modified from Yates and Le Cozannet (2012).

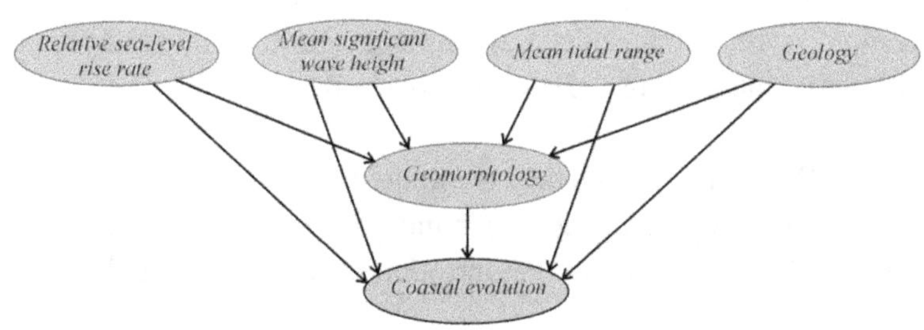

"Coastal Evolution". First, we consider "coast" the zone below the strip of the seafloor penetrated by average waves at low tide and the strip of the land reached by the average surf at high tide.

Evolution is assumed to be of four possible types. One, often called erosion in the literature, should be termed more correctly as abrasion, in our view. Another is accretion. There is yet another possible outcome, inundation, when the shoreline just retreats and the former coast is covered by water without notable abrasion. Finally, none of these may happen. Thus, "Coastal Evolution" as a variable in BBN is attributed four states: (i) abrasion, (ii) accretion, (iii) inundation, (iv) none. Note that accretion may be achieved in two different ways, by accumulation of sediments or by bare retreat of sea.

"Geology" was kept as is, including "hard" and "soft" sediments. However, the term "sediment" was replaced by a wider analog "material", because abrasion implies washout of clasts offshore, and these clasts unlikely should be termed "sediment". Any clasts, both of soft and hard material, are soft material themselves.

"Geomorphology" has been perhaps the most heterogeneous and least understood variable in the BBN. Four types of coastal landforms recognized as states of this variable included: rocky cliffs and platforms, abradable (e.g., chalk) cliffs, beaches and wetlands. Nevertheless, physiographically, these types far from describe the diversity of coastal relief reported in cornerstone works on geomorphology – compare, e.g., with the ria, Dalmatian, fjord, lagoon, scherm, Aral, firth, marsh, atoll and buried ice coast types reported by Schukin (1960; vol. 3).

The same time, these four states of the "Geomorphology" variable appear, at a closer look, totally dependant on the two types of material identified as the states of "Geology". "Rocky cliffs and platforms" are the same as the coast made of hard material, and the rest represent three (supposedly not all) different kinds of coast made of soft material. Furthermore, the distinction between the three, though obviously great by itself, unlikely is relevant in concern of the object of this study, coastal evolution. Indeed, once the material in the coastal zone is soft, it will be removed anyway, be it at the base of a cliff or at the edge of a beach. Hence, the diversity of near-shore geomorphological processes operating in different types of coast creates or maintains the diversity of these environments but the impact of all of them on the shoreline behavior is determined only by the mechanical properties of material, hard vs. soft, in the narrow coastal zone. Therefore, we decided to remove the variable "Geomorphology" at all.

Conversely, the variable "Mean significant wave height", in our opinion, unites and smoothes the results of quite diverse processes having strikingly different impact on the coast, the surf and the storms. The states of both could be provisionally put as "yes/no" (occur/do not occur), but with date the former state may be further specified.

"Relative sea level rise rate" should be treated wider, just as "Sea level", which includes three states, "Sea level rises", "Sea level drops" and none.

As for "Mean tidal range", the relation between tides and the coast abrasion/accretion/inundation should be better understood qualitatively. Therefore, we exclude this issue from the BBN and leave it for future studies.

In addition, we deem necessary to introduce two concepts more, "Along-shore currents" and "Flows of clasts of material", their states being defined exactly in the same way as in the "Mean significant wave height".

It is noteworthy that BBN variables listed here are related to vastly different time scales. For example, sea level changes represent long-term trends taking hundreds of years, whereas storms potentially cause more erosion in few hours. Meanwhile, attempts to introduce time in BBN usually result just in a multitude of copies of the basic network (Pshenichny et al., 2009). For this, as was stated by these researchers, the event bush is potentially better option.

PROCESSING OF MEANINGS

Having specified the meanings, one should set the relations between them. The methodology of formalization of meanings and setting the relations between them was described by Pshenichny (2017; see Chapter 2 of this book). In theory, this should be done by the predmetka, which represents a chart disclosing the relations between meanings in some finite context. To obtain it, an intermediary in a form of incomplete event bush is very helpful (operating with incomplete records of events that are allowed to have "…" instead of some meanings), and then the events in complete form should be returned by the Novorossian inference – special procedure of transforming the incomplete into complete by a set of fixed rules (Pshenichny, 2017). From this perfectly described context, a few mature event bushes may be obtained, each having only completely formulated events. The event bush or another tool of representation of dynamic knowledge is necessary to explicitly show the successions of changes, or *scenarios*, in the environment. These cannot be explicitly shown in the predmetka.

However, if the meanings are not numerous and their interrelations are quite transparent, their complete form is rather obvious, and one may proceed right to completely formulated event bush(es) from the beginning.

Of the events specified above, two, "Sea level rises" and "Sea level drops", represent long-term trends, which, in fact, determine a coastal environment. This is their abrupt distinction from all others. Therefore, we consider not reasonable to include change of the sea level trend in our event bush, but rather, consider two independent event bushes with the same semantics, one, under the assumption of sea level rise, and the other, sea level drop. The "none" option would be automatically included in both.

Based on the theoretical knowledge, we made the following assumptions on the possibility of cause-effect relation:

- No accretion of coast given the sea level rise;
- No abrasion if there are neither surf nor storms;
- Only storm may probably abrade hard material.

We do not exclude that other assumptions are reasonable but at present we use these three only.

The results of the modeling are represented in Fig. 2. Full formulations of all events is given in Table 1. Tertiary events are not listed because they are formulated in the same way as the primary internal or secondary events of the same index they were produced by. By the syntactic rules set out by Pshenichny (2017), neither the formulation of individual event (in the table), nor their ornament (in the plot) may not be simplified. This makes both hardly readable. Diviacco (2017; see Chapter 4 of this book) has given an exhausting account of this problem of conflict of formalization and representation. Still, longing for formalization, we sacrifice intuitive clarity.

Table 1. Formulations of events in Figure 2

Event	Description
iaE_1	Coasts made of soft material and not of hard material are not affected by sea level rise, are not affected by surf, are not affected by storms, are not abraded, are not accreted, are not inundated
iaE_2	Coasts made of hard material and not of soft material are not affected by sea level rise, are not affected by surf, are not affected by storms, are not abraded, are not accreted, are not inundated
ibE_3	Sea level rises
ibE_4	Surf is on near the shore
ibE_5	Storms occur near the shore
ibE_6	Flows of clasts of material go along shore
ibE_7	Along-shore currents occur
iiE_8	Coasts made of soft material or of hard material are not affected by sea level rise, are affected by surf, are not affected by storms, are not abraded, are not accreted, are not inundated
iiE_9	Coasts made of soft material and not of hard material are not affected by sea level rise, are not affected by surf, are affected by storms, are not abraded, are not accreted, are not inundated
iiE_{10}	Coasts made of soft material and not of hard material are not affected by sea level rise, are not affected by surf, are affected by storms, are abraded, are not accreted, are not inundated
iiE_{11}	Coasts made of soft material and not of hard material are not affected by sea level rise, are not affected by surf, are affected by storms, are not abraded, are accreted, are not inundated
iiE_{12}	Clasts of soft or hard material occur in water, do not flow and do not settle down
iiE_{13}	Clasts of soft or hard material occur in water, do not flow and settle down
iiE_{14}	Clasts of soft or hard material occur in water, flow and do not settle down
iiE_{15}	Flow of clasts of material goes along shore
iiE_{16}	Coasts made of soft material and not of hard material are affected by sea level rise, are not affected by surf, are not affected by storms, are not abraded, are not accreted, are not inundated
iiE_{17}	Coasts made of soft material and not of hard material are affected by sea level rise, are not affected by surf, are not affected by storms, are not abraded, are not accreted, are inundated
iiE_{18}	Coasts made of soft material and not of hard material are affected by sea level rise, are affected by surf, are not affected by storms, are not abraded, are not accreted, are inundated
iiE_{19}	Coasts made of soft material and not of hard material are affected by sea level rise, are not affected by surf, are affected by storms, are not abraded, are not accreted, are inundated
iiE_{20}	Coasts made of soft material and not of hard material are affected by sea level rise, are affected by surf, are affected by storms, are abraded, are not accreted, are inundated
iiE_{21}	Coasts made of hard material and not of soft material are not affected by sea level rise, are not affected by surf, are affected by storms, are not abraded, are not accreted, are not inundated
iiE_{22}	Coasts made of hard material and not of soft material are not affected by sea level rise, are not affected by surf, are affected by storms, are abraded, are not accreted, are not inundated
iiE_{23}	Coasts made of soft material and of hard material are not affected by sea level change, are not affected by surf, are affected by storms, are not abraded, are accreted, are not inundated
iiE_{24}	Coasts made of hard material and not of soft material are affected by sea level rise, are not affected by surf, are not affected by storms, are not abraded, are not accreted, are not inundated
iiE_{25}	Coasts made of hard material and not of soft material are affected by sea level rise, are not affected by surf, are not affected by storms, are not abraded, are not accreted, are inundated
iiE_{26}	Coasts made of hard material and not of soft material are affected by sea level rise, are affected by surf, are not affected by storms, are not abraded, are not accreted, are inundated
iiE_{27}	Coasts made of hard material and not of soft material are affected by sea level rise, are not affected by surf, are affected by storms, are not abraded, are not accreted, are inundated

continued on following page

Table 1. Continued

Event	Description
iiE_{28}	Coasts made of hard material and not of soft material are affected by sea level rise, are not affected by surf, are affected by storms, are abraded, are not accreted, are inundated
ibE_{29}	Sea level drops
iiE_{30}	Coasts made of soft material and not of hard material are affected by sea level drop, are not affected by surf, are not affected by storms, are not abraded, are not accreted, are not inundated
iiE_{31}	Coasts made of soft material and not of hard material are affected by sea level drop, are not affected by surf, are not affected by storms, are not abraded, are accreted, are not inundated
iiE_{32}	Coasts made of hard material and not of soft material are not affected by sea level drop, are not affected by surf, are affected by storms, are not abraded, are not accreted, are not inundated
iiE_{33}	Coasts made of hard material and not of soft material are not affected by sea level drop, are not affected by surf, are affected by storms, are abraded, are not accreted, are not inundated
iiE_{34}	Coasts made of hard material and not of soft material are affected by sea level drop, are not affected by surf, are not affected by storms, are not abraded, are not accreted, are not inundated
iiE_{35}	Coasts made of hard material and not of soft material are affected by sea level drop, are not affected by surf, are not affected by storms, are not abraded, are accreted, are not inundated
iiE_{36}	Coasts made of hard material and not of soft material are affected by sea level drop, are affected by surf, are not affected by storms, are not abraded, are not accreted, are not inundated
iiE_{37}	Coasts made of hard material and not of soft material are affected by sea level drop, are not affected by surf, are affected by storms, are not abraded, are not accreted, are not inundated

DISCUSSION

Compared to the Bayesian belief network shown in Fig. 1, the pair of event bushes presented in Fig. 2 gives a display of scenarios of evolution of coastal zones much closer to the nature and, hence, much more impartial than do the data models used in the known databases. Nevertheless, the following questions arise immediately:

- Can these be used as a basis for probabilistic (including Bayesian) computation and
- How they can be related to the databases.

Answering the former question, the transformation of event bush into BBN was discussed by Pshenichny et al. (2009). They suggested to use the bush directly to perform probabilistic computation. For this, the connectives of the bush should be attributed probabilistic sense, *p* being the probability value:

- Flux Without Influx – Translation of probability value E_i *Flux* $E_j \Rightarrow p(E_i)=p(E_j)$;
- Flux With Influx – Multiplication of probabilities at influx: E_i, E_j *Flux* $E_k \Rightarrow p(E_k)=p(E_i)p(E_j)$ and deduction at the corresponding flux E_i *Flux* $E_m \Rightarrow p(E_m)=1-p(E_i)p(E_j)$;
- Conflux had not been suggested in the early version of the event bush but now it can be treated probabilistically in the same way as influx: E_1, E_2, ..., E_{n-1} *Conflux* $E_n \Rightarrow p(E_n)=p(E_1)p(E_2)...p(E_{n-1})$;
- Split (originally it was suggested for the furcation, which is now considered as fluxes in various event bushes and not a connective itself) – division of probability by *n-1*, *n-1* being the number of resulting events (and *n*, the total number of events involved in one split), under assumption that all

Figure 2a. The event bushes showing the evolution of coastal environment under the condition of no sea level rise. For formulation of events, see Table 1.

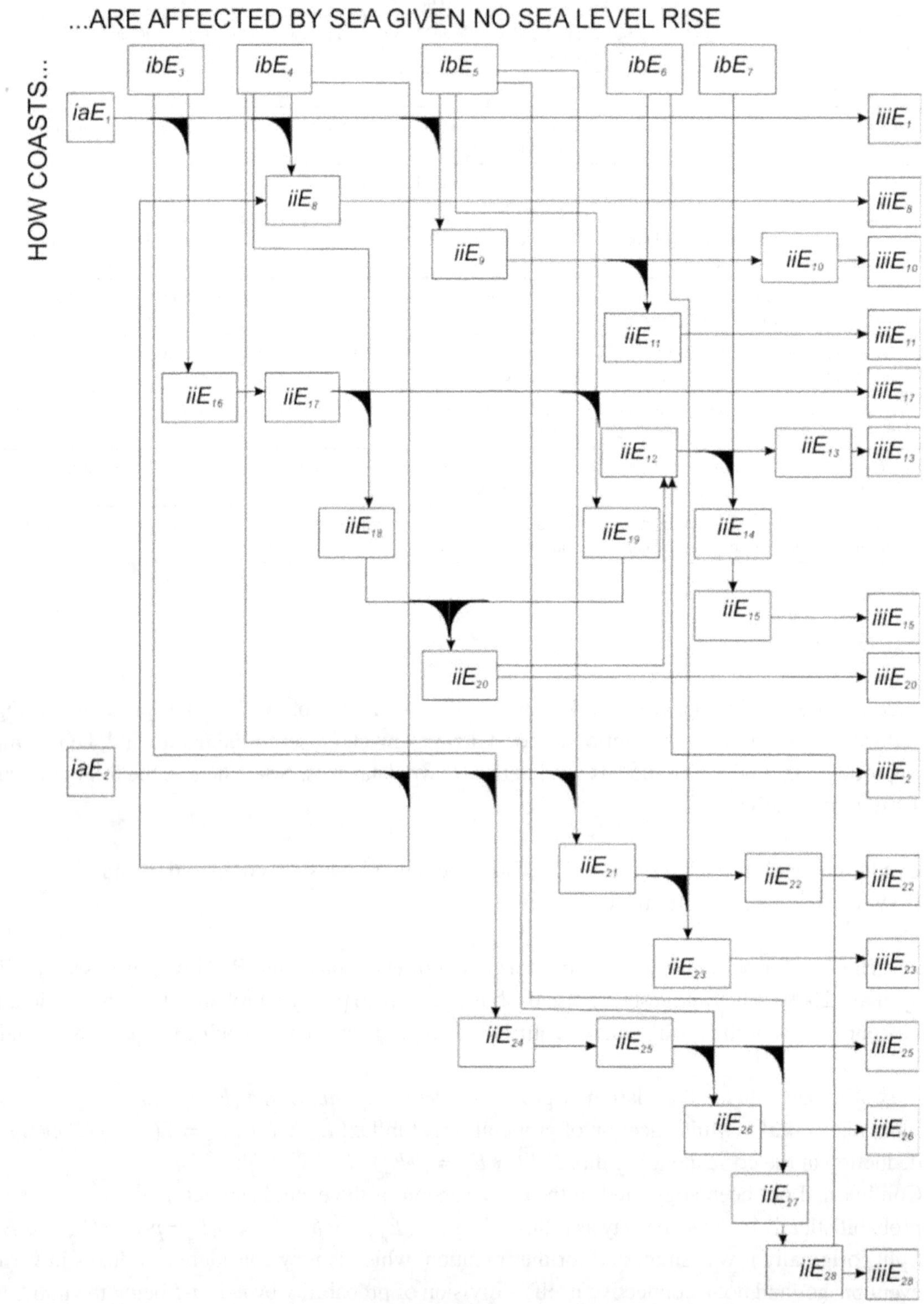

Figure 2b. The event bushes showing the evolution of coastal environment under the condition of no drop. For formulation of events, see Table 1.

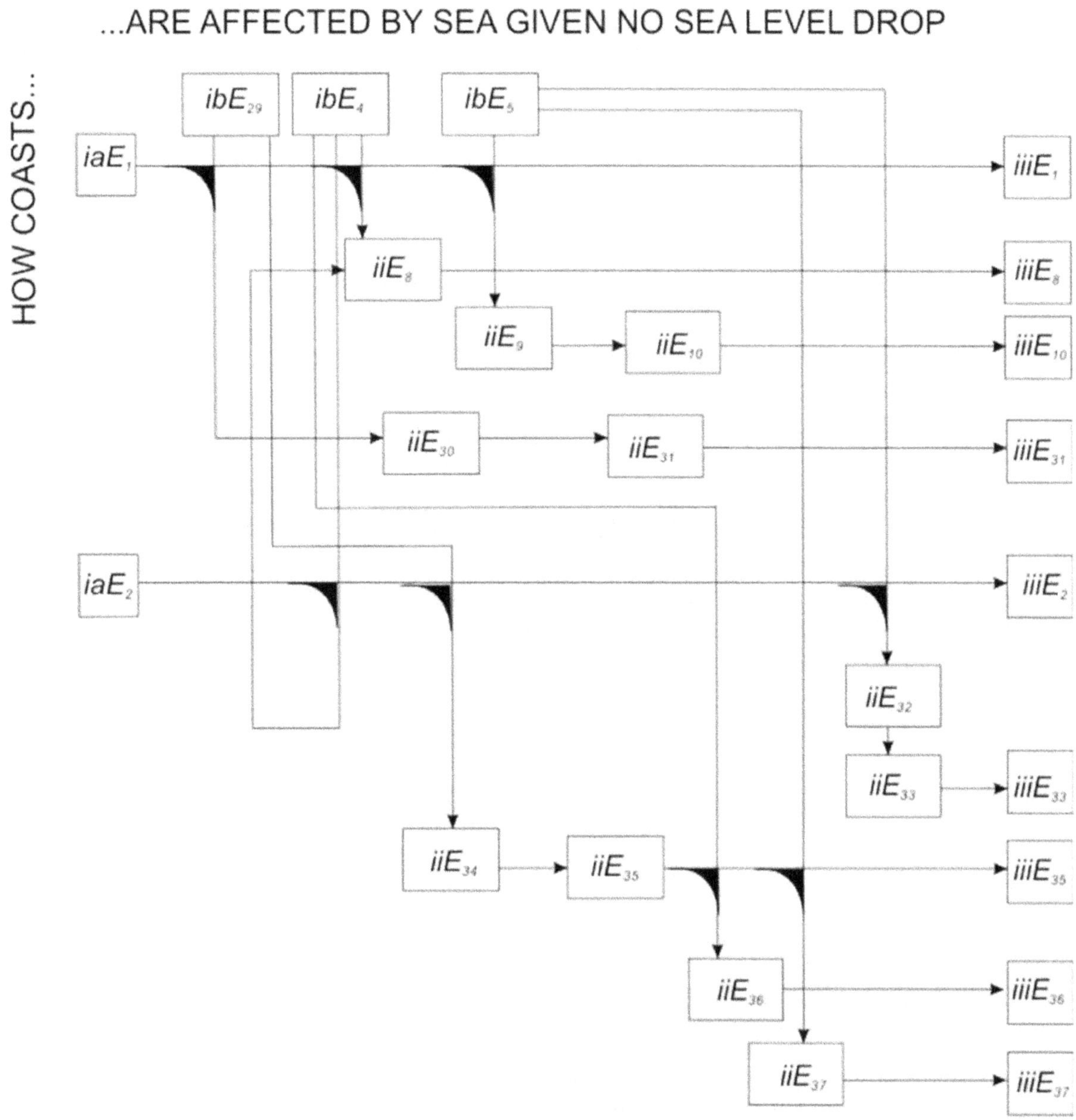

these are equiprobable (if not, several influxes should be used rather than split) – E_1 *Split* E_2, E_3, ..., $E_n \Rightarrow p(E_2)=p(E_3)=...=p(E_3)=p(E_1)/(n-1)$.

Such model was termed as probability bush by Pshenichny et al. (2009).

Alternatively, aiming to account for incompleteness of knowledge and to allow its update, Bayesian inference can be performed on the event bush(es). For this, they can be either treated separately (that would be our own choice) or merged in one BBN, taking the events for variables with two states: "yes" and "no", except the events "Sea level rises", "Sea level drops". For these, one more variable shall be introduced, "Sea level is doing something", the states of which would be exactly these two events and one more: "Sea level is stable".

Still, the semantic information preserved in the connectives of the event bush (e.g., influx versus conflux) would be lost, for BBN does not have means to express it. Unlikely more than two states, "yes" and "no", may be reasonably added to other variables because any transition between them is actually expressed by the number assigned to them, and any distinction that can be captured qualitatively should augment to the verbal formulation of a variable, while the kit of variable states remains the same: "yes" and "no". This is why, in our opinion, even performing the Bayesian formalization of the event bush, one should not just convert the bush into BBN but keep it as is, just adding Bayesian sense to its nodes and edges. Such transformed event bush may be called Bayesian bush, by analogy with the probability bush of Pshenichny et al. (2009).

In the Bayesian bush, conditional probabilities of consequences given the causes should be specified, and prior probabilities of at least some of these (some causes and/or some consequences, set). Causes and consequences are the states of variables. Linking the events by an arc in event bush we imply that the conditional probability of state "yes" of the causal event and state "no" of the resulting event is zero. For instance, in Fig. 2a, consider an event iiE_{12} "Clasts of soft or hard material occur in water, do not flow and do not settle down". Its state "yes" has two semi-independent causes, state "yes" of the event iiE_{20} "Coasts made of soft material and not of hard material are affected by sea level rise, are affected by surf, are affected by storms, are abraded, are not accreted, are inundated", and state "yes" of the event iiE_{28} "Coasts made of hard material and not of soft material are affected by sea level rise, are not affected by surf, are affected by storms, are abraded, are not accreted, are inundated".

Hence, the following conditional probabilities apply to it: $p(yes\text{-}iiE_{12}|yes\text{-}iiE_{20})$, $p(yes\text{-}iiE_{12}|no\text{-}iiE_{20})$, $p(yes\text{-}iiE_{12}|yes\text{-}iiE_{28})$, and $p(yes\text{-}iiE_{12}|no\text{-}iiE_{28})$, and the full probability of iiE_{12} is:

$$p(yes\text{-}iiE_{12})=p(yes\text{-}iiE_{12}|yes\text{-}iiE_{20})+p(yes\text{-}iiE_{12}|no\text{-}iiE_{20})+p(yes\text{-}iiE_{12}|yes\text{-}iiE_{28})+p(yes\text{-}iiE_{12}|no\text{-}iiE_{28}).$$

Then, setting a prior probability of any of the cause, say, $p(yes\text{-}iiE_{20})$, the aposteriori probability of this cause given the consequence $p(yes\text{-}iiE_{12})$, by the Bayes formula, is:

$$p(yes\text{-}iiE_{20}|yes\text{-}iiE_{12})= p(yes\text{-}iiE_{12}|yes\text{-}iiE_{20})\ p(yes\text{-}iiE_{20})/p(yes\text{-}iiE_{12}).$$

The suggested event-bush-based models of coast evolution should be verified by existing databases and/or statistically representative expert judgment. This itself represents a research task, as it may require development of event-bush-based expert knowledge elicitation protocols and technology of linking of event bushes to relational databases. Still, as stated by Pshenichny and Kanzheleva (2011), the intention n of the event bush is to simulate the functionation of modeled objects, and this makes one reasonably expect that such protocols would be more feasible for knowledge engineers and understandable by experts than, say, similar protocols for BBNs that are already in use (Felpeto et al., 2005). The other task, linking of the event bush technique to the relational database design, is addressed by Mandrik and Arustamov (2017; see Chapter 11 of this book).

CONCLUSION

Qualitative meanings seem to provide better insight into the nature and mechanisms of coast evolution than measurable parameters or indices derived from them. The method of event bush is formal enough to operate with these meanings in a strict and impartial way and thus strongly decrease the dependence of probabilistic assessments on the data model accepted in the database from which the data for computation were taken. Moreover, the event bush is readily transformable into Bayesian belief network (thus becoming Bayesian bush) and can be used as such. Still, the issues of verification of Bayesian bush-based models including linking them to expert judgment elicitation and database design are still ahead.

REFERENCES

Behncke, B., & Pshenichny, C. A. (2009). Modeling unusual eruptive behavior of Mt. Etna, Italy, by means of event bush. *Journal of Volcanology and Geothermal Research*, *185*(3), 157–171. doi:10.1016/j.jvolgeores.2009.04.020

Codd, E. F. (1970). A relational model of data for large shared data banks. *Communications of the ACM*, *13*(6), 377–387. doi:10.1145/362384.362685

Coelho, C., Silva, R., Gomes, F. V., & Pinto, F. T. (2006). A Vulnerability Analysis Approach for the Portuguese West Coast. *Transactions of the Wessex Institute*, *12*. doi:10.2495/RISK060241

Diviacco, P. (2017). On Formalization and Representation in Collaborative Research. In C. A. Pshenichny, P. Diviacco, & D. I. Mouromtsev (Eds.), *Dynamic Knowledge Representation in Scientific Domains*. IGI Global.

European Commission. (2004). *EUROSION: Living with coastal erosion in Europe: Sediment and Space for Sustainability, Part IV – A guide to coastal erosion management practices in Europe: Lessons Learned.* Technical Report 27. Author.

Felpeto, A. & Marti, J. (2005). GIS-based automatic system for volcanic hazard assessment. In *Geophysical Research Abstracts* (vol. 7). European Geosciences Union (EGU) 2nd General Assembly, Vienna, Austria, EGU05-A-04020.

Gornitz, V. M., Daniels, R. C., White, T. W., & Birdwell, K. R. (1994). The Development of a Coastal Risk Assessment Database: Vulnerability to Sea-Level Rise in the U.S. Southeast. *Journal of Coastal Research*, *12*, 327–338.

Gutierrez, B. T., Plant, N. G., & Thieler, E. R. (2011). A Bayesian network to predict coastal vulnerability to sea level rise. *Journal of Geophysical Research*, *116*(F2), F02009. doi:10.1029/2010JF001891

Hapke, C., & Plant, N. (2010). Predicting coastal cliff erosion using a Bayesian probabilistic model. *Marine Geology*, *278*(1-4), 140–149. doi:10.1016/j.margeo.2010.10.001

Jensen, F. V. (2001). *Bayesian Networks and Decision Graphs*. New York: UCL Press. doi:10.1007/978-1-4757-3502-4

Karunarathna, H., & Reeve, D. (2007). Predicting morphodynamic response of a coastal plain estuary using a Boolean model. *Proceedings of the 5th IAHR Symposium on River, Coastal and Estuarine Morphodynamics*, 1109–1115. doi:10.1201/NOE0415453639-c139

Karunarathna, H., & Reeve, D. (2008). A Boolean approach to prediction of long-term evolution of estuary morphology. *Journal of Coastal Research*, *24*(2B), 51–61. doi:10.2112/05-0542.1

Le Cozannet, G., Garcin, M., & Pshenichny, C. A. (2012). Reconsidering data models in coastal databases. *KEMG-2012 Proceedings*. Retrieved from geognosis.ifmo.ru

Mandrik, M. V., & Arustamov, S. A. (2017). Biography as Scenario: Representing Historical Knowledge by the Event Bush. In Dynamic Knowledge Representation in Scientific Domains. IGI Global.

Pin, P., & Chen, S. (1976). The entity-relationship model—toward a unified view of data. *ACM Transactions on Database Systems, 1*(1), 9-36.

Plant, N. G., & Holland, K. T. (2011). Prediction and assimilation of surfzone processes using a Bayesian network, Part I: Forward models. *Coastal Engineering*, *58*, 119–130. doi:10.1016/j.coastaleng.2010.09.003

Pshenichny, C. A. (2017). Qualitative and Quantitative Calculi in the Theory of Multitudes. In C. A. Pshenichny, P. Diviacco, & D. I. Mouromtsev (Eds.), *Representation of Dynamic Knowledge in Scientific Domains*. IGI Global.

Pshenichny, C. A., & Kanzheleva, O. M. (2011). Theoretical foundations of the event bush method. In K. Sinha, L. Gundersen, J. Jackson, & D. Arctur (Eds.), Societal Challenges and Geoinformatics (pp. 139-165). Boulder, CO: GSA Special Paper 482. doi:10.1130/2011.2482(12)

Pshenichny, C. A., Nikolenko, S. I., Carniel, R., Vaganov, P. A., Khrabrykh, Z. V., Moukhachov, V. P., & Rezyapkin, A. A. et al. (2009). The event bush as a semantic-based numerical approach to natural hazard assessment (exemplified by volcanology). *Computers & Geosciences*, *35*(5), 1017–1034. doi:10.1016/j.cageo.2008.01.009

Schukin, I. S. (1960). *Obschaya geomorphologia [General Geomorphology]* (Vols. 1–3). Moscow State University Publishers. (in Russian)

Shaw, J., Taylor, R. B., Solomon, S., Christian, H. A., & Forbes, D. L. (1998). Potential Impacts of Global Sea-Level Rise on Canadian Coasts. *The Canadian Geographer*, *42*(4), 365–379. doi:10.1111/j.1541-0064.1998.tb01352.x

Thieler, E. R., & Hammar-Klose, E. S. (1999). *National assessment of coastal vulnerability to future sea-level rise: Preliminary results for the U.S. Atlantic Coast.* Technical Report, Open-File Report 99-953, US Geological Survey.

Yates, M. L., & Le Cozannet, G. (2012). Brief communication 'Evaluating European Coastal Evolution using Bayesian Networks. *Natural Hazards and Earth System Sciences*, *12*(4), 1173–1177. doi:10.5194/nhess-12-1173-2012

Chapter 9
Modeling of Geological Evolution of the Gulf of Mannar Area, South India, by the Event Bush Method

Biju Longhinos
College of Engineering, Trivandrum, India

Cyril Pshenichny
ITMO University, Russia

Vladimir Anokhin
Russian Academy of Sciences, Institute of Limnology, Russia

Jijoy Joseph
University College, Trivandrum, India

Sophie Koneva
Karpinsky All-Russian Geological Research Institute (VSEGEI), Russia

Teena Chauhan
University of Bergen, Norway

ABSTRACT

The geological record is never completely available for observation. However, we have to query it – first to suggest or select and then to verify our hypotheses of the geological history, present day, and future of the region in question. So far, this job has been done either intuitively, or on the contrary, by quantitative modeling. Still, the former looks insufficient, especially if the case is contemporary tectonics or other potentially hazardous processes, and the latter gives reliable result only if involves abundant data – and still gives no warranty that is adequate enough to the modeled issue. Therefore, an intermediate solution is desired for regional geology, able to give a reliable result based on available data. The information modeling by means of the event bush method looks promising. In this chapter, the method of event bush is applied to verify the wrench tectonics hypothesis for the neo- and contemporary tectonic regime of the Gulf of Mannar region in the southernmost part of the Hindustan peninsula.

DOI: 10.4018/978-1-5225-5261-1.ch009

INTRODUCTION

The basin of Gulf of Mannar (Mannar Basin) lies between the south-eastern India and western Sri Lanka (6-9°N and 78-80°E), forming a part of extension of Laccadive Sea of the Indian Ocean to the north (Figure 1).

Early workers treated the Mannar Basin as southern extension of a larger Cauvery Basin followed from Muthupattinam at the south to Portonovo at the north (Baillie et al. 2002). Later, Rana et al. (2008) considered Mannar as a separate basin, which only shows some structural similarity to Cauvery Basin, mostly by its horst-graben tectonics. Based on the stratigraphic data, these scientists state that in general the Mannar basin remained a separate graben all through its history between the Sri Lankan and Indian cratons (Figure 2).

The Indian coast of Gulf of Mannar, from Mandapam at the north to Kanyakumari at the south, has received great attention due to its physiographical diversity observed all along its length of 440 km. The coastal landforms are quite diverse, including large strand plains, deltaic marshes, huge sand dune fields, and ridges built of sedimentary rocks and isolated patches of granulitic rocks. Contemporary

Figure 1. Location of the Gulf of Mannar (blue shade) and the red box indicates the study area

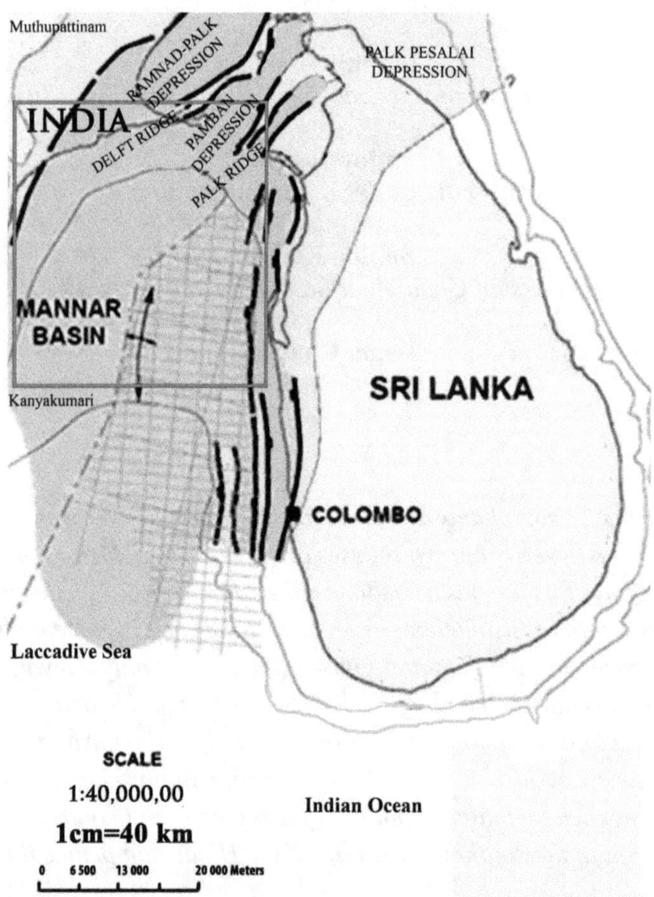

Figure 2. Tectonic structure of the northern part of the Mannar basin and adjacent territory

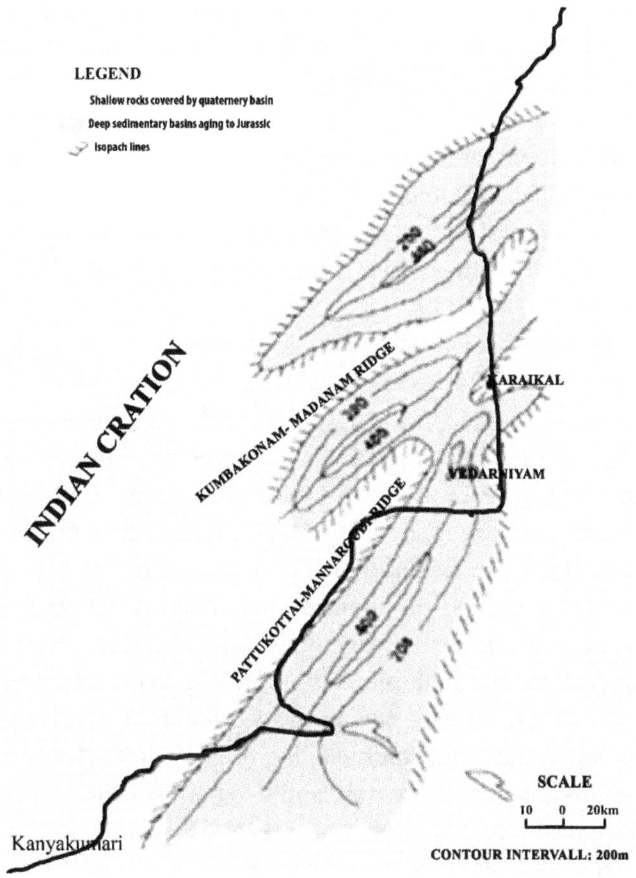

tectonic movements have been felt in the area in 1948, including the 6m subsidence of a village near Dhanushkodi (Vaz et al., 2006). Anomalously high heat flow (100 to 180 mw per sq m, quite comparable to the Himalayan hotsprings) was identified in the Ramanad area. The Mannar Basin is known for its highest heat flow value in the peninsular Precambrian region (Longhinos et al., 2010). Towards the south, closer to Kanyakumari, dikes and bodies of olivine basalts of supposedly Late Mesozoic age are reported (Ramasamy et al., 1994; Ramasamy, 1995; Longhinos et al., 2010). The same time, huge (sometimes over 60 m thick) cover of migrating sand, at places semi-consolidated, indicates an ongoing phase of aeolian deposition and redeposition since Pleistocene.

This, along with other processes of vigorous weathering and denudation, represents the main problem in interpretation of the tectonic history and current geodynamic regime of the territory. From one side, the area must represent a perfect case of marginal environment of rift activation, with prominent peripheral half-grabens and other typical associated features, including the nowadays processes. From the other, the sand cover spreading along the coast limits exposure of geological units and may have obscured events that the coast underwent even recently. Only few appropriate exposures can be found there, and the geologist is supposed rather to infer than to observe the neotectonic and contemporary movements in the area. Although, locally there is preserved a direct geological evidence of ongoing uplift seen along the coast. It is expressed in the advance and retreat of sea and vertical displacements of the beach. The same time, huge dunes may imply a retreat of the sea, which may be caused either by uplift of land or by downsagging of the basin coupled with strengthening of wind blowing fine-grained coastal sands into the hinterland. Therefore, a complex interpretation is needed with a very accurate treatment of all cause-effect relations between the hypothesized tectonic "driving force" and observed stratigraphic results.

For this, first, all possible consequences of the suggested movements should be tracked given the (i) geological sequence that is believed to have had existed and (ii) principal physical-geological agents that had to exist in the area and/or exist now (i.e., sun+air, sea, probably vegetation). Then, every known outcrop or another geologically peculiar object in the area must be related to this framework to either support or refute the suggested hypothesis.

Taking into account the well-known looseness and contextuality of geological knowledge, its dependence on the personal and social aspects of research (Frodeman, 1995; Diviacco et al., 2012, and references therein), this should be done impartially, preferably on a strict formal basis.

This entirely falls in the scope of relatively new area of research, information modeling of geological environments (Pshenichny & Kanzheleva, 2011), whose theoretical foundation in the semantic science is dynamic knowledge representation (Pshenichny, 2017 – Chapter 1 of this book). Pshenichny and Kanzheleva suggest a classification of environments, according to which the considered case can be regarded as directed alternative change environment, i.e., an environment in which (i) evolution may be considered as having clear and definable beginning and end and (ii) there may be multiple starting events, multiple resulting events and multiple paths from one event to another. The method deemed so far the most appropriate for this kind of environment is the event bush (Pshenichny & Kanzheleva, 2011). The present work aims at information modeling of geological evolution of the Gulf of Mannar area by the event bush method. As a result, we are going to build a strict and consistent verbal model of tectonic evolution of the area with full set of possible scenarios and full set of possible types of resulting geological sequence given the assumptions on preexistent structure and acting agents.

In particular, this implies the following tasks:

- Summarization of existing geological data on the area,
- Explicit formulation of tectonic hypothesis,
- Presentation of the event bush method,
- Construction of information model,
- Verification of information model by the geological data.

These tasks are addressed in corresponding sections of the chapter.

REGIONAL GEOLOGY

Tectonic Framework

The region occurs within the Achankovil shear zone, which is regarded as transcontinental by some researchers in the sense that it can be traced into now-separate crustal fragments of Madagascar and Sri Lanka (Drury et al. 1984; Harris & Santosh, 1993; Harris et al., 1994; Santosh et al., 1992, 2005; Braun

& Kriegsman, 2003; Guru-Rajesh & Chetty, 2006; Sreejith & Ravindra Kumar, 2013) whereas others consider it to be a high-strain zone within Southern Granulite Terrain (Ghosh et al., 2004).

The Gulf of Mannar is bordered by Cenozoic sedimentary rocks and the Proterozoic Pandiyan mobile belt (followed both in India and Sri Lanka). The Pandiyan belt is built mostly of metamorphic rocks of gneissic to granulitic grade, aging (roughly) Precambrian and intruded by granites and dolerites. Towards the eastern coast of the Indian peninsula, the outcrops of hard crystalline rocks become rare. These rocks are either overlain by Mesozoic sedimentary rocks with sparse intrusions of Late Cretaceous basalts, or by Cenozoic sedimentary rocks. The batch of basement rocks at the south-eastern coast of India is believed to represent a highly differentiated horst and graben morphology typical for rift extension.

According to Raiverman (1992), except the transcontinental joints stretching from the equator across the Himalayas up to the present territory of Russia, the Mannar basin has not been influenced by a single structural lineament that controlls the tectonics of both Cauvery and Palar basins. Instead, it remained a broad depression all through its history between the Sri Lankan and Indian peninsular massifs (cratons). Curray (1984) compared the structure of the Mannar basin to aulacogen and considered its origin as early stage of continental rifting that began around the Late Jurassic. The oil exploration in the 2000s has demarked the domains of magmatism in the Mannar basin, dating back to the Late Cretaceous. Satellite-derived gravity data show very low gravity values off the Kanyakumari coast at the south and off the Sayakudi coast at the north, though high-density basaltic rocks are identified in the reflection pattern, seismic profiling and drilling.

The crustal lineaments and faults developed in the Mannar basin can be classified by strike into four main groups: (i) NNE-SSW to NE-SW, (ii) NW-SE, (iii) N-S and (iv) E-W. All four were identified on land but also continue in sea. Though these systems are quite ancient, ranging, by existing indirect estimates, from ~3000 Ma to ~100 Ma, all of them have been repeatedly reactivating and affect even the present-day sediments. The NNE-SSW-striking fault between India and Sri Lanka in the Palk Bay – Gulf of Mannar region inferred from the convex flexure in the Thalai Mannar sector of the Adams Bridge and in the group of islands off the Jaffna area of Sri Lanka shows recent left-lateral (sinistral type) strike-slip movements (Katz, 1978).

The basin of Gulf of Mannar is a deposition center with a thickness of sedimentary fill varying from 1 to 3 km. The oldest sedimentary rocks in this region date back to the Albain (Early Mid-Cretaceous), but in the Mannar basin they occur only at the seafloor.

On land, within the basin, sedimentary rocks are exposed at the Indian coast between Kanyakumari at the south and Pamban at the north. Sedimentation is believed to be a result of several tectono-eustatic events.

At present, the basin is fed by terrestrial sediments from India and Sri Lanka, mainly by river Tamaraparani and partly by river Vaiga from Indian sub-continent and heavily sediment loaded rivers Mahaoya, Kelani, Kalu and Gin from the Sri Lankan side. These rivers flow almost completely through Proterozoic calc-alkaline granulitic and gneissic terrains, devoid of synchronous mafic and ultramafic rocks. From the sub-continent, there are four seasonal easterly streams: Vaipar, Nambiar, Hanumand Ar and Uppar flow south of river Tamaraparani, having estuaries sealed by perennial sand dune migration. Therefore, they supply very little amount of sediments to the basin.

Stratigraphy and Composition of Rocks

The lithostratigraphy of the rock units in this Indian coastal tract of Gulf of Mannar comprises a metamorphosed Proterozoic granulitic and gneissic rocks overlain by a thin veil of sedimentary rocks (Paramasivam & Srinivasan, 1980). The former include charnockites, quartz-feldspathic gneiss, and granitic gneiss – all together categorised as granulites in the literature. The latter comprise fossiliferous limestone and calcareous sandstone, which are overlain by the Holocene to recent soil cover.

The sedimentary rocks include partly fragmented material of dominantly carbonate composition with various faunal contents such as molluscs, bryozoans, foraminifers, and fragments of red and blue-green algae (Armstrong-Altrin et al. 2001). These rocks were deposited directly over the Proterozoic rocks, when a considerable proportion of the land along the coast was occupied by the sea as a result of transgression during the Mio-Pliocene. These sedimentary rocks referred to as Kudangulam Formation were correlated with the Karaikal beds of Late Miocene to Pliocene age (Ramanathan 1979; Paramasivam & Srinivasan 1980). On the eastern flank of the Mannar basin, in Sri Lanka, the Jaffna Formation is most probably the equivalent to Kudangulam, on the basis of the fossil assemblages and facies characteristics. The presence of microfossils like *Austritrillina howchini* and *Taberina malabarica* places the Jaffna Formation in the upper part of the upper Miocene, specifically in the Burdigalian stage (Sahani 1979; Cooray 1984). The facies association and the sediment composition show that the carbonates were deposited in a shallow marine near-shore environment (Ramasamy et al., 1996; Ramasamy & Armstrong-Altrin, 1998). A preliminary stable-isotope study on the Kudangulam limestone reveals that freshwater circulation played a major role in the diagenesis (Armstrong-Altrin et al., 2001). Chauhan et al. (2008) noted two phases of sand dune stabilization in the Late Pleistocene, leading to the formation of calcrete inliers, later being covered by the next phase of moving sand. To the extreme south of the study area, around Kanyakumari, the limestone is absent and Quaternary sandstone lays directly on the granulites.

Ramasamy et al. (1994) and Longhinos et al. (2010) reported the dike swarms parallel to the coast and intruding the deposits of the Mannar basin. The dikes strike fairly 60-75°N, with nearly vertical dip

Table 1. Stratigraphy of Gulf of Mannar Basin (compiled from ONGC, 1993a-c)

Time in million years	Geologic Time scale	Offshore Formation (rock type/ thickness)	Inland Formation (rock type/ thickness)
Recent to 2	Quaternary	Marine sediments (marine clay/ 0.3 to 5m)	Teri (fine Aeolian sand / 0.5 to 60m)
2 -5	Unconformity		Panambarai (Sandstone/ 1o to 15 m)
5 -10	Neogene Paleogene	Tittacherry Limestone	Kudangulam (Limestone / 5 to 120m)
10 -53		Vedaraniyam Sandstone	
53- 60		Kamalapuram Limestone	
60-66.5	Unconformity		
66.5 – 70	Cretaceous	Mannar 1 Basalt	
70- 88		Komarakshi Shale	
88 – 92		Mannar 2 Basalt	
Above 560	Proterozoic	Granulites *(Garnet Biotite Gneiss to Garnet Hornblende Granulite)*	

(whether they occur at the west of the area, in the Pathanamthitta district of Kerala, or at the east, in the Tirunelveli district of Tamil Nadu). Longhinos & Ramasarma (2008) found similar mafic bodies in the hinterland. Such dikes occur in different parts of the Achankovil shear zone but have become observable only at recent quarrying, probably for this reason they have not received any geological attention. Longhinos et al. (2010) deciphered magnetic anomaly variations across the terrain of Kudangulam and correlated them to the faults and subvolcanic necks and dikes. Ar-Ar dating of a few basaltic intrusives of the Munnar basin gives the age of *c.* 82 Ma (Rathore et al., 2007).

The stratigraphy of the offshore geological sequence has been best characterized by the oil exploratory drilling data (ONGC, 1993a, b, c; Maitra et al., 2007), and the inland geology, reported in numerous works of Armstrong-Altrin et al. (2001, 2003, 2004, 2005). Based on both, the main stratigraphic units in the region are shown in Table 1.

Recent and Contemporary Endogenous Activity at the Southeastern Coast of India

The main sets of deep ancient crustal lineaments and faults (NNE-SSW to NE-SW, NW-SE, N-S and E-W) continue to affect the present-day sediments. Geothermal anomaly is reported in the Palk Bay – Gulf of Mannar region. Hot water flows there from boreholes ranging in depth from 200 to 600 m. Kanishkan and Regnathan (1990) estimated its temperature as 47°C to 90°C, the highest being recorded at Thiruthuraipoondi. Gopalakrishnan and Varadan (1996) list the following possible reasons for the increased geothermal activity in the region: heat produced by friction of rocks being displaced along faults, recent intrusions or subvolcanic bodies, "mantle plume", or "hot spot".

Seismic activity along the Tamil Nadu coast is known for quite a long time during the last century (Ramalingeswara Rao, 1992). A large number of earthquake epicenters from Chennai to Kodiakkarai were identified since 1967. One epicenter of magnitude 3 to 4 is located within the Palk Bay – Gulf of Mannar basin. Gopalakrishnan (1996), Project Vasundara, GSI (1994) and Subramanian and Gopalakrishnan (2002) consider that these earthquakes were caused by reactivation and movement along different faults at the southeast coast of Tamil Nadu, where the present study area falls in. Reddy (1995) ascertains that areas showing high gravity anomalies and high heat flow are prone for seismic activity as well.

Also, vigorous denudation may not only obscure but, on the contrary, unveil the tectonic activity. Specific abrasion features evidencing from rapid tilting of coastal sedimentary rocks are observed along the coast of Manappad. In other places similar structures may be obscured by dunes.

Study Area

The whole western coast of Gulf of Mannar, from Mandapam at the north to Kanyakumari at the south and up to 20 km inland (westward), has been considered in the present study. Geological expeditions were made in 2009-2015 and covered all this area. The coast here has coastal plains and cliffs made of either crystalline or sedimentary rocks. The basement rocks are exposed closer to the cliffs. The plains extend up to 10-15 km off the shoreline. We believe that the character of rock exposures reflects complex basement structures, striking perpendicular to the coast and representing torqued shorter half-grabens at the south and large linear depressions at the north of the study area. The topography of the area is shown in Figure 3, and geological structure based on our own and published field data, in Figure 4.

Figure 3. The topography of south-west coast of the Mannar basin

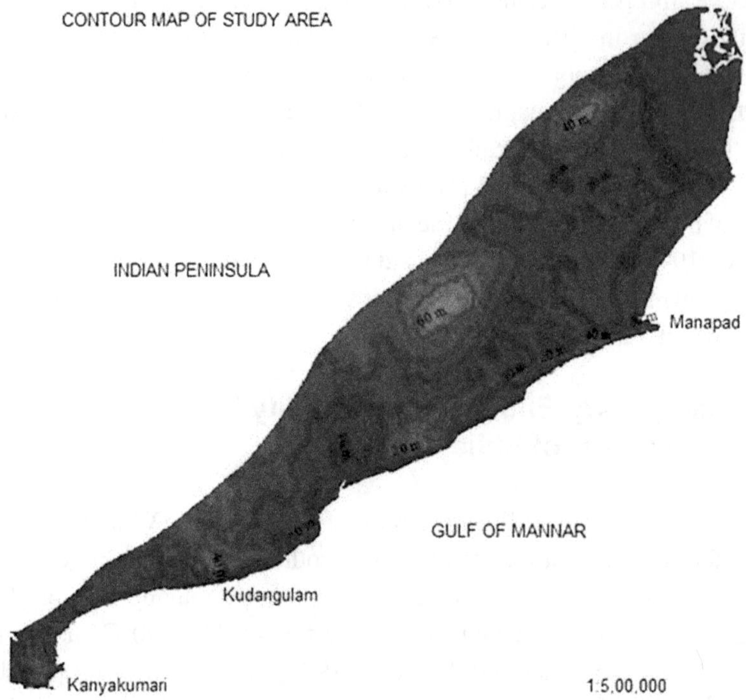

Figure 4. The geology of western coast of the Mannar basin

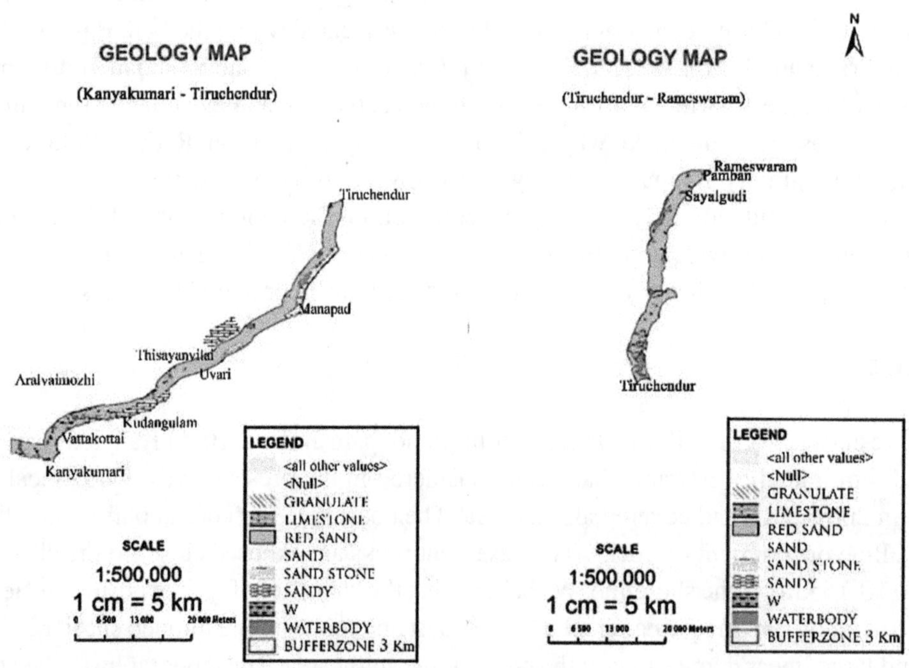

The coast at Kanyakumari has a cliff of granulitic and gnessic rocks, bordered by beach rocks of Quaternary age. Towards the north, along the coast, the crystalline cliff disappears but the coast remains rocky. In two locations, basaltic dikes were seen to cut the gneiss. These dikes (striking 70°-80°N with nearly vertical dip) are traced further into the hinterland off the coast. About 3 km north towards Leepuram, outcrops of calcareous sandstone overlying granulites with marked unconformity are well preserved at the coast. The calcareous sandstone shows clear stratification with a seaward dip < 10°.

In another 1.5 km north of Leepuram, in a place known as Circular Fort, a prominent cliff abuts 10 m into the sea. The cliff is made of shell limestone with 20% medium-grained sandy sediments. The shells are quite recent in age and not well cemented in the rock.

The coast takes an eastward turn about 100 m north of the Circular Fort promontory, where cliffs disappear and the coast is occupied by long beaches with coastal and aeolian accumulative landforms. About 500 m away into the hinterland, gneisses and granulitic rocks are exposed usually 15-20 m above sea level. Basaltic bodies are plugged into the gneisses, a few of them are exposed in wells at depth of 30 m or more and others are inferred from geophysical studies (Longhinos & Ramasarma, 2008). Two seasonal estuaries (of heavily dammed Uppar and Hanumand Ar streams, flowing from the eastern slopes of Western Ghats) are located in this part of the coastal plain. Aeolian dunes extend 500 m into the hinterland, eventually waning. East of the Hanumand Ar stream the coast becomes cliffy again with dunes obscuring the underlying rock.

About 300 m east of the estuary, a limestone-capped cliff of 54 m height runs parallel to the shore for about 5-6 km, thereby disappearing beneath the Aeolian dunes north of the Vijayapathi mines. Between the coast and the cliff, there is a coastal plain underlain by weathered limestone, gneiss and charnockite. Road cuttings into the cliff show the following sequence (from bottom to top): gneiss intruded by basaltic dikes – fossiliferous limestone – calcareous sandstone. Sandstone cliff appears along the coastal part of Vijayapathi (near the Idinthakari village). North of Vijayapathi, the limestone exposures are seen further in the hinterland (Tyssiyanvilai, Udangudi and Edyangudi localities) where sandstone crops out in cliffs parallel to the coast up to Mandappu in north. A calcareous submerged reef is found along the Periathalai – Ovari coast. Further to the north, marine calcareous sandstones, rich in quartz and feldspar, with embedded mollusk shells are exposed in Manappad and Tiruchendur in the high tide and berm zones. In Manappad, the sandstone cap rises up to 60 m above sea level. Between Kulashekarapattinam, located northern side of Manappad creek, and Tiruchendur, the sandstone is exposed at 15 meters above the mean tide level. Narayanaswamy (1947) suggests that the sea level in the Holocene stabilized in this region after the cementation of sand.

TECTONIC MODEL

As far as we know, researchers have not delved deep in specific peculiarities of recent and contemporary tectonism in the region of Gulf of Mannar. Geological and geophysical characteristics of this region suggest (i) horst-graben framework, (ii) abundant episodes of local extension marked by dike intrusion and (iii) highly differentiated vertical movements (both uplift and downsagging within several kilometers). The third point should be specified further:

- Downsagged blocks rapidly change to uplifted ones more or less in longitudinal direction;
- Downsagged blocks eventually change to uplifted ones more or less in latitudinal direction;
- Downsagged granulites (where observable) appear more gneissic than uplifted ones;
- Slicken slides are encountered in the field.

We believe that the mechanism that best complies with these facts and limitations is strike-slip displacement along sub-parallel 60-75°N striking faults with secondary interchanging zones of compression and strain arising in rock blocks between these faults. In our understanding, when horizontal tension develops, it may result not necessarily in one but in a series of subparallel strike-slip faults, and between any two neighboring faults elongated blocks of rock would be stuck (Figure 5). Any such block must experience, by inertia, oppositely directed motion of rock mass at its sides. If the block is narrow enough, these two motions would repeatedly curl forming structures looking like eddies in plane. In these eddies, stress and uplift occur, and in between, strain and downsagging do. Then, one may expect that in the strained localities secondary faults may develop favoring intrusion of dikes and/or any kind of disintegration and splitting of the initial elongated block into shorter portions. Thereafter, these portions appear divided by the localities of strain. Within a portion, two sub-blocks can be identified, each more or less distinctly separated from the strained butt and colliding with its counterpart at the other (stressed) butt, i.e., approximately in the middle of the portion. Then, as collision progresses, each sub-block begins to tilt from the stressed buttend to the strained one, that is, along the strike of the initial strike-slip faults. Hence, the initial block appears more or less split into paired mutually colliding sub-blocks, tilted from stress to strain. This more or less fits the concept of transcurrent wrench tectonics sensu van der Pluijm (2016), or pull-apart strike-slip structures of Sylvester (1958).

This idea, being, by van der Pluijm (2016), purely geometric and free from any genetic, tectonic or size connotation, should be discerned from the concept of global wrench tectonics of Storetvedt (2003) who proposed similar mechanism for the entire earthcrust. Wrench tectonics as a possible mechanism operating in the considered region was mentioned by Chandra et al (2006), and Damodaran et al. (2016).

Figure 5. An idealized scheme of initial stage of the wrench tectonics mechanism (modified from Sylvester, 1958)

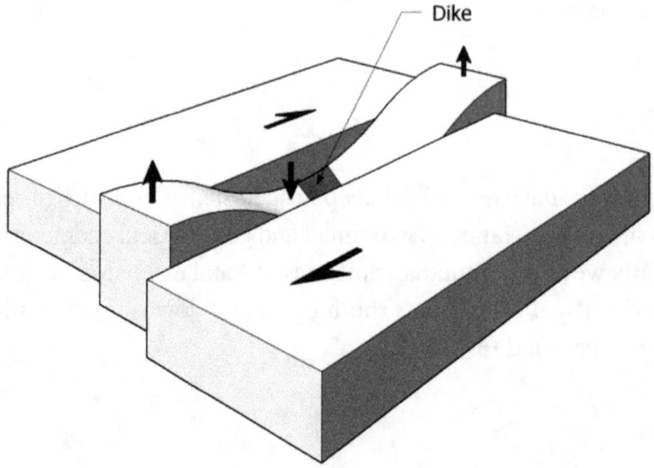

However, like in most regions of the world, due to principal incompleteness of the geological record and inability of continuous direct observation of tectonic movements, neither the wrench tectonics nor any other possible competing hypothesis can be completely proven based on the data. Rather, the question should be put as whether the existing data support (not prove!) or refute given hypothesis, i.e., whether they are in contradiction with it or not yet. This issue has been resolved largely intuitively in geology, without application of logic or logic-like inference tools. Such intuitive approach can work for simple models and limited data, but in large-scale regional studies like the one presented here, when data are abundant and the model, quite complex, a method is needed to impartially assess the compliance of existing data with the proposed model. As for the dikes intruding the basement and sedimentary rocks, it is quite obvious intuitively that they do fit the proposed hypothesis, using the extension zones between transcurrent faults. However, as for the rock sequences supposedly formed under the processes of wrench tectonics, it is not obvious at all. In this work, mainly the data on field relations of rock units will be considered. Its compliance with the wrench tectonic model will be examined below by the method of event bush. Modeling of knowledge and data by knowledge representation tools to resolve problems in particular field was termed information modeling by Pshenichny and Kanzheleva (2011).

INFORMATION MODELING OF THE STUDY AREA GEOLOGICAL EVOLUTION

Outline of the Event Bush Method

The knowledge engineering method of event bush intends to give a strict and finite but extendable display of an area of reality and corresponding domain of knowledge. From now on it will be described mainly based on the works of Pshenichny et al. (2008; 2009) and Pshenichny and Kanzheleva (2011). Importantly, it does not intend to paint an objective and true picture of reality based solely on formal grounds. The purpose of the event bush is to more or less impartially structure and shape up the subjective information gathered by a scientist. Like any other formalization, it does not say itself what is true and what is false, or what is relevant and what is not, for the studied environment. Nevertheless, it helps organize what scientist considers relevant, allowing to formulate everything that can happen given a list of premises. It imposes some requirements on these to ensure the most complete and objective inference, but the choice of premises is totally at the scientist's discretion. In a discourse or polemics, the tool of event bush may help us formally express and compare contrasting standpoints. Meanwhile, formalization itself often urges a scientist to revise knowledge, find gaps and strands of "wooly" reasoning, terminological and conceptual intricacy and the like. Such imperfections can be tackled and to some degree cured by the event bush. But like any other weapon, in principle it can be used alternatively, e.g., to produce a beautiful, formally perfect nonsense.

The method of event bush rests on assumption that some areas of reality (e.g., formation of a geological sequence in an area given the initial geological structure and wrench tectonics) can be represented as shown in Figure 6. In the considered area of reality, the following events are identified.

(*ia*) Primary internal events. These are primary, not overlapping and non-unique inputs (basic objects, processes or tendencies) – e.g., "Body of granulite overlain by limestone and sandstone is unaffected by varying horizontal force". Such inputs, according to the concept of event bush, would determine any further course of events ("happenings").

Figure 6. Structure of the event bush

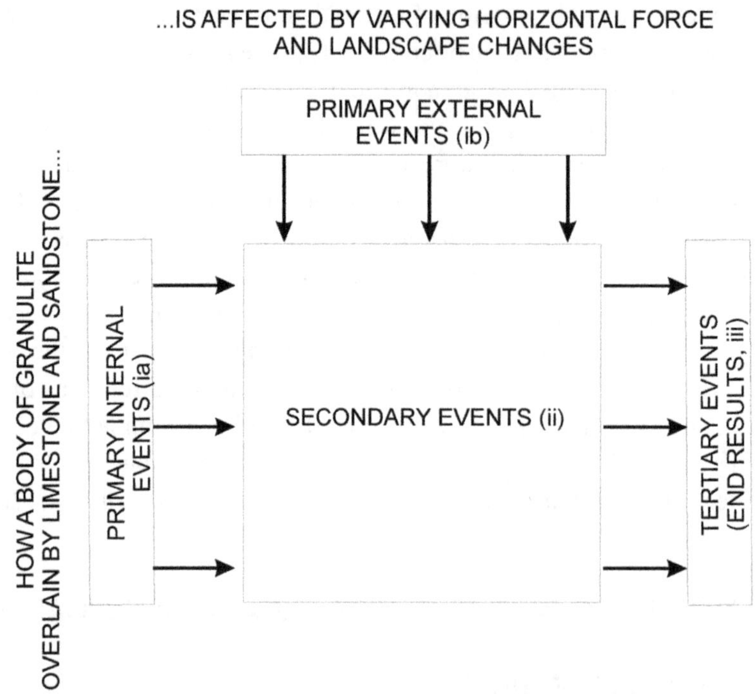

(*ib*) Primary external events meaning the circumstances that come from the environment. They indicate the way the environment may affect basic inputs or influence their further, indirect manifestations, thus "shaping up" different "happenings". For instance, an event "Varying horizontal force applies" is expected to affect the primary geological structure (i.e., somehow transform the body of granulite overlain by limestone and sandstone).

(*ii*) Secondary events (processes or objects) that result from primary inputs with or without the contribution of incoming circumstances – the "happenings" proper (e.g., "Horizontal shears form in body of granulite overlain by limestone and sandstone affected by varying horizontal force") formulated in a strict concise way indicating their core features determined by the causes, following the principle "one more cause – one more property".

(*iii*) Tertiary events that denote end results, or products, generated either by primary internal or by secondary events, with or without primary external ones. Tertiary events document the completed "happenings". In the considered case, it looks reasonable to take as end results the structural patterns that may be generated in the considered environment.

Other types of events (e.g., those describing some other type of results – say, quaternary ones) can be added, but the presence of the (*i*) - (*iii*) is mandatory. Their general interrelation is shown by an example in Figure 7 (a, b)

In accordance with the syntax of the event bush presented in Figure 6, the relations between the (*i*) - (*iii*) can be set:

Figure 7a. Connectives in the event bush model – flux and influx. Bold lines mean scenario that is actually observed. See other comments in the text.

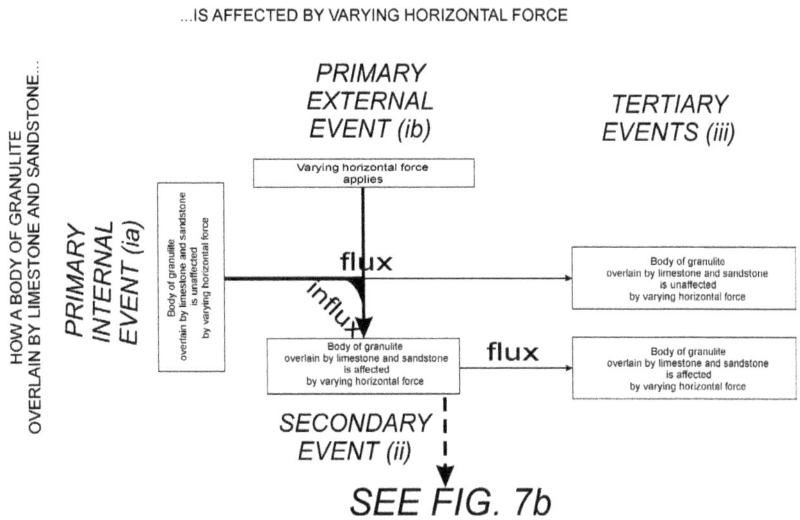

Figure 7b. Connectives in the event bush model – flux, influx and furcation. Bold lines mean scenario that is actually observed. See other comments in the text. (The bush (b) is from bush (a))

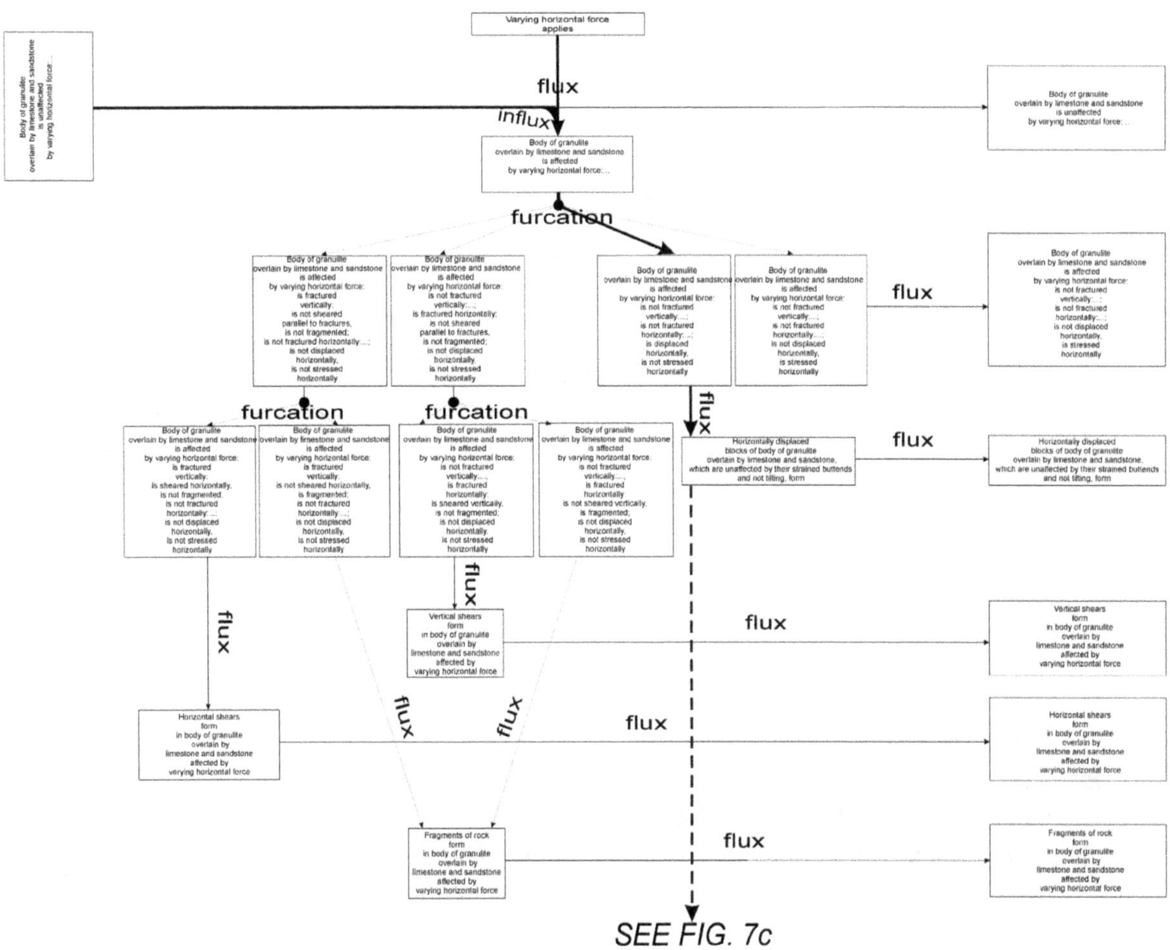

Figure 7c. Connectives in the event bush model – flux, influx, furcation and conflux. Bold lines mean scenario that is actually observed. See other comments in the text. (The bush (c) is from bush (b)).

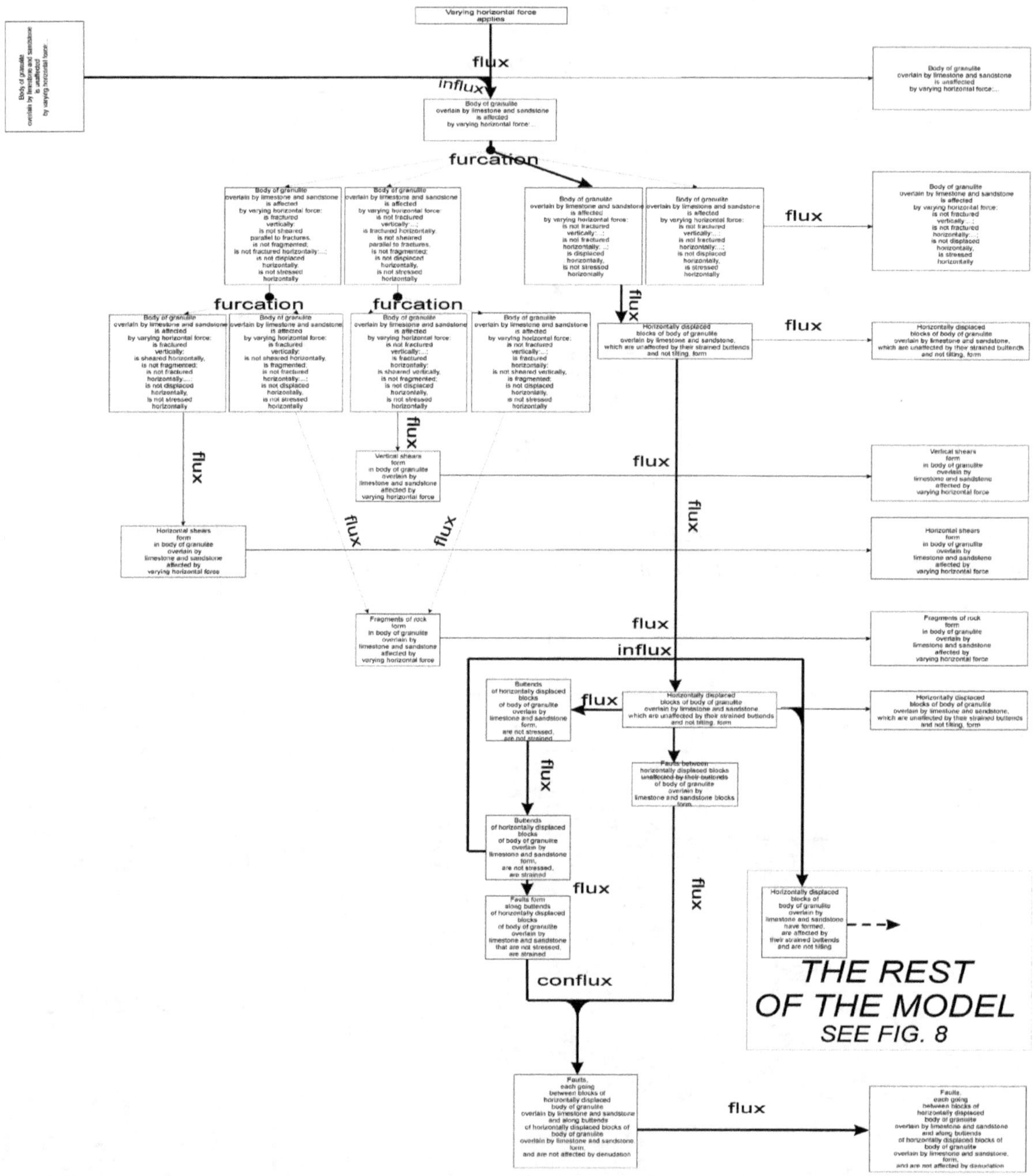

1. Events (*ia*) and (*ib*) must not lead to other (*ia*) or (*ib*);
2. Event (*ib*) may lead to an event only together with (*ia*) or (*ii*);
3. Event (*ii*) must not lead to (*ia*) or (*ib*);
4. Event (*iii*) must not lead to (*ia*), (*ib*), (*ii*), and another (*iii*).

These relations are enforced by the connectives of the event bush, each having a graphic designation. Pshenichny and Kanzheleva (2011) define four connectives, flux, influx, conflux, and furcation, of which the first two, flux and influx, are mandatory for an event bush.

If one event leads to another, this is defined as flux:

E_i *Flux* E_j

and denoted by simple arrow. For instance, in Figure 7, obviously,

Body of granulite overlain by limestone and sandstone is unaffected by varying horizontal force (E_i)

FLUX

Body of granulite overlain by limestone and sandstone is unaffected by varying horizontal force (E_j).

If two or more events lead to one, this refers either to influx (denoted by a rounded crossing, or "right turn" graphic sign) or to conflux (denoted by "double right turn") connectives.

Putting the right turn, we express an opinion (or an obvious possibility) that two events together may lead to another event, so that one, above the right turn sign (E_j), "flows in", or modifies, another event E_i (to the left of it), resulting in the event below the said sign (E_k):

E_i, E_j *Influx* E_k.

For example, in Figure 7a,

Body of granulite overlain by limestone and sandstone is unaffected by varying horizontal force (E_i), Varying horizontal force applies (E_j)

INFLUX

Body of granulite overlain by limestone and sandstone is unaffected by varying horizontal force (E_k).

Obviously, all events (*ib*) act as modifiers. The subject of the resulting event is inherited from the main premise, and the predicate is formed by the modifying one. For intuitive clarity, we adopt that the "right turn" points always down, from the modifying event to the resulting one, which denotes the influx. The main and the modifying events may change places, only if this makes sense in the considered domain of knowledge and involves only the events of type (*ii*).

Once an event (*ia*) or (*ib*) is introduced or event (*ii*) appears in a bush, one should subsequently coincide it with all previously introduced events to find out whether any new event can result from each pair.

Along with the flux, influx is mandatory for the event bush, for solely the influx involves the events ib in the inference.

Furcation (graphically denoted as circle; Figure 7b) represents the case that one event (E_i, left of the circle) necessarily leads to one of incompatible alternatives (E_{i+1}, E_{i+2}, …, E_n., right of it):

E_i *Furcation* $E_{i+1}, E_{i+2}, \ldots, E_n$.

This can be exemplified by the following inference shown in Figure 7b:

Body of granulite overlain by limestone and sandstone is affected by varying horizontal force:

is not fractured vertically, is not fractured horizontally, is not displaced horizontally, is not stressed horizontally (E_i)

FURCATION

Body of granulite overlain by limestone and sandstone is affected by varying horizontal force:

is fractured vertically, is not fractured horizontally, is not displaced horizontally, is not stressed horizontally (E_{i+1}),

Body of granulite overlain by limestone and sandstone

is affected by varying horizontal force:

is not fractured vertically, is fractured horizontally, is not displaced horizontally, is not stressed horizontally (E_{i+2}),

Body of granulite overlain by limestone and sandstone

is affected by varying horizontal force:

is not fractured vertically, is not fractured horizontally, is displaced horizontally, is not stressed horizontally (E_{i+3}),

Body of granulite overlain by limestone and sandstone

is affected by varying horizontal force:

is not fractured vertically, is not fractured horizontally, is not displaced horizontally, is stressed horizontally (E_{i+4}).

The connective of conflux (see Figure 7c) implies that two or more events (E_i, E_{i+1}, E_{i+2}, …, E_{n-1}) "conflow" into one (E_n) playing similar (symmetric) role:

E_i, E_{i+1}, E_{i+2}, …, E_{n-1} *Conflux* E_n.

The conflux is marked in the event bush as a "double right turn" and will be exemplified below in this chapter. The event bush does not admit that more than two events lead to one and play non-symmetric role (as in influx).

Wishing to express a case, in which different crisply defined options are compatible, one may put several arcs going out of one node, i.e., use several flux connectives. This means that an event can lead to any number of other events, in influx/conflux with others or by itself.

The rules of composition of and inference in the event bush do not envisage an opportunity of producing several events from several events in one single step, so the listed inference options expressed by the connectives of the event bush exhaust the entire variety of possibilities.

Change of events is a correctly built expression including one connective of any type (Pshenichny & Kanzheleva, 2011). Change has one causal part (that includes the events, one or more, left of the connective) and one effect part (the events to the right of the connective).

Flow, or scenario, is a finite totally ordered set (Schröder, 2003) of changes, in which the left part of the first change is an *ia* event, the right part of the last change is a *iii* event, and the right part of every preceding change is the left part of the following change. If there is a change with influx in it, the modifier of this influx (*ib* or *ii*) is included, but not the events that caused it, unless there are successive influx changes.

If there is a scenario, in which at least one event participates in causal parts of two changes (e.g., one, being flux, and another, furcation), there is another scenario, which overlaps with the former one in the interval from the beginning (*ia*) to this very event, and then ramifies. Furthermore, if there is a scenario without the ii events, first, it may only consist of one change, namely the flux from *ia* to *iii*, and then, there also must be a scenario that overlaps with this one in the beginning (i.e., at the *ia* event) and includes at least one influx. This influx of *ia* and another event, *ib* or *ii*, will produce an ii event, which will then route this scenario further. There must not be an event in a bush not included in a scenario. For more detail on the methodology of event bush the reader is referred to the papers by Pshenichny et al. (2009) and Pshenichny and Kanzheleva (2011).

Modeling

The Model

As a result of the geological studies reported above, three interrelated points have been formulated: (i) present-day geological sequence is described, (ii) pre-existing geological structure is hypothesized (as a body of granulite overlain by limestone and sandstone and breached by basaltic dikes), and (iii) the mechanism of wrench tectonics is suggested to proceed from the pre-existing to present-day geological framework.

To reconstruct the scenarios, which could, based on the current knowledge, lead from the past to present geologic structure by the wrench tectonic mechanism, an event bush modeling was performed. Herewith, we omit the presence of dikes as they do not add to relations between the stratified bodies, but may considerably add to the complexity of the model.

Based on the rules of construction of the event bush presented by Pshenichny et al. (2008; 2009), the modeling starts with formulation of the modeling task, which was, as shown in Figure 6, defined as to understand how a body of granulite overlain by limestone and sandstone is affected by varying horizontal force and landscape changes. This determines the choice of primary internal and primary external events. These are: primary internal, "Body of granulite overlain by limestone and sandstone is unaffected by varying horizontal force", and primary externals, "Varying horizontal force applies" and several events describing denudation, "Air denudates", "Sea denudates" and "External sand is brought by sea".

Primary internal event (and, hence, all other events having the same subject, or predmet, by Pshenichny, 2017 – Chapter 1 of this book) are instantiated as "grainy fill" and mean particular outcrops documented at observation sites. All other events in the model are considered to be lump-filled, i.e., instantiated by single phenomenon (i.e., tectonic force, air, sea or sand).

Influx of the first primary external event, "Varying horizontal force applies", to the primary internal event, leads to the event "Body of granulite overlain by limestone and sandstone is affected by varying horizontal force" (see Figure 7b). The latter, based on general knowledge in the field of structural geology, can be continued as furcation quoted in the previous subsection. Of the four resulting events, only one is the case in the study area, "Body of granulite overlain by limestone and sandstone is affected by varying horizontal force, is not fractured vertically, is not fractured horizontally, is displaced horizontally, is not stressed horizontally". The corresponding scenario is marked bold in Figure 7b. Other three events resulting from that furcation might take place but their role is minor and is not considered here. Then there is no need to follow the scenarios of influx of primary external events describing the denudation in them, and solely the scenarios related to the only one furcated event that is relevant (see Figure 7c)

Figure 8. Sketch of event bush modeling the knowledge about formation of the stratified rock sequence of the Gulf of Mannar area given the wrench tectonics. Bold lines mean scenarios that are believed to actually take place in the said area. See other comments in the text.

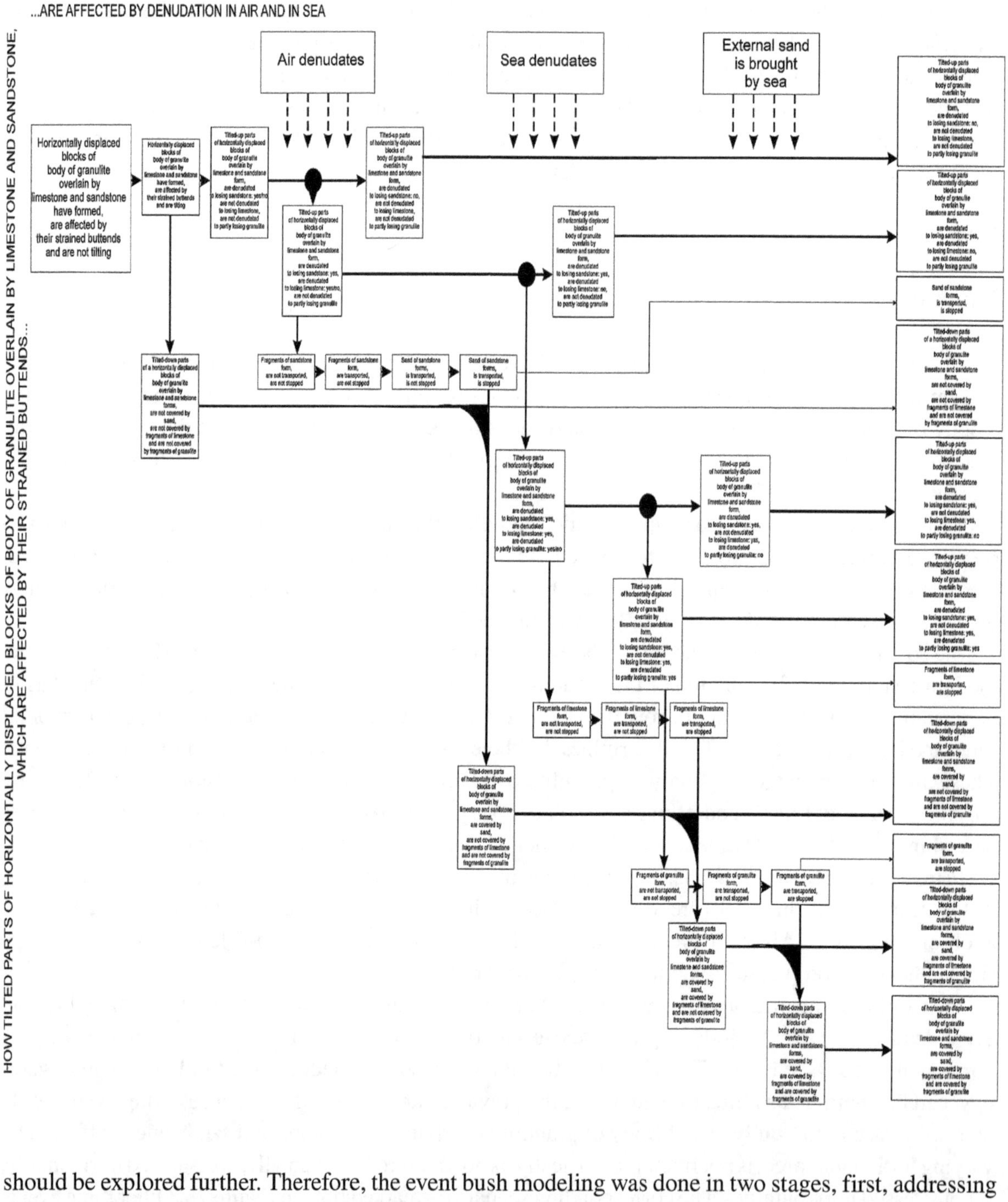

should be explored further. Therefore, the event bush modeling was done in two stages, first, addressing the general task, "How a body of granulite overlain by limestone and sandstone is affected by varying horizontal force" (Figure 6) and then, narrowing it in the area shown by a frame in Figure 8 to, "How horizontally displaced blocks of granulite body overlain by limestone and sandstone affected by their strained buttends and not tilting are affected by landscape changes". The term "denudation" was chosen instead of specific terms deflation for air and abrasion for sea to universally describe the destructive job done both by air and sea in sediments and rocks. Then, the semantic rules of event bush require subject be put ahead of predicates. Thus the events "Air denudates" and "Sea denudates" were obtained. Although these expressions sound not conventional for the geoscientists, formally they seem to be correct.

Complete version of the event bush schematically plotted in Figure 8 is too big to be shown in graphic form; therefore it is given in analytic form (in formulae) in Appendices 1 and 2.

The following formal assumptions have been made to build this event bush adequate to the considered environment.

Assumption 1: First all study area was continuously occupied by granulite, then denudation of the granulite occurred, then the area was continuously overlain by limestone, then continuously overlain by sandstone, each rock unit (granulite, limestone, sandstone) lying at similar height, and then another denudation occurred.

Corollaries from this assumption: if sea retreats, in air can be denudated only the sand it leaves, i.e., dunes will form. Tilted-down parts are assumed to be not denudating at all.

Assumption 2: Granulite can be lost only partly (not entirely) and only if sandstone and limestone are denudated; limestone can be denudated only sandstone is; if sand is on top, any rock unit can be denudated only if the sand is denudated.

Assumption 3: Fragments of sandstone turn into sand at transportation.

Note: If eluvium forms (i.e., sandstone is covered by fragments of sandstone, limestone, by fragments of limestone, granulite, by fragments of granulite, this is a fuzzy object belonging to two crisp ones (sensu Zadeh, 1965): "to be not denudated in air" and "to be denudated to losing sandstone/limestone/part of granulite".

Assumption 4: Either a rock unit is covered by external sand brought by sea, or by fragments of its own rock, or by sand of fragments of sandstone mixed with external sand brought by sea, or by nothing.

Note: modern soil is considered to be "nothing" in this context.

Assumption 5: Fragments of rock (and sand of fragments of rock) are always transported downward until stopping in the lowermost point (tilted-down parts of blocks).

Corollary from assumption 5: Fragments of rock may not be brought to tilted-up parts of block.

Assumption 6: Tectonism (faulting, block formation, wrenching and sub-block tilting) occurs only once (though, possibly, continuously) at each site; no part of any block was first tilted up, then down, or vice versa.

Corollaries from Assumption 6: If sub-block tilting may occur only once, a tilted-down part may not be tilted up then, and hence, it is not possible that, once denudated in air, the rock is then denudated in sea and then brought back in subaerial conditions. If within a tilted-down part of a block some fragments were mixed with external sand brought by sea, i.e., this part was covered by sea, it will not become exposed to air again and the fragments of rocks denudated later, i.e., bedded below the rock that has given said fragments, if ever present in this part's section, will also be mixed with external sand brought by sea.

Note: downsagging is not synonymic to sea transgression. Sea may advance for purely eustatic reason.

Assumption 7: Fragments of rock coarser than sand (henceforth fragments) may not travel up in any way (including wind that is believed to be normally capable of bringing only sand or dust, not gravel and pebbles).

Corollaries from this assumption:

Fragments of rocks may be bedded in any form (e.g., in sand layers) only where deposition takes place;

All sections of elevated parts of earthcrust blocks will have no such fragments.

Assumption 8: only fragments of sandstone may produce sand; fragments of other rocks (limestone and granulite) contribute to in situ sand negligibly; external (far-traveling) sand may consist of fragments of whatever rock types.

Assumption 9: the same rock unit (e.g., sandstone, limestone or event the available part of granulite) in the same scenario cannot be denudated twice. Either it is denudated by air, or by sea, or not denudated at all, but not by air and sea together.

This said, we shall proceed to modeling.

Results

In the model, tertiary events mean the geological sections formed as a result of this or that tectonic regime. Importantly, various regimes may lead to formation of the same sequence. The general model is too wide, and full group of tertiary events was inferred only for the narrowed model, which is schematically visualized in Figure 8 and presented entirely in Appendices 1 and 2. From these events, those were selected that correspond to scenarios that we suggest for the Gulf of Mannar area. Totally, there appeared to be 47 such events, each corresponding to particular scenario of evolution of geological sequence given the wrench tectonics. Then, these events were plotted as stratigraphic columns to identify the geologic sequence each of them corresponds to, and it was found that commonly more than one (from two to six) such events lead to the same type of section. These types are ten in number. All of them are illustrated, described and exhaustingly exemplified below. Description of sequences is made bottom-up. In descriptions of events, the main exogenic genetic properties characterizing the event are given in parentheses in the end.

Type 1

Granulite, unconformity, limestone, sandstone (denudated or not on top – Figure 9).

Figure 9. Type 1 section with varieties: (a) – top not denudated (events $iiiE_6$, $iiiE_{18}$, $iiiE_{21}$), (b) - top denudated (events $iiiE_7$, $iiiE_9$).

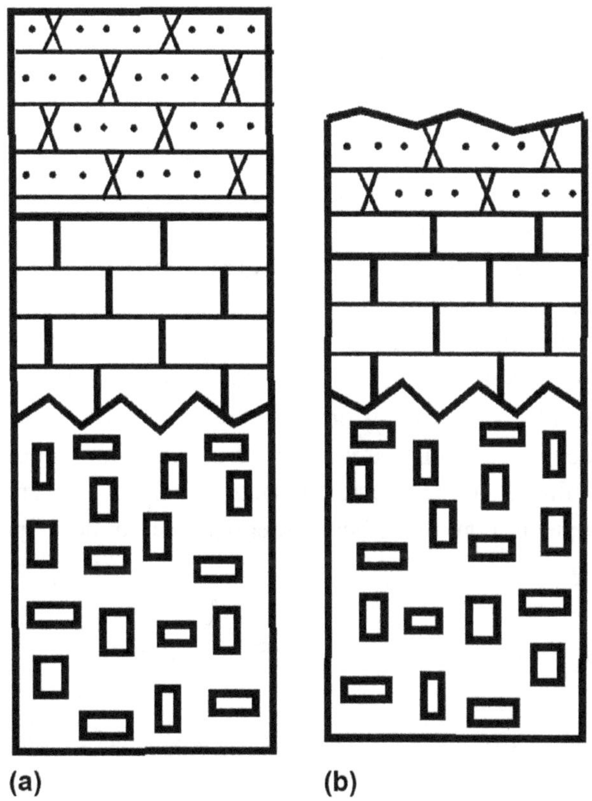

Tertiary Events:

$iiiE_6$: Tilted-up parts of horizontally displaced blocks of body of granulite overlain by limestone and sandstone are elevated, are not denudated in air to losing sandstone, are not denudated in air to losing limestone, are not denudated in air to partly losing granulite, are not denudated in sea to losing sandstone, are not denudated in sea to losing limestone, are not denudated in sea to partly losing granulite, are not covered by external sand brought by sea (no subaerial denudation).

$iiiE_7$: Tilted-up parts of horizontally displaced blocks of body of granulite overlain by limestone and sandstone are elevated, are denudated in air to losing sandstone, are not denudated in air to losing limestone, are not denudated in air to partly losing granulite, are not denudated in sea to losing sandstone, are not denudated in sea to losing limestone, are not denudated in sea to partly losing granulite, are not covered by external sand brought by sea (subaerial denudation).

$iiiE_9$: Tilted-up parts of horizontally displaced blocks of body of granulite overlain by limestone and sandstone are elevated, are not denudated in air to losing sandstone, are not denudated in air to losing limestone, are not denudated in air to partly losing granulite, are denudated in sea to losing sandstone, are not denudated in sea to losing limestone, are not denudated in sea to partly losing granulite, are not covered by external sand brought by sea (subaqueous denudation).

$iiiE_{18}$: Tilted-down parts of horizontally displaced blocks of body of granulite overlain by limestone and sandstone are downsagged, are not covered by sand of fragments of sandstone unmixed with external sand brought by sea, are not covered by sand of fragments of sandstone mixed with external sand brought by sea, are not covered by fragments of limestone unmixed with external sand brought by sea, are not covered by fragments of limestone mixed with external sand brought by sea, are not covered by fragments of granulite unmixed with external sand brought by sea, are not covered by fragments of granulite mixed with external sand brought by sea, are not covered by external sand brought by sea (No subaqueous deposition).

$iiiE_{21}$: Tilted-down parts of horizontally displaced blocks of body of granulite overlain by limestone and sandstone are downsagged, are not covered by sand of fragments of sandstone unmixed with external sand brought by sea, are not covered by sand of fragments of sandstone mixed with external sand brought by sea, are not covered by fragments of limestone unmixed with external sand brought by sea, are not covered by fragments of limestone mixed with external sand brought by sea, are not covered by fragments of granulite unmixed with external sand brought by sea, are not covered by fragments of granulite mixed with external sand brought by sea, are covered by external sand brought by sea (subaqueous conditions but neither denudation, nor deposition).

Type 2

Granulite, unconformity, limestone, sandstone, sand (Figure 10).

Figure 10. Type 2 section with varieties: (a) – just fine-grained sand on top (events $iiiE_{10}$, $iiiE_{36}$), (b) – coarse- and then fine-grained sand on top (event $iiiE_{48}$).

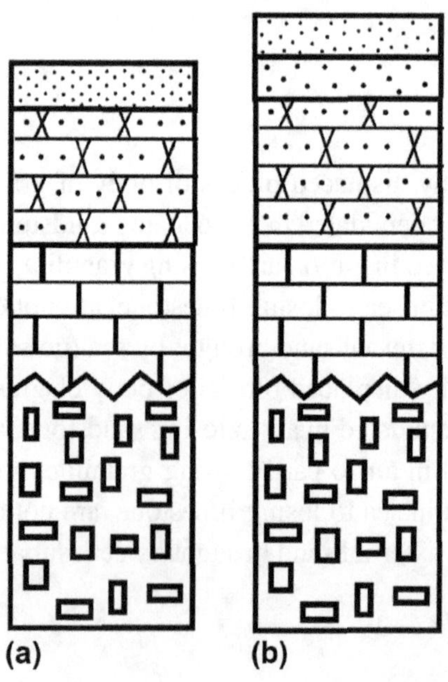

Tertiary Events:

iiiE$_{10}$: Tilted-up parts of horizontally displaced blocks of body of granulite overlain by limestone and sandstone are elevated, are not denudated in air to losing sandstone, are not denudated in air to losing limestone, are not denudated in air to partly losing granulite, are not denudated in sea to losing sandstone, are not denudated in sea to losing limestone, are not denudated in sea to partly losing granulite, are covered by external sand brought by sea (subaqueous deposition).

iiiE$_{36}$: Tilted-down parts of horizontally displaced blocks of body of granulite overlain by limestone and sandstone are downsagged, are covered by sand of fragments of sandstone unmixed with external sand brought by sea, are not covered by sand of fragments of sandstone mixed with external sand brought by sea, are not covered by fragments of limestone unmixed with external sand brought by sea, are not covered by fragments of limestone mixed with external sand brought by sea, are not covered by fragments of granulite unmixed with external sand brought by sea, are not covered by fragments of granulite mixed with external sand brought by sea, are not covered by external sand brought by sea (subaqueous deposition).

iiiE$_{48}$: Tilted-down parts of horizontally displaced blocks of body of granulite overlain by limestone and sandstone are downsagged, are covered by sand of fragments of sandstone unmixed with external sand brought by sea, are not covered by sand of fragments of sandstone mixed with external sand brought by sea, are not covered by fragments of limestone unmixed with external sand brought by sea, are not covered by fragments of limestone mixed with external sand brought by sea, are not covered by fragments of granulite unmixed with external sand brought by sea, are not covered by fragments of granulite mixed with external sand brought by sea, are covered by external sand brought by sea (subaqueous deposition).

Type 3

Granulite, unconformity, limestone, unconformity, sand (Figure 11).

Figure 11. Type 3 section.

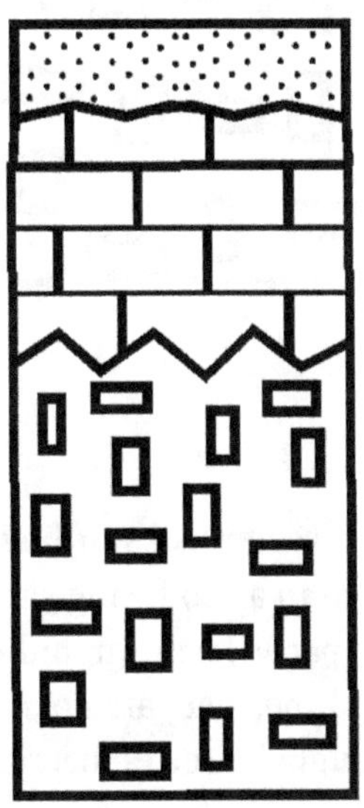

Tertiary Events:

$iiiE_{11}$: Tilted-up parts of horizontally displaced blocks of body of granulite overlain by limestone and sandstone are elevated, are not denudated in air to losing sandstone, are not denudated in air to losing limestone, are not denudated in air to partly losing granulite, are denudated in sea to losing sandstone, are not denudated in sea to losing limestone, are not denudated in sea to partly losing granulite, are covered by external sand brought by sea (subaqueous denudation and deposition).

$iiiE_{15}$: Tilted-up parts of horizontally displaced blocks of body of granulite overlain by limestone and sandstone are elevated, are denudated in air to losing sandstone, are not denudated in air to losing limestone, are not denudated in air to partly losing granulite, are not denudated in sea to losing sandstone, are not denudated in sea to losing limestone, are not denudated in sea to partly losing granulite, are covered by external sand brought by sea (subaerial denudation and subaqueous deposition).

Type 4

Granulite, unconformity, limestone, unconformity (Figure 12).

Figure 12. Type 4 section.

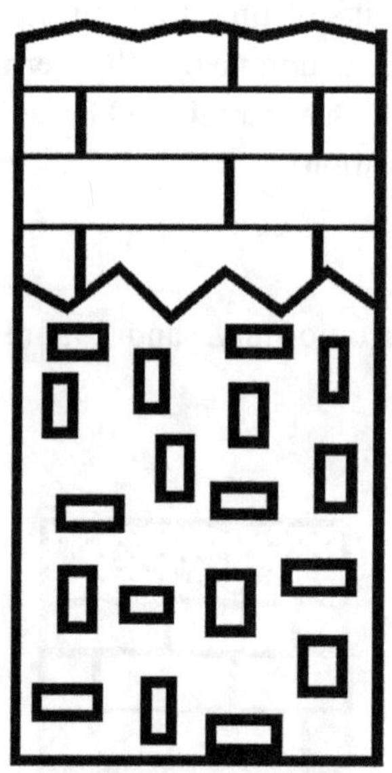

Tertiary Event:

$iiiE_{16}$: Tilted-up parts of horizontally displaced blocks of body of granulite overlain by limestone and sandstone are elevated, are denudated in air to losing sandstone, are not denudated in air to losing limestone, are not denudated in air to partly losing granulite, are not denudated in sea to losing sandstone, are denudated in sea to losing limestone, are not denudated in sea to partly losing granulite, are not covered by external sand brought by sea (subaerial denudation or subaqueous denudation).

Type 5

Granulite, unconformity (Figure 13).

Figure 13. Type 5 section.

Tertiary Events:

iiiE$_{19}$: Tilted-up parts of horizontally displaced blocks of body of granulite overlain by limestone and sandstone are elevated, are denudated in air to losing sandstone, are not denudated in air to losing limestone, are not denudated in air to partly losing granulite, are not denudated in sea to losing sandstone, are denudated in sea to losing limestone, are denudated in sea to partly losing granulite, are not covered by external sand brought by sea (subaerial and then subaqueous denudation).

iiiE$_{23}$: Tilted-up parts of horizontally displaced blocks of body of granulite overlain by limestone and sandstone are elevated, are not denudated in air to losing sandstone, are not denudated in air to losing limestone, are not denudated in air to partly losing granulite, are denudated in sea to losing sandstone, are denudated in sea to losing limestone, are denudated in sea to partly losing granulite, are not covered by external sand brought by sea (subaqueous denudation).

iiiE$_{29}$: Tilted-up parts of horizontally displaced blocks of body of granulite overlain by limestone and sandstone are elevated, are denudated in air to losing sandstone, are denudated in air to losing limestone, are denudated in air to partly losing granulite, are not denudated in sea to losing sandstone, are not denudated in sea to losing limestone, are not denudated in sea to partly losing granulite, are not covered by external sand brought by sea (subaerial denudation).

iiiE$_{35}$: Tilted-up parts of horizontally displaced blocks of body of granulite overlain by limestone and sandstone are elevated, are denudated in air to losing sandstone, are denudated in air to losing limestone, are not denudated in air to partly losing granulite, are not denudated in sea to losing sandstone, are not denudated in sea to losing limestone, are denudated in sea to partly losing granulite, are not covered by external sand brought by sea (subaerial and then subaqueous denudation).

iiiE$_{39}$: Tilted-up parts of horizontally displaced blocks of body of granulite overlain by limestone and sandstone are elevated, are denudated in air to losing sandstone, are denudated in air to losing limestone, are denudated in air to partly losing granulite, are not denudated in sea to losing sandstone, are not denudated in sea to losing limestone, are denudated in sea to partly losing granulite, are not covered by external sand brought by sea (subaerial and then subaqueous denudation).

From these events, event *iiiE*$_{19}$ looks more probable to us than all the rest because others would require too high eustatic sea level rise, which must have been documented in numerous localities along the coast.

Type 6

Granulite, unconformity, sand (Figure 14).

Figure 14. Type 6 section.

Tertiary Events:

$iiiE_{20}$: Tilted-up parts of horizontally displaced blocks of body of granulite overlain by limestone and sandstone are elevated, are denudated in air to losing sandstone, are not denudated in air to losing limestone, are not denudated in air to partly losing granulite, are not denudated in sea to losing sandstone, are denudated in sea to losing limestone, are not denudated in sea to partly losing granulite, are covered by external sand brought by sea (first subaerial denudation, then subaqueous denudation, then subaqueous deposition).

$iiiE_{26}$: Tilted-up parts of horizontally displaced blocks of body of granulite overlain by limestone and sandstone are elevated, are not denudated in air to losing sandstone, are not denudated in air to losing limestone, are not denudated in air to partly losing granulite, are denudated in sea to losing sandstone, are denudated in sea to losing limestone, are not denudated in sea to partly losing granulite, are covered by external sand brought by sea (subaqueous denudation, then subaqueous deposition).

$iiiE_{32}$: Tilted-up parts of horizontally displaced blocks of body of granulite overlain by limestone and sandstone are elevated, are denudated in air to losing sandstone, are denudated in air to losing limestone, are not denudated in air to partly losing granulite, are not denudated in sea to losing sandstone, are not denudated in sea to losing limestone, are not denudated in sea to partly losing granulite, are covered by external sand brought by sea (subaerial denudation, then subaqueous deposition).

$iiiE_{38}$: Tilted-up parts of horizontally displaced blocks of body of granulite overlain by limestone and sandstone are elevated, are denudated in air to losing sandstone, are denudated in air to losing limestone, are not denudated in air to partly losing granulite, are not denudated in sea to losing sandstone, are not denudated in sea to losing limestone, are denudated in sea to partly losing granulite, are covered by external sand brought by sea (subaqueous denudation, then subaqueous deposition).

$iiiE_{40}$: Tilted-up parts of horizontally displaced blocks of body of granulite overlain by limestone and sandstone are elevated, are not denudated in air to losing sandstone, are not denudated in air to losing limestone, are not denudated in air to partly losing granulite, are denudated in sea to losing sandstone, are denudated in sea to losing limestone, are denudated in sea to partly losing granulite, are covered by external sand brought by sea (subaqueous denudation, then subaqueous deposition).

$iiiE_{60}$: Tilted-up parts of horizontally displaced blocks of body of granulite overlain by limestone and sandstone are elevated, are denudated in air to losing sandstone, are denudated in air to losing limestone, are not denudated in air to partly losing granulite, are not denudated in sea to losing sandstone, are not denudated in sea to losing limestone, are denudated in sea to partly losing granulite, are covered by external sand brought by sea (subaerial denudation, then subaqueous denudation, then subaqueous deposition).

Type 7

Granulite, unconformity, limestone, sandstone, sand with fragments of granulite (Figure 15).

Figure 15. Type 7 sections: (a) – with fragments of granulite throughout the sand layer of heterogeneous grain size, (events $iiiE_{22}$, $iiiE_{24}$, $iiiE_{25}$, $iiiE_{27}$), (b) – with fragments of granulite in the upper part of the sand layer, basal sub-layer being of heterogeneous grain size, and upper, coarse-depleted (events $iiiE_{28}$, $iiiE_{30}$).

Tertiary Events:

iiiE_{22}: Tilted-down parts of horizontally displaced blocks of body of granulite overlain by limestone and sandstone are downsagged, are not covered by sand of fragments of sandstone unmixed with external sand brought by sea, are covered by sand of fragments of sandstone mixed with external sand brought by sea, are not covered by fragments of limestone unmixed with external sand brought by sea, are not covered by fragments of limestone mixed with external sand brought by sea, are not covered by fragments of granulite unmixed with external sand brought by sea, are not covered by fragments of granulite mixed with external sand brought by sea, are not covered by external sand brought by sea (subaqueous deposition).

iiiE_{24}: Tilted-down parts of horizontally displaced blocks of body of granulite overlain by limestone and sandstone are downsagged, are not covered by sand of fragments of sandstone unmixed with external sand brought by sea, are not covered by sand of fragments of sandstone mixed with external sand brought by sea, are not covered by fragments of limestone unmixed with external sand brought by sea, are not covered by fragments of limestone mixed with external sand brought by sea, are not covered by fragments of granulite unmixed with external sand brought by sea, are covered by fragments of granulite mixed with external sand brought by sea, are not covered by external sand brought by sea (subaqueous deposition).

iiiE_{25}: Tilted-down parts of horizontally displaced blocks of body of granulite overlain by limestone and sandstone are downsagged, are not covered by sand of fragments of sandstone unmixed with external sand brought by sea, are covered by sand of fragments of sandstone mixed with external sand brought by sea, are not covered by fragments of limestone unmixed with external sand brought by sea, are not covered by fragments of limestone mixed with external sand brought by sea, are not covered by fragments of granulite unmixed with external sand brought by sea, are covered by fragments of granulite mixed with external sand brought by sea, are not covered by external sand brought by sea (subaqueous deposition).

iiiE_{27}: Tilted-down parts of horizontally displaced blocks of body of granulite overlain by limestone and sandstone are downsagged, are not covered by sand of fragments of sandstone unmixed with external sand brought by sea, are not covered by sand of fragments of sandstone mixed with external sand brought by sea, are not covered by fragments of limestone unmixed with external sand brought by sea, are covered by fragments of limestone mixed with external sand brought by sea, are not covered by fragments of granulite unmixed with external sand brought by sea, are not covered by fragments of granulite mixed with external sand brought by sea, are not covered by external sand brought by sea (subaqueous deposition).

iiiE_{28}: Tilted-down parts of horizontally displaced blocks of body of granulite overlain by limestone and sandstone are downsagged, are not covered by sand of fragments of sandstone unmixed with external sand brought by sea, are covered by sand of fragments of sandstone mixed with external sand brought by sea, are not covered by fragments of limestone unmixed with external sand brought by sea, are covered by fragments of limestone mixed with external sand brought by sea, are not covered by fragments of granulite unmixed with external sand brought by sea, are not covered by fragments of granulite mixed with external sand brought by sea, are not covered by external sand brought by sea (subaqueous deposition).

$iiiE_{30}$: Tilted-down parts of horizontally displaced blocks of body of granulite overlain by limestone and sandstone are downsagged, are not covered by sand of fragments of sandstone unmixed with external sand brought by sea, are covered by sand of fragments of sandstone mixed with external sand brought by sea, are not covered by fragments of limestone unmixed with external sand brought by sea, are covered by fragments of limestone mixed with external sand brought by sea, are not covered by fragments of granulite unmixed with external sand brought by sea, are covered by fragments of granulite mixed with external sand brought by sea, are not covered by external sand brought by sea (subaqueous denudation, then subaqueous deposition).

Type 8

Granulite, unconformity, limestone, sandstone, sand, sand with fragments of underlying rocks (Figure 16).

Figure 16. Type 8 sections: (a) – with the top of the sequence: layer of coarse-grained sand, layer of fragments of granulite, layer of fragments of limestone (event $iiiE_{49}$), (b) – with the top of the sequence: layer of coarse-grained sand, layer of fragments of limestone, layer of fragments of granulite (events $iiiE_{50}$, $iiiE_{56}$), (c) – with the top of the sequence: layer of coarse-grained sand, layer of fragments of limestone (event $iiiE_{51}$).

Tertiary Events:

$iiiE_{49}$: Tilted-down parts of horizontally displaced blocks of body of granulite overlain by limestone and sandstone are downsagged, are covered by sand of fragments of sandstone unmixed with external sand brought by sea, are not covered by sand of fragments of sandstone mixed with external sand brought by sea, are covered by fragments of limestone unmixed with external sand brought by sea, are not covered by fragments of limestone mixed with external sand brought by sea, are not covered by fragments of granulite unmixed with external sand brought by sea, are not covered by fragments of granulite mixed with external sand brought by sea, are not covered by external sand brought by sea (subaqueous deposition).

$iiiE_{50}$: Tilted-down parts of horizontally displaced blocks of body of granulite overlain by limestone and sandstone are downsagged, are covered by sand of fragments of sandstone unmixed with external sand brought by sea, are not covered by sand of fragments of sandstone mixed with external sand brought by sea, are covered by fragments of limestone unmixed with external sand brought by sea, are not covered by fragments of limestone mixed with external sand brought by sea, are covered by fragments of granulite unmixed with external sand brought by sea, are not covered by fragments of granulite mixed with external sand brought by sea, are not covered by external sand brought by sea (subaqueous deposition).

$iiiE_{51}$: Tilted-down parts of horizontally displaced blocks of body of granulite overlain by limestone and sandstone are downsagged, are covered by sand of fragments of sandstone unmixed with external sand brought by sea, are not covered by sand of fragments of sandstone mixed with external sand brought by sea, are covered by fragments of limestone unmixed with external sand brought by sea, are not covered by fragments of limestone mixed with external sand brought by sea, are not covered by fragments of granulite unmixed with external sand brought by sea, are not covered by fragments of granulite mixed with external sand brought by sea, are not covered by external sand brought by sea (subaqueous deposition).

$iiiE_{56}$: Tilted-down parts of horizontally displaced blocks of body of granulite overlain by limestone and sandstone are downsagged, are covered by sand of fragments of sandstone unmixed with external sand brought by sea, are not covered by sand of fragments of sandstone mixed with external sand brought by sea, are covered by fragments of limestone unmixed with external sand brought by sea, are not covered by fragments of limestone mixed with external sand brought by sea, are covered by fragments of granulite unmixed with external sand brought by sea, are not covered by fragments of granulite mixed with external sand brought by sea, are not covered by external sand brought by sea (subaqueous deposition).

Type 9

Granulite, unconformity, limestone, sandstone, coarse-grained sand, finer-grained sand with fragments of underlying rocks (Figure 17).

Figure 17. Type 9 sections: (a) – with the top of the sequence: coarse-grained sand, finer-grained sand with fragments of limestone, finer-grained sand with fragments of granulite (events $iiiE_{54}$, $iiiE_{59}$), (b) – with the top of the sequence: coarse-grained sand, finer-grained sand with fragments of limestone (events $iiiE_{53}$, $iiiE_{58}$).

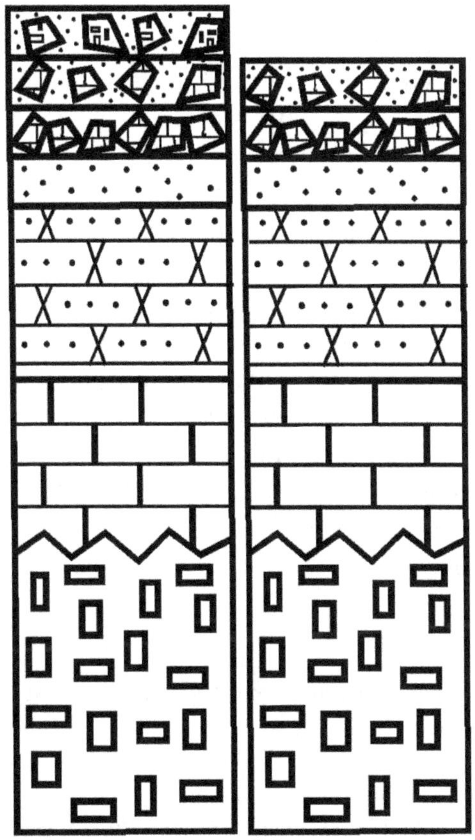

Tertiary Events:

$iiiE_{53}$: Tilted-down parts of horizontally displaced blocks of body of granulite overlain by limestone and sandstone are downsagged, are covered by sand of fragments of sandstone unmixed with external sand brought by sea, are not covered by sand of fragments of sandstone mixed with external sand brought by sea, are covered by fragments of limestone unmixed with external sand brought by sea, are covered by fragments of limestone mixed with external sand brought by sea, are not covered by fragments of granulite unmixed with external sand brought by sea, are not covered by fragments of granulite mixed with external sand brought by sea, are not covered by external sand brought by sea (subaqueous deposition).

$iiiE_{54}$: Tilted-down parts of horizontally displaced blocks of body of granulite overlain by limestone and sandstone are downsagged, are covered by sand of fragments of sandstone unmixed with external sand brought by sea, are not covered by sand of fragments of sandstone mixed with external sand brought by sea, are covered by fragments of limestone unmixed with external sand brought by sea, are covered by fragments of limestone mixed with external sand brought by sea, are not covered by fragments of granulite unmixed with external sand brought by sea, are covered by fragments of granulite mixed with external sand brought by sea, are not covered by external sand brought by sea (subaqueous deposition).

$iiiE_{58}$: Tilted-down parts of horizontally displaced blocks of body of granulite overlain by limestone and sandstone are downsagged, are covered by sand of fragments of sandstone unmixed with external sand brought by sea, are not covered by sand of fragments of sandstone mixed with external sand brought by sea, are not covered by fragments of limestone unmixed with external sand brought by sea, are covered by fragments of limestone mixed with external sand brought by sea, are not covered by fragments of granulite unmixed with external sand brought by sea, are not covered by fragments of granulite mixed with external sand brought by sea, are not covered by external sand brought by sea (subaqueous deposition).

$iiiE_{59}$: Tilted-down parts of horizontally displaced blocks of body of granulite overlain by limestone and sandstone are downsagged, are covered by sand of fragments of sandstone unmixed with external sand brought by sea, are not covered by sand of fragments of sandstone mixed with external sand brought by sea, are not covered by fragments of limestone unmixed with external sand brought by sea, are covered by fragments of limestone mixed with external sand brought by sea, are not covered by fragments of granulite unmixed with external sand brought by sea, are covered by fragments of granulite mixed with external sand brought by sea, are not covered by external sand brought by sea (subaqueous deposition).

Type 10

Granulite, unconformity, limestone, sandstone, coarse-grained sand, sand with fragments of underlying rocks, fine-grained sand (Figure 18).

Figure 18. Type 10 sections: (a) – with a layer of fragments of limestone between coarse- and fine-grained sand layers in the upper part (event $iiiE_{55}$), (b) – with a layer of fragments of limestone and a layer of fragments of granulite between coarse- and fine-grained sand layers in the upper part (event $iiiE_{57}$).

Tertiary Events:

iiiE_{55}: Tilted-down parts of horizontally displaced blocks of body of granulite overlain by limestone and sandstone are downsagged, are covered by sand of fragments of sandstone unmixed with external sand brought by sea, are not covered by sand of fragments of sandstone mixed with external sand brought by sea, are covered by fragments of limestone unmixed with external sand brought by sea, are not covered by fragments of limestone mixed with external sand brought by sea, are not covered by fragments of granulite unmixed with external sand brought by sea, are not covered by fragments of granulite mixed with external sand brought by sea, are covered by external sand brought by sea (subaqueous deposition).

iiiE_{57}: Tilted-down parts of horizontally displaced blocks of body of granulite overlain by limestone and sandstone are downsagged, are covered by sand of fragments of sandstone unmixed with external sand brought by sea, are not covered by sand of fragments of sandstone mixed with external sand brought by sea, are covered by fragments of limestone unmixed with external sand brought by sea, are not covered by fragments of limestone mixed with external sand brought by sea, are covered by fragments of granulite unmixed with external sand brought by sea, are not covered by fragments of granulite mixed with external sand brought by sea, are covered by external sand brought by sea (subaqueous deposition).

Now we shall examine how these inferred sections correspond with the stratigraphic data on the study area.

DISCUSSION

Linking of the Event Bush Model to Local Field Data

Like in any model, the kit of sections suggested above is an idealization. Real outcrops offer much more complicated and fuzzy picture than those inferred in the event bush. In addition, the outcrops commonly exhibit only part of the sequence, and therefore, observing, say, sandstone conformably overlain by coarse-grained sand, with the lower and upper parts of the section covered by vegetation or obscured by scree (or removed artificially – say, at construction works), one has enough reason to relate this section to types 2, 7, 8 or 9, and only additional clearing may decrease or eliminate this uncertainty. However, when cleared up enough, any observed section in the study area will be related uniquely to particular type of inferred section.

Still, in many cases we have enough ground to certainly relate known outcrops to this or that type of inferred sections. This relation is presented in Figure 19.

The relation presented in Figure 19 can be illustrated by particular sections, allowing for inevitable uncertainty discussed above. In many cases, one may cope with this uncertainty by prior geological knowledge used in the event bush model. For instance, an outcrop at Vattakottai: shell limestone (visible thickness 3-4 m), soil (0.5 m), even though granuite is not exposed there, should be definitely referred to type 4. Likewise, another site, at Aruvaimozhi (Teri sand, 4-6 m, soil, 1-2 m), is reliably regarded as type 6. The section at Kudangulam (limestone 10-12 m, Teri sand 8-10 m, soil 0.5 m) may be classified as type 1, 2 or 3, 3 being the most probable. The same types may pretend to interpret the sequence at

Tysyanviali (limestone of visible thickness of about 2 m, carbonate sandstone of 1-2 m, Teri sand of 3-4 m and soil less than 0.5 m), and again type 3 looks more plausible judging by field impression. In the Ovari – Tiruchendur stretch, limestone of unknown thickness occurs (visible thickness at least 1 m) overlain by 3-4 m-thick sandstone covered, in turn, by Teri sand dunes reaching 40-55 m height and only locally topped by negligible soil layer (less than 0.5 m). This sequence is also interpreted as likely type 3 but possibly 1 or 2. Even bare sand dunes 7-10 m high observed in the Sayakudi – Pamban area may be related to types 2, 3, 6 or maybe even 10, but of course much more precise field studies are required.

Figure 19. Occurrence of inferred section types (numbered circles at the left) in localities of the study area (listed on the right)

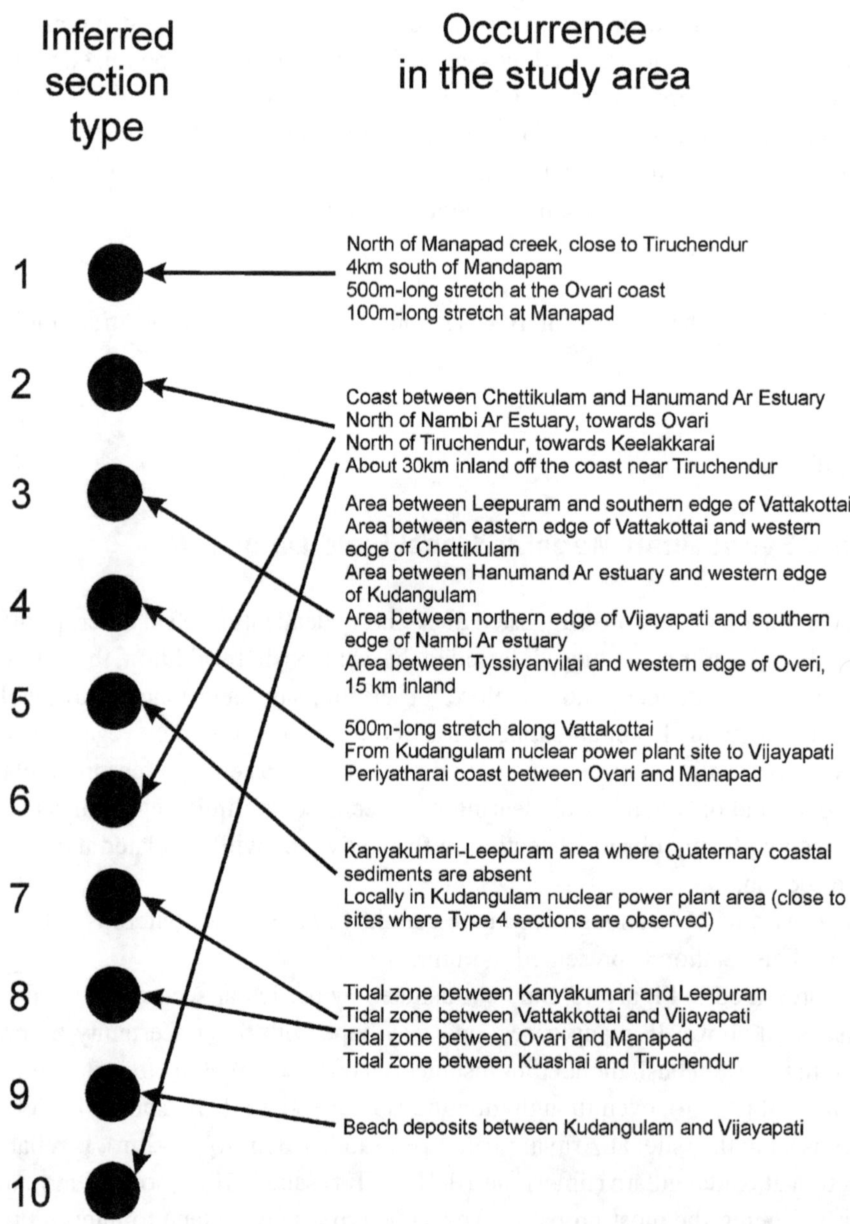

Still, nowhere in the study area the sequences were found that are uncovered by the proposed model (i.e., representing sections other than inferred). Hence, the main question of this work, whether the hypothesis of wrench tectonics is supported by existing field data, should be answered positively. Of course, abundant fieldwork is needed to further constrain the interpretation of regional stratigraphy and decrease the related uncertainty, and final solution should include contouring the elongated blocks stuck between paired strike-slip faults and tilted-up and tilted-down parts of sub-blocks of these blocks. Importantly, not necessarily one (as is accepted in geology), but a few versions of hypothesized blocks with extension-collision zones within them could be plotted if these appear equally well supported by the data. Theoretically, though not necessary in the considered case, data on dikes and geophysical anomalies also can be included in the event bush model.

Future Work

In this study, the regional context is used to look for solution of a methodological query, whether a hypothesis in regional geology can be impartially verified by existing field data, however uncertain and incomplete, by a knowledge representation/information modeling technique. The main question put forth by the results presented above is, what experience can be gained for other regions and other types of geological structure. This question has three aspects: (i) methodology, (ii) information technology and (iii) fieldwork technology.

Methodologically, the model developed at stage two (Figure 8) could be formulated in a more concise and universal way, as a set of general rules, from which the events describing evolution of a granulite-limestone-sandstone body would follow as a particular case.

Then, the development of the general model plotted in Figure 8c would give a framework for lithotectonic studies throughout the world able to test (and perhaps correct) the tectonic hypotheses against field data.

Nonetheless, practical feasibility of this approach largely depends on the computer-based tools to build event bushes. At present, this is being done by vector graphic software not specific for this task and offering no opportunity to test the correctness and completeness of syntax and semantics and exclude repetitions, as well as merge, store, exchange and query the event bushes.

When such software is developed, this would inevitably lead to a change of philosophy of geological fieldwork predicted, in yet larger scale, by a pioneer of application of information technologies in geology (Loudon, 2000).With an event bush editor in hands, geologist would be able to prepare before the fieldwork, along with traditional primary materials, an event bush or, better, a kit of vent bushes describing alternative hypotheses of geological history of the study area. In the field, each new piece of data would be not only recorded and stored by existing means but also put in an event bush framework to support or refute prior hypotheses or, possibly, to correct them right on the way. Similarly, the program of field campaign could be promptly modified (in case of marine work especially) if new data require to do so. Finally, the combination of event bush and geographical information system (GIS) would enable not only conceptual but spatially-focused support or refutation of hypotheses.

CONCLUSION

The event bush modeling has shown that the hypothesis of wrench tectonics is compliant with the geological structure of the region of Gulf of Mannar. This experience can be used in other regions and for other tectonic or paleogeographic hypotheses. Nevertheless, for reliable and wide application of this technique specific software tools are required. When such software is developed, this may strongly influence the philosophy of geological fieldwork.

ACKNOWLEDGMENT

The authors are deeply obliged to Denis Leontiev for plotting figures and technical assistance in the event bush modeling. Their help was invaluable, and this work could not take place otherwise.

REFERENCES

Armstrong-Altrin, J. S., Madhavaraju, J., Asir, N. G. G., & Ramasamy, S. (2005). *Provenance and depositional history of sandstones from the Upper Miocene Kudankulam Formation*. Tamil Nadu.

Armstrong-Altrin, J. S., Ramasamy, S., & Makhnach, A. (2001). Stable Isotope Geochemistry and Evidence for Meteoric Diagenesis in Kudankulam Formation,Tamil Nadu. *Journal of the Geological Society of India*, *57*, 39–48.

Armstrong-Altrin, J. S., Verma, S. P., Lee, Y. I. I., & Ramasamy, S. (2004). Geochemistry of sandstones from the upper Miocene Kudankulam Formation, southern India: Implications for provenance, weathering, and tectonic setting. *Journal of Sedimentary Research*, *74*(2), 285–297. doi:10.1306/082803740285

Armstrong-Altrin, J. S., Verma, S. P., Madhavaraju, J., Lee, Y. I. I., & Ramasamy, S. (2003). Geochemistry of Upper Miocene Kudankulam Limestones, Southern India. *International Geology Review*, *45*(1), 16–26. doi:10.2747/0020-6814.45.1.16

Baillie, P. W., Shaw, R. D., Liyanaarachchi, D. T. P., & Jayaratne, M. G. (2002). *A New Mesozoic Sedimentary Basin, Offshore Sri Lanka.* EAGA 64th Conference & Exhibition, Florence, Italy.

Braun, I., & Kriegsman, L. M. (2003). Proterozoic crustal evolution of southernmost India and Sri Lanka. In M. Yoshida & B. F. Windley (Eds.), Geological Society, London, Special Publications: Vol. 206. Proterozoic East Gondwana: Supercontinents Assembly and Breakup (pp. 169–203). Academic Press. doi:10.1144/GSL.SP.2003.206.01.10

Chandra, S., Gideon, N. D., Chaturvedi, C., & Rana, M. S. (2006). *6th International Conference & Exposition on Petroleum Geophysics*. Retrieved from https://www.spgindia.org/conference/6thconf_kolkata06/161.pdf

Chauhan, T., Longhinos, B., & Seralathan, P. (2008). Characterization of carbonate substratum in and around Chettikulam-Shrirangnarayanapuram villages, Radhapuram Taluk, Tirunelveli, Tamilnadu, India. *Proceedings of the XXV Annual Convention of Indian Association of Sedimentologists and National Seminar on Sedimentary basins of India - Their geological significance and Economic Prospects,* 49-50.

Cooray, P. G. (1984). Geology with special reference to the Precambrian. In C. H. Fernando (Ed.), *Ecology and Biogeography of Sri Lanka* (pp. 1–34). The Hague, The Netherlands: Dr. W. Junk B.V. Publishers. doi:10.1007/978-94-009-6545-4_1

Curray, J. R. (1984). Sri Lanka: Is it a mid-plate platelet? *J. Nara*, *31*, 30–50.

Damodaran, T., Srinivasa Rao, G., Radhakrishna, M., & Murthy, K. S. R. (2016). Crustal structure and rift tectonics across the Cauvery–Palar basin, Eastern Continental Margin of India based on seismic and potential field modeling. *Journal of Earth System Science*, *125*(2), 329–342. doi:10.1007/s12040-016-0669-y

Diviacco, P., Juan, A. A., Daradoumis, T., Roca, M., Grasman, S. E., & Faulin, J. (2012). Addressing Conflicting Cognitive Models in Collaborative E-Research: A Case Study in Exploration Geophysics in Collaborative and Distributed E-Research. In Collaborative and Distributed E-Research: Innovations in Technologies, Strategies and Applications. IGI Global.

Drury, S. A., Harris, N. B. W., Holt, R. W., Reeves-Smith, G. J., & Wightman, R. T. (1984). Pre-Cambrian Tectonics and Crustal Evolution in South India. *Journal of Geophysics*, *92*, 3–20.

Frodeman, R. (1995). Geological reasoning: Geology as an interpretive and historical science. *Geological Society of America Bulletin*, *107*(8), 960–968. doi:10.1130/0016-7606(1995)107<0960:GRGAAI>2.3.CO;2

Ghosh, J. G., de Wit, M. J., & Zartman, R. E. (2004). Age and Tectonic Evolution of Neoproterozoic Ductile Shear Zones in the Southern Granulite Terrain of India, with implications for Gondwana studies. *Tectonics*, 23.

Gopalakrishnan, K., & Varadan, G. N. (1996). Geothermal Manifestations and their relation to Geotectonic features along Tamil Nadu-Pondicherry Coast. Geothermal Energy in India. Geological Survey of India, 45, 25-37.

Guru-Rajesh, K., & Chetty, T. R. K. (2006). Structure and tectonics of the Achankovil shear zone, Southern India. *Gondwana Research*, *10*(1-2), 86–98. doi:10.1016/j.gr.2005.11.018

Harris, N. B. W., & Santosh, M. (1993). Geological Society of India Memoir: Vol. 25. Chronological constraints on granulite formation in southern India and Sri Lanka. Academic Press.

Harris, N. B. W., Santosh, M., & Taylor, P. N. (1994). Crustal Evolution in South India: Constraints from Nd Isotopes. *The Journal of Geology*, *102*(2), 139–150. doi:10.1086/629659

Kanishkan, B., & Renganathan, M. (1990). *Study of the Tertiary formations to the south of Coleroon river to locate possible lignite bearing areas in Tamil Nadu*. Geological Survey of India, Unpubished Report.

Katz, M. B. (1978). Sri Lanka in Gondwanaland and the Evolution of Indian Ocean. *Geological Magazine*, *115*(4), 237–244. doi:10.1017/S0016756800037146

Longhinos, B., Anand, S. P., & Rajaram, M. (2010). Physical Geology of Subvolcanic Systems: Laccolith, Sills and Dykes. *Proceedings, LASI 4 Conference*, 22-26.

Longhinos, B., & Ramasarma, M. (2008). *Seismo-tectonic signatures in and around Kudankulam, Tirunelveli, India. In Proceedings in Nuclear Power safety*. Kudangulam.

Loudon, T. V. (2000). Geoscience after IT. *Computers & Geosciences*, 26.

Maitra, A., Chakrobarty, Chandan, Rana, M.S. & Chandra, S. (2007). Tectono-sedimentation model and hydrocarbon prospectivity of Gulf of Mannar, Cauvery Basin. *KDMIPE, ONGC Unpub. Report*, 17.

Narayanaswami, S. (1947). *The geology and mineral resources of the Tiruchendur and southern and eastern portions of the Nanguneri taluk, Tirunelvelli District.* Unpublished Report, GSI P.R.

ONGC. (1993a). *Lithostratigraphy of the Indian Petroliferous Basins; document VII-Cauvery Basin. KDM Instt. Petro. Expl.* Dehra Dun, India: Oil and Natural Gas Commission.

ONGC. (1993b). *Fault systems of part of coastal belt of Tamil Nadu; document VII-Cauvery Basin. KDM Instt. Petro. Expl.* Dehra Dun, India: Oil and Natural Gas Commission.

ONGC. (1993c). *Composite Bouguer Gravity Map of Cauvery basin, Tamil Nadu; document VII-Cauvery Basin. KDM Instt. Petro. Expl.* Dehra Dun, India: Oil and Natural Gas Commission.

Paramasivam, T. S., & Srinivasan, R. (1980). Sedimentary Limestones of Tisaiyanvilai Area, Tirunelveli District, Tamil Nadu. *Oil And Natural Gas Commission of India Bulletin, 17*, 183–186.

Project Vasundara, GSI. (1994). *Seismotectonic map*. Geological Survey of India. T-VI.

Pshenichny, C. A., & Kanzheleva, O. M. (2011). Theoretical foundations of the event bush method. In K. Sinha, L. Gundersen, J. Jackson, & D. Arctur (Eds.), Geological Society of America Special Paper: Vol. 482. Societal Challenges and Geoinformatics (pp. 139–165). Academic Press. doi:10.1130/2011.2482(12)

Pshenichny, C. A., Nikolenko, S. I., Carniel, R., Sobissevitch, A. L., Vaganov, P. A., Khrabrykh, Z. V., . . . Gusev, E. A. (2008). The event bush as a potential complex methodology of conceptual modelling in the geosciences. *Proceedings, iEMSs – International Congress on Environmental Modelling and Software, 2*, 900-912.

Pshenichny, C. A., Nikolenko, S. I., Carniel, R., Vaganov, P. A., Khrabrykh, Z. V., Moukhachov, V. P., & Rezyapkin, A. A. et al. (2009). The event bush as a semantic-based numerical approach to natural hazard assessment (exemplified by volcanology). *Computers & Geosciences, 35*(5), 1017–1034. doi:10.1016/j.cageo.2008.01.009

Raiverman, V. (1992). Trans-Asiatic lineaments and Himalayan orogeny. In A. K. Sinha (Ed.), Himalayalan orogen and global tectonics. The International Lithosphere Programme.

Ramalingeswara Rao, B. (1992). Seismicity and Geodynamics of the low- to high-grade transition zone of Peninsular India. *Tectonophysics, 201*(1-2), 175–185. doi:10.1016/0040-1951(92)90182-6

Ramanathan, S. (1979). Tertiary Formations of South India. *Miscellaneous Publication - Geological Survey of India, 45*, 165–180.

Ramasamy, S. (1995). Petrography and Major Element Geochemistry of Sedimentary Rocks exposed around Kudankulam, Tamil Nadu. In *Proceedings, XIth Convention of Indian Association of Sedimentologists*. Roorkee University.

Ramasamy, S., & Armstrong-Altrin, J. S. (1998). Inferences on Rhodoids from Neogene arbonates of Kudankulam, Tamil Nadu. *Journal of the Geological Society of India*, *52*, 341–344.

Ramasamy, S., Armstrong-Altrin, J. S., & Chandrasekar, K. (1996). Heavy Mineral Distribution around Manavalakurichi, Kanyakumari District, Tamil Nadu, India. Indian Association of Sedimentologists.

Ramasamy, S., Paul, B. A. J., & Madhavaraju, J. (1994). Stratigraphy, petrography and sedimentation history of Kudankulam formation along the south Tamil Nadu coast, India. *Bulletin of Pure and Applied Sciences*, *13*, 43–58.

Rana, M. S., Chakraborty, C., Sharma, R., & Giridhar, M. (2008). Mannar Volcanics - Implications for Madagascar Breakup. *Proceedings, 7th International Conference & Exposition on Petroleum Geophysics*.

Rathore, S. S., Bansal, M., Vijan, A. R., & Pangtey, P. S. (2007). Ar-Ar dating of intrusives from Gulf of Mannar sub-basin of Cauvery Basin. *KDMIPE, ONGC Unpublished Report*, 17.

Reddy, A. G. B. (1995). A Geophysical approach to the problems of seismicity in the Indian shield. *Journal of the Geological Society of India*, *45*, 5–17.

Sahani, A. (1979). Miocene Vertebrates from the Coastal Tertiary Rocks of Peninsular India and Sri Lanka. *Miscellaneous Publication - Geological Survey of India*, *45*, 197–205.

Santosh, M., Kagami, H., Yoshida, M., & Nanda-Kumar, V. (1992). Pan-African charnockite formation in East Gondwana: Geochronologic (Sm-Nd and Rb-Sr) and petrogenetic constraints. *Bulletin of Indian Geological Association*, *25*, 1–10.

Santosh, M., Tanaka, K., Yokoyama, K., & Collins, A. S. (2005). Late Neoproterozoic–Cambrian feisic magnetism along transcrustal shear zones in southern India: U–Pb electron microprobe ages and implications for the amalgamation of the Gondwana supercontinent. *Gondwana Research*, *8*(1), 31–42. doi:10.1016/S1342-937X(05)70260-4

Schröder, B. S. W. (2003). *Ordered Sets: An Introduction*. Boston: Birkhäuser. doi:10.1007/978-1-4612-0053-6

Sreejith, C., & Ravindra Kumar, G. R. (2013). Petrogenesis of high-K metagranites in the Kerala Khondalite Belt, southern India: A possible magmatic-arc link between India, Sri Lanka, and Madagascar. *Journal of Geodynamics*, *63*, 69–82. doi:10.1016/j.jog.2012.10.002

Storetvedt, K. M. (2003). *Global wrench tectonics: theory of earth evolution*. Bergen: Fagbokforlaget.

Storetvedt, K. M., & Longhinos, B. (2012) The Atlantic and its bordering continents – a wrench tectonic analysis: lithospheric deformation, basin histories and major hydrocarbon provinces. *New Concepts in Global Tectonics Newsletter, 64*.

Subramanian, K. S., & Gopalakrishnan, K. (2002). Earth tremors in the coastal belt of Tamil Nadu and Pondicherry in September, 2001 – A Geological Perspective. *Journal of the Geological Society of India*, *60*, 691–694.

Sylvester, A. G. (1958). Strike-slip Faults. *Geological Society of America Bulletin*, *100*(11), 1666–1703. doi:10.1130/0016-7606(1988)100<1666:SSF>2.3.CO;2

van der Pluijm, B. (2016). Wrench Tectonics: Strike-slip Boundaries. *Processes in Structural Geology & Tectonics*. Retrieved from http://psgt.earth.lsa.umich.edu/PDF/14_WrenchTectonics_2016.pdf

Vaz, G., Subbarao, V., & Ravikumar, V. (2006). Thermal Springs in Indian Coastal Areas of the Palk Bay: Their implication in Relation to Lineaments, Coastal Morphology and Seismicity. *Journal of the Geological Society of India*, *68*, 593–596.

APPENDIX 1

List of Events in the Event Bush Model of Wrench Tectonics Mechanism

- iaE_1: Horizontally displaced blocks of body of granulite overlain by limestone and sandstone, which are affected by their strained buttends, form and are not tilting.
- ibE_2: Air denudates.
- ibE_3: Sea denudates.
- ibE_4: External sand is brought by sea.
- iiE_5: Horizontally displaced blocks of body of granulite overlain by limestone and sandstone, which are affected by their strained buttends, form and are tilting.
- iiE_6: Tilted-up parts of horizontally displaced blocks of body of granulite overlain by limestone and sandstone are elevated, are not denudated in air to losing sandstone, are not denudated in air to losing limestone, are not denudated in air to partly losing granulite, are not denudated in sea to losing sandstone, are not denudated in sea to losing limestone, are not denudated in sea to partly losing granulite, are not covered by external sand brought by sea, are not covered by sand of fragments of sandstone mixed with external sand brought by sea.
- iiE_7: Tilted-up parts of horizontally displaced blocks of body of granulite overlain by limestone and sandstone are elevated, are denudated in air to losing sandstone, are not denudated in air to losing limestone, are not denudated in air to partly losing granulite, are not denudated in sea to losing sandstone, are not denudated in sea to losing limestone, are not denudated in sea to partly losing granulite, are not covered by external sand brought by sea, are not covered by sand of fragments of sandstone mixed with external sand brought by sea.
- iiE_8: Fragments of sandstone form, are not transported to tilted-down parts, are not stopped in tilted-down parts.
- iiE_9: Tilted-up parts of horizontally displaced blocks of body of granulite overlain by limestone and sandstone are elevated, are not denudated in air to losing sandstone, are not denudated in air to losing limestone, are not denudated in air to partly losing granulite, are denudated in sea to losing sandstone, are not denudated in sea to losing limestone, are not denudated in sea to partly losing granulite, are not covered by external sand brought by sea, are not covered by sand of fragments of sandstone mixed with external sand brought by sea.
- iiE_{10}: Tilted-up parts of horizontally displaced blocks of body of granulite overlain by limestone and sandstone are elevated, are not denudated in air to losing sandstone, are not denudated in air to losing limestone, are not denudated in air to partly losing granulite, are not denudated in sea to losing sandstone, are not denudated in sea to losing limestone, are not denudated in sea to partly losing granulite, are covered by external sand brought by sea, are not covered by sand of fragments of sandstone mixed with external sand brought by sea.
- iiE_{11}:Tilted-up parts of horizontally displaced blocks of body of granulite overlain by limestone and sandstone are elevated, are not denudated in air to losing sandstone, are not denudated in air to losing limestone, are not denudated in air to partly losing granulite, are denudated in sea to losing sandstone, are not denudated in sea to losing limestone, are not denudated in sea to partly losing granulite, are covered by external sand brought by sea, are not covered by sand of fragments of sandstone mixed with external sand brought by sea.

- iE_{12}: Fragments of sandstone form, are transported to tilted-down parts, are not stopped in tilted-down parts.
- iiE_{13}: Sand of fragments of sandstone forms, is transported to tilted-down parts, is not stopped in tilted-down parts, is not mixed with external sand brought by sea.
- iiE_{14}: Sand of fragments of sandstone forms, is transported to tilted-down parts, is stopped in tilted-down parts, is not mixed with external sand brought by sea.
- iiE_{15}: Tilted-up parts of horizontally displaced blocks of body of granulite overlain by limestone and sandstone are elevated, are denudated in air to losing sandstone, are not denudated in air to losing limestone, are not denudated in air to partly losing granulite, are not denudated in sea to losing sandstone, are not denudated in sea to losing limestone, are not denudated in sea to partly losing granulite, are covered by external sand brought by sea, are not covered by sand of fragments of sandstone mixed with external sand brought by sea.
- iiE_{16}: Tilted-up parts of horizontally displaced blocks of body of granulite overlain by limestone and sandstone are elevated, are denudated in air to losing sandstone, are not denudated in air to losing limestone, are not denudated in air to partly losing granulite, are not denudated in sea to losing sandstone, are denudated in sea to losing limestone, are not denudated in sea to partly losing granulite, are not covered by external sand brought by sea, are not covered by sand of fragments of sandstone mixed with external sand brought by sea.
- iiE_{17}: Sand of fragments of sandstone forms, is transported to tilted-down parts, is stopped in tilted-down parts, is mixed with external sand brought by sea.
- iiE_{18}: Tilted-down parts of horizontally displaced blocks of body of granulite overlain by limestone and sandstone are down sagged, are not covered by sand of fragments of sandstone unmixed with external sand brought by sea, are not covered by sand of fragments of sandstone mixed with external sand brought by sea, are not covered by fragments of limestone unmixed with external sand brought by sea, are not covered by fragments of limestone mixed with external sand brought by sea, are not covered by fragments of granulite unmixed with external sand brought by sea, are not covered by fragments of granulite mixed with external sand brought by sea, are not covered by external sand brought by sea.
- iiE_{19}: Tilted-up parts of horizontally displaced blocks of body of granulite overlain by limestone and sandstone are elevated, are denudated in air to losing sandstone, are not denudated in air to losing limestone, are not denudated in air to partly losing granulite, are not denudated in sea to losing sandstone, are denudated in sea to losing limestone, are denudated in sea to partly losing granulite, are not covered by external sand brought by sea, are not covered by sand of fragments of sandstone mixed with external sand brought by sea.
- iiE_{20}: Tilted-up parts of horizontally displaced blocks of body of granulite overlain by limestone and sandstone are elevated, are denudated in air to losing sandstone, are not denudated in air to losing limestone, are not denudated in air to partly losing granulite, are not denudated in sea to losing sandstone, are denudated in sea to losing limestone, are not denudated in sea to partly losing granulite, are covered by external sand brought by sea, are not covered by sand of fragments of sandstone mixed with external sand brought by sea.
- iiE_{21}: Tilted-down parts of horizontally displaced blocks of body of granulite overlain by limestone and sandstone are down sagged, are not covered by sand of fragments of sandstone unmixed with external sand brought by sea, are not covered by sand of fragments of sandstone mixed with

external sand brought by sea, are not covered by fragments of limestone unmixed with external sand brought by sea, are not covered by fragments of limestone mixed with external sand brought by sea, are not covered by fragments of granulite unmixed with external sand brought by sea, are not covered by fragments of granulite mixed with external sand brought by sea, are covered by external sand brought by sea.

- iiE_{22}: Tilted-down parts of horizontally displaced blocks of body of granulite overlain by limestone and sandstone are down sagged, are not covered by sand of fragments of sandstone unmixed with external sand brought by sea, are covered by sand of fragments of sandstone mixed with external sand brought by sea, are not covered by fragments of limestone unmixed with external sand brought by sea, are not covered by fragments of limestone mixed with external sand brought by sea, are not covered by fragments of granulite unmixed with external sand brought by sea, are not covered by fragments of granulite mixed with external sand brought by sea, are not covered by external sand brought by sea.
- iiE_{23}: Tilted-up parts of horizontally displaced blocks of body of granulite overlain by limestone and sandstone are elevated, are not denudated in air to losing sandstone, are not denudated in air to losing limestone, are not denudated in air to partly losing granulite, are denudated in sea to losing sandstone, are denudated in sea to losing limestone, are not denudated in sea to partly losing granulite, are not covered by external sand brought by sea, are not covered by sand of fragments of sandstone mixed with external sand brought by sea.
- iiE_{24}: Tilted-down parts of horizontally displaced blocks of body of granulite overlain by limestone and sandstone are down sagged, are not covered by sand of fragments of sandstone unmixed with external sand brought by sea, are not covered by sand of fragments of sandstone mixed with external sand brought by sea, are not covered by fragments of limestone unmixed with external sand brought by sea, are not covered by fragments of limestone mixed with external sand brought by sea, are not covered by fragments of granulite unmixed with external sand brought by sea, are covered by fragments of granulite mixed with external sand brought by sea, are not covered by external sand brought by sea.
- iiE_{25}: Tilted-down parts of horizontally displaced blocks of body of granulite overlain by limestone and sandstone are down sagged, are not covered by sand of fragments of sandstone unmixed with external sand brought by sea, are covered by sand of fragments of sandstone mixed with external sand brought by sea, are not covered by fragments of limestone unmixed with external sand brought by sea, are not covered by fragments of limestone mixed with external sand brought by sea, are not covered by fragments of granulite unmixed with external sand brought by sea, are covered by fragments of granulite mixed with external sand brought by sea, are not covered by external sand brought by sea.
- iiE_{26}: Tilted-up parts of horizontally displaced blocks of body of granulite overlain by limestone and sandstone are elevated, are not denudated in air to losing sandstone, are not denudated in air to losing limestone, are not denudated in air to partly losing granulite, are denudated in sea to losing sandstone, are denudated in sea to losing limestone, are not denudated in sea to partly losing granulite, are covered by external sand brought by sea, are not covered by sand of fragments of sandstone mixed with external sand brought by sea.
- iiE_{27}: Tilted-down parts of horizontally displaced blocks of body of granulite overlain by limestone and sandstone are down sagged, are not covered by sand of fragments of sandstone unmixed

with external sand brought by sea, are not covered by sand of fragments of sandstone mixed with external sand brought by sea, are not covered by fragments of limestone unmixed with external sand brought by sea, are covered by fragments of limestone mixed with external sand brought by sea, are not covered by fragments of granulite unmixed with external sand brought by sea, are not covered by fragments of granulite mixed with external sand brought by sea, are not covered by external sand brought by sea.

- iiE_{28}: Tilted-down parts of horizontally displaced blocks of body of granulite overlain by limestone and sand stone are down sagged, are not covered by sand of fragments of sandstone unmixed with external sand brought by sea, are covered by sand of fragments of sandstone mixed with external sand brought by sea, are not covered by fragments of limestone unmixed with external sand brought by sea, are covered by fragments of limestone mixed with external sand brought by sea, are not covered by fragments of granulite unmixed with external sand brought by sea, are not covered by fragments of granulite mixed with external sand brought by sea, are not covered by external sand brought by sea.
- iiE_{29}: Tilted-up parts of horizontally displaced blocks of body of granulite overlain by limestone and sandstone are elevated, are denudated in air to losing sandstone, are denudated in air to losing limestone, are not denudated in air to partly losing granulite, are not denudated in sea to losing sandstone, are not denudated in sea to losing limestone, are not denudated in sea to partly losing granulite, are not covered by external sand brought by sea.
- iiE_{30}: Tilted-down parts of horizontally displaced blocks of body of granulite overlain by limestone and sandstone are down sagged, are not covered by sand of fragments of sandstone unmixed with external sand brought by sea, are covered by sand of fragments of sandstone mixed with external sand brought by sea, are not covered by fragments of limestone unmixed with external sand brought by sea, are covered by fragments of limestone mixed with external sand brought by sea, are not covered by fragments of granulite unmixed with external sand brought by sea, are covered by fragments of granulite mixed with external sand brought by sea, are not covered by external sand brought by sea.
- iiE_{31}: Fragments of limestone form, are not transported to tilted-down parts, are not stopped in tilted-down parts, are not mixed with external sand brought by sea.
- iiE_{32}: Tilted-up parts of horizontally displaced blocks of body of granulite overlain by limestone and sandstone are elevated, are denudated in air to losing sandstone, are denudated in air to losing limestone, are not denudated in air to partly losing granulite, are not denudated in sea to losing sandstone, are not denudated in sea to losing limestone, are not denudated in sea to partly losing granulite, are covered by external sand brought by sea.
- iiE_{33}: Fragments of limestone form, are transported to tilted-down parts, are not stopped in tilted-down parts are not mixed with external sand brought by sea.
- iiE_{34}: Fragments of limestone form, are transported to tilted-down parts, are stopped in tilted-down parts, are not mixed with external sand brought by sea.
- iiE_{35}: Tilted-up parts of horizontally displaced blocks of body of granulite overlain by limestone and sandstone are elevated, are denudated in air to losing sandstone, are denudated in air to losing limestone, are not denudated in air to partly losing granulite, are not denudated in sea to losing sandstone, are not denudated in sea to losing limestone, are denudated in sea to partly losing granulite, are not covered by external sand brought by sea.

- iiE_{36}: Tilted-down parts of horizontally displaced blocks of body of granulite overlain by limestone and sandstone are down sagged, are covered by sand of fragments of sandstone unmixed with external sand brought by sea, are not covered by sand of fragments of sandstone mixed with external sand brought by sea, are not covered by fragments of limestone unmixed with external sand brought by sea, are not covered by fragments of limestone mixed with external sand brought by sea, are not covered by fragments of granulite unmixed with external sand brought by sea, are not covered by fragments of granulite mixed with external sand brought by sea, are not covered by external sand brought by sea.
- iiE_{37}: Fragments of limestone form, are transported to tilted-down parts, are stopped in tilted-down parts, are mixed with external sand brought by sea.
- iiE_{38}: Tilted-up parts of horizontally displaced blocks of body of granulite overlain by limestone and sandstone are elevated, are denudated in air to losing sandstone, are denudated in air to losing limestone, are not denudated in air to partly losing granulite, are not denudated in sea to losing sandstone, are not denudated in sea to losing limestone, are denudated in sea to partly losing granulite, are covered by external sand brought by sea.
- iiE_{39}: Tilted-up parts of horizontally displaced blocks of body of granulite overlain by limestone and sandstone are elevated, are denudated in air to losing sandstone, are denudated in air to losing limestone, are denudated in air to partly losing granulite, are not denudated in sea to losing sandstone, are not denudated in sea to losing limestone, are not denudated in sea to partly losing granulite, are not covered by external sand brought by sea.
- iiE_{40}: Tilted-up parts of horizontally displaced blocks of body of granulite overlain by limestone and sandstone are elevated, are not denudated in air to losing sandstone, are not denudated in air to losing limestone, are not denudated in air to partly losing granulite, are denudated in sea to losing sandstone, are denudated in sea to losing limestone, are denudated in sea to partly losing granulite, are not covered by external sand brought by sea, are not covered by sand of fragments of sandstone mixed with external sand brought by sea.
- iiE_{41}: Tilted-down parts of horizontally displaced blocks of body of granulite overlain by limestone and sandstone are down sagged, are covered by sand of fragments of sandstone unmixed with external sand brought by sea, are not covered by sand of fragments of sandstone mixed with external sand brought by sea, are not covered by fragments of limestone unmixed with external sand brought by sea, are not covered by fragments of limestone mixed with external sand brought by sea, are not covered by fragments of granulite unmixed with external sand brought by sea, are covered by fragments of granulite mixed with external sand brought by sea, are not covered by external sand brought by sea.
- iiE_{42}: Tilted-down parts of horizontally displaced blocks of body of granulite overlain by limestone and sandstone are down sagged, are covered by sand of fragments of sandstone unmixed with external sand brought by sea, are covered by sand of fragments of sandstone mixed with external sand brought by sea, are not covered by fragments of limestone unmixed with external sand brought by sea, are not covered by fragments of limestone mixed with external sand brought by sea, are not covered by fragments of granulite unmixed with external sand brought by sea, are not covered by fragments of granulite mixed with external sand brought by sea, are not covered by external sand brought by sea.

- iiE_{43}: Tilted-up parts of horizontally displaced blocks of body of granulite overlain by limestone and sandstone are elevated, are denudated in air to losing sandstone, are denudated in air to losing limestone, are denudated in air to partly losing granulite, are not denudated in sea to losing sandstone, are not denudated in sea to losing limestone, are not denudated in sea to partly losing granulite, are covered by external sand brought by sea.
- iiE_{44}: Fragments of granulite form, are not transported to tilted-down parts, are not stopped in tilted-down parts, are not mixed with external sand brought by sea.
- iiE_{45}: Fragments of granulite form, are transported to tilted-down parts, are not stopped in tilted-down parts, are not mixed with external sand brought by sea.
- iiE_{46}: Fragments of granulite form, are transported to tilted-down parts, are stopped in tilted-down parts, are not mixed with external sand brought by sea.
- iiE_{47}: Fragments of granulite form, are transported to tilted-down parts, are stopped in tilted-down parts, are mixed with external sand brought by sea.
- iiE_{48}: Tilted-down parts of horizontally displaced blocks of body of granulite overlain by limestone and sandstone are down sagged, are covered by sand of fragments of sandstone unmixed with external sand brought by sea, are not covered by sand of fragments of sandstone mixed with external sand brought by sea, are not covered by fragments of limestone unmixed with external sand brought by sea, are not covered by fragments of limestone mixed with external sand brought by sea, are not covered by fragments of granulite unmixed with external sand brought by sea, are not covered by fragments of granulite mixed with external sand brought by sea, are covered by external sand brought by sea.
- iiE_{49}: Tilted-down parts of horizontally displaced blocks of body of granulite overlain by limestone and sandstone are down sagged, are covered by sand of fragments of sandstone unmixed with external sand brought by sea, are not covered by sand of fragments of sandstone mixed with external sand brought by sea, are covered by fragments of limestone unmixed with external sand brought by sea, are not covered by fragments of limestone mixed with external sand brought by sea, are not covered by fragments of granulite unmixed with external sand brought by sea, are not covered by fragments of granulite mixed with external sand brought by sea, are not covered by external sand brought by sea.
- iiE_{50}: Tilted-down parts of horizontally displaced blocks of body of granulite overlain by limestone and sandstone are down sagged, are covered by sand of fragments of sandstone unmixed with external sand brought by sea, are not covered by sand of fragments of sandstone mixed with external sand brought by sea, are covered by fragments of limestone unmixed with external sand brought by sea, are not covered by fragments of limestone mixed with external sand brought by sea, are covered by fragments of granulite unmixed with external sand brought by sea, are not covered by fragments of granulite mixed with external sand brought by sea, are not covered by external sand brought by sea.
- iiE_{51}: Tilted-down parts of horizontally displaced blocks of body of granulite overlain by limestone and sandstone are down sagged, are covered by sand of fragments of sandstone unmixed with external sand brought by sea, are not covered by sand of fragments of sandstone mixed with external sand brought by sea, are covered by fragments of limestone unmixed with external sand brought by sea, are not covered by fragments of limestone mixed with external sand brought by sea, are not covered by fragments of granulite unmixed with external sand brought by sea, are not

covered by fragments of granulite mixed with external sand brought by sea, are not covered by external sand brought by sea.

- iiE_{52}: Tilted-up parts of horizontally displaced blocks of body of granulite overlain by limestone and sandstone are elevated, are not denudated in air to losing sandstone, are not denudated in air to losing limestone, are not denudated in air to partly losing granulite, are denudated in seat losing sandstone, are denudated in seat losing limestone, are denudated in seat partly losing granulite, are covered by external sand brought by sea, are not covered by sand of fragments of sandstone mixed with external sand brought by sea.
- iiE_{53}: Tilted-down parts of horizontally displaced blocks of body of granulite overlain by limestone and sandstone are down sagged, are covered by sand of fragments of sandstone unmixed with external sand brought by sea, are not covered by sand of fragments of sandstone mixed with external sand brought by sea, are covered by fragments of limestone unmixed with external sand brought by sea, are covered by fragments of limestone mixed with external sand brought by sea, are not covered by fragments of granulite unmixed with external sand brought by sea, are not covered by fragments of granulite mixed with external sand brought by sea, are not covered by external sand brought by sea.
- iiE_{54}: Tilted-down parts of horizontally displaced blocks of body of granulite overlain by limestone and sandstone are down sagged, are covered by sand of fragments of sandstone unmixed with external sand brought by sea, are not covered by sand of fragments of sandstone mixed with external sand brought by sea, are covered by fragments of limestone unmixed with external sand brought by sea, are covered by fragments of limestone mixed with external sand brought by sea, are not covered by fragments of granulite unmixed with external sand brought by sea, are covered by fragments of granulite mixed with external sand brought by sea, are not covered by external sand brought by sea.
- iiE_{55}: Tilted-down parts of horizontally displaced blocks of body of granulite overlain by limestone and sandstone are down sagged, are covered by sand of fragments of sandstone unmixed with external sand brought by sea, are not covered by sand of fragments of sandstone mixed with external sand brought by sea, are covered by fragments of limestone unmixed with external sand brought by sea, are not covered by fragments of limestone mixed with external sand brought by sea, are not covered by fragments of granulite unmixed with external sand brought by sea, are not covered by fragments of granulite mixed with external sand brought by sea, are covered by external sand brought by sea.
- iiE_{56}: Tilted-down parts of horizontally displaced blocks of body of granulite overlain by limestone and sandstone are down sagged, are covered by sand of fragments of sandstone unmixed with external sand brought by sea, are not covered by sand of fragments of sandstone mixed with external sand brought by sea, are covered by fragments of limestone unmixed with external sand brought by sea, are not covered by fragments of limestone mixed with external sand brought by sea, are covered by fragments of granulite unmixed with external sand brought by sea, are not covered by fragments of granulite mixed with external sand brought by sea, are not covered by external sand brought by sea.
- iiE_{57}: Tilted-down parts of horizontally displaced blocks of body of granulite overlain by limestone and sandstone are down sagged, are covered by sand of fragments of sandstone unmixed with external sand brought by sea, are not covered by sand of fragments of sandstone mixed with

external sand brought by sea, are covered by fragments of limestone unmixed with external sand brought by sea, are not covered by fragments of limestone mixed with external sand brought by sea, are covered by fragments of granulite unmixed with external sand brought by sea, are not covered by fragments of granulite mixed with external sand brought by sea, are covered by external sand brought by sea.

- iiE_{58}: Tilted-down parts of horizontally displaced blocks of body of granulite overlain by limestone and sandstone are down sagged, are covered by sand of fragments of sandstone unmixed with external sand brought by sea, are not covered by sand of fragments of sandstone mixed with external sand brought by sea, are not covered by fragments of limestone unmixed with external sand brought by sea, are covered by fragments of limestone mixed with external sand brought by sea, are not covered by fragments of granulite unmixed with external sand brought by sea, are not covered by fragments of granulite mixed with external sand brought by sea, are not covered by external sand brought by sea.
- iiE_{59}: Tilted-down parts of horizontally displaced blocks of body of granulite overlain by limestone and sandstone are down sagged, are covered by sand of fragments of sandstone unmixed with external sand brought by sea, are not covered by sand of fragments of sandstone mixed with external sand brought by sea, are not covered by fragments of limestone unmixed with external sand brought by sea, are covered by fragments of limestone mixed with external sand brought by sea, are not covered by fragments of granulite unmixed with external sand brought by sea, are covered by fragments of granulite mixed with external sand brought by sea, are not covered by external sand brought by sea.
- $iiiE_6$: Tilted-up parts of horizontally displaced blocks of body of granulite overlain by limestone and sandstone are elevated, are not denudated in air to losing sandstone, are not denudated in air to losing limestone, are not denudated in air to partly losing granulite, are not denudated in sea to losing sandstone, are not denudated in sea to losing limestone, are not denudated in sea to partly losing granulite, are not covered by external sand brought by sea, are not covered by sand of fragments of sandstone mixed with external sand brought by sea.
- $iiiE_7$: Tilted-up parts of horizontally displaced blocks of body of granulite overlain by limestone and sandstone are elevated, are denudated in air to losing sandstone, are not denudated in air to losing limestone, are not denudated in air to partly losing granulite, are not denudated in sea to losing sandstone, are not denudated in sea to losing limestone, are not denudated in sea to partly losing granulite, are not covered by external sand brought by sea, are not covered by sand of fragments of sandstone mixed with external sand brought by sea.
- *iiiE*: Fragments of sandstone form, are not transported to tilted-down parts, are not stopped in tilted-down parts.
- $iiiE_{9L}$: Tilted-up parts of horizontally displaced blocks of body of granulite overlain by limestone and sandstone are elevated, are not denudated in air to losing sandstone, are not denudated in air to losing limestone, are not denudated in air to partly losing granulite, are denudated in sea to losing sandstone, are not denudated in sea to losing limestone, are not denudated in sea to partly losing granulite, are not covered by external sand brought by sea, are not covered by sand of fragments of sandstone mixed with external sand brought by sea.
- $iiiE_{10}$: Tilted-up parts of horizontally displaced blocks of body of granulite overlain by limestone and sandstone are elevated, are not denudated in air to losing sandstone, are not denudated in air

to losing limestone, are not denudated in air to partly losing granulite, are not denudated in sea to losing sandstone, are not denudated in sea to losing limestone, are not denudated in sea to partly losing granulite, are covered by external sand brought by sea, are not covered by sand of fragments of sandstone mixed with external sand brought by sea.

- *iii*E_{11}: Tilted-up parts of horizontally displaced blocks of body of granulite overlain by limestone and sandstone are elevated, are not denudated in air to losing sandstone, are not denudated in air to losing limestone, are not denudated in air to partly losing granulite, are denudated in sea to losing sandstone, are not denudated in sea to losing limestone, are not denudated in sea to partly losing granulite, are covered by external sand brought by sea, are not covered by sand of fragments of sandstone mixed with external sand brought by sea.
- *iii*E_{14}: Sand of fragments of sandstone forms, is transported to tilted-down parts, is stopped in tilted-down parts, is not mixed with external sand brought by sea.
- *iii*E_{15}: Tilted-up parts of horizontally displaced blocks of body of granulite overlain by limestone and sandstone are elevated, are denudated in air to losing sandstone, are not denudated in air to losing limestone, are not denudated in air to partly losing granulite, are not denudated in sea to losing sandstone, are not denudated in sea to losing limestone, are not denudated in sea to partly losing granulite, are covered by external sand brought by sea, are not covered by sand of fragments of sandstone mixed with external sand brought by sea.
- *iii*E_{16}: Tilted-up parts of horizontally displaced blocks of body of granulite overlain by limestone and sandstone are elevated, are denudated in air to losing sandstone, are not denudated in air to losing limestone, are not denudated in air to partly losing granulite, are not denudated in sea to losing sandstone, are denudated in sea to losing limestone, are not denudated in sea to partly losing granulite, are not covered by external sand brought by sea, are not covered by sand of fragments of sandstone mixed with external sand brought by sea.
- *iii*E_{17}: Sand of fragments of sandstone forms, is transported to tilted-down parts, is stopped in tilted-down parts, is mixed with external sand brought by sea.
- *iii*E_{18}: Tilted-down parts of horizontally displaced blocks of body of granulite overlain by limestone and sandstone are down sagged, are not covered by sand of fragments of sandstone unmixed with external sand brought by sea, are not covered by sand of fragments of sandstone mixed with external sand brought by sea, are not covered by fragments of limestone unmixed with external sand brought by sea, are not covered by fragments of limestone mixed with external sand brought by sea, are not covered by fragments of granulite unmixed with external sand brought by sea, are not covered by fragments of granulite mixed with external sand brought by sea, are not covered by external sand brought by sea.
- *ii*E_{19}: Tilted-up parts of horizontally displaced blocks of body of granulite overlain by limestone and sandstone are elevated, are denudated in air to losing sandstone, are not denudated in air to losing limestone, are not denudated in air to partly losing granulite, are not denudated in sea to losing sandstone, are denudated in sea to losing limestone, are denudated in sea to partly losing granulite, are not covered by external sand brought by sea, are not covered by sand of fragments of sandstone mixed with external sand brought by sea.
- *iii*E_{20}: Tilted-up parts of horizontally displaced blocks of body of granulite overlain by limestone and sandstone are elevated, are denudated in air to losing sandstone, are not denudated in air to losing limestone, are not denudated in air to partly losing granulite, are not denudated in sea to

losing sandstone, are denudated in sea to losing limestone, are not denudated in sea to partly losing granulite, are covered by external sand brought by sea, are not covered by sand of fragments of sandstone mixed with external sand brought by sea.

- *iiiE*$_{21}$: Tilted-down parts of horizontally displaced blocks of body of granulite overlain by limestone and sandstone are down sagged, are not covered by sand of fragments of sandstone unmixed with external sand brought by sea, are not covered by sand of fragments of sandstone mixed with external sand brought by sea, are not covered by fragments of limestone unmixed with external sand brought by sea, are not covered by fragments of limestone mixed with external sand brought by sea, are not covered by fragments of granulite unmixed with external sand brought by sea, are not covered by fragments of granulite mixed with external sand brought by sea, are covered by external sand brought by se;a.
- *iiiE*$_{22}$: Tilted-down parts of horizontally displaced blocks of body of granulite overlain by limestone and sandstone are down sagged, are not covered by sand of fragments of sandstone unmixed with external sand brought by sea, are covered by sand of fragments of sandstone mixed with external sand brought by sea, are not covered by fragments of limestone unmixed with external sand brought by sea, are not covered by fragments of limestone mixed with external sand brought by sea, are not covered by fragments of granulite unmixed with external sand brought by sea, are not covered by fragments of granulite mixed with external sand brought by sea, are not covered by external sand brought by sea.
- *iiiE*$_{23}$: Tilted-up parts of horizontally displaced blocks of body of granulite overlain by limestone and sandstone are elevated, are not denudated in air to losing sandstone, are not denudated in air to losing limestone, are not denudated in air to partly losing granulite, are denudated in sea to losing sandstone, are denudated in sea to losing limestone, are not denudated in sea to partly losing granulite, are not covered by external sand brought by sea, are not covered by sand of fragments of sandstone mixed with external sand brought by sea.
- *iiiE*$_{24}$: Tilted-down parts of horizontally displaced blocks of body of granulite overlain by limestone and sandstone are down sagged, are not covered by sand of fragments of sandstone unmixed with external sand brought by sea, are not covered by sand of fragments of sandstone mixed with external sand brought by sea, are not covered by fragments of limestone unmixed with external sand brought by sea, are not covered by fragments of limestone mixed with external sand brought by sea, are not covered by fragments of granulite unmixed with external sand brought by sea, are covered by fragments of granulite mixed with external sand brought by sea, are not covered by external sand brought by sea.
- *iiiE*$_{25}$: Tilted-down parts of horizontally displaced blocks of body of granulite overlain by limestone and sandstone are down sagged, are not covered by sand of fragments of sandstone unmixed with external sand brought by sea, are covered by sand of fragments of sandstone mixed with external sand brought by sea, are not covered by fragments of limestone unmixed with external sand brought by sea, are not covered by fragments of limestone mixed with external sand brought by sea, are not covered by fragments of granulite unmixed with external sand brought by sea, are covered by fragments of granulite mixed with external sand brought by sea, are not covered by external sand brought by sea.
- *iiiE*$_{26}$: Tilted-up parts of horizontally displaced blocks of body of granulite overlain by limestone and sandstone are elevated, are not denudated in air to losing sandstone, are not denudated in air

to losing limestone, are not denudated in air to partly losing granulite, are denudated in sea to losing sandstone, are denudated in sea to losing limestone, are not denudated in sea to partly losing granulite, are covered by external sand brought by sea, are not covered by sand of fragments of sandstone mixed with external sand brought by sea.

- $iiiE_{27}$: Tilted-down parts of horizontally displaced blocks of body of granulite overlain by limestone and sandstone are down sagged, are not covered by sand of fragments of sandstone unmixed with external sand brought by sea, are not covered by sand of fragments of sandstone mixed with external sand brought by sea, are not covered by fragments of limestone unmixed with external sand brought by sea, are covered by fragments of limestone mixed with external sand brought by sea, are not covered by fragments of granulite unmixed with external sand brought by sea, are not covered by fragments of granulite mixed with external sand brought by sea, are not covered by external sand brought by sea.
- $iiiE_{28}$: Tilted-down parts of horizontally displaced blocks of body of granulite overlain by limestone and sand stone are down sagged, are not covered by sand of fragments of sandstone unmixed with external sand brought by sea, are covered by sand of fragments of sandstone mixed with external sand brought by sea, are not covered by fragments of limestone unmixed with external sand brought by sea, are covered by fragments of limestone mixed with external sand brought by sea, are not covered by fragments of granulite unmixed with external sand brought by sea, are not covered by fragments of granulite mixed with external sand brought by sea, are not covered by external sand brought by sea.
- $iiiE_{29}$: Tilted-up parts of horizontally displaced blocks of body of granulite overlain by limestone and sandstone are elevated, are denudated in air to losing sandstone, are denudated in air to losing limestone, are not denudated in air to partly losing granulite, are not denudated in sea to losing sandstone, are not denudated in sea to losing limestone, are not denudated in sea to partly losing granulite, are not covered by external sand brought by sea.
- $iiiE_{30}$: Tilted-down parts of horizontally displaced blocks of body of granulite overlain by limestone and sandstone are down sagged, are not covered by sand of fragments of sandstone unmixed with external sand brought by sea, are covered by sand of fragments of sandstone mixed with external sand brought by sea, are not covered by fragments of limestone unmixed with external sand brought by sea, are covered by fragments of limestone mixed with external sand brought by sea, are not covered by fragments of granulite unmixed with external sand brought by sea, are covered by fragments of granulite mixed with external sand brought by sea, are not covered by external sand brought by sea.
- $iiiE_{31}$: Fragments of limestone form, are not transported to tilted-down parts, are not stopped in tilted-down parts, are not mixed with external sand brought by sea.
- $iiiE_{32}$: Tilted-up parts of horizontally displaced blocks of body of granulite overlain by limestone and sandstone are elevated, are denudated in air to losing sandstone, are denudated in air to losing limestone, are not denudated in air to partly losing granulite, are not denudated in sea to losing sandstone, are not denudated in sea to losing limestone, are not denudated in sea to partly losing granulite, are covered by external sand brought by sea.
- $iiiE_{34}$: Fragments of limestone form, are transported to tilted-down parts, are stopped in tilted-down parts, are not mixed with external sand brought by sea.

- *ii*E_{35}: Tilted-up parts of horizontally displaced blocks of body of granulite overlain by limestone and sandstone are elevated, are denudated in air to losing sandstone, are denudated in air to losing limestone, are not denudated in air to partly losing granulite, are not denudated in sea to losing sandstone, are not denudated in sea to losing limestone, are denudated in sea to partly losing granulite, are not covered by external sand brought by sea.
- *iii*E_{36}: Tilted-down parts of horizontally displaced blocks of body of granulite overlain by limestone and sandstone are down sagged, are covered by sand of fragments of sandstone unmixed with external sand brought by sea, are not covered by sand of fragments of sandstone mixed with external sand brought by sea, are not covered by fragments of limestone unmixed with external sand brought by sea, are not covered by fragments of limestone mixed with external sand brought by sea, are not covered by fragments of granulite unmixed with external sand brought by sea, are not covered by fragments of granulite mixed with external sand brought by sea, are not covered by external sand brought by sea.
- *iii*E_{37}: Fragments of limestone form, are transported to tilted-down parts, are stopped in tilted-down parts, are mixed with external sand brought by sea.
- *iii*E_{38}: Tilted-up parts of horizontally displaced blocks of body of granulite overlain by limestone and sandstone are elevated, are denudated in air to losing sandstone, are denudated in air to losing limestone, are not denudated in air to partly losing granulite, are not denudated in sea to losing sandstone, are not denudated in sea to losing limestone, are denudated in sea to partly losing granulite, are covered by external sand brought by sea.
- *iii*E_{39}: Tilted-up parts of horizontally displaced blocks of body of granulite overlain by limestone and sandstone are elevated, are denudated in air to losing sandstone, are denudated in air to losing limestone, are denudated in air to partly losing granulite, are not denudated in sea to losing sandstone, are not denudated in sea to losing limestone, are not denudated in sea to partly losing granulite, are not covered by external sand brought by sea.
- *iii*E_{40}: Tilted-up parts of horizontally displaced blocks of body of granulite overlain by limestone and sandstone are elevated, are not denudated in air to losing sandstone, are not denudated in air to losing limestone, are not denudated in air to partly losing granulite, are denudated in sea to losing sandstone, are denudated in sea to losing limestone, are denudated in sea to partly losing granulite, are not covered by external sand brought by sea, are not covered by sand of fragments of sandstone mixed with external sand brought by sea.
- *iii*E_{41}: Tilted-down parts of horizontally displaced blocks of body of granulite overlain by limestone and sandstone are down sagged, are covered by sand of fragments of sandstone unmixed with external sand brought by sea, are not covered by sand of fragments of sandstone mixed with external sand brought by sea, are not covered by fragments of limestone unmixed with external sand brought by sea, are not covered by fragments of limestone mixed with external sand brought by sea, are not covered by fragments of granulite unmixed with external sand brought by sea, are covered by fragments of granulite mixed with external sand brought by sea, are not covered by external sand brought by sea.
- *iii*E_{42}: Tilted-down parts of horizontally displaced blocks of body of granulite overlain by limestone and sandstone are down sagged, are covered by sand of fragments of sandstone unmixed with external sand brought by sea, are covered by sand of fragments of sandstone mixed with external sand brought by sea, are not covered by fragments of limestone unmixed with external

sand brought by sea, are not covered by fragments of limestone mixed with external sand brought by sea, are not covered by fragments of granulite unmixed with external sand brought by sea, are not covered by fragments of granulite mixed with external sand brought by sea, are not covered by external sand brought by sea.

- $iiiE_{43}$: Tilted-up parts of horizontally displaced blocks of body of granulite overlain by limestone and sandstone are elevated, are denudated in air to losing sandstone, are denudated in air to losing limestone, are denudated in air to partly losing granulite, are not denudated in sea to losing sandstone, are not denudated in sea to losing limestone, are not denudated in sea to partly losing granulite, are covered by external sand brought by sea.
- $iiiE_{44}$: Fragments of granulite form, are not transported to tilted-down parts, are not stopped in tilted-down parts, are not mixed with external sand brought by sea.
- $iiiE_{46}$: Fragments of granulite form, are transported to tilted-down parts, are stopped in tilted-down parts, are not mixed with external sand brought by sea.
- $iiiE_{47}$: Fragments of granulite form, are transported to tilted-down parts, are stopped in tilted-down parts, are mixed with external sand brought by sea.
- $iiiE_{48}$: Tilted-down parts of horizontally displaced blocks of body of granulite overlain by limestone and sandstone are down sagged, are covered by sand of fragments of sandstone unmixed with external sand brought by sea, are not covered by sand of fragments of sandstone mixed with external sand brought by sea, are not covered by fragments of limestone unmixed with external sand brought by sea, are not covered by fragments of limestone mixed with external sand brought by sea, are not covered by fragments of granulite unmixed with external sand brought by sea, are not covered by fragments of granulite mixed with external sand brought by sea, are covered by external sand brought by sea.
- $iiiE_{49}$: Tilted-down parts of horizontally displaced blocks of body of granulite overlain by limestone and sandstone are down sagged, are covered by sand of fragments of sandstone unmixed with external sand brought by sea, are not covered by sand of fragments of sandstone mixed with external sand brought by sea, are covered by fragments of limestone unmixed with external sand brought by sea, are not covered by fragments of limestone mixed with external sand brought by sea, are not covered by fragments of granulite unmixed with external sand brought by sea, are not covered by fragments of granulite mixed with external sand brought by sea, are not covered by external sand brought by sea.
- $iiiE_{50}$: Tilted-down parts of horizontally displaced blocks of body of granulite overlain by limestone and sandstone are down sagged, are covered by sand of fragments of sandstone unmixed with external sand brought by sea, are not covered by sand of fragments of sandstone mixed with external sand brought by sea, are covered by fragments of limestone unmixed with external sand brought by sea, are not covered by fragments of limestone mixed with external sand brought by sea, are covered by fragments of granulite unmixed with external sand brought by sea, are not covered by fragments of granulite mixed with external sand brought by sea, are not covered by external sand brought by sea.
- $iiiE_{51}$: Tilted-down parts of horizontally displaced blocks of body of granulite overlain by limestone and sandstone are down sagged, are covered by sand of fragments of sandstone unmixed with external sand brought by sea, are not covered by sand of fragments of sandstone mixed with external sand brought by sea, are covered by fragments of limestone unmixed with external sand

brought by sea, are not covered by fragments of limestone mixed with external sand brought by sea, are not covered by fragments of granulite unmixed with external sand brought by sea, are not covered by fragments of granulite mixed with external sand brought by sea, are not covered by external sand brought by sea.

- $iiiE_{52}$: Tilted-up parts of horizontally displaced blocks of body of granulite overlain by limestone and sandstone are elevated, are not denudated in air to losing sandstone, are not denudated in air to losing limestone, are not denudated in air to partly losing granulite, are denudated in seat losing sandstone, are denudated in seat losing limestone, are denudated in seat partly losing granulite, are covered by external sand brought by sea, are not covered by sand of fragments of sandstone mixed with external sand brought by sea.
- $iiiE_{53}$: Tilted-down parts of horizontally displaced blocks of body of granulite overlain by limestone and sandstone are down sagged, are covered by sand of fragments of sandstone unmixed with external sand brought by sea, are not covered by sand of fragments of sandstone mixed with external sand brought by sea, are covered by fragments of limestone unmixed with external sand brought by sea, are covered by fragments of limestone mixed with external sand brought by sea, are not covered by fragments of granulite unmixed with external sand brought by sea, are not covered by fragments of granulite mixed with external sand brought by sea, are not covered by external sand brought by sea.
- $iiiE_{54}$: Tilted-down parts of horizontally displaced blocks of body of granulite overlain by limestone and sandstone are down sagged, are covered by sand of fragments of sandstone unmixed with external sand brought by sea, are not covered by sand of fragments of sandstone mixed with external sand brought by sea, are covered by fragments of limestone unmixed with external sand brought by sea, are covered by fragments of limestone mixed with external sand brought by sea, are not covered by fragments of granulite unmixed with external sand brought by sea, are covered by fragments of granulite mixed with external sand brought by sea, are not covered by external sand brought by sea.
- $iiiE_{55}$: Tilted-down parts of horizontally displaced blocks of body of granulite overlain by limestone and sandstone are down sagged, are covered by sand of fragments of sandstone unmixed with external sand brought by sea, are not covered by sand of fragments of sandstone mixed with external sand brought by sea, are covered by fragments of limestone unmixed with external sand brought by sea, are not covered by fragments of limestone mixed with external sand brought by sea, are not covered by fragments of granulite unmixed with external sand brought by sea, are not covered by fragments of granulite mixed with external sand brought by sea, are covered by external sand brought by sea.
- $iiiE_{56}$: Tilted-down parts of horizontally displaced blocks of body of granulite overlain by limestone and sandstone are down sagged, are covered by sand of fragments of sandstone unmixed with external sand brought by sea, are not covered by sand of fragments of sandstone mixed with external sand brought by sea, are covered by fragments of limestone unmixed with external sand brought by sea, are not covered by fragments of limestone mixed with external sand brought by sea, are covered by fragments of granulite unmixed with external sand brought by sea, are not covered by fragments of granulite mixed with external sand brought by sea, are not covered by external sand brought by sea.

- $iiiE_{57}$: Tilted-down parts of horizontally displaced blocks of body of granulite overlain by limestone and sandstone are down sagged, are covered by sand of fragments of sandstone unmixed with external sand brought by sea, are not covered by sand of fragments of sandstone mixed with external sand brought by sea, are covered by fragments of limestone unmixed with external sand brought by sea, are not covered by fragments of limestone mixed with external sand brought by sea, are covered by fragments of granulite unmixed with external sand brought by sea, are not covered by fragments of granulite mixed with external sand brought by sea, are covered by external sand brought by sea.
- $iiiE_{58}$: Tilted-down parts of horizontally displaced blocks of body of granulite overlain by limestone and sandstone are down sagged, are covered by sand of fragments of sandstone unmixed with external sand brought by sea, are not covered by sand of fragments of sandstone mixed with external sand brought by sea, are not covered by fragments of limestone unmixed with external sand brought by sea, are covered by fragments of limestone mixed with external sand brought by sea, are not covered by fragments of granulite unmixed with external sand brought by sea, are not covered by fragments of granulite mixed with external sand brought by sea, are not covered by external sand brought by sea.
- $iiiE_{59}$: Tilted-down parts of horizontally displaced blocks of body of granulite overlain by limestone and sandstone are down sagged, are covered by sand of fragments of sandstone unmixed with external sand brought by sea, are not covered by sand of fragments of sandstone mixed with external sand brought by sea, are not covered by fragments of limestone unmixed with external sand brought by sea, are covered by fragments of limestone mixed with external sand brought by sea, are not covered by fragments of granulite unmixed with external sand brought by sea, are covered by fragments of granulite mixed with external sand brought by sea, are not covered by external sand brought by sea.

APPENDIX 2

The Event Bush Model of Wrench Tectonics Mechanism

iaE_1 *Flux* iiE_5

ibE_2, iiE_6 *Influx* iiE_7

ibE_2, iiE_7 *Influx* iiE_{29}

ibE_2, iiE_{29} *Influx* iiE_{39}

ibE_3, iiE_6 *Influx* iiE_9

ibE_3, iiE_7 *Influx* iiE_{16}

ibE_3, iiE_9 *Influx* iiE_{23}

ibE_{3}, iiE_{16} *Influx* iiE_{19}

ibE_{3}, iiE_{23} *Influx* iiE_{40}

ibE_{3}, iiE_{29} *Influx* iiE_{35}

ibE_{3}, iiE_{36} *Influx* iiE_{48}

ibE_{4}, iiE_{6} *Influx* iiE_{10}

ibE_{4}, iiE_{7} *Influx* iiE_{15}

ibE_{4}, iiE_{9} *Influx* iiE_{11}

ibE_{4}, iiE_{14} *Influx* iiE_{17}

ibE_{4}, iiE_{16} *Influx* iiE_{20}

ibE_{4}, iiE_{18} *Influx* iiE_{21}

ibE_{4}, iiE_{23} *Influx* iiE_{26}

ibE_{4}, iiE_{29} *Influx* iiE_{32}

ibE_{4}, iiE_{34} *Influx* iiE_{37}

ibE_{4}, iiE_{35} *Influx* iiE_{38}

ibE_{4}, iiE_{39} *Influx* iiE_{43}

ibE_{4}, iiE_{43} *Influx* iiE_{52}

ibE_{4}, iiE_{46} *Influx* iiE_{47}

ibE_{4}, iiE_{49} *Influx* iiE_{55}

ibE_{4}, iiE_{50} *Influx* iiE_{57}

iiE_{5} *Flux* iiE_{6}

iiE_{5} *Flux* iiE_{18}

iiE_{6} *Flux* $iiiE_{6}$

iiE_{7} *Flux* iiE_{8}

iiE_7 *Flux* $iiiE_7$

iiE_8, iiE_{18} *Influx* iiE_{12}

iiE_8 *Flux* $iiiE_8$

iiE_9 *Flux* iiE_8

iiE_9 *Flux* $iiiE_9$

iiE_{10} *Flux* $iiiE_{10}$

iiE_{11} *Flux* $iiiE_{11}$

iiE_{12} *Flux* iiE_{13}

iiE_{13} *Flux* iiE_{14}

iiE_{14}, iiE_{18} *Influx* iiE_{36}

iiE_{14} *Flux* $iiiE_{14}$

iiE_{15} *Flux* $iiiE_{15}$

iiE_{16} *Flux* $iiiE_{16}$

iiE_{17}, iiE_{18} *Influx* iiE_{22}

iiE_{17}, iiE_{36} *Influx* iiE_{42}

iiE_{17} *Flux* $iiiE_{17}$

iiE_{18} *Flux* $iiiE_{18}$

iiE_{19} *Flux* $iiiE_{19}$

iiE_{20} *Flux* $iiiE_{20}$

iiE_{21} *Flux* $iiiE_{21}$

iiE_{22} *Flux* $iiiE_{22}$

iiE_{23} *Flux* iiE_{31}

iiE_{23} *Flux* $iiiE_{23}$

iiE_{24} *Flux* $iiiE_{24}$

iiE_{25} *Flux* $iiiE_{25}$

iiE_{26} *Flux* $iiiE_{26}$

iiE_{27} *Flux* $iiiE_{27}$

iiE_{28} *Flux* $iiiE_{28}$

iiE_{29} *Flux* iiE_{31}

iiE_{29} *Flux* $iiiE_{29}$

iiE_{30} *Flux* $iiiE_{30}$

iiE_{31} *Flux* $iiiE_{31}$

iiE_{32} *Flux* $iiiE_{32}$

iiE_{33} *Flux* iiE_{34}

iiE_{34}, iiE_{36} *Influx* iiE_{49}

iiE_{34} *Flux* $iiiE_{34}$

iiE_{35} *Flux* $iiiE_{35}$

iiE_{36}, iiE_{31} *Influx* iiE_{33}

iiE_{36}, iiE_{37} *Influx* iiE_{58}

iiE_{36}, iiE_{47} *Influx* iiE_{41}

iiE_{36} *Flux* $iiiE_{36}$

iiE_{37}, iiE_{18} *Influx* iiE_{27}

iiE_{37}, iiE_{22} *Influx* iiE_{28}

iiE_{37}, iiE_{49} *Influx* iiE_{53}

iiE_{37} *Flux* $iiiE_{37}$

iiE_{38} *Flux* $iiiE_{38}$

iiE_{39} *Flux* iiE_{44}

iiE_{39} *Flux* $iiiE_{39}$

iiE_{40} *Flux* iiE_{44}

iiE_{40} *Flux* $iiiE_{40}$

iiE_{41} *Flux* $iiiE_{41}$

iiE_{42} *Flux* $iiiE_{42}$

iiE_{43} *Flux* $iiiE_{43}$

iiE_{44} *Flux* $iiiE_{44}$

iiE_{45} *Flux* iiE_{46}

iiE_{46}, iiE_{49} *Influx* iiE_{50}

iiE_{46} *Flux* $iiiE_{46}$

iiE_{47}, iiE_{18} *Influx* iiE_{24}

iiE_{47}, iiE_{22} *Influx* iiE_{25}

iiE_{47}, iiE_{27} *Influx* iiE_{30}

iiE_{47}, iiE_{49} *Influx* iiE_{51}

iiE_{47}, iiE_{50} *Influx* iiE_{56}

iiE_{47}, iiE_{53} *Influx* iiE_{54}

iiE_{47}, iiE_{58} *Influx* iiE_{59}

iiE_{47} *Flux* $iiiE_{47}$

iiE_{48} *Flux* $iiiE_{48}$

iiE_{49}, iiE_{44} *Influx* iiE_{45}

iiE_{49} *Flux* $iiiE_{49}$

iiE_{50} *Flux* $iiiE_{50}$

iiE_{51} *Flux* $iiiE_{51}$

iiE_{52} *Flux* $iiiE_{52}$

iiE_{53} *Flux* $iiiE_{53}$

iiE_{54} *Flux* $iiiE_{54}$

iiE_{55} *Flux* $iiiE_{55}$

iiE_{56} *Flux* $iiiE_{56}$

iiE_{57} *Flux* $iiiE_{57}$

iiE_{58} *Flux* $iiiE_{58}$

iiE_{59} *Flux* $iiiE_{59}$

Chapter 10
Modeling Spatial Evolution:
Review of Methods and Its Significance

G. Jayanthi
Pondicherry University, India

V. Uma
Pondicherry University, India

ABSTRACT

Geographic features in the real world are represented by spatial entities such as point, line, and area in two-dimensional surfaces. These features tend to evolve in time, thereby characterizing change in their physical identity, evolution into new species, thus describing geomorphological change of geographic features. These phenomena can be formalized using spatio-temporal relations. Formal representation of changing geographic (spatial) features is the interest of this chapter. Formal methods for representing the event and process that causes geomorphological change are presented. The formalization of geographic entities that are temporally and spatially related in a two-dimensional plane using the interval logic and spatial logic would facilitate the understanding of how modeling of space-time using spatio-temporal relations represents spatial evolution over time. Representation of temporal dynamism can be accomplished using various models. Modeling using spatio-temporal graph is more apt as it contributes to the cause-effect analysis.

INTRODUCTION

Human cognition by nature systematically deduces knowledge that would pave the way for incidental solution. Cognitive thinking towards understanding of physical phenomena on earth such as epidemic disease spread, drought, deforestation, evacuation, migration of settlements etc., imparts knowledge to human beings. Information gathered to attain this knowledge transforms an individual to domain knowledge expert. Artificial Intelligence (AI) is a subject of cognitive science that governs the development of computational algorithms and tools that would act like human beings in solving complex tasks. Automated reasoning through inference engine, a knowledge based approach, is the foundation of such systems (Elaine Rich, 2009). Knowledge based systems contribute to Geographical Information System

DOI: 10.4018/978-1-5225-5261-1.ch010

(GIS) and can increase the computational efficiency of GIS. Thus, engineering real-world fact can be done by employing systematic Knowledge Engineering (KE) processes (Patterson, 2005) namely (1) Data Acquisition (2) Data pre-processing and transformation (3) Domain knowledge Representation and Reasoning. The final outcome of such is an expert system. The source of knowledge for automated reasoning systems like spatio-temporal knowledge based system consists of spatial and temporal facts that are coded using formal principles of KE. These systems are called expert system as they automate the discovery of knowledge that would further update the facts in the knowledge base to ensure real-world dynamism. The resources on the earth are enormous in its type, physical appearance, size, variety and are geographically distributed in wide range of space. Each of these resources is identified in a location of space at distinct time period (seasonal), within specified time interval or in a moment of time. These resources tend to change over time. Thus, space-time evolution of earth is the motivation behind this study.

Modeling space-time is the foundation for both knowledge-based reasoning and probabilistic reasoning approaches, in which spatial locations like land use, land cover regions, forests, urban settlements and their temporal activities are parameterized, empirically formulated to qualitatively reason spatio-temporal events like deforestation, forest fire, migration of habitats etc., (Jayanthi and Uma, 2016b). Moreover, this approach has motivated researchers towards reasoning about the events that cause phenomenal changes such as earthquake, landslides, drought etc., (Jayanthi and Uma, 2016a; 2016b; 2016c; 2016d and 2017). Thus, modeling of evolution considering the spatial and temporal factors is the interest of this chapter. Spatial entities representing geographic features are identified by their spatial and aspatial properties. Universally, geographical entities are related with respect to space and time. The physical locations of such features and their spatial extension over time in various forms are related in a moment of time, over an interval or at regular intervals (time period) (Campelo, 2013; Hornsby and Egenhofer, 2000 and Mondo, 2010). Thus, temporal, spatial and spatio-temporal relations that are intrinsically formed, dynamically associate or disassociate spatial entities over time (Mondo, 2010). Geographic feature captured in a moment of time is a static scenario, which is a snapshot of real-world entity. A distinguishable change as a resultant of process is identified when captured over an interval (Campelo, 2013, Hornsby and Egenhofer, 2000; Jayanthi and Uma, 2016d; 2017). In this perspective, researchers have shown interest towards modeling of spatio-temporal evolution using graph structure (Mondo, 2010) in which nodes represents spatial locations and edges are links that relate spatial location at different time instants. The spatial analytics based on dependencies between spatial locations at various temporal scales is applied in forecasting weather, rainfall, traffic flow etc., (Box, 2012; Cohn, 1995 and 1997; Egenhofer, 1994a and 1994b; Galton, 2008 and 2009).

Evolution of spatial structures over time emerges into a connected spatio-temporal network, a graph structure formed through association of spatio-temporal relations. The structural properties of graph structure are (1) reflexive and its complement irreflexive (2) symmetry and (3) transitive. These properties are applied in analyzing membership, homogeneity, heterogeneity and relational hierarchy (Mondo, 2010). Thus, modeling of evolution considering the spatial and temporal factors using graph structures has its significance in dynamic knowledge representation wherein connectivity and reachability of geographic features on geo-spatial network are analyzed using spatial topological relations formalized through RCC-8 (Randell, 1992) and temporal relations (Allen, 1981 and 1983). Nevertheless, potential researches are in progress towards the optimization of computational methods wherein linear complexity is highly demanded along with highly expressive knowledge based systems.

The objective of the proposed chapter is to explain the formalization of geographic entities that are temporally and spatially related in a two-dimensional plane using the interval logic and spatial logic.

Considering the spatial evolution, this chapter presents a discussion on representation of temporal dynamics of spatial objects using various models and highlights the significance of theory of multitudes and spatio-temporal process graph in modeling evolution.

KNOWLEDGE REPRESENTATION

The systematic transformation of real-world facts to symbolic facts coded with syntax and semantics using formal mathematical logic is *Knowledge Representation (KR).* It is a field of Artificial Intelligence that forms the base for expert systems to automate tasks. For example, in industrial applications controlling the speed of motors in motion at some point of time, controlling a manufacturing process by measuring boiling and cooling temperatures to activate an action are the tasks of expert system. They are formally represented as rules of actions. Symbolic representation of facts using first order predicate logic is one of the primitive ways of formal representation of facts in KR. The quantifiers and symbolic connectives used and their respective logical definitions are listed in (Table 1). Formal representation of facts using predicate logic has been applied in development of Medical Diagnosis Systems, Speech Recognition Systems, Agent based Planning Systems etc. Predicate logic based primitive formal representation of facts are done using symbolic quantifiers and connectives shown in (Table 1). These primitives alone do not suffice the need of codifying temporal and spatial knowledge. The increasing demand for highly expressive logical language of KR to formalize temporal and spatial knowledge, has introduced two major frameworks (1) Temporal KR and (2) Spatial KR. These two frameworks are formally evaluated based on principles of KR using the axioms of interval logic and spatial logic (Allen, 1981 and Randell, 1992). The syntax and semantics of temporal and spatial relations are detailed in (Table 2a, 2b and 3). The fundamental theory of interval logic (Allen, 1981), spatial logic and connections (Randell, 1992) are the building blocks of temporal and spatial KR framework respectively.

Temporal Relations

Temporal Knowledge Representation is the logic based formulation of action description whose occurrence is implied by time. Thus, every action in the real-world is characterized by time. For example, consider an announcement *"The Sports Day Meet is to be scheduled before 20th August 2017".* The description of the activity is characterized by time, which is a event to be held within stipulated duration. Allen, 1983 has introduced qualitative reasoning of action which has paved way for logic based formalism of time.

Table 1. Symbols and semantics

Symbols	Semantics	Symbols	Semantics
$\forall$	Universal quantifier (*for all*)	$\Rightarrow$	Logical Implication
$\exists$	Existential quantifier (*there exists*)	$\rightarrow$	Conditional
$\wedge$	Conjunction	$\leftrightarrow$	Bi-conditional
$\vee$	Disjunction		
$\neg$	Negation		
$\Leftrightarrow$	Logical Equivalence		

Table 2a. Interval relations of temporal logic

Syntactic Relations	Logical Definition
before(λ_1, λ_2)	$\lambda_1.End < \lambda_2.Start$
overlaps(λ_1, λ_2)	$\lambda_1.Start < \lambda_2.Start \wedge \lambda_1.End < \lambda_2.End$
starts(λ_1, λ_2)	$\lambda 1.End < \lambda_2 .End \wedge \lambda_1.Start = \lambda_2.Start$
during(λ_1, λ_2)	$\lambda_2.Start < \lambda_1.Start \wedge \lambda_1.End < \lambda_2.End$
meets(λ_1, λ_2)	$\lambda_1.End = \lambda_2.Start$ at point P
finishes(λ_1, λ_2)	$\lambda_2.Start < \lambda_1.Start \wedge \lambda_1.End = \lambda_2.End$
equal(λ_1, λ_2)	λ_1, λ_2 occurs in same interval such that $\lambda_1.Start = \lambda_2.Start \wedge \lambda_1.End = \lambda_2.End$

Note: λ_1 and λ_2 are events defined in an Interval.

The two fundamental outcomes of this formal theory are ‘event’ and ‘process’. A process is a continuant of a event either it be recurring or non – recurring within a span of time. For example, academic report preparation for “*Annual day meet 2017*” is a process, whereas the report is the compilation of events those have occurred since the beginning of the academic year.

λ_1, λ_2 are events that occur in an interval with *Start* and *End* points. For example, consider the *before* predicate in which the *End* time of the event λ_1 is less than *Start* time of the event λ_2. Obviously, the logical definitions of other predicates are listed in (Table 2a). The time points β and α can be logically related as shown in (Table 2b). β.time and α.time denotes the time of occurrence of the events β and α respectively. The schematic representation of point and interval relations in Figure 1 shows the event intervals and their interactions. A point in a temporal relation describes an instant. Duration of time in which an activity begins and ends can be defined as an interval. ‘Start’ and ‘End’ of an interval are the two components by which an action is inferred quantitatively. Event occurs in an interval. Interval has start and end points during which an action can be described. Thus, point and interval relations are the building blocks of automated reasoning system over time. Temporal logic is a formal way of reasoning actions that are interrelated, independent and concurrent (Allen, 1984). Reasoning actions is a prospective research area and the prime concepts are event, process and reasoning knowledge over uncertain intervals using fuzzy logic. Knowledge discovery by temporal reasoning and information retrieval by modeling temporal data have contributed to the development of cognitive systems, robot navigation, agent technology (Pani and Bhattacharjee, 2001). Logical languages for KR originated with Situation calculus (Mc Charthy and Hayes 1969) and interval logic (Allen, 1981; 1983; 1984 and 1994; Bennett, 2004). Event Calculus approach was introduced by Kowalski and Sergot in 1986. Research on temporal logic has been streamlined in various directions such as semi interval logic (Freksa, 1991; 1992), approximate intervals (Bittner, 2002) that are addressing imprecise knowledge.

Spatial Relations

The logic based formalism of spatial geometry and their topological structures bounded by connected set of points is *Spatial Knowledge Representation.* The physical identity of spatial objects and their interconnections in a two-dimensional plane are represented using first order logic based formalism termed as spatial logic. The theory of spatial logic defines a spatial location as connected set of points. The intersection of two connected spatial locations is topologically related using logical definition of

Table 2b. Point relations of temporal logic

Syntactic Relations	Logical Definition
before (α,β)	α < β
after (α,β)	α > β
equal(α,β)	α = β

Note: α and β are time points.

Figure 1. Temporal logic point and Interval relations – Schematic definition

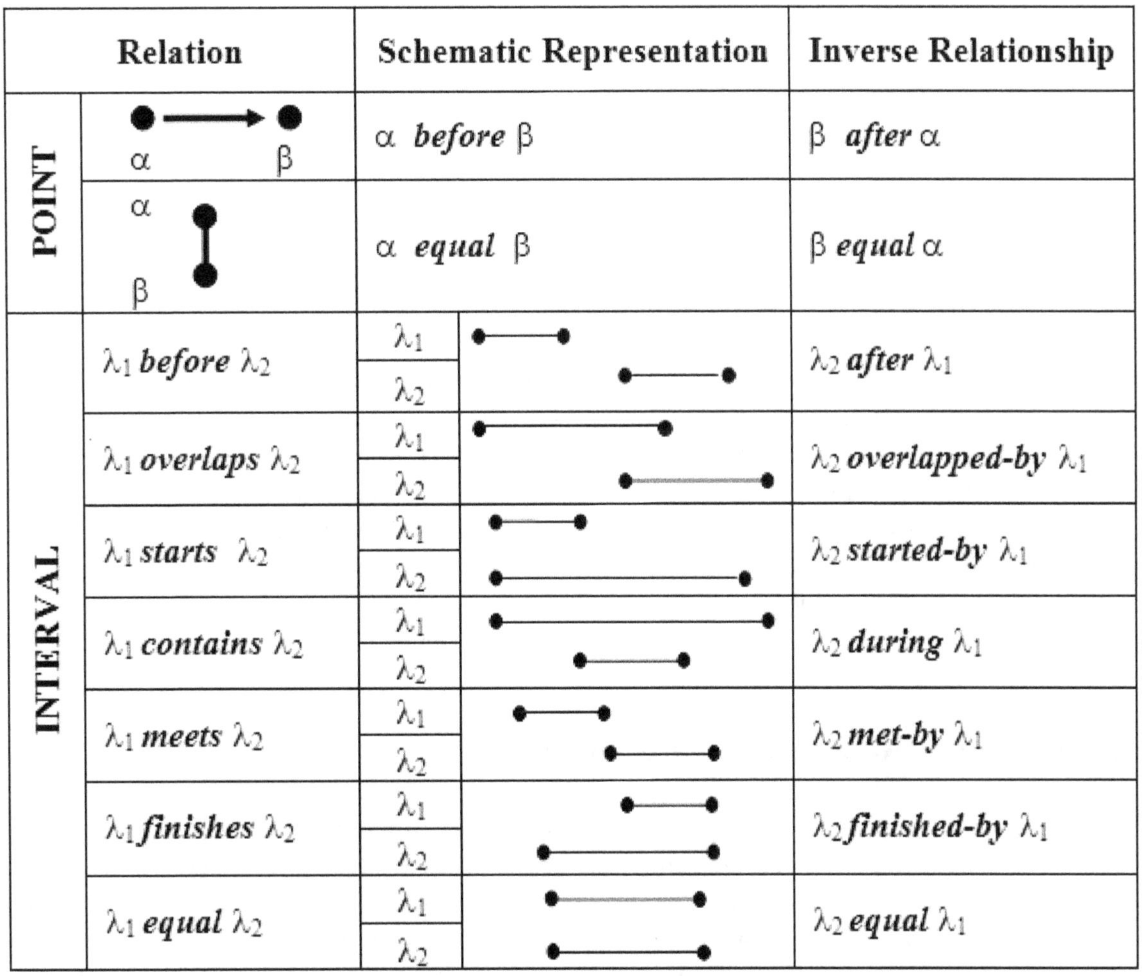

predicates illustrated in (Table 3). The base and composite relations of spatial logic are shown in Figure 2. Novel approaches to spatial knowledge representations have contributed to geo-spatial dynamics (Egenhofer and Franzosa, 1991; Randell et al., 1992; Cui et al., 1993; Cohn et al., 1997).

A hierarchy of spatial relations DR_{xy}, O_{xy} and PP_{xy} are inferred when DC_{xy} is composed with EC_{xy}; P_{xy} is composed with PO_{xy} and TPP_{xy} composed with $NTPP_{xy}$, where x and y denote spatial regions R_1 and R_2 respectively. Thus, composite relations are derived using base relations as shown in Figure 2 (b),

Table 3. Spatial logic relations

Topological Relation	Syntactic Representations
Connected (C_{xy}) "x", "y" are spatial regions	$\forall xy[C_{xy} \rightarrow C_{yx}]$
Disconnected (DC_{xy})	$\neg C_{xy}$
Part (P_{xy})	$\forall z [C_{zx} \rightarrow C_{zy}]$
Proper Part (PP_{xy})	$P_{xy} \wedge \neg P_{xy}$
Overlaps (O_{xy})	$\exists z[P_{zx} \wedge P_{zy}]$
Partial Overlap (PO_{xy})	$O_{xy} \wedge \neg P_{xy} \wedge \neg P_{yx}$
Discrete (DR_{xy})	$\neg O_{xy}$
Externally Connected (EC_{xy})	$\neg O_{xy} \wedge C_{xy}$
Tangential Proper Part (TPP_{xy})	$\exists z[EC_{zy} \wedge EC_{zx}] \wedge PP_{xy}$
Non-Tangential Proper part ($NTPP_{xy}$)	$\neg\exists z[EC_{zy} \wedge EC_{zx}] \wedge PP_{xy}$
Non-Tangential Proper Part Inverse ($NTPP^{-1}{}_{xy}$)	$NTPP_{yx}$
Tangential Proper part Inverse ($TPP^{-1}{}_{xy}$)	TPP_{yx}
Proper Part Inverse ($PP^{-1}{}_{xy}$)	PP_{yx}
Part Inverse ($P^{-1}{}_{xy}$)	P_{yx}

(d) and (f). The origin of spatial logic formulation based on principles of set theory was introduced by Clarke in 1981. He extended the formal theory in 1985. His formalism is under laid by the concept of Mereology - Parts and Wholes, the physical geography of earth. Later, Point set topology was introduced by Egenhofer in 1991 to improve the expressiveness of spatial logic. This has paved way for Region Connection Calculus (RCC) formulation by Randell in 1992, and his work has contributed eight topological relations to spatial logic framework. The other major works are 9-intersection Calculus, reasoning about directions in two-dimensional plane by Frank, 1991; 1992 and 1996, Boolean Connection Algebra (BRCC) introduced by Stell in 2000, RCC-8c and RCC-8c° (Egenhofer 1994a and 1994b; Stell 2000 and 2004; Kontchakov et al., 2014).

Spatio-Temporal Knowledge Representation

Events are discrete components whether it be temporal or spatial. Aggregation of such events constitutes process. Therefore, association of spatial and temporal factors as a whole or part in describing location based events is termed as spatio-temporal process. KR with spatio-temporal facts involves composition of spatial and temporal logics and their properties are explored when relating spatial objects at different temporal scales. Thus, interval relations and topological relations are major components of spatio-temporal knowledge based systems (Campelo, 2012 and 2013).

Spatio-temporal knowledge representation based expert systems are predominantly used in agent based systems such as evacuation planning in disaster management, eradication of infectious disease, ecological and environmental planning systems, monitoring trajectory path navigation etc. The significance of spatio-temporal KR in real-world problem solving is explained using the following illustrative examples.

Figure 2. (a), (c), (e) shows the base relations of spatial logic using RCC-8 topological relations; (b), (d) and (f) show the inferred composite relations

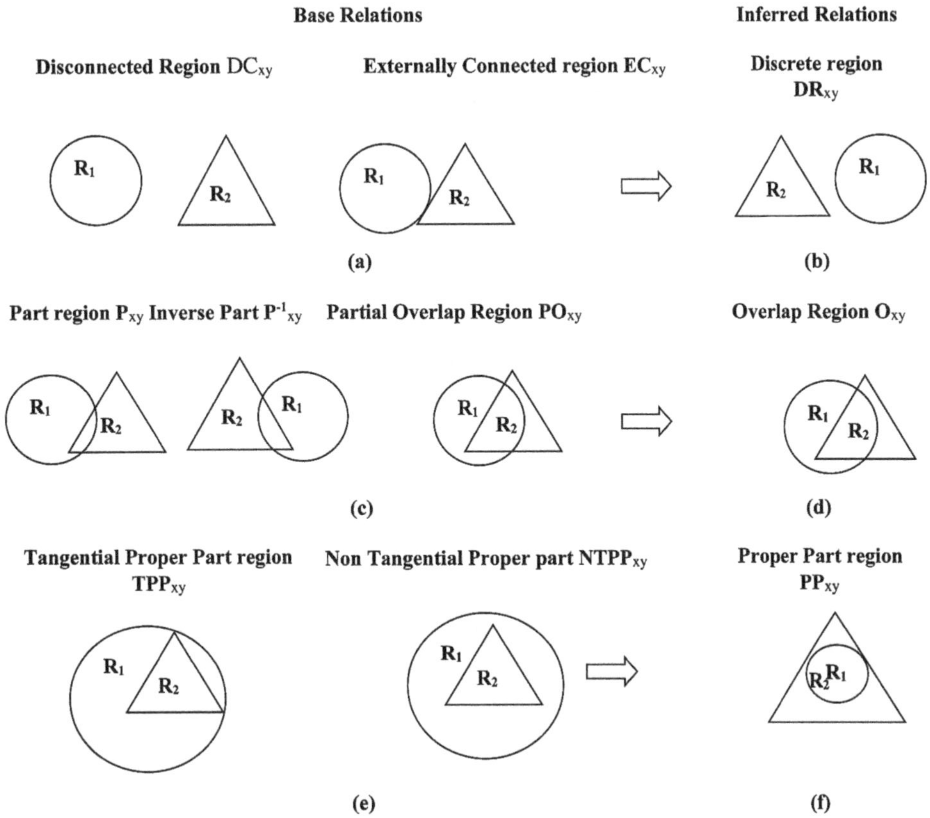

Example: Temporal and Spatial Relations in Task Planning Systems – Illustration 1

Consider a logistics planning system wherein the task of loading the cargo at source station and unloading it at the destination station can be well formulated by attributing event with spatial entities. The temporal and spatial state of the task is illustrated in Figure 3. The actions a_1, a_2 and a_3 define tasks for agent. Topological relations between spatial objects A and B are defined in (Table 4).

The task of this logistic problem is to load the Plane A with Cargo B at the source airport and unload the Cargo B from Plane A at the destination airport. The sequence of task involved is formalized into actions a_1, a_2 and a_3 based on preconditions. The relations that would be in effect after each corresponding

Table 4. Spatial relation between spatial objects

Action	Precondition Relation	Effect Relation
a_1:load_cargo	is_disjoin(Plane A, Cargo B)	is_contain(Plane A, Cargo B)
a_2:fly_plane	is_contain(Plane A, Cargo B)	is_onto(Cargo B, Plane A)
a_3:unload_cargo	is_onto(Cargo B, Plane A)	is_disjoin(Plane A, Cargo B)

Figure 3. Interval relations in transportation and logistics planning

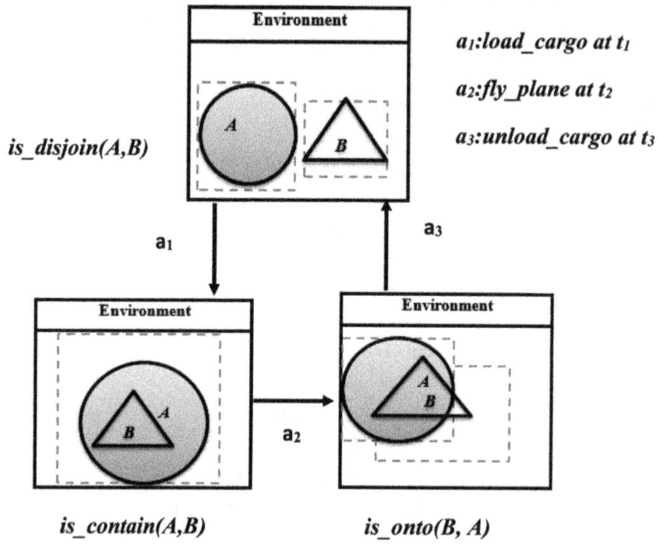

Table 5. Task of agent and temporal relation

Action and Relation	Interval Relation	Direction
is_before (load_cargo, fly_plane)	a_1 *before* a_2	$A(obj_1) \rightarrow B(obj_2)$
is_after (fly_plane, unload_cargo)	a_2 *after* a_3	$B(obj_2) \rightarrow A(obj_1)$

action are shown in (Table 4). The predicate *"is_disjoin", "is_contain", "is_onto"* follows the definition given in (Table 3) for DC_{xy}*(disconnected)*, PP_{xy} *(Proper Part)*, O_{xy} *(Overlaps)* respectively. The significance of temporal relations is made known from actions defined by predicates *"is_before", "is_meets" and "is_after"* that follows the definition given in (Table 2a) for *"before", "meets" and "after"* relations respectively in interval logic. The temporal relations between actions are shown in (Table 5).

The spatio-temporal relations inferred from this problem solving approach are explained with the temporal sequence of tasks viz., (1) *load_cargo* at source followed by (2) *fly_plane* and (3) *unload_cargo* at destination. The three tasks are qualitatively inferred using interval relations with predicate *"is_before" and "is_after"* as defined in (Table 5). Thus, it is significant that aggregation of temporal events defined by actions would result in process here in this example the process is sequence of actions reaching the target as illustrated in the Figure 3.

Example: Illustration-2 Block Stacking Problem

Block "B" and "A" are placed at different locations. The goal is to stack block B on top of block A. The temporal and spatial relations in sequencing the actions a_1 and a_2 are illustrated in (Table 6 and 7). The snapshot of action sequence is shown in Figure 4. The sequential event in moving the blocks is described by the precondition and effect of action shown in (Table 6). *Move_B* is the action triggered when block

Figure 4. Block stacking problem

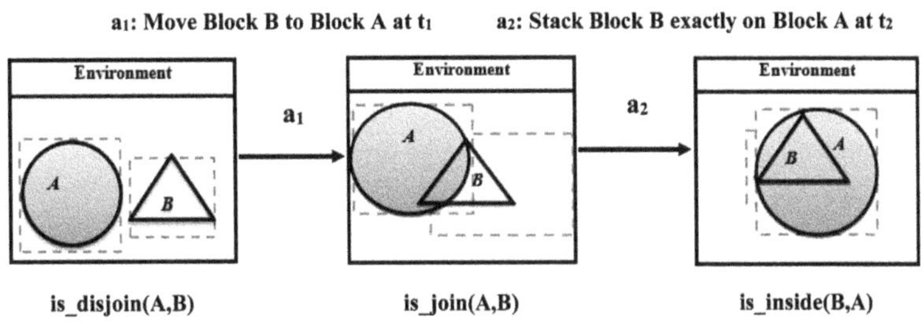

Table 6. Spatial relation between spatial object

Action	Precondition Relation	Effect Relation
a_1:Move_B	is_disjoin(Block A, Block B)	is_join(A,B)
a_2:Stack_B	is_join (Block B, Block A)	is_inside(B,A)

Table 7. Task of agent and temporal relation

Action and Relation	Temporal Relation	Direction
is_after (stack_B, Move_B)	a_1 *before* a_2	$B(obj_2) \rightarrow A(obj_1)$

Figure 5. Path planning

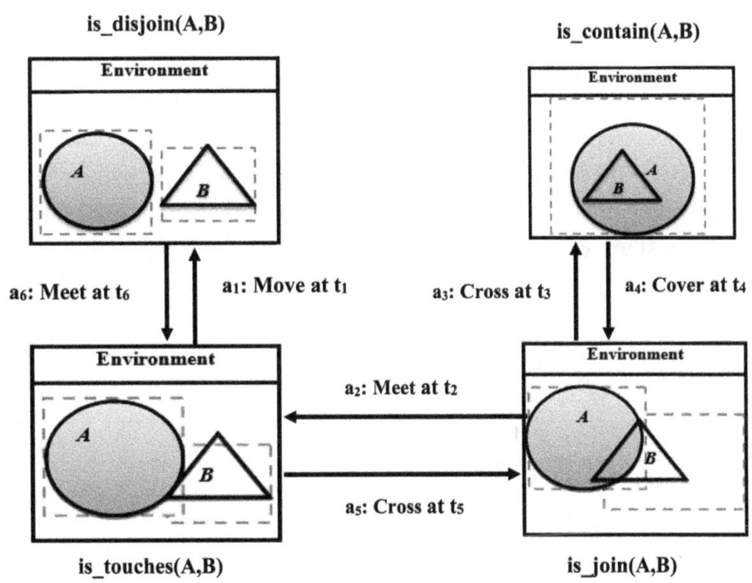

A and B are apart. The effect of the action results in *"is_join(A,B)"*. The next action *Stack_B* is triggered as Block B is in join relation with Block A.

The predicates *"is_disjoin", "is_join", "is_inside"* follow the definition of DC_{xy}*(disconnected)*, P_{xy} *(Part)*, O_{xy} *(Overlaps)* respectively. The significance of temporal relations is made known from actions defined by predicates *"is_before", and "is_after"* that follows the definition of *"before" and "after"* relations in interval logic and it is shown in (Table 7).

Example: Path Navigation of Robot - Illustration 3

The path navigation of robot is to trace the path moved by the robot in reaching the target. In Figure 5, A is robot and B is an object. The robot A has to navigate and pick the object B when it meets at some point of time in its movement along the path. Figure 5 illustrates the forward and backward move of robot at various time points. The predicates *"is_disjoin", "is_touches", "is_join" and "is_contain"* follow the definitions of DC_{xy}*(disconnected)*, O_{xy} *(Overlaps)*, P_{xy}*(Part) and* PP_{xy} *(Proper part)* respectively. The significance of temporal relations is made known from actions defined by predicates *"is_before", and "is_after"* that follows the definition of *"before", "meets" and "after"* relations in interval logic and it is shown in (Table 9).

The path sequence in forward direction towards the target is: At time $t_{1,}$ the object B and robot A are apart. Robot A moves and touches the object B at time t_2. The robot picks the object at time t_3. The fetched object is placed down at time t_4. Robot leaves the object at time t_5 and it is apart from the object B at t_6 in backward path sequence. The interpretation of temporal and spatial relations by sequencing the actions in path navigation of robot is illustrated in Figure 5. The descriptions of actions based on precondition relation and effect of relation after the corresponding action are defined in (Table 8). Action sequencing is done based on the interval relation shown in (Table 9).

Thus, the significance of interval relations and spatial topological relations in representing temporal and spatial knowledge in various problem solving tasks is illustrated in this section. The methods discussed in this section confines to dynamic changes occurring on two dimensional plane and its interaction with neighboring spatial structures (point, line, polygon and its topological structures). For example, a bounded region is a connected set of spatial points called as sub-regions r_1 and r_2 at time τ_1 (temporal granule) in a geographic region of interest $\Re$ as shown in Figure 6. A bounded region that undergoes deformation over time either in shape or structure can be a landscape, river boundary, forested region etc. it can become non-existent at time τ_2 which is clearly a dynamic phenomena. The change in topological relation at different times τ_1, τ_2, τ_3 is inferred using RCC-8 topological relations and temporal

Table 8. Task of agent and spatial relations

Action	Precondition Relation	Effected Relation
a_1:move	is_disjoin (A, B)	is_touches(B, A)
a_2:meet	is_touches(B, A)	is_join(B,A)
a_3:cross	is_join(B,A)	is_contain(B,A)
a_4:cover	is_contain(B,A)	is_join(A,B)
a_5:cross	is_join(A,B)	is_touches(A,B)
a_6:meet	is_touches(A,B)	is_disjoin (A, B)

Table 9. Temporal relations between spatial objects

Action and Relation	Temporal Relations	Direction
is_before(move,meet)	a_1 *before* a_2	$A(obj_1) \rightarrow B(obj_2)$
is_before(meet,cross)	a_2 *before* a_3	$B(obj_2) \rightarrow A(obj_1)$
is_meet(cross,cover)	a_3 meets a_4	$B(obj_2) \rightarrow A(obj_1)$
is_after(cross,cover)	a_5 *after* a_4	$A(obj_1) \rightarrow B(obj_2)$
is_after(meet,cross)	a_6 *after* a_5	$A(obj_1) \rightarrow B(obj_2)$

Figure 6. Sub regions connected in a geographic region R

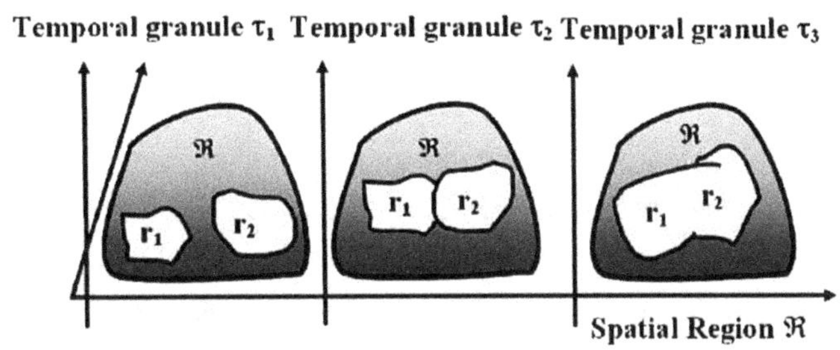

aspect of the spatial dynamism is inferred using Allen's Interval relations. The event causing geo-spatial dynamism is the interest of scientific studies and such events can be inferred through coding of facts and formulations of rules in a knowledge base. The profound theory of time introduced by Allen's interval relations and spatial topological relations formulated by Randell's RCC-8 are adopted to illustrate the importance of spatio-temporal KR in scientific applications.

DISCUSSION: MODELING SPATIAL EVOLUTION

Spatial evolution is the study of geographic location over time. Every spatial location related at time *t* would be related in near future time say *t+1*. The identity of geographic features also varies in such cases. Hence, geographic features are related at different times. A spatial structure of geographic location incurs dynamic change as time passes and temporal study on such spatial structures investigates various spatial relations that have formed during the phenomenal change. This section introduces various methods in spatio-temporal modeling of evolution. Event and Process are interrelated activities. Process is aggregation of sequential, non-sequential and sub sequential events. For example, ecology cultivation is a process that occurs as a result of sequence of events (growth of species) over time. Whereas rainfall at a particular spatial location is not the only source for growth of biological species, thus it is a non-sequential event. The dependency between these two events like growth of plants, grass lands, trees induces immigration of habitats which is a sub sequential event. Thus, aggregation of events results in a process.

Temporal Characteristics of Spatial Change

Temporal characteristics of spatial change thus describe a process which is a function of time. This concept can be understood from Figure 7 in which the spatial objects are apart by a distance d (Discrete Relation DR_{xy}) at time t. The objects S_1 or S_2 tend to move from their respective locations at consecutive time t+1 and subsequently the objects S_1 and S_2 meet at t+k (Externally connected EC_{xy}) such that distance d becomes zero. A process is aggregation of events occurring at consecutive time intervals. Spatial topological relations at each time instant when aggregated into a time interval, spatial change is inferred qualitatively using topological relations of RCC-8 as discussed in section 2. Allen's Interval relations when applied to spatial interconnections define a event. Events at successive time scale would confer a process as explained in Figure 7. The evolving structure of the space is shown in dotted lines. Thus, spatial evolution is a function of time as defined by Equation (1):

$$\text{Evolving Space } \delta = f(S, T) \quad (1)$$

where S is a two-dimensional geographic space on earth consisting of spatial regions S_i's. The relation between spatial regions is described at different time points *t, t+1, t+2,...., t + k* of time T such that $\sum_{k=0}^{n} t + k = T$ where, *t* is a temporal granule and *k* is stepwise increment factor of t.

Figure 7. Temporal and spatial components in evolving process

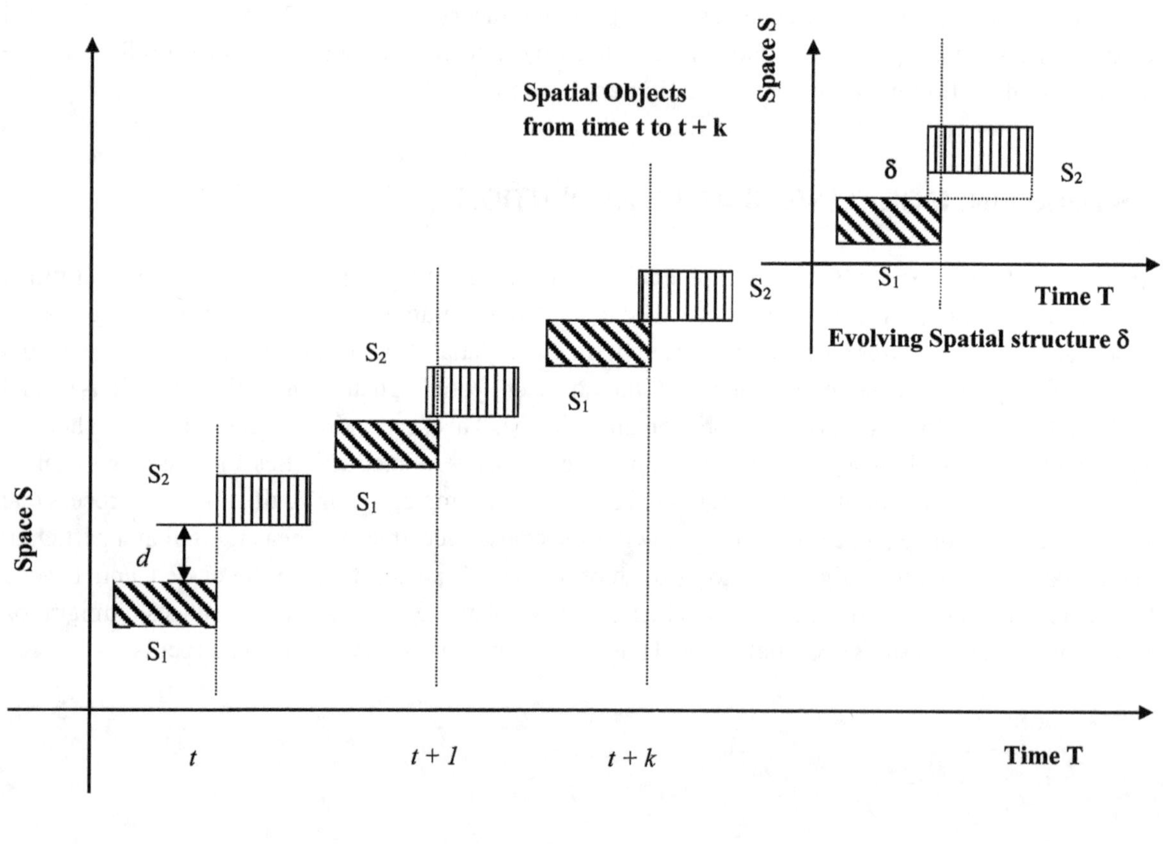

Table 10. Components of KR framework

KR Framework		Temporal KR		Spatial KR		Spatio-Temporal KR	
Inference		Event ^{T}	Process ^{T}	Event ^{S}	Process ^{S}	Event ST	Process ST
Relations	Temporal	{P, I}	{I}	-	-	{P, I}	{I}
	Spatial	-	-	{C}	{IC}	{C}	{IC}
	Spatio-Temporal	{P, I}	{I}	{C}	{IC}	{P, I, C}	{I, IC}

Note: T: Temporal; S: Spatial; ST: Spatio-Temporal; P: Point; I: Interval; C: Connected set of points is an event; IC: Interconnected set of points at successive interval is process.

Table 11. RCC-8 neighborhood relations

Base Relation	Neighborhood Relation Sets
DC_{xy}	$\{EC_{xy}\}$
EC_{xy}	$\{PO_{xy}\} \cup \{DC_{xy}\}$
PO_{xy}	$\{TPP_{xy}\} \cup \{TPPI_{xy}\} \cup \{C_{xy}\} \cup \{DC_{xy}\}$
TPP_{xy}	$\{PO_{xy}\} \cup \{NTPP_{xy}\} \cup \{C_{xy}\}$
$TPPI_{xy}$	$\{PO_{xy}\} \cup \{NTPPI_{xy}\} \cup \{C_{xy}\}$
$NTPP_{xy}$	$\{TPP_{xy}\} \cup \{C_{xy}\}$
$NTPPI_{xy}$	$\{TPPI_{xy}\} \cup \{C_{xy}\}$
C_{xy}	$\{PO_{xy}\} \cup \{TPP_{xy}\} \cup \{TPPI_{xy}\} \cup \{NTPP_{xy}\} \cup \{NTPPI_{xy}\}$

Point and interval are the components of temporal KR framework using which a set of relations are defined in (Tables 2a and 2b). The significance of point and interval is suppressed within spatial KR framework as regions and its connections are considered as important components to qualitatively infer spatial locations like land parcel using the predicates *near_by, close_to, interior, exterior* etc. Thus, connection between two spatial regions derives a set of relations as defined in (Table 3). The illustrative examples in section 2 have shown the importance of both temporal and spatial components. {P, I, C, IC} are defined in (Table 10) to qualitatively infer spatio-temporal process which is a collection of events defined by actions and temporal states.

The importance of KR framework detailed in (Table 10) for inferring event and process has increased the demand for the development of Geographical Information Systems and its variant like Temporal GIS, Spatio-Temporal GIS etc., for real time applications such as disaster management, environmental risk assessment, evacuation planning etc.

Representation of Temporal Dynamics of Spatial Object: Review of Spatio-Temporal Models

In recent years, the growth of knowledge based systems encompasses mathematical foundation of logic with set theory, relations algebra owing to KR framework. Point set theory and its intersections were discovered by Egenhofer in 1994b. They are used in illustrating the importance of spatial predicates in description of spatial relations based on related and unrelated geometrical shapes. The 9-intersection calculi support Geographical Information Systems (GIS) in knowledge discovery.

Figure 8. RCC-8 neighborhood – Schematic definition

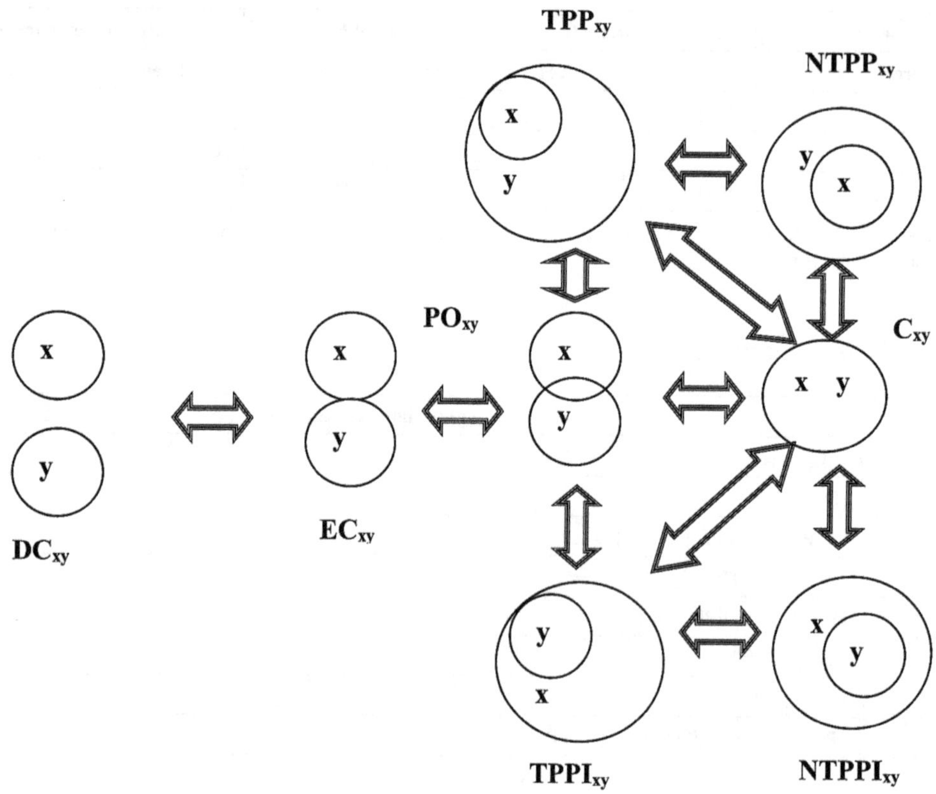

Spatial Neighborhood Regions

Egenhofer's 9-intersection framework includes binary topological relations with various geometries like *polygon, lines and points* to comparatively analyze the interconnections between interior, exterior and boundaries of spatial objects. RCC- 8 topological relation based on spatial logic have their significance in representing continuity of space. Muller, 1998a and 1998b had shown formal methods to qualitatively infer space-time history as continuant of space over time. It further derives topological neighborhood relations as shown in Figure 8 (Muller, 2002; Goodday, 1994). The spatial regions x and y are apart (DC_{xy}) that conceptually connects regions externally (EC_{xy}). The partially overlapping regions formed from EC_{xy} may conceptually change into other topological relations as shown in Figure 8. The spatial topological connection and their conceptual neighbors are formed by set union of base relations defined by Figure 8. Thus, continuity of space derives conceptual neighbors as listed in (Table 11). Event describes static aspect of the world, while process inculcates collection of events which could be a state space of activities thus, describing space-time history (Galton, 2003; 2004; 2008 and 2009). The continuity of a process is derived from continuity of event, thus event is integral part of process. Galton and Worboy in 2005 introduced dynamic geo-network to illustrate the persistence of relationship between events in knowledge discovery. Dynamic geographic network has illustrated the use of event and process in Geographical Information Systems (GIS). Formal representation of identity based geo-spatial dynamism has attracted the use of graph structures. The spatial object whose identity continue to appear in future time

Figure 9. Entities evolving over time

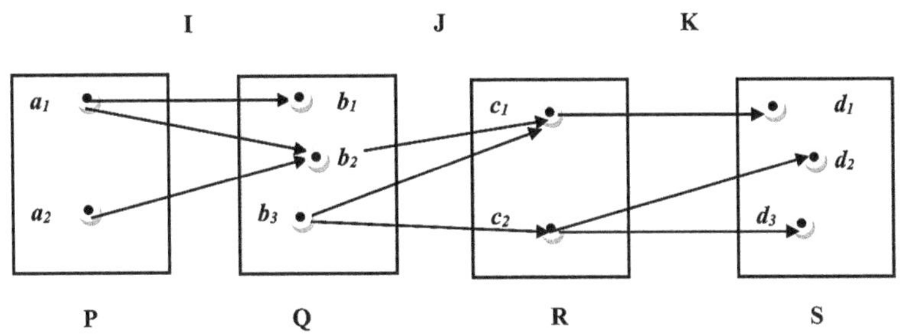

or transform in its identity seen at later time is modeled using bigraphs. It is a bare graph connecting more than one node by a common edge representing a link in association with identity of a object or its continuity, subject to transformation over time. This bigraph dynamism was introduced by Stell et al., in 2011 to aid in algebraic analyzes of geo-spatial dynamism. The existing models illustrated in this section have given importance to evolution but not reported about the forecasting of this dynamism in future time.

Event and Process are interrelated activities. Identification of a feature by its existence and non-existence can be studied using two main domain knowledge representations (1) Temporal KRR and (2) Spatial KRR. Dynamic knowledge representation underlies the principles of KRR. The origin of spatial KRR is found by Whitehead. The state of physical change in a geographic location is formally expressed as parts and whole by Clarke in 1981. The complexity in expressing the physical state of geography lies with geometrical shape and directional information. This has later paved way to formulate logical axioms on spatial geometry named as topological structures by Randell in 1992. Formalization of physical geography based on topological structure is qualitative as it uses predicates like *nearby, close_to, crosses, touches, interior, and exterior etc.,* to express the identification of physical entity in a geographic space.

Objects at different geographic spaces can be characterized by temporal relations. A spatial activity whether man made or natural in a moment of time or between two time points is an event. When such events are aggregated the resultant is a process. Thus, dynamism in a geographic space is a function of time by which existence and non-existence of an object can be inferred. Moreover, it is continuant. As stated in most of the prospective research works, aggregation of geographic space using point and interval attributes of time is qualitative representation of temporal dynamism of space (Cohn et al., 1994 and 1997; Hazarika and Cohn, 2001; Hazarika, 2005). Space-time history is the continuity of space over consecutive time points. UML extensions have also been introduced to differentiate the semantics of events and objects in reasoning spatial process (Worboys, 2004 and 2005; Bennett, 2004). A process on geographic space can be inferred by characterizing temporal relations. Event is a function of spatial topology; Process is a function of time and when interrelated spatio - temporal process is inferred. Spatial structures of geographic location incur dynamic change as time passes and temporal study on such spatial structures investigates various spatial relations that have formed during the phenomenal change (Stell, 2004; 2008 and 2011). In this perspective, we would like to propose modeling of dynamic spatio-temporal patterns and spatio-temporal relations in those evolving structures. The significance in scientific domain applications is to be investigated.

Figure 10. Bare bigraph for relation I

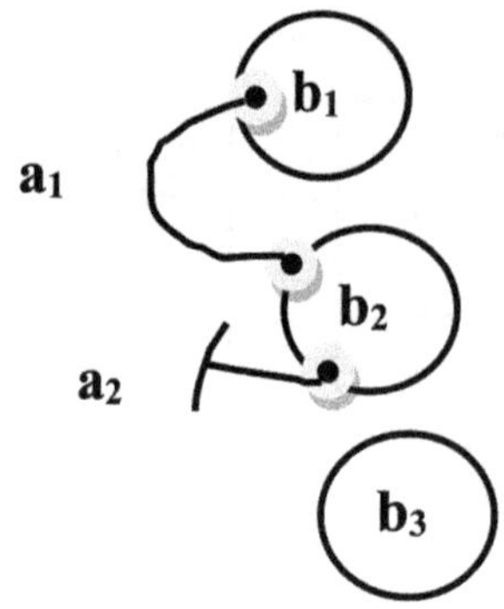

Figure 11. Composite relations

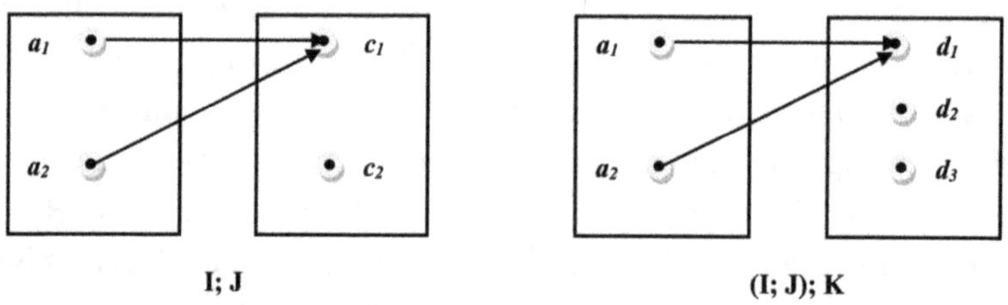

Bare Bigraph Dynamism

The change of spatial entities at different times is modeled using bigraph by Stell, 2011. The entities and their relationships with each other at different times P, Q, R and S respectively are shown in Figure 9. The interpretation of the relation I indicated in Figure 9 is that two different spatial regions may join or continue to exist or split into a separate region say growth of specific species group identified at different times. The existence of the identity of the spatial object is described by relating a_1 and b_1. Separation of a_1 into two entities as b_1 and b_2 shows the distribution of one entity into two different entities. The existence of spatial object at different times in some connection with spatial structure is illustrated by bare bigraph shown in Figure 10. The relation I in Figure 9 is shown in bare graph in which a_1 and a_2 are spatially disjoint entities. At a later time a_1 changes in to b_1 and b_2. Bare bigraph, a model for spatial evolution is illustrated in Figure 11 with composite relations that devise the evolution of spatial entities a_1 and a_2 into d_1.

Spatio-Temporal Process Graph

The dynamic assertion and updation of knowledge in concurrence with change in real-world is dynamic KR. The use of dynamic KR can be well understood by reasoning events and process from spatio-temporal graph as a model of evolution. The discussion in section 3 has shown spatial structure δ is function of time. The dynamic update of knowledge base is denoted by KB′ shown in Equation (2). The update of knowledge base involving temporal and spatial KR framework is detailed in (Table 12):

Table 12. Dynamic KR in spatio-temporal domain

KR Framework	Relationship Set
Temporal KR	{*before, equal, meets, starts, overlaps, contains, after, met-by, finished-by, started-by, during*}
Spatial KR	$\{C_{xy}, DC_{xy}, DR_{xy}, P_{xy}, PP_{xy}, O_{xy}, EC_{xy}, TPP_{xy}, NTPP_{xy}, NTPP^{-1}_{xy}, TPP^{-1}_{xy}, PP^{-1}_{xy}, P^{-1}_{xy}\}$
Dynamic Knowledge Representation	KB′ = KB∪δ[**{C,DC,DR,P,PP,O,EC,TPP,NTPP},{** *before, equal, meets, starts, overlaps, contains, after, met-by, finished-by, started-by, during* }] KB_T ∪ Spatial activity at time t and consecutive time t+1

$$KB' = KB \cup \delta \quad (2)$$

The dynamic knowledge that needs update in knowledge base involves composite relations formed with temporal and spatial KR framework that would aid in inferring process. In this perspective, a process graph is introduced in this section that signifies the importance of temporal and spatial relations in KR framework discussed in detail in section 2 of this chapter. Consider the spatial regions R_1, R_2, R_3 and R_4 at different times say t, t′and t″ shown in Figure 12. The time t has the component τ which is a sub part or a granule, such that τ ∈t. The spatial interconnections of R_1, R_2, R_3 and R_4 and dynamic change of their interconnections are described in snapshot t, t′ and t″ respectively. The update of knowledge base in accordance with t, t′ and t″ can be well explained using graph structure. Hence, process graph is introduced to illustrate the representation of change in spatial and temporal structures. Spatio-Temporal Process Graph **ST_G_P (V, E)** consists of {V}, the set of vertices where each vertex represents topological connections between two nearby spatial regions x and y. {E} is the set of edges that links topologically connected neighbors at different time t thereby representing spatial and temporal event relations on graph.

Significance of Multitude

The theory of multitudes can be used to represent spatial evolution. It is considered that the theory of multitude offers much finer vision of the world than the predicate logic. The spatial regions shown in Figure 12 can be represented using the theory. In Figure 12 the multitude of region is indexed at various time instances. Thus, representation of spatial evolution using theory of multitude is effective in reasoning temporal and spatial knowledge. The evolving spatial structure shown in Figure 12 at different temporal scale can be inferred from graph shown in Figure 13.

Temporal and Spatial Structures of a Spatial Process

1. 1. The initial fact of spatial knowledge is shown in Figure 12: The suffix 'x' and 'y' in Figure 13 corresponds to spatial relations R_1, R_2, R_3 and R_4 in Figure 12 these variables are used interchangeably to model the graph shown in Figure 13. At time t, the spatial relations formed are:

Disconnected (R_1, R_2) and Disconnected (R_3, R_4) and Disconnected (R_1, R_3) and Disconnected (R_1, R_4) and Disconnected (R_2, R_3) and Disconnected (R_2, R_4)

2. 2. The update of temporal knowledge base KB'_T is [t ∪ {**τ1,τ2,τ3**}]. The fact of Spatial KB is updated by the spatial relation formed after t which is t′ = t+τ:

Figure 12. Spatial regions (a) Initial time t (b) near future time t+τ (c) periodic or non–periodic interval [t, t′+τ]

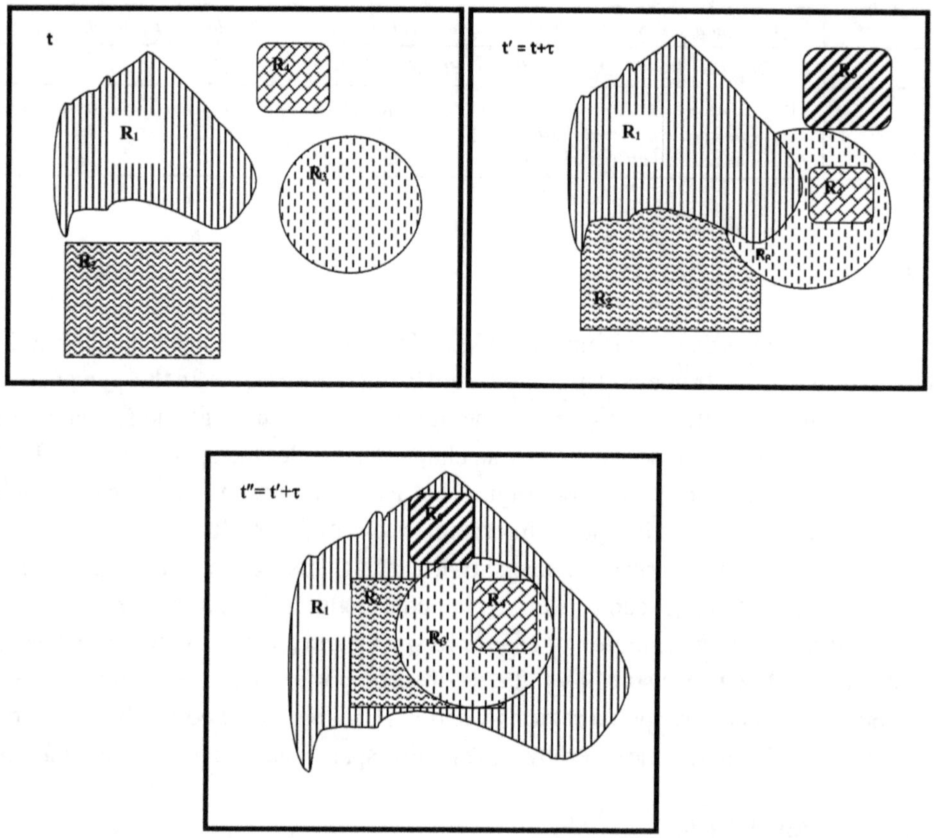

Figure 13. Spatio-temporal graph

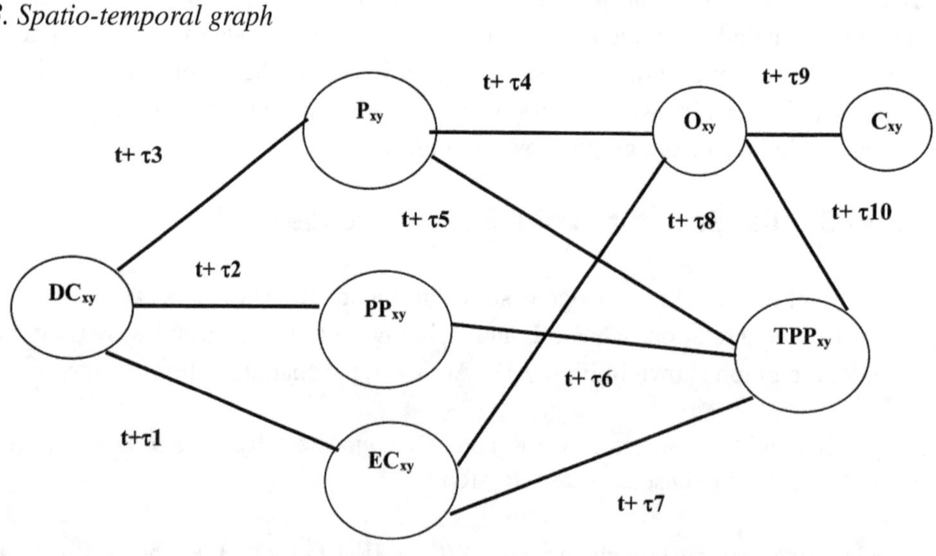

Part (R_1, R_3) and Proper part (R_4, R_3) and Externally connected (R_5, R_3)

3. 3. The update of temporal knowledge base KB'_T is [t ∪ {**τ1,τ2,τ3**}∪{**τ4,τ5,τ6**}]. The fact of Spatial KB is updated by spatial relation formed after **τ1 through τ6 which** is t″= t′+τ:

Tangential proper part (R_3, R_2) and Overlap (R_2, R_1) and Overlaps (R_3, R_1) and Overlaps(R_4, R_1)and Overlaps(R_5, R_1)

Temporal and spatial relations that would update the knowledge base in accordance with step 1 through 3 is defined in (Table 12). The analysis of spatial process which is a function of time can be done using two dimensional arrays wherein the adjacency of spatially connected regions is represented. The edges denote the temporal description of the spatial topology. The adjacency matrix S_{xy} for the ST_Gp process graph is shown in Figure 14 where k varies from 1 to x and m varies from 1 to y. The intersection of kth row and m^{th} column spatial relation contains the temporal description of spatial topological change.

Analysis of Spatio-Temporal Process

The diagonal represent the reflexivity of spatial relations. The symmetry is analyzed from row and column intersection. The spatial geometries are linearly related at different times. In the matrix shown above any row and column describes a process. The edge set {{T+τ_3}∪ {T+τ_2} ∪ {T+τ_1}} describes a spatial process in the graph shown in Figure 13. Vertices denotes spatially connected regions, edge links the topologically connected regions at different times. The adjacency matrix of spatial process S_{xy} contains temporal edges relating the spatial structure. Thus, evolving spatial region and its description at different time (τ) could be inferred from array structure shown in eq. (3) and (4).

$S_{xy} = \{t + \tau i; i \neq 0, i = 1, \ldots, N, N \in Z, k \neq m;$ (3)

Figure 14. Adjacency Matrix S_{xy} of the process graph ST_G_P

	DC	P	PP	EC	O	TPP	C
DC	-	T+τ_3	T+τ_2	T+τ_1	-	-	-
P	T+τ_3	-	-	-	T+τ_4	T+τ_5	-
PP	T+τ_2	-	-	-	-	T+τ_6	-
EC	T+τ_1	-	-	-	T+τ_8	T+τ_7	-
O	-	T+τ_4	-	T+τ_8	-	T+τ_{10}	T+τ_9
TPP	-	T+τ_5	T+τ_6	T+τ_7	T+τ_{10}	-	-
C	-	-	-	-	T+τ_9	-	-

Figure 15. Spatial and Temporal relation of process graph

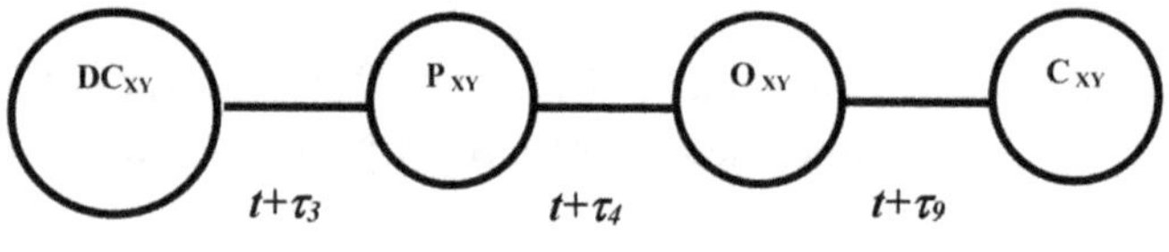

$S_{xy} = \{\phi; i = 0;$ (4)

k and m are row and column indices of spatio-temporal graph. The spatial structure that extends, appears with transformation say in topological forms like *{DC, PP, TPP, TPPI, NTPP, NTPPI}* based on RCC-8 neighborhood relations can be inferred and qualitatively analyzed using temporal description given by edge links along with spatial connections. The inherent process contained in the process graph is shown in Table 15.

$$\{\{t+\tau_3\} \cup \{t+\tau_4\} \cup \{t+\tau_9\}\} \Rightarrow Sp \quad (5)$$

$$\text{before}(t+\tau_3, t+\tau_4) \wedge \text{before}(t+\tau_4, t+\tau_9) \Rightarrow \text{before}(t+\tau_3, t+\tau_9) \quad (6)$$

The start and end of the interval relation shown in Equation (6) is characterized by $t+\tau_i$. The spatially connected regions formed at different temporal granule τ_i say DC_{xy}, P_{xy}, O_{xy}, and C_{xy} is inferred from Equation (6). The end time of DC_{xy} is less than start time of P_{xy} similarly with $O_{xy,}$ *and* C_{xy}. The union of temporal links which are edges connecting the spatially connected regions implies a process as shown in Equation (5). The spatial relations that are temporally related at time t and its granule components τ can be inferred using Allen's relation *before* and *starts* as shown in Equation (6). The continuity of spatially connected regions at consecutive time is thus, characterized by $t+\tau_i$.

Reasoning spatial process is illustrated in Equation (6) with Allen's *before* relation. The disconnected spatial regions x and y at $t+\tau_3$ occurs before the formation of regions x and y into part relation at $t+\tau_4$ which becomes overlapping regions at $t+\tau_9$. This sequence leads to connected regions at later time after $t+\tau_9$. Thus, logical conjunction of spatial relations $DC_{xy} \wedge P_{xy} \wedge O_{xy}$ at $t+\tau_3 \wedge t+\tau_4 \wedge t+\tau_9$ transitively implies connected region C_{xy} after $t+\tau_9$. Eventually, the evolving spatial structures can be applied in qualitative prediction of foregoing spatial process using the adjacency matrix shown in Figure 13 as spatial regions in DC_{xy} may form C_{xy} by any of the adjacent path shown in graph Figure 13. Hence, the adjacency matrix consisting of temporal description of spatial relations would considerably infer spatial events at every interval and process over consecutive intervals.

The formal theory of space-time in KR illustrated in modeling of spatial evolution is conceptualized using graph structures. The dynamic spatial structures which would evolve over time can be qualitatively inferred at an instant and in near future time too. For example, river boundary touching a land parcel at later time due to natural events such as seasonal rainfall, floods or construction of bridge nearby river which was not found nearby river at former time. The dynamic extension of water resources due to which geological structures deforms over time can be qualitatively inferred through the formal theory of space-time with qualitative knowledge representation. Moreover, evolution of new species throughout a specific time period, the cause-effect of such evolution can be qualitatively inferred and even predicted in near future time.

CONCLUDING REMARKS

Formal approaches to knowledge representation are presented to illustrate the significance of Artificial Intelligence as a tool for Knowledge Engineering (KE). The formal representation of time defined by Allen and Spatial region connection defined by Randell are the prime elements in construction of knowledge based systems that has wide range of domain applications in Geographical Information Systems.

The significance of temporal and spatial relations in representing knowledge is illustrated in this chapter with real time scenarios such as Transportation and logistics planning, path planning, block stacking etc. Thus, spatial and temporal relations can be associated to represent dynamism of real-world as described in the section Knowledge Representation of this chapter.

A review of spatio temporal models is presented to illustrate the significance of knowledge representation framework in identifying physical entities that are evolving over time. The importance of temporal and spatial component in modeling spatial evolution is explained, one such example illustrated in this chapter is Bare Bigraph as a model for evolution. Towards the end, the significance of Spatio Temporal Process Graph as dynamic knowledge representation is conceptualized using graph structure. This method of knowledge representation would apply in cause-effect analysis of evolving geographical features.

REFERENCES

Allen, J. F. (1981). An interval-based representation of temporal knowledge. In *International Conference of Artificial Intelligence*, Morgan Kaufmann.

Allen, J. F. (1983). Maintaining knowledge about temporal intervals. *Communications of the ACM*, *26*(11), 832–843. doi:10.1145/182.358434

Allen, J. F. (1983). Maintaining knowledge about temporal intervals. *Communications of the ACM*, *26*(11), 832–843. doi:10.1145/182.358434

Allen, J. F. (1984). Towards a general theory of action and time. *Artificial Intelligence*, *23*(2), 123–154. doi:10.1016/0004-3702(84)90008-0

Allen, J. F., & Ferguson, G. (1994). Actions and events in interval temporal logic. *Journal of Logic and Computation*, *4*(5), 531–579. doi:10.1093/logcom/4.5.531

Bennett, B., & Galton, A. P. (2004). A unifying semantics for time and events. *Artificial Intelligence*, *153*(1–2), 13–48. doi:10.1016/j.artint.2003.02.001

Bittner, T. (2002). Approximate qualitative temporal reasoning. *Annals of Mathematics and Artificial Intelligence*, *36*(1–2), 39–80. doi:10.1023/A:1015899702951

Box, P., Jenkins, G., Reinsel, G., & Ljung, G. M. (2012). *Time Series Analysis Forecasting and Control*. Wiley Series in Probability and Statistics.

Campelo, C. E. C. (2013). *Representing and Reasoning about Changing Spatial Extensions of Geographic Features* (Doctoral thesis). University of Leeds.

Campelo, C. E. C., Bennett, A. P., & Dimitrova, V. (2011). Identifying geographical processes from time-stamped data. *Lecture Notes in Computer Science, 6631*(5), 70–87. doi:10.1007/978-3-642-20630-6_5

Campelo, C. E. C., Bennett, B., & Dimitrova, V. (2012). From polygons and timestamps to dynamic geographic features: Grounding a spatio-temporal geo-ontology. *Lecture Notes in Computer Science, 7518*(10), 251–260. doi:10.1007/978-3-642-33999-8_31

Campelo, C. E. C., Bennett, B., & Dimitrova, V. (2013). Identifying geographic events and processes in spatio-temporal data. *Frontiers of Artificial Intelligence and Applications., 241*(8), 47–58.

Claramunt, C., & Theriault, M. (1995). Managing time in GIS: An event-oriented approach. *Proceedings Recent Advances in Temporal Databases, 1*(9), 23–42.

Claramunt, C., & Theriault, M. (1996). *Towards semantics for modelling spatio-temporal processes within GIS. In Advances in GIS Research* (pp. 27–43). Taylor and Francis.

Clarke, B. L. (1981). A calculus of individuals based on connection. *Notre Dame Journal of Formal Logic, 22*(3), 204–218. doi:10.1305/ndjfl/1093883455

Clarke, B. L. (1985). Individuals and points. *Notre Dame Journal of Formal Logic, 26*(1), 61–75. doi:10.1305/ndjfl/1093870761

Cohn, A. G., Bennett, B., Gooday, J., & Gotts, N. (1997). Qualitative spatial representation and reasoning with the region connection calculus. *GeoInformatica, 1*(3), 275–316. doi:10.1023/A:1009712514511

Cohn, A. G., Gooday, J. M., & Bennett, B. (1994). A comparison of structures in spatial and temporal logics. Proceedings of Philosophy and the Cognitive Sciences, 1–14.

Cui, Z., Cohn, A., & Randell, D. A. (1993). Qualitative and topological relationships in spatial databases. Lect. Notes Comput. Sci., Advances in Spatial Databases, 692(5), 296–315. doi:10.1007/3-540-56869-7_17

Egenhofer, M. J. (1991). Reasoning about binary topological relations. *Advanced Spatial Databases, 525*(8), 141–160. doi:10.1007/3-540-54414-3_36

Egenhofer, M. J., Clementini, E., & Di Felice, P. (1994a). Topological relations between regions with holes. *International Journal of Geographical Information Science, 8*(2), 129–142. doi:10.1080/02693799408901990

Egenhofer, M. J., & Franzosa, R. D. (1991). Point-set topological spatial relations. *International Journal of Geographical Information Systems, 5*(2), 161–174. doi:10.1080/02693799108927841

Egenhofer, M. J., Mark, D. M., & Herring, J. (1994b). *The 9-intersection: Formalism and Its Use for Natural-Language Spatial Predicates.* Technical Report 94-1, National Center for Geographic Information and Analysis, Santa Barbara, CA.

Frank, A. U. (1991). Qualitative spatial reasoning about cardinal directions. *International Journal of Geographical Information Systems*, (10): 1–10.

Frank, A. U. (1992). Theories and methods of spatio-temporal reasoning in geographic space. *International Conference GIS – From Space to Territory, Proceedings Lecture Notes Computer Science*, 639. doi:10.1007/3-540-55966-3

Frank, A. U. (1996). Qualitative spatial reasoning: Cardinal directions as an example. *International Journal of Geographical Information Systems, 10*(3), 269–290. doi:10.1080/02693799608902079

Freksa, C. (1991). Qualitative spatial reasoning. *Cognitive and Linguist Aspect of Geographic Space, 63*, 361–372. doi:10.1007/978-94-011-2606-9_20

Freksa, C. (1992). Temporal reasoning based on semi-intervals. *Artificial Intelligence, 54*(2), 199–227. doi:10.1016/0004-3702(92)90090-K

Galton, A. (2003). Desiderata for a spatio-temporal geo-ontology. Spatial Information Theory. *Foundation of Geographic Information Science, 2825*(2), 1–12.

Galton, A. (2004). Fields and objects in space, time and space-time. *Spatial Cognition and Computation, 4*(1), 15–37. doi:10.1207/s15427633scc0401_4

Galton, A. (2008). Experience and history: Processes and their relation to events. *Journal of Logic and Computation, 18*(3), 323–340. doi:10.1093/logcom/exm079

Galton, A. (2009). Spatial and temporal knowledge representation. *Earth Science Informatics, 2*(3), 169–187. doi:10.1007/s12145-009-0027-6

Galton, A., & Worboys, M. (2005). Processes and events in dynamic geo-networks. Lecture Notes in Computer Science, 3799(11), 45–59. doi:10.1007/11586180_4

Goodday, J. M., & Cohn, A. G. (1994). *Conceptual neighborhoods in temporal and spatial reasoning. ECAI-94 Spatial Temporal Reasoning Workshop.*

Hazarika, S. M. (2005). *Qualitative Spatial Change: Space-Time Histories and Continuity*. Thesis.

Hazarika, S. M., & Cohn, A. G. (2001). *Qualitative spatio-temporal continuity*. Spatial Information Theory.

Hornsby, K., & Egenhofer, M. J. (2000). Identity-based change: A foundation for spatio-temporal knowledge representation. *International Journal of Geographical Information Science, 14*(3), 207–224. doi:10.1080/136588100240813

Jayanthi, G, Uma, V. (2016a). Event Attributed Spatial Entity Knowledge (EASE) Based Spatio-Temporal Reasoning to Infer Geographic Processes. *Proceeding of IEEE(ICGHPC.16).*

Jayanthi, G., & Uma, V. (2016b). Reasoning About Events Causing Geographic Process Using Spatial and Temporal Relations. *Proceedings of the IEEE, ICCCI*, 16.

Jayanthi, G., Uma, V.(2016d). Automated Planning Considering Event Based Spatial Object Interactions Using Spatio-Temporal Reasoning. *Proceeding of IEEE(ICRTIT.16).*

Jayanthi, G., & Uma, V. (2017). A qualitative inference method for prediction of geographic process using spatial and temporal relations. *International Journal of Artificial Intelligence and Soft Computing, 6*(1), 1–23. doi:10.1504/IJAISC.2017.084215

Jayanthi, G., & Uma, V. (under review). submitted on June 2016c). Reasoning Temporal Description of Spatial Entity Knowledge Towards Qualitative Geographic Process Prediction. *International Journal of Knowledge and Information Systems.*

Kowalski, R., & Sergot, M. (1986). A logic based calculus of events. *New Generation Computing*, *4*(1), 67–95. doi:10.1007/BF03037383

McCarthy, J., & Hayes, P. J. (1969). *Some philosophical problems from the standpoint of artificial intelligence. In Readings in Nonmonotonic Reasoning* (pp. 26–45). Morgan Kaufmann Publishers Inc.

Mondo, D., Stell, J. G., Claramunt, C., & Thibaud, R. (2010). A Graph Model for Spatio-Temporal Evolution. *Journal of Universal Computer Science*, *16*(11), 1452–1477.

Muller, P. (1998a). A qualitative theory of motion based on spatio-temporal primitives. Principles of Knowledge Representation and Reasoning, 131–143.

Muller, P. (1998b). Space-time as a primitive for space and motion. *Proceedings of FOIS*, 63–76.

Muller, P. (2002). Topological spatio-temporal reasoning and representation. *Computational Intelligence*, *18*(3), 420–450. doi:10.1111/1467-8640.00196

Pani, A. K., & Bhattacharjee, G. P. (2001). Temporal representation and reasoning in artificial intelligence: A review. *Mathematical and Computer Modelling*, *34*(1–2), 55–80. doi:10.1016/S0895-7177(01)00049-8

Patterson, D. W. (2005). *Introduction to Artificial Intelligence and Expert Systems*. PHI learning.

Randell, D. A., Cui, Z., & Cohn, A. G. (1992). *A spatial logic based on regions and connection*. Proceedings Knowledge Representation and Reasoning.

Rich, Knight, & Nair. (2009). *Artificial Intelligence*. Tata McGraw Hill.

Stell, J., & Worboys, M. (2008). A theory of change for attributed spatial entities. *Lecture Notes in Computer Science*, *5266*(9), 308–319. doi:10.1007/978-3-540-87473-7_20

Stell, J. G. (2000). Boolean connection algebras: A new approach to the region-connection calculus. *Artificial Intelligence*, *122*(1), 111–136. doi:10.1016/S0004-3702(00)00045-X

Stell, J. G. (2004). Part and complement: Fundamental concepts in spatial relations. *Annals of Mathematics and Artificial Intelligence*, *41*(1), 1–17. doi:10.1023/B:AMAI.0000018575.09226.ff

Stell, J. G., Mondo, D., Thibaud, R., & Claramunt, C. (2011). Spatio-Temporal Evolution as Bigraph Dynamism. *COSIT, LNCS.*, *6899*, 148–167.

Whitehead, A. N. (1929). Process and Reality: An Essay in Cosmology. Macmillan.

Worboys, M. (2005). Event oriented approaches to geographic phenomena. *International Journal of Geographical Information Science*, *19*(1), 1–28. doi:10.1080/13658810412331280167

Worboys, M., & Hornsby, K. (2004). From objects to events: GEM, the geospatial event model. *GIScience Lecture Notes in Computer Science*, *3234*(10), 1–17.

KEY TERMS AND DEFINITIONS

Event: Temporal description of an action, activity, or scenario in the real world.

Process: Aggregation of events over interval.

Spatial Knowledge Representation: The logic-based formalism of spatial geometry and its topological structures bounded by a connected set of points.

Spatio-Temporal Process Graph: Spatio-Temporal Process Graph $ST\_G_p(V, E)$ consists of {V}, the set of vertices where each vertex represents topological connections between two nearby spatial regions x and y. {E} is the set of edges that links topologically connected neighbors at different time t thereby representing spatial and temporal event relations on graph.

Temporal Knowledge Representation: The logic-based formulation of action description whose occurrence is implied by time.

Section 3

Dynamic Knowledge Representation in Other Domains

Chapter 11
Biography as Scenario:
Representing Historical Knowledge by the Event Bush

Maria Mandrik
Archive of the Russian Academy of Sciences, Russia

Sergey Arustamov
ITMO University, Russia

ABSTRACT

Purely descriptive knowledge appears to have a structure that can be formalized using the newly developed tool of dynamic knowledge representation, the event bush. In its framework, a singular biography can be transformed into a knowledge-based system with an opportunity of creation of a database. However, the very approach to the database design may undergo quite a change thereafter. This change may concern the concept of primary key, composition of data scheme, and other fundamental issues of database design.

INTRODUCTION

Recent advances in dynamic knowledge engineering and representation allow us to enter that particular area of descriptive knowledge, which has been considered totally inappropriate for formalization. In this area, perhaps the least prone to formalization has been considered the study of history. Meanwhile, exactly this discipline is strongly needed for political analysis and decision-making on regional and global scale. This urges one to look for impartial and bias-free way of representation of historical knowledge. The present paper addresses one of the typical patterns of descriptive historical knowledge, biography. Below we will examine whether it is possible to formalize it by means of the principal method of dynamic knowledge engineering, the event bush.

DOI: 10.4018/978-1-5225-5261-1.ch011

SOURCE TEXT

As an example, any kind of biography may be appropriate, even a synthetic biography of an imagined character. However, let us select a biography of well-known personality, whose impact on the world's history is needless to assert. A good candidate may be Napoleon Bonaparte (1769–1821). To see the applicability of the event bush, one may take a part of his life (the whole story may require a space that does not fit the paper format). For our purposes the criteria for selection of a biographic text are simplicity, brevity and brightness of narration (to easily recognize and extract distinct events). The following description taken from Napoleon Bonaparte (2014) fits these criteria entirely.

"Early Years. Considered one of the world's greatest military leaders, Napoleon Bonaparte was born on August 15, 1769, in Ajaccio, Corsica, France. He was the fourth, and second surviving, child of Carlo Buonaparte, a lawyer, and his wife, Letizia Ramolino.

By the time around Napoleon's birth, Corsica's occupation by the French had drawn considerable local resistance. Carlo Buonaparte had at first supported the nationalists siding with their leader, Pasquale Paoli. But after Paoli was forced to flee the island, Carlo switched his allegiance to the French. After doing so he was appointed assessor of the judicial district of Ajaccio in 1771, a plush job that eventually enabled him to enroll his two sons, Joseph and Napoleon, in France's College d'Autun.

Eventually, Napoleon ended up at the military college of Brienne, where he studied for five years, before moving on to the military academy in Paris. In 1785, while Napoleon was at the academy, his father died of stomach cancer. This propelled Napoleon to take the reins as the head of the family. Graduating early from the military academy, Napoleon, now second lieutenant of artillery, returned to Corsica in 1786.

Back home Napoleon got behind the Corsican resistance to the French occupation, siding with his father's former ally, Pasquale Paoli. But the two soon had a falling-out, and when a civil war in Corsica began in April 1793, Napoleon, now an enemy of Paoli, and his family relocated to France, where they assumed the French version of their name: Bonaparte."

EVENT BUSH METHOD

The method of event bush was developed by Pshenichny et al. (2009) to reveal the behavior of information in dynamic environments of directed alternative changes (Pshenichny and Kanzheleva, 2011). It is gaining increasing importance in a variety of fields, from the geosciences to marketing and history (Solomin and Mandrik, 2015). For detailed description of the method, the reader is referred to these publications.

In brief, the event bush represents an environment of directed alternative changes as a set of events of four types: primary internal (*ia*), primary external (*ib*), secondary (*ii*) and tertiary (*iii*). Event is defined as a statement of general form $S - P_1, P_2, \ldots, P_i$, where S is subject, and $P_1, P_2, \ldots, P_i$ are predicates, with or without negation. The following restrictions are put herewith, (i) subject may be only one in any event, (ii) subject may not be negated, (iii) all predicates may not be negated in an event, at least one must be without negation (e.g., an event $S - \sim P$, ~ being the negation, is not allowed).

The following pairwise relations between event types *(i)* - *(iii)* are set in the event bush:

- Not a single event may lead to *(ia)* or *(ib)*;
- Event *(ia)* without an event *(ib)* or *(ii)* may lead only to one event *(ii)* or *(iii)*;
- Event *(ia)* may not lead to any event together with another *(ia)*;
- Event *(ib)* may lead only to an event *(ii)* and only together with *(ia)* or another *(ii)*;
- Event *(iii)* may not lead to any other event.

These relations are enforced by the connectives of the event bush (Pshenichny and Kanzheleva, 2011): flux (E_1 *Flux* E_2), influx (E_1, E_2 *Influx* E_3), conflux (E_1, E_2, ..., E_{n-1} *Conflux* E_n) and furcation (E_1 *Furcation* E_2, ..., E_n), E being individual event. The first two are mandatory for the event bush, others, optional. Connectives allow one to build *scenarios*, which, by rules of composition of the event bush, must start from *ia*, end up with *iii* and include direct or indirect input of *ib*.

CONSTRUCTION OF EVENT BUSH

General rules of composition of event bush were presented by Pshenichny et al. (2008). Based on these, the context of the Napoleon's biography is modeled as an ensemble of three event bushes, of which two can be regarded as rules partially controlling the structure of the third.

The formulations of events of all bushes are given in Table 1. These formulations have been subsequently updated in process of construction of the bush as discussed by Pshenichny et al. (2008), and in final form some of them look as follows:

iaE_5: Buonaparte (S_1) grows up (P_1), does not side with those who are close to him ($\sim P_2$), does not side with his opponents ($\sim P_3$), does not feel good with those who are close to him ($\sim P_4$), does not make career with his opponents ($\sim P_5$).

iaE_7: Buonaparte (S_1) grows up (P_1), sides with those who are close to him (P_2), does not feel good with those who are close to him ($\sim P_3$), does not side with his opponents ($\sim P_4$), does not make career with his opponents ($\sim P_5$).

Such formulations are hard to read and track the change (e.g., $\sim P_2$ in iaE_5 to P_2 in iaE_7). Table 1 presents short formulations showing the changes.

Importantly, by the rules of composition of the bush, events include predicates in present tense only. Time is expressed by the passage from one event to another, or *change*, in terms of the event bush method. Each change is marked by connective. Each event, therefore, includes everything that pertains to the subject in the present state – e.g., Napoleon is born, does not die (i.e., did not die, or has not died) in infancy, is an officer, but is not (yet) a head of family, and so on. If any of these circumstances change, then there should be a different event, and a connective should be chosen to enforce a passage from one to the other.

Table 1. Events of the Napoleon biography context

General Bush 1		
1	iaE_1	Buonaparte is born, does not grow up, does not die in infancy
2	iiE_1	Buonaparte is born, does not grow up, does not die in infancy
3	ibE_2	Unknown disaster happens that threatens life of a newborn
4	iiE_3	Buonaparte is born, does not grow up, dies in infancy
5	$iiiE_3$	Buonaparte is born, does not grow up, dies in infancy
6	iiE_4	Buonaparte is born, grows up, does not die in infancy
7	$iiiE_4$	Buonaparte is born, grows up, does not die in infancy
General Bush 2		
8	iaE_5	Buonaparte grows up
9	$iiiE_5$	Buonaparte grows up
10	ibE_6	Those who are close to Buonaparte exist
11	iiE_7	Buonaparte sides with those who are close to him
12	iiE_8	Buonaparte feels good with those who are close to him
13	$iiiE_8$	Buonaparte feels good with those who are close to him
14	ibE_9	Political wind is unfavorable for those who are close to Buonaparte
15	iiE_{10}	Buonaparte sides with his opponents
16	iiE_{11}	Buonaparte makes career with his opponents
17	$iiiE_{11}$	Buonaparte makes career with his opponents
Particular Bush (Napoleon Biography Proper)		
18	iaE_{12}	Napoleon Buonaparte is born on August 15, 1769, in Ajaccio, Corsica, does not die in infancy
19	iiE_{12}	Napoleon Buonaparte is born on August 15, 1769, in Ajaccio, Corsica, does not die in infancy
20	$iiiE_{12}$	Napoleon Buonaparte is born on August 15, 1769, in Ajaccio, Corsica, does not die in infancy
21	iiE_{13}	Napoleon Buonaparte has a father who makes career
22	iiE_{14}	Napoleon Buonaparte studies in France's College d'Autun
23	iiE_{15}	Napoleon Buonaparte studies in military college of Brienne
24	$iiiE_{15}$	Napoleon Buonaparte studies in military college of Brienne
25	ibE_{16}	Death of Napoleon's father occurs
26	iiE_{17}	Napoleon Buonaparte has experienced father's death
27	iiE_{18}	Napoleon Buonaparte is an officer
28	$iiiE_{18}$	Napoleon Buonaparte is an officer
29	ibE_{19}	Civil war in Corsica occurs
30	iiE_{20}	Napoleon Buonaparte does not side with those who are close to him
31	iiE_{21}	Napoleon Buonaparte is the head of family
32	iiE_{22}	Napoleon Buonaparte moves family from Corsica to continental France
33	$iiiE_{22}$	Napoleon Buonaparte moves family from Corsica to continental France

The following bushes are built from the specified events.

General bush (rule) 1:

$(iaE_1\ \mathit{Flux}\ iiE_1); (iaE_1, ibE_2\ \mathit{Influx}\ iiE_3); (iiE_1\ \mathit{Flux}\ iiE_4); (iiE_3\ \mathit{Flux}\ iiiE_3); (iiE_4\ \mathit{Flux}\ iiiE_4)$

General bush (rule) 2:

$(iaE_5\ \mathit{Flux}\ iiiE_5); (iaE_5, ibE_6\ \mathit{Influx}\ iiE_7); (iiE_7\ \mathit{Flux}\ iiE_8); (iiE_8\ \mathit{Flux}\ iiiE_8); (iiE_8, ibE_9\ \mathit{Influx}\ iiE_{10}); (iiE_{10}\ \mathit{Flux}\ iiE_{11}); (iiE_{11}\ \mathit{Flux}\ iiiE_{11})$

The bush representing the Napoleon's biography:

$(iaE_{12}\ \mathit{Flux}\ iiE_{12}); (iiE_{12}\ \mathit{Flux}\ iiiE_{12}); (iiE_{12}, iiE_{11}\ \mathit{Influx}\ iiE_{13}); (iiE_{13}\ \mathit{Flux}\ iiE_{14}); (iiE_{14}\ \mathit{Flux}\ iiE_{15}); (iiE_{15}\ \mathit{Flux}\ iiiE_{15}); (iiE_{15}, ibE_{16}\ \mathit{Influx}\ iiE_{17}); (iiE_{17}\ \mathit{Flux}\ iiE_{18}); (iiE_{18}\ \mathit{Flux}\ iiiE_{18}); (iiE_{18}, ibE_{19}\ \mathit{Influx}\ iiE_{20}); (iiE_{20}\ \mathit{Flux}\ iiE_{21}); (iiE_{21}\ \mathit{Flux}\ iiE_{22}); (iiE_{22}\ \mathit{Flux}\ iiiE_{22})$

They correspond directly to the source text, as can be seen from comparison of the latter with the above formulae and Table 1.

DISCUSSION

The following observations can be made from the above bushes. An important property of event bush is that, by rules of its composition, aiming to build one scenario that corresponds to the modeled case, a full set of incompatible scenarios are being built. Each of the three bushes represents a multitude of scenarios, together forming a *multiflow structure* (Pshenichny et al., 2009). In the third ("biographic") bush, one scenario corresponds to actual biography of Napoleon:

$(iaE_{12}\ \mathit{Flux}\ iiE_{12}); (iiE_{12}, iiE_{11}\ \mathit{Influx}\ iiE_{13}); (iiE_{13}\ \mathit{Flux}\ iiE_{14}); (iiE_{14}\ \mathit{Flux}\ iiE_{15}); (iiE_{15}, ibE_{16}\ \mathit{Influx}\ iiE_{17}); (iiE_{18}, ibE_{19}\ \mathit{Influx}\ iiE_{20}); (iiE_{20}\ \mathit{Flux}\ iiE_{21}); (iiE_{21}\ \mathit{Flux}\ iiE_{22}); (iiE_{22}\ \mathit{Flux}\ iiiE_{22})$

while others, ending with $iiiE_{12}$, $iiiE_{15}$, $iiiE_{12}$ and $iiiE_{18}$, stand for those options that Napoleon could take if some or all of the circumstances expressed by the *ibE* events had not occurred (perhaps one of these scenarios may refer to Napoleon's brother, Joseph). Also, alternative scenarios can be observed in the general bushes (the rules), and one of those in the rule 1 clearly describes the tragic destiny of two other children of Carlo Buonaparte who died in infancy. Thus one may think of a kind of relational database sensu Codd (1970) listing and linking (i) scenarios of evolution of the class denoted by the subject of the *ia* event (e.g., class "Buonaparte offspring") in considered context (environment) and (ii) instances that follow these scenarios. This may have application well outside of the humanities.

Herewith, it should be noted that event bush describing a biography may reasonably have only one primary internal event (*ia*) and, hence, only one primary internal subject. If the task is to describe solely the biography of the person (not the results of his/her activity), then this subject ("Buonaparte offspring") will remain singular for all secondary and tertiary events of the bush. Another specificity of "biographic" bush is that its first change is *(iaE Flux iiE)* that is rather untypical for event bushes, normally starting with a couple *(iaE Flux iiiE)* and *(iaE, ibE Influx iiE)*. This change expresses the fact that someone is just born and nothing else has happened.

One more specific feature that, to our mind, would relate to any biography is presence of general bushes (rules) that reflect the general tendencies of the context conditioning the lives of people of given class (e.g., the Buonapartes). The way these rules condition the inference in a particular bush still need to be studied but some remarks can be already made. In the considered case, Rule 1 controls inference in the "biographic" bush by the class-subclass relation of subjects ("Buonaparte" being the class, and "Napoleon Buonaparte", subclass) and at least some of the predicates ("to be born" and "to be born on August 15, 1769, in Ajaccio, Corsica") of the *ia* events (iaE_1 and iaE_{12}). Rule 2 is effected via the same relation between the subjects of some of the *ii* events and primary external events (ibE_9 and ibE_{19}). Interestingly, the "biographic" bush is affected by another event from the rule 2, iiE_{11}, which implies a change in life of Carlo Buonaparte controlled by the same rule (betray those who are close to him if political wind changes – perhaps later this very rule reflecting a family feature helped Napoleon many times in Egypt, Paris and many other places).

The choice of these rules remains at the scientist's discretion and is governed by his/her subjectivity, but, when chosen, they are represented in a strict, concise and transparent way. Such "strictness in subjectivity" is, generally speaking, a virtue of the event bush.

Moreover, these general tendencies are not necessarily known – e.g., we do not know why two of four Buonaparte offsprings died in infancy but we have to put that "Unknown disaster (*S*) happens that threatens life of a newborn (*P*)" in Buonaparte family.

These features give ground to state that biography can be taken for a *formal pattern of knowledge* described by specific structure of the event bush, and as such may be found in a number of fields, including possibly some technical applications.

Among these, a perspective option could be database design. Existing types of database (relational – Codd, 1970, object-oriented – Kim, 1990, or any other) are rooted in the interrelated formalisms of first-order predicate logic and set theory. Either a database describes properties of each element of some set or interrelated sets (so that one string corresponds to one element in a table, and relations between sets are expressed as relations between the tables in the data scheme), or, at best, it relates data to some objects, but objects themselves are considered purely static, neither interacting with each other nor evolving.

The event bush offers an opportunity to relate data to an evolving environment. Diviacco et al. (2014) already demonstrated how to refer any piece of data or judgment of any kind to any fragment of event bush in the field of geoscience.

In the archivistics, we see the following application of the above results. Suppose that there is a document relating to the childhood of Napoleon Buonaparte – e.g., a letter of Carlo Buonaparte to Letizia Ramolino (regardless of whether it tells about their glorious son or not). Obviously, this letter can be tagged to the event iiE_4 "Buonaparte is born, grows up, does not die in infancy" as well as to the tertiary event with the same meaning, $iiiE_4$. Generally speaking, any relevant document may be related to *n* events in the bush listed either via AND (if it is definitely relevant to all of them, like in the above case: iiE_4AND$iiiE_4$) or via OR if there is no certainty what particular events the document relates to. Stating that a piece of data or judgment may relate to a particular part of event (subject/predmet by Pshenichny, 2017a, or predicate/primeta by Pshenichny, 2017a), Diviacco et al. (2014), in fact, declared that the said piece of data or judgment may relate to all events having this part given they take place (i.e., are filled, by Pshenichny, 2017b). If it refers to a particular change between two or more events, it relates to all events participating in this change and listed via AND, and if the reference is scenario, it does to all events of this scenario listed via AND. If the reference is several scenarios of the bush or the bush per se, all related events must be united by AND. Uncertainty about the relevance of a document to given event or group of events may be expressed by listing these events via OR.

Alternatively, the relevance of any document to any event of the event bush may be expressed by means of a multivalued logic, e.g., "relevant/irrelevant/uncertain" or, as suggested, e.g., by Anoprienko and Ivanitsa (2001), "completely relevant/ completely irrelevant/partly relevant, partly not/uncertain (neither relevant nor not)". Their quadruple logic seems to be sensitive enough for the specificity of archivistics, as in the same document (e.g., letter) different things may be discussed, something may be stated indirectly, as a hint, guess or rumor, and so forth.

Then, an expression uniting all the relevant events (excluding only completely irrelevant ones) makes up a *name* of a document and, the same time, its identifier. Any two documents cannot be exactly the same, and hence, they cannot relate to exactly the same events. Still, the difference in meanings may be so small that it is unreasonable to express it as a specific event – e.g., a letter of Carlo Buonaparte to Letizia Ramolino and her response the other day, which tells about exactly the same matters. Herewith, it may be unreasonable to introduce specific events such as, "Carlo writes a letter" and "Letizia writes a letter". Instead, these letters may have the same name and a numerical index: iiE_4AND$iiiE_4$-1, iiE_4AND$iiiE_4$-2. Then their identifiers become absolutely unique given the event bush and can be used as primary keys for documents in a database based on event bush.

Obviously, not all events in the event bush may take place in reality, or, following the terminology of Pshenichny (2017b), be filled. Query, can inexistent event be reference for real documents? If a document contains deliberate disinformation, or failed forecast, or just considers a spectrum of possible consequences, it does relate to scenarios of event bush that do not take place.

Importantly, documents in an event-bush-based database could be easily queried by name and, in fact, do not need any additional organization and format. Technically, they could be arrange as a simple list of names in any text editor linked to particular events of the event bush. Still, this poses the necessity of development of a special software to create, store and edit the event bushes that has been stressed by researchers (e.g., Pshenichny & Kanzheleva, 2011); however, yet today such databases can be created by existing asynchronous tools of support of collaborative research (Diviacco et al., 2014).

It should be stressed that the suggested approach in no way refers to the record group and does not doubt its integrity. Rather, it refers to optimization of search of digitalized (and even not necessarily digitalized) documents, which should not be physically regrouped. Even non-digitalized documents may be provided with paper labels indicating the scenarios these documents relate to, and recorded as such in paper registries.

However, for digitalized documents, the optimization of search by tagging the electronic copies of documents to some metadata is carried out since 1999. Special ontologies, such as RIC (Records in Contexts…, 2016) and EAD (Encoded Archival Description, 2017), have been suggested in XML environment. Comparing the suggested methodology with existing standards, one may notice that special attribute, *bioghist*, is created in the EAD standard to tag the records containing biographical or historical information about the creator(s) of the archived materials. However, the approach in EAD is reverse to that proposed by us: having a material, look for a general knowledge to mark relate it to. What we propose is to organize knowledge into some construct (say, event bush) and then to look whether there are any archive materials available that relate to this knowledge.

However, the most important distinction between any ontological approach and the one based on dynamic knowledge engineering has been considered by Pshenichny and Mouromtsev (2015) and Rogalchuk et al. (2017, see Chapter 13 of this book). According to these researchers, ontologies, conceptual graphs and other traditional tools of knowledge engineering/representation operate with objects considered either as classes (i.e., types: concept types or relation types) or as individuals, while dynamic knowledge representation (i) operates not with individual meanings but with their compositions and (ii) these meanings only distantly resemble concepts in the logical sense. The latter, as evidenced by the quoted studies, is much more suitable for representation of the knowledge about changing world, an example of which is any biography. Therefore, the suggested approach seems to have a wide area of application in archivistics and other fields dealing with storing and querying data in any form.

CONCLUSION

1. An event bush that describes biography has only one primary internal event *iaE*, and the subject of this event is the subject of all *ii* and *iii* events;
2. Unlike the majority scenarios in event bushes, biography is characterized by necessary *(iaE Flux iiE)* change in the beginning. This can be considered as a formal property of biography in terms of event bush;
3. Another formal property of biography is presence of general rules (in the form of other event bushes) that govern it entirely or partly;
4. General rules are put formally but the very choice of these rules reflects the subjectivity unavoidably present in descriptive knowledge;
5. Semantics of events in the event bush does not allow past tense but past and present are easily distinguishable from future in full event formulations because predicates that should be activated in the future are negated;
6. Multiflow structure of the event bush represents, along with the scenario that has actually taken place and become the biography of person in question, other options, which may become biographies of someone else;
7. Event bush representation makes biography a formal pattern of dynamic knowledge, which can be extended to many subjects (not necessarily humans but, say, countries, political movements and so forth) and even transformed into a database.

REFERENCES

Anoprienko, A. Ya., & Ivanitsa, S. V. (2001). Postbinarnyi kompiuting i interval'nye vychisleniya v kontekste kodo-logicheskoi evolutsii [Post-Binary Computing and Interval Computation in the Context of Code-Logical Evolution]. Donetsk National Technical University – UNITEX Publishers.

Codd, E. F. (1970). A relational model of data for large shared data banks. *Communications of the ACM*, *13*(6), 377–387. doi:10.1145/362384.362685

Encoded Archival Description. (2017). Library of Congress. Retrieved from http://loc.gov/ead/

Kim, W. (1990). *Introduction to Object-Oriented Databases*. The MIT Press.

Napoleon Bonaparte. (2014). Retrieved from http://www.biography.com/people/napoleon-9420291

Pshenichny, C. A., & Kanzheleva, O. M. (2011). Theoretical foundations of the event bush method. Societal Challenges and Geoinformatics, 482, 139-165. doi:10.1130/2011.2482(12)

Pshenichny, C. A., & Mouromtsev, D. I. (2015). Grammar of Dynamic Knowledge for Collaborative Knowledge Engineering and Representation. In Collaborative Knowledge in Scientific Research Networks. IGI Global. doi:10.4018/978-1-4666-6567-5.ch016

Pshenichny, C. A., Nikolenko, S. I., Carniel, R., Sobissevitch, A. L., Vaganov, P. A., Khrabrykh, Z. V., . . . Gusev, E. A. (2008). The event bush as a potential complex methodology of conceptual modelling in the geosciences. *Proceedings, iEMSs – International Congress on Environmental Modelling and Software*, *2*, 900-912.

Pshenichny, C. A., Nikolenko, S. I., Carniel, R., Vaganov, P. A., Khrabrykh, Z. V., Moukhachov, V. P., & Rezyapkin, A. A. et al. (2009). The Event Bush as a Semantic-based Numerical Approach to Natural Hazard Assessment (Exemplified by Volcanology). *Computers & Geosciences*, *35*(5), 1017–1034. doi:10.1016/j.cageo.2008.01.009

Records in Contexts. A Conceptual Model for Archival Description. (2016). International Council on Archives, Experts Group on Archival Description. Retrieved from https://www.ica.org/sites/default/files/RiC-CM-0.1.pdf

Solomin, K., & Mandrik, M. (2015). Constructing historical knowledge through graphic boundary objects. In Collaborative Knowledge in Scientific Research Networks. IGI Global. doi:10.4018/978-1-4666-6567-5.ch017

Chapter 12

Dynamic Knowledge Representation as a Formalization Conveyor for Manmade Systems With Useful Impulse

Andrey Naumov
ITMO University, Russia

Ilya Popov
ITMO University, Russia

Igor Bondarenko
ITMO University, Russia

Boris Krylov
ITMO University, Russia

Roman Timonin
OpenWay Group, Russia

Ivan Ofitserov
Independent Researcher, Russia

ABSTRACT

Many manmade systems, from tea clippers to banking database management tools, exhibit an important common feature that a useful impulse is produced and functionates in them, being affected by favorable or unfavorable circumstances. While some of these systems are highly formalized, others remain operated or investigated quite intuitively. Their structural similarity urges one to think of conveying of experience of formalization from most to least formalized systems, but this process itself needs formal grounds to avoid the errors of inadequate formalization. In this chapter, a solution in the dynamic knowledge representation method and underlying theory of multitudes are sought.

DOI: 10.4018/978-1-5225-5261-1.ch012

INTRODUCTION

In a wide range of man-made systems (technical, social, economic and probably others) there is a peculiar class, in which indivisible or desirably indivisible pieces of useful material, energy or information are being transmitted from specified sender to specified recipient, and different circumstances are present that may affect this transmission in a good or bad way. Henceforth, the transmission of a singular piece from sender to recipient (with all possible adventures on the way) will be referred to as *impulse* (with intuitive rather than specifically physical connotation). Depending on the type of system, this *impulse* may be: delivery of any material stuff; money transfer from one account to another; sale and purchase of a material good; processing of a data packet; one utterance in communication between two people, and the like. By the way, if to refer to physics, impulse is a key notion of *transactional* interpretation in quantum mechanics (Cramer, 1986).

In the computer age, an eminent example of such systems is database management systems (DBMS). In them, the impulse carrying useful information is termed *transaction*, or, precisely speaking, *simple* (single) *transaction* – an indivisible sequence of data processing operations (Vassiliou, 1980). Gray (1981, p. 1) defined it as "a transformation of state which has the properties of atomicity (all or nothing), durability (effects survive failures) and consistency (a correct transformation)." In 1983, Reuter and Harder added one more property, Isolation (how transaction integrity is visible to other users and systems), and put forth an acronym ACID as shorthand for these four key properties (Reuter & Harder, 1983). ACID requirements ensure the security of data. To meet them, the transaction should be accomplished at once, not in parts, and transfer the database from one correct state to another, or, if for some reason some actions in the transaction are impossible, the database must return to the initial state (transaction rollback).

Modern banking is based on financial DBMS, in which transactions mean money transfers. For them, the issues of data (i.e., money) security are especially urgent. Financial institutions are one of those who suffer the biggest losses from misuse of computer-based tools. Analysis of crimes shows that although the overall number of thefts and robberies generally decreases worldwide, the number of wholesale money thefts increases (HSE, 1996-2017). By this report, more than one million of bank transactions, by which cash is withdrawn and illegally transferred onto offshore accounts, take place every day, total daily loss amounting up to 37 million dollars (Harris, 2009). This emphasizes the urgency of transaction safety issues, which, in turn, depend on the transaction design. The latter is largely governed by the formalisms used to encode the DBMS architecture (i.e., the transaction calculus). These formalisms, desirably, should be applicable to the entire class of systems with useful impulse and thus may serve many more purposes than they were originally developed for.

However, to refocus these formalisms on other fields, it is insufficient just to state that the systems they are now scheduled for are of the same class as before. A precise relation between relevant components of the systems is needed that would enable formalization first at the qualitative, and then, supposedly, at the quantitative level. To inquire whether such relation exists, both systems should be represented in one formal language, adequate enough to capture both relevant similarity and relevant distinction between them.

Mathematically, transaction is commonly treated as mapping of database table from its environment to some other programs (e.g., compensable programs by He, 2010). However, to make such framework fit any kind of system with useful impulse, one needs a general instruction, what mathematical objects to coincide with the environment of impulse, that is, the circumstances (actions) that take place in the system and possibly affect the action taken on the impulse traveling in it. Still, mathematics per se gives no such instruction. Importantly, trying to imagine a universal mathematical framework, one should take into account that some impulses may be tightly related to the carrier – e.g., the money may be not only transferred electronically but also carried by a person. In the latter case, the carrier would be prone to many more types of influence and perhaps even have free will. In such event, unlikely its conceptualization as mapping would be adequate. Similarly, the visualization of transaction used by the researchers (Gray, 1981; He, 2009, and others) emphasized their structure (linear, ramifying, nested) but did not formalize what happens at each step of transaction. Instead, informal denotations were used, such as, "GET MSG", "DEBIT", "CREDIT", "POST", "PUT MSG", or "ORDER", "BUILD", "CREDIT?", "OK", "SHIP", "BUILD", "GET PAYMENT" (Gray, 1981, Figure 1 in p. 3). Likewise, in practice, denotations could be "BEGIN", "STOP", "ROLLBACK" and so forth, i.e., describe what happens to transaction, but their brevity and informality does not allow an integer *scenario*, in which the transaction would be the main "actor".

Another solution is to build a transaction calculus based on the classical first-order predicate logic. In fact, although not declaring this, this has been done by Jagannathan et al. (2005) who suggested the Transactional Featherweight Java Calculus, in which particular classes (class Updater, class Runner, class Transactor) act as specific *n*-ary predicates defining exchange of values between different entities, e.g., "class Updater has two fields, n and v and an update method which assigns v to n's val field" (p. 169). Still, these predicates, in fact, imply that value is transferred automatically, i.e., say very little about the transaction itself. Hence, they, even being detailed enough, also do not allow to build a scenario of transaction behavior because do not assume the transaction to be an actor. In fact, they represent a set of static relations bound by logical connectives (mainly implication), but without an entity that could "sue them up" in a scenario.

Yet another approach to the transaction calculus may be based on the temporal logic. There are different systems of temporal logic, some being an extension of propositional logic (e.g., Venema, 2001), and some, built on top of the first-order predicate logic (Hodkinson & Reynolds, 2007). The latter, in our view, could be the basis for transaction protocols, but with the same peculiarity as above.

Still, among the approaches demonstrated by Gray, one, focused on the time-domain addressing, referred values $V_0, V_1, V_2, \ldots$, to the entity E in time intervals $[T_0,T_1), [T_1,T_2), [T_2,*)$:

$$E{:}\langle V_0, [T_0,T_1)\rangle, \langle V_1, [T_1,T_2)\rangle, \langle V_2, [T_2,*)\rangle$$

With slight modification and omitting the time intervals, one readily obtains:

$$E{:}\langle V_0, \sim V_1, \sim V_2, \ldots, \sim V_n\rangle, \langle \sim V_0, V_1, \sim V_2, \ldots, \sim V_n\rangle, \ldots, \langle \sim V_0, \sim V_1, \sim V_2, \ldots, V_n\rangle$$

where ~ is negation. This looks pretty much like definition of a family of events in the theory of multitudes (Pshenichny, 2017a, see Chapter 1 of this book):

$$S\text{–}P^u{}_1, \ldots, P^u{}_m, (\sim)P_{m+1}, (\sim)P_{m+2}, \ldots, (\sim)P_n$$

where S is predmet, $P^u{}_1, \ldots, P^u{}_m$ are universa (finite in number, never negated), and $(\sim)P_{m+1}, \ldots, (\sim)P_n$ are non-universum primetas, also finite in number, which may be negated or not that is denoted as $(\sim)$. Predmets are analogous, but not equal, to subjects, and primetas, to predicates of the predicate logic. For exact definition and explanation of these terms, the reader is referred to Pshenichny (2017a). In the above description of transaction suggested by Gray, E should be predmet (E=S), values, non-universum primetas (V_0= P_{m+1}; V_1= P_{m+2}; ...), universum primeta (s) needed, time intervals omitted, and additional constraint introduced that one, must, and two, may not, of the non-universum primetas be asserted in each pair of brackets. Still, as shown by Pshenichny (2017a), introduction of universum and time is not a problem at all, and the abovementioned constraint just means exclusion axioms for all other cases. Therefore, the notation of Gray fairly well fits the suggested syntax of the theory of multitudes. This formalism, according to Pshenichny (2017b), allows further quantification and is thus applicable to model the dynamics of systems.

Armed with this methodology, we may put one of predmets in a system, S_1, the transaction proper (keeping in mind that its meaning can be extended to the impulse, perhaps with the carrier), and proceed to modeling of some simple cases of transaction destiny.

Therefore, the purpose of this work is to find out whether the methods of dynamic knowledge engineering may convey qualitative formalization approaches from the domain of banking DBMS to other domains, in which systems with useful impulse are present.

This purpose implies the following tasks:

1. Choosing a situation in banking DBMS operation and appropriate methods for modeling;
2. Constructing the models;
3. Comparison of these models with other systems with useful impulse.

SELECTION OF METHODS

One of the main requirements for flawless operation of DBMS is reliability of databases. This requirement means, in particular, the possibility of recovering a consistent state of a database after any kind of hardware on software failure. Obviously, some additional information is required for this.

Recovery after sudden loss of Random Access Memory (RAM) contents (mild failure) may occur at emergency shutdown of power supply, at fatal processor failure (e.g., RAM control actuation) or other circumstances. The situation is characterized by losing part of information that was stored in the main memory buffers at the moment of failure.

Rough failure is characterized by damage to external storage media. Recovery of the main external media after the rough failure is not likely but still should be taken into account. Since the rough failure is not followed by loss of the main memory buffers, it is possible to restore the database to a level allowing to continue executing even incomplete transactions (Chen, 1976).

To start with, let us conceptualize one of the simplest yet important occurrences of DBMS functionating – the rough failure and concomitant system recovery. The basis for recovery is data redundancy provided by the transaction log. Similarly to the pages from the database, data from the log are not recorded on the disk at once, but are preliminary buffered in the main memory (Courts, 1997).

The basic principle of ejection of the log buffer and database page buffers is that the indication of change of a database object must come to the external memory of the database before the changed object comes there itself. It is necessary to observe the following rules: maximal speed of transaction, as well as the guarantee of restoring the last consistent state of the database. Additional condition for buffers pushing is the limited amount of database buffers and the transaction log. Periodically, and also at the onset of the specified event (e.g., when the number of free pages in the buffer has decreased and

Figure 1. Possible variants of transaction states by the moment of passing the last control point and by possible failure. Modified from Weikum (2002).

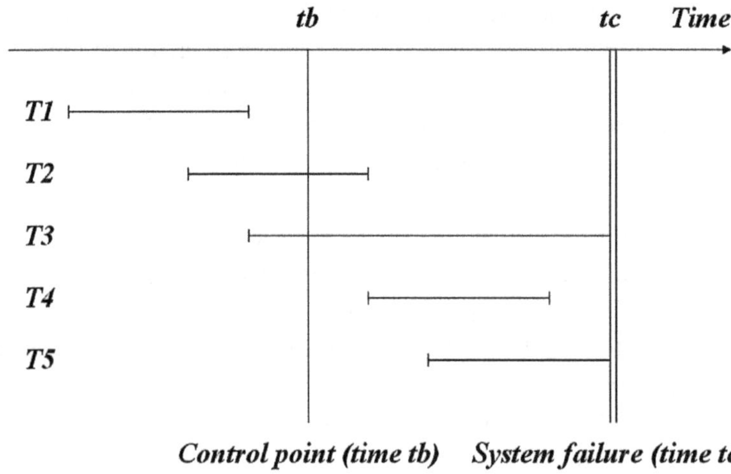

reached a critical value), the system passes the control point. At this point, the database buffer contents is ejected to the external memory and special physical record is made, which is the list of all transactions being executed by this moment since the onset of the event or the past control point. Therefore, the minimal requirement that ensures the recovery of the last consistent state of the database is ejecting the log of all records of the changes made in the database to the external memory when the transaction is fixed at the control point.

The last moment when changed pages have been confidently popped is the moment of the last control point. There are five options of transaction proceeding relative to the last control point and possible failure (Figure 1).

If the last control point was passed at the moment *tb*, and the rough failure happened at the moment *tc*, five options (T1-T5) are possible for a *transaction*:

T1: Transaction was successfully completed before the control point. All data of this transaction are saved in the permanent memory – log records and data pages modified by this transaction. No operations are needed to recover the T1 transaction.

T2: Transaction was started before the control point and was successfully completed after the control point, but before the failure. Records of transaction log which relate to this transaction are popped into the external memory. Data pages changed by this transaction are popped into the external memory only partially. For this transaction it is necessary to repeat the operations that were executed after the control point acceptance.

T3: Transaction was started before the control point and was not completed because of failure. Such transaction should be rollbacked. Some data pages changed by this transaction are already in the external memory – namely, those pages that were updated before the control point. The changes made after the control point were not documented in the database. The records of transaction log made before the control point are popped into the external memory, but there are no records after it.

T4: Transaction was started after the control point and successfully completed before the system failure. RECORDS of transaction log which relate to this transaction are popped into the log external memory. There are no changes in the database that were made by this transaction in the external memory of the database.

T5: Transaction was started after the control point and was not completed because of failure. There are no traces of this transaction either in the external memory of the database. No action should be taken for such transaction, as if there were no such transaction at all.

These options should be represented by means of dynamic knowledge engineering (Pshenichny, 2017a; b; Chapter 1 and Chapter 2 of this book).

For most of the practical applications, the key concepts of dynamic knowledge engineering can be introduced as follows:

- Meaning that relates to a thing per se is termed predmet (somewhat similar to but not the same as subject in the traditional Aristotelian logic);
- Meaning that relates to a property is termed primeta (close to predicate in the Aristotelian and predicate logics);
- Meaning that relates to a thing with some properties is termed event (having some parallelism with judgment, or statement, or proposition, in the logic);
- Event that may have examples is termed multitude (an object resembling set in the set theory or concept in the logic).

The examples, or fill, of multitudes, may migrate from one multitude to another. Succession of events passing fill to each other is termed scenario.

Among the methods, which theoretically may operate with these objects (predmets, primetas, events and scenarios), there is a wide range of UML behavior diagrams (Booch, 1999), event trees, event bushes, Bayesian and influence diagrams and so forth (Pshenichny & Mouromtsev, 2015). Still, while some of these methods (namely, the event bush) were developed ad hoc to operate with them by using the syntax of the multitude theory, other methods largely await their incorporation in its framework.

Two alternative methods of knowledge representation, activity diagram and event bush, have been applied to represent the same environment.

The method of event bush was suggested by Pshenichny et al. (2008; 2009) and further developed by Pshenichny and Kanzheleva (2011). The fundamentals of this method are presented by Pshenichny (2017; Chapter 2 of this book).

Activity diagrams are one of the five diagrams in the UML for modeling the dynamic aspects of systems (Booch, 1999). Every activity diagram is a presentation of algorithms of some actions that are executed in the system, which lead to desired result. Still, unlike the traditional flowchart, it shows concurrency as well as branches of control. Dynamics refers to the performance of the system as a whole or parts of it. Activity diagrams allow modeling complicated life cycle of object with *transitions* from one state (activity) to another one. This kind of diagram can be used also for describing the dynamics of a collection of objects. It is possible to draw several activity diagrams for the same system, describing different system attributes and showing different actions performed in it.

Activity diagram consists of the following elements (Figure 2, a-f):

1. Control node is an abstract activity node that coordinates workflow;
2. Activity initial node is a control node, in which the flow(s) starts upon calling this activity from the outside. There may be only one activity node in a diagram;
3. Action is an elementary component of activity. Actions are displayed in the form of rounded rectangle;
4. Decision node is intended to define branching rules and different variants of further development. Displayed as a rhombus with condition marks near transitions;
5. Separation point and merge point provide splitting of one flow into several parallel flows and their synchronization;
6. Activity final node (or final activity state) is a control node that stops all flows of this activity diagram. There can be more than one final node in the diagram.

Flows in the activity diagram may merge and separate, but not in the same node (Figure 2).
Separation point is characterized by the following (see Figure 3):

1. Only one flow enters; and
2. Two or more flows go out, executed simultaneously.

Merge point mirrors the separation one with one, yet important, specificity:

1. Two or more flows enter, executed simultaneously; and
2. Only one flow goes out. Incoming parallel flows are synchronized, i.e. each one of them waits till all others reach this point, and then the execution is continued as one flow.

Figure 2. Elements of activity diagrams. See comments in the text.

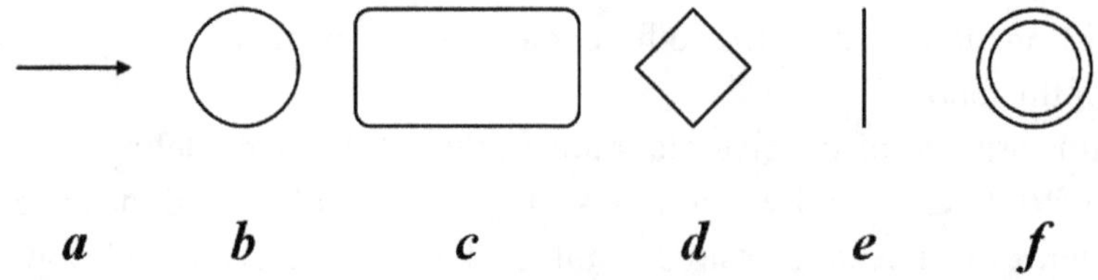

Figure 3. Separation point and merge point in activity diagram. Modified from White (2004).

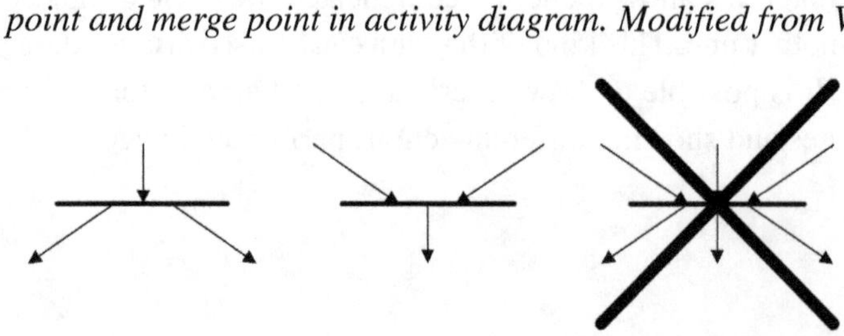

In addition, an element of the activity diagram may be:

- An object under action. This is not an imperative diagram element, but in some cases, it is necessary to show the object initiating execution of actions or being their result;
- A comment with the help of which it is possible to explain the contents of the action or object.

There are many more different elements of activity diagram and variations of its syntax which description is not given in the subsection due to the absence of practical interest towards them in this investigation. Some of the will be reviewed hereafter as required.

The activity diagrams have quite a few advantages compared to other UML behavior diagrams. They explicitly show the relations between the processes, such as compatibility, incompatibility, execution in turn and others. One of the main disadvantages of this method, in our mind, is that it does not offer strict rules of formulation of entities in its nodes. Although clear by intuition, expressions like "Woke up", "Shower" or "Coffee" bear the same imperfection as the formulations used by Gray (1981) in visualization of transactions. Nevertheless, this disadvantage can be removed if to put the activity diagram method in the framework of the multitude theory that offer a syntax for formulation of events described by Pshenichny (2017a). This will be done below, in process of modeling the transaction behavior.

SIMULATION OF TRANSACTION BEHAVIOR BY SELECTED METHODS

We will describe the different cases of transaction behavior in relation to the system failure and control point (Figure 1) by the activity diagram. Intending to develop a framework for the general case of systems with useful impulse, we will give the transaction a wider meaning and treat it as impulse.

The cases *T1* – *T5* of transaction behavior depicted in Figure 1 can be represented as follows in the activity diagram (Figure 4). Every state of transaction is given exhausting verbal formulation in Figure 4a. To proceed to strict notation in terms of the theory of multitudes (Figure 4b), let us introduce the following: S_1 – Impulse; S_2 – Control point; S_3 – System crash; S_4 – Copy of an impulse up to control point; P_1 – To flow; P_2 – To pass a control point; P_3 – To come across a system crash; P_4 – To disappear completely; P_5 – To disappear before passing a control point; P_6 – To take place; P_7 – To be sent to archive.

Accepting the grammar of the theory of multitudes, we obtain an individual and unique *event* in every box of the activity diagram. Succession of events makes up a scenario. All scenarios in the diagram emanate from the same initial state. Figure 4 shows four distinct incompatible scenarios, each ending up with a final node. Importantly, two scenarios, *T2* and *T3*, share the same thread. One may argue that Figure 1, giving a concise and clear representation of transaction behavior, is much better option than the conceptualization presented in Figure 4. To address this claim, special nodes referring to the *T1* – *T5* scenarios of Figure 1 were introduced here before the end nodes of each scenario. With these, it is seen that the activity diagram discloses the meanings of *T1* – *T5* showing the sequence of atomic events in each scenario (including the actions since the moment of occurrence of a hard failure of the system until the successful completion of the impulse). The activity diagram covers all five scenarios but, as is seen in Figure 4, scenarios *T1* and *T4* do not differ in it. Still, if to analyze them, one may conclude that they, being formally discerned in Figure 1 (one completely before, the other completely after the hard failure), in fact, do not differ in any relevant detail. Hence, unlikely it is reasonable to recognize them as different

Figure 4. Activity diagram for impulse traveling in a system with possible hard system crash: (a) – events are formulated verbally, (b) – symbolically

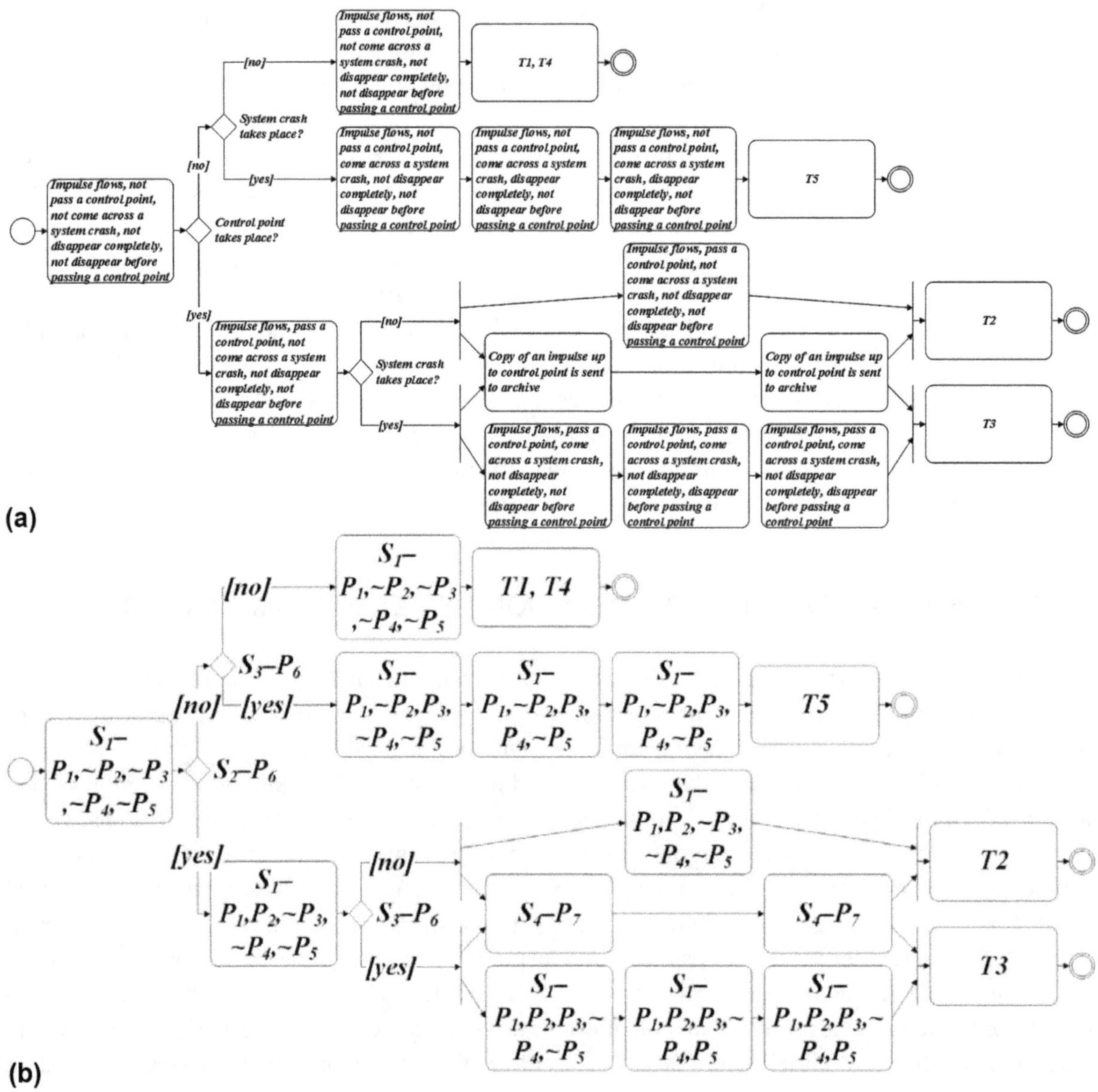

scenarios. This alone evidences for appropriateness of conceptualization by the activity diagram based on the theory of multitudes grammar.

Unlikely the hard system failure can be totally excluded. However, at the first decision node querying whether the control point is passed, the system should be able to initiate control to answer "yes" to this query. The role of consolidation nodes is just to unite parallel processes and lead them to a common termination expressed as final node.

Of course, the very composition of events in an activity diagram is a matter of personal vision. Figure 5 presents another vision of the same context, the event bush. The modeling task can be formulated as

Figure 5. Event bush model of a hard system crash: (a) – verbal, (b) – symbolic formulation

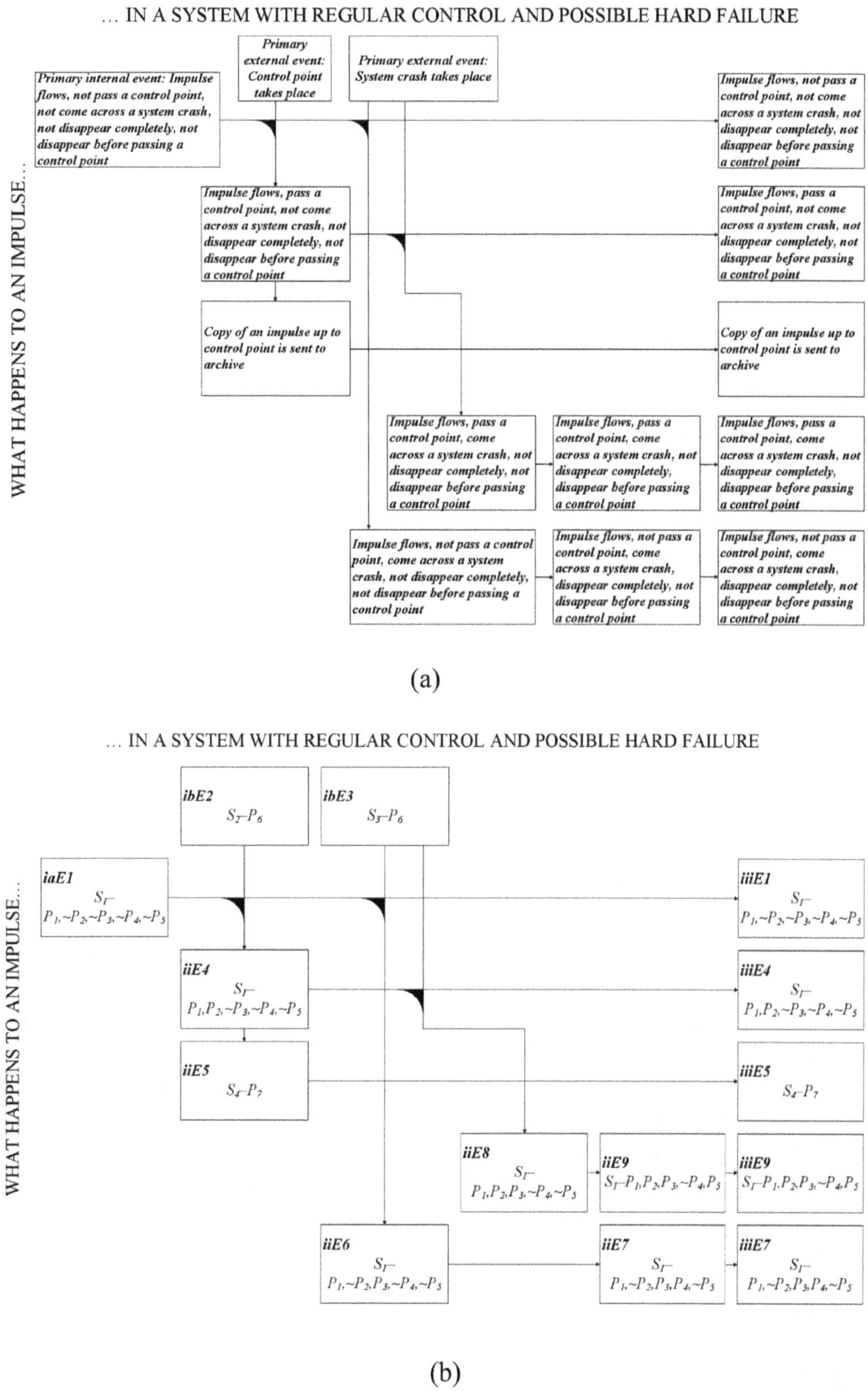

"What happens to an impulse in a system with regular control and possible hard failure". This task determines the choice of primary internal and primary external events. Omitting the process of construction of event bush that was reported and discussed in several publications (see, e.g., Pshenichny et al., 2008; Mouromtsev et al., 2014), we proceed to the final result. For clarity, tertiary events in Figure 5a are replaced by the indices of scenarios from Figure 1. Note that scenarios *T1* and *T4* are indiscernible again.

In the event bush model, primary internal event (*ia*) is:

Iae_1: Impulse (S_1) flows (P_1), does not pass a control point ($\sim P_2$), does not come across a system crash ($\sim P_3$), does not disappear completely ($\sim P_4$), does not disappear before passing a control point ($\sim P_5$), where ~ is negation.

Primary external events (*ib*) are:

Ibe_2: Control point (S_2) takes place (P_6).
Ibe_3: System crash ($\underline{S}3$) take place (P_6).

Initial change is:

$((ia)\ S_1 - P_1, \sim P_2, \sim P_3, \sim P_4, \sim P_5)\ Flux\ ((iii)\ S_1 - P_1, \sim P_2, \sim P_3, \sim P_4, \sim P_5)$

These are included in influxes:

$((ia)\ S_1 - P_1, \sim P_2, \sim P_3, \sim P_4, \sim P_5), ((ib)\ S_2 - P_6)\ Influx((ii)\ S_1 - P_1, P_2, \sim P_3, \sim P_4, \sim P_5)$

$((ii)\ S_4 - P_7), ((ib)\ S_3 - P_6)\ Influx((ii)\ S_1 - P_1, P_2, P_3, \sim P_4, \sim P_5)$

$((ia)\ S_1 - P_1, \sim P_2, \sim P_3, \sim P_4, \sim P_5), ((ib)\ S_1 - P_6)\ Influx((ii)\ S_1 - P_1, \sim P_2, P_3, \sim P_4, \sim P_5)$

Despite that this model is also based on personal vision, the rules of composition of the event bush suggest that, having plotted the scenario that corresponds to the modeler's vision, the one must (i) finish other scenarios that formally follow, given the connectives used in this scenario, and then (ii) examine the resulting bush aiming to suppose what other scenarios can bind the existing events. If this inevitably requires introduction of new events (i.e., new boxes in the event bush), these new events are also examined to find out what other, alternative relations to preexisting events they may have, and so forth. There is no formal limit for construction of the event bush. Like any other modeling, it formalizes the scientist's view and the same time strongly relies on the scientist's intuition and a will to produce an adequate and concise conceptualization.

The syntax of the event bush allows to formally build scenarios from events using special connectives (*Flux, Influx, Conflux and Outflow*). Thus, the most desired scenario that corresponds to the intuitive solution in the proposed bush is:

$(iaE_1)\ Flux\ (iiiE_1)$

$(iaE_1), (ibE_2) Influx\ (iiE_4)$

(iiE_4) *Flux* (iiE_5)

(iiE_4) *Flux* $(iiiE_4)$

(iiE_4), (ibE_3)*Influx* (iiE_3)

(iiE_5) *Flux* $(iiiE_5)$

(iaE_1), (ibE_3)*Influx* (iiE_6)

(iiE_6) *Flux* (iiE_7)

(iiE_7) *Flux* $(iiiE_7)$

(iiE_8) *Flux* (iiE_9)

(iiE_9) *Flux* $(iiiE_9)$

Similar syntax can be applied to the activity diagrams, taking its starting, decision, separation and merge nodes as connectives such that each of them is a type of some connective of the theory of multitudes, as has been done for the connectives of the event bush (Pshenichny, 2017b). One would be able to record scenarios from the activity diagrams in the same way as from the event bush and compare the two. At present, however, this can be done only intuitively.

At the intuitive level, activity diagrams may look more handy because they display only the scenarios in question, while the event bush, by rules of its composition, must include the "nothing happens" scenario (iaE_1) *Flux* $(iiiE_1)$. Nonetheless, exactly this scenario makes the representation of the system formally complete. Another peculiarity of the activity diagram is transparent demonstration of parallel processes, which are not that eminent in the event bush. Still, this feature requires to repeat the chains of events (Figure 4a-b) that makes the plot notable larger (even though not harder to follow). In general, the event bush seems to better serve the theoretical, and the activity diagram, the practical purposes of knowledge representation.

REFOCUSING THE MODELS

The suggested conceptualizations, developed for a particular case of banking DBMS, may be readily extended onto many other cases of systems with useful impulse, which allow creating replicas. For instance, publishing a weird news on the Internet (i.e., creating P_1 an impulse, S_1), it is reasonable to recommend readers to perform copying it on a hard disk upon reading, i.e., to make the copying, S_2, happen, P_6 (control point takes place – $S_2 - P_6$) in case the blocking of the web source (S_3) takes place, P_6 (hard failure occurs – $S_3 - P_6$). Then, S_4 would be "Copy of the news" (copy is put in a folder – $S_4 - P_7$), P_2 – "to get copied", P_3 – "to suffer blocking of the web source", P_4 – "to be lost completely", P_4 – "to be lost retrievably".

The same model would apply to seasonal production of agricultural commodities in a country. After harvesting, the volume of grain production (impulse) is accounted (control point) and some amount is stored by the state for strategic reserve (a copy is put in the folder). Then, if there happens disaster that ruins the country's economy (hard failure), famine is avoided by the stored grain (impulse is complete). Nevertheless, here an important qualitative constraint appears: once information (transaction or news) can be copied entirely, the harvest cannot – it would be necessarily divided. This would be expressed by another primeta (*P*) "to be decreased". Another issue to be accounted for, as was mentioned above, is possible relation of the impulse to its carrier – that differs, say, wire transfer from carrying money in the wallet. To express this relation, another predmet "career" will be needed, S_5, that gives birth to a primeta "to be linked to career", which will be one more universum for S_1 (impulse or its analog), and a bunch of primary external events (each with its own predmet, but possibly all with the same peimtea, "to take place", P_6), to denote the adventures that may happen to the career. Wherever anything happens to him, this may result in influx modifying the events, in which S_1 (impulse or its analog) is predmet.

Still, this model covers only one aspect of transaction behavior. Turning back to the Introduction, we may say that the models presented above develop the visual approach of Gray (1981) disclosing his laconic labels like "GET MSG", "DEBIT", "CREDIT" and so forth and using another concept suggested by him in concern of time-domain addressing, which appeared remarkably coherent with the definition of event in the theory of multitudes. In this approach, the only "actor" is transaction (impulse), which is thought to be singular in every considered case. In terms of the multitude theory, this should be understood as that impulse is a multitude with a lump fill. We believe that in this way we will also capture the fundamental properties of transactions (atomicity, consistency, durability and independence), as well as different threads of transaction (singular, multiple ramifying, multiple merging, nested and more complex).

Nevertheless, another approach to transaction calculus advocated by Jagannathan et al. (2005) is that transaction is made by a succession of classes that act as *n*-ary predicates defining exchange of values between different entities. This seems to fit fairly well the transitions between predmets and events having grainy fill, allowing to directly express the quantitative measure of impulse (transaction), such as amount of information, money, or anything else (Pshenichny, 2017b). In this respect, impulse would be directly expressed by connectives of the multitude theory. Predmetka chart allows to bring together the concept of lump-filled transaction developed in this work with this understanding of transaction as succession of changes between grainy-filled multitudes governed by connectives. Although, one more formal representation of transaction, that as mapping from its environment to other programs (e.g., He, 2010), still looks not so straightforwardly convertible in terms of the multitude theory.

Finally, modeling the same context (even so trivial as that in Figure 1) by various methods may bring useful insights into their properties in representation of given type of task. The two methods used in this study could be compared merely in terms of practical use (ease versus expressivity), but bringing more methods (say, other behavior diagrams of UML) and involving more techniques of transaction modeling (Gai et al., 2016, Gladkaya et al., 2017, Ritov et al., 2017, Stasiuk et al., 2017) in the same terms of the multitude theory would give enough ground for their comparison and grouping by formal properties expressed in terms of the said theory. Still, a prerequisite for such analysis is interpretation of linking elements of each method (such as initial, separation, merge and final points of the activity diagrams) as types of connectives of the multitude theory.

CONCLUSION

In the course of research, modeled was one of the simplest yet fundamental aspects of transaction in DBMS, the relation of transaction to the control point and possible hard failure in the system. Two methods of dynamic knowledge engineering, the activity diagram and event bush, were used to build alternative models. Still, the same semantics and syntax were used for both, the syntax being defined in the theory of multitudes that is thought to underlie both methods. Nevertheless, only in one of these, the event bush, the rules of building scenarios (say, upper-level syntactic rules) directly follow from those of the theory of multitudes, while the activity diagram these relations remain intuitive. Nevertheless, the co-application of these methods has given compatible results, and both methods highlighted some redundancy in the existing conceptualization of transaction behavior. At the intuitive level, activity diagrams may look more handy because they display only the scenarios in question, while the event bush, by rules of its composition, must include the "nothing happens" scenario. Nonetheless, exactly this scenario makes the representation of the system formally complete. Another peculiarity of the activity diagram is transparent demonstration of parallel processes, which are not that eminent in the event bush. Still, this feature requires to repeat the chains of events that makes the plot notable larger (even though not harder to follow). In general, the event bush seems to better serve the theoretical, and the activity diagram, the practical purposes of knowledge representation. Joint application of the activity diagrams and event bushes is highly favorable to use the advantages of both methods in modeling of alternative directed change environments.

The results of modeling show few but key regularities of behavior of systems with useful impulse. First, they apply to a particular, yet wide and cross-disciplinary, type of such systems, in which replica of impulse is generated. Then, in these systems, as evidenced by both methods used, the lucky scenario represents two threads leading, at best, to two separate results, the completed impulse and its intact replica. In case of system failure, only replica survives if produced prior to failure. Finally, despite the intuitive desire to recognize an impulse generated and completed entirely before and entirely after both control point and possible hard failure as two distinct scenarios, they show no distinction and should be considered one, and also non-desired, scenario.

REFERENCES

Booch, G. (1999). *The Unified Modeling Language User Guide.* Pearson Education.

Chen, P. P.-S. (1976). The Entity-Relationship Model: Toward a Unified View of Data. *ACM Transactions on Database Systems, 1*(1), 9–36. doi:10.1145/320434.320440

Courts, H. R., & Keith, G. B. (1997). *Method for preventing overwriting cache buffer transaction entries until corresponding log buffer entries have been copied to log partition of the disk.* U.S. Patent No. 5,636,360. 3.

Cramer, J. G. (1986). The transactional interpretation of quantum mechanics. *Reviews of Modern Physics, 58*(3), 647–688. doi:10.1103/RevModPhys.58.647

Gai, K., Qiu, M., & Elnagdy, S. A. (2016). Security-Aware Information Classifications Using Supervised Learning for Cloud-Based Cyber Risk Management in Financial Big Data. *IEEE 2nd International Conference on Big Data Security on Cloud (BigDataSecurity), IEEE International Conference on High Performance and Smart Computing (HPSC), and IEEE International Conference on Intelligent Data and Security (IDS).* New York: IEEE. Doi:10.1109/BigDataSecurity-HPSC-IDS.2016.66

Gladkaya, E. A., Egorova, E. N., & Torsunova, E. R. (2017). *Using Economic and Mathematical Methods During Formation of Construction Cluster.* Integration and Clustering for Sustainable Economic Growth. doi:10.1007/978-3-319-45462-7_26

Gray, J. (1981). The Transaction Concept: Virtues and Limitations. In *Seventh International Conference on Very Large Databases.* Tandem Computers Incorporated.

Haerder, T., & Reuter, A. (1983). Principles of transaction-oriented database recovery. *ACM Computing Surveys, 15*(4), 287. doi:10.1145/289.291

Harris, M. (2009). The ten most expensive online crimes ever. *CNET.* Retrieved from https://www.cnet.com/news/the-ten-most-expensive-online-crimes-ever/

He, J. (2010) Transaction calculus. In *UTP'08 Proceedings of the 2nd international conference on Unifying theories of programming.* Dublin, Ireland: Springer-Verlag.

Health and Safety Executive (HSE). (1996-2017). *HSE Offshore: Offshore topics. Statistics, Health and Safety Executive.* Retrieved from http://www.hse.gov.uk/offshore/statistics.htm

Hodkinson, I., & Reynolds, M. (2007). Temporal Logic. In Handbook of Modal Logic. Elsevier.

Jagannathan, S., Vitek, J., Welc, A., & Hosking, A. (2005). A transactional object calculus. *Science of Computer Programming, 57*(2), 164–186. doi:10.1016/j.scico.2005.03.001

Pshenichny, C. A. (2014). Knowledge engineering in volcanology: Practical claims and general approach. *Journal of Volcanology and Geothermal Research, 286*, 78–92. doi:10.1016/j.jvolgeores.2014.07.012

Pshenichny, C. A., & Kanzheleva, O. M. (2011). Theoretical foundations of the event bush method. In K. Sinha, L. Gundersen, J. Jackson, & D. Arctur (Eds.), Geological Society of America Special Paper: Vol. 482. Societal Challenges and Geoinformatics (pp. 139–165). Academic Press. doi:10.1130/2011.2482(12)

Pshenichny, C. A., & Mouromtsev, D. I. (2015). Grammar of Dynamic Knowledge for Collaborative Knowledge Engineering and Representation. In Collaborative Knowledge in Scientific Research Networks. IGI Global. doi:10.4018/978-1-4666-6567-5.ch016

Pshenichny, C. A., Nikolenko, S. I., Carniel, R., Sobissevitch, A. L., Vaganov, P. A., Khrabrykh, Z. V., . . . Gusev, E. A. (2008). The event bush as a potential complex methodology of conceptual modelling in the geosciences. *Proceedings, iEMSs – International Congress on Environmental Modelling and Software, 2*, 900-912.

Pshenichny, C. A., Nikolenko, S. I., Carniel, R., Vaganov, P. A., Khrabrykh, Z. V., Moukhachov, V. P., & Rezyapkin, A. A. et al. (2009). The event bush as a semantic-based numerical approach to natural hazard assessment (exemplified by volcanology). *Computers & Geosciences, 35*(5), 1017–1034. doi:10.1016/j.cageo.2008.01.009

Ritov, M. Y., Gorlov, A. P., & Eryomenko, V. T. (2017). Mathematical modeling of the assessing process the of security level of the complex organizational and technical systems of industrial enterprises. *2nd International Conference on Industrial Engineering, Applications and Manufacturing (ICIEAM).* Doi:10.1109/ICIEAM.2016.7911697

Stasiuk, A. I., Hryshchuk, R. V., & Goncharova, L. L. (2017, May). A Mathematical Cybersecurity Model of a Computer Network for the Control of Power Supply of Traction Substations. *Cybernetics and Systems Analysis*, *53*(3), 476–484. doi:10.1007/s10559-017-9949-z

Vassiliou, Y., & Lochovsky, F. H. (1980). DBMS Transaction Translation. *Information Systems Working Papers Series.*

Venema, T. (2001). Temporal Logic. In L. Goble (Ed.), *The Blackwell Guide to Philosophical Logic* (pp. 203–223). Malden, MA: Blackwell Publishers.

Weikum, G., & Vossen, G. (2002). *Transactional Information Systems: Theory, Algorithms, and the Practice of Concurrency Control and Recovery*. Morgan Kaufmann Publishers.

White, S.A. (2004). Process Modeling Notations and Workflow Patterns. *BPTrends,* 1-20.

Chapter 13
Petri Net vs. Event Bush in Description of Environments of Alternative Directed Changes

Vyacheslav Rogalchuk
St. Petersburg State Transport University, Russia

Sergey Arustamov
ITMO University, Russia

Nikita Solovyov
ITMO University, Russia

Aleksandra Malisheva
Universite d'Angers, France

ABSTRACT

Two alternative methods of knowledge representation, Petri net and event bush, have been applied to represent the same environments of alternative directed changes. The chapter examines the interrelation between the two methods and considers the ways of their mutual optimization and co-application. It discusses the balance between conceptual thoroughness and formal transparency from one side and readability, representation power, and common-sense clarity from the other. The interrelation and interplay of these two methods may bring fruitful revelations about how to treat dynamic knowledge, what strategy to choose to represent scenarios depending on the task in mind. This is seen as an important contribution to a new field of research, representation of dynamic knowledge, which may have wide application in a variety of fields, from history to technical design.

INTRODUCTION

Knowledge base is an essential part of any expert system. When formulated as a set of rules, it gives ground for the inference, which draws conclusions from data entered by the user (Giarratano & Riley, 2005). This type of inference is specific for expert systems and provides advice and decision support (see,

DOI: 10.4018/978-1-5225-5261-1.ch013

e.g., Arendt & van Uden, 2011) as well as systematic capture, transfer and reuse of knowledge (Quintana-Amate et al., 2015). Still, in contrast to databases that have various and well-developed standards of data structuring (e.g., those developed in relational or object-oriented framework), the knowledge base, in common practice, until now often represents a simple repository "containing articles that relate to fault fixing, how-to documents, user guides, white papers and so on" with an ad-hoc retrieval system (Serio Ltd., 2014). However, even when organized as a set of rules, or productions, the knowledge base often lacks proper tools of knowledge engineering to adequately incorporate the human expertise and account for the peculiarities of meanings. In fact, the only knowledge-engineering approach that is being used is ontology (taxonomy, hierarchy, classification)-related.

The examples of this approach in knowledge base design include, but are not exhausted by, ontologies and hierarchies per se (Kulon et al., 2006), structural models in the DANE method (Goel et al., 2012; DANE 2.0 Users Guide, 2011), UML class diagrams (La Rocca, 2012; da Silva et al., 2014), class and object frames (La Rocca, 2012), IDL code (La Rocca, 2012), "holistic models" (Syairudin et al., 2015), or "view models" of Akasaka et al. (2012) that allow to perform analogical reasoning and generate new functions of modeled systems. Product tree (La Rocca, 2012) is another ontology-related approach that resembles event tree but its "parent-child" relation maps not the "cause-effect" but the "whole-part" relation, thus performing an instantiation of a product model. "Customer satisfaction tree" (Akasaka et al., 2012) has inverse composition but maps the same relation, parent corresponding to part and child, to whole. Tree-shaped conceptual design graphs and their modifications presented by Calkins et al. (2000) are also based on part-whole relation and therefore represent another example of the same ontology-related approach.

Nevertheless, ontologies, conceptual graphs and other traditional tools of knowledge engineering/representation operate with objects considered either as classes (i.e., types: concept types or relation types) or as individuals (Martin, 2002). This has proven to be efficient for a wide range of tasks like building data models or making classifications and descriptions of those domains in which things, their properties and relations are considered to be fixed – e.g., production units of a factory or biological species. Nevertheless, there are some tasks in natural-scientific and technical domains that require consideration of changing properties and relations and claim for formal description of scenarios and patterns of evolution. The knowledge relating to such tasks has been termed *dynamic*, the methods of its processing, *dynamic knowledge representation (engineering, management* and so forth), and the contexts that include them, *dynamic environments* (Pshenichny & Kanzheleva, 2011).

Importantly, dynamic knowledge representation should be understood herewith as representation of dynamic knowledge, not as dynamic representation of knowledge. There is quite a research done in the fields of the so-called dynamic ontologies and ontology evolution (Stojanovic, 2004; Zablith, 2008; Murdock et al., 2010, Noy et al., 2006, and others), dynamic knowledge representation using agent-based modeling (An, 2009), and even adaptive fuzzy Petri nets for dynamic knowledge representation and inference (Li and Lara-Rosano, 2000), but the approach in this field is quite different and focuses on how to adapt ontology or another, essentially static, knowledge model to new knowledge about phenomena, not to capture the changes of the phenomena themselves. Speaking about "evolution", "processes", "scenarios" and related issues, the researchers in this field mean the change of the model proper, not of the world it describes, and look for solutions to keep ontology self-consistent despite the changing world rather than to model the changes of the latter. Thus, dynamic ontologies describe the same static knowledge but assume its incompleteness. So do the dynamic graphs (Demetrescu et al., 2005) and dynamic attributed graphs (Desmier et al., 2013), which perhaps can be applied to represent dynamic knowledge but this is not their virtue denoted as "dynamic". Similarly to ontologies, "dynamic" in this case means "editable" and refers not to the knowledge, but to the model that describes it and to the method to create such model.

Examples of dynamic knowledge can be found virtually in any field – in history (in research of evolution of cultures, nations, states), in policy (in monitoring of the change of political situation, forecasting and advice practices), in medicine (in studying the reaction of organisms to changing conditions, getting sick and recovering), in technical design (in analyzing the functioning of equipment under changing conditions or partial breakdown), in geosciences (in reconstruction of evolution of landscapes or geological structures, assessment of natural hazards) and so forth.

The key concept of dynamic knowledge engineering and representation is *event*. Event has been recently defined as a statement of general form $S – P_1, P_2, \ldots, P_i$, where S is subject, and $P_1, P_2, \ldots, P_i$ are predicates, with or without negation. The following restrictions are put herewith, (i) subject may be only one in any event, (ii) subject may not be negated, (iii) all predicates may not be negated in an event, at least one must be without negation (e.g., an event $S – \neg P$, $\neg$ being the negation, is not allowed). For instance, "Birds (S) fly (P_1), sing (P_2) and are not scared ($\neg P_3$)" (Pshenichny and Mouromtsev, 2015). At the level of graph-based visualization this means that nodes in a graph that represents dynamic knowledge do not mean individual subjects and predicates; rather, the events per se (i.e., the combinations of a subject with predicates, possibly with negation where allowed) make up the nodes of such graph (Figure 1).

Events are arranged in successions by a set of *connectives* generally defined in the framework of unified grammar of dynamic knowledge (Pshenichny & Mouromtsev, 2015). The simplest succession, having one connective and being a change of one non-empty set of events to another, is understood as *process*. A chain of processes makes *flow* or, in case of distinct beginning and end, *scenario*. A non-empty group of interrelated scenarios is either a description or a model of a *dynamic environment*.

Pshenichny & Kanzheleva (2011) and then Pshenichny & Mouromtsev (2015) classified environments as shown in Figure 2.

Environment must include objects and may include events. Environments of the former type are, e.g., classifications or data models and are regarded as object-based, or static. Those, which include events, are event-based, or dynamic, correspondingly. This distinction appears valuable also for the methods, which are or can be used in knowledge engineering and representation. Regardless of underlying formalisms, all these methods appear to admit a graphic expression as a network or graph-like structure; still, the meaning of nodes and edges may vary.

Figure 1. Difference between static and dynamic environments. Static environments imply fixed relations between objects. Dynamic environment can be represented as a succession of static ones, possibly including mutually exclusive events ("Syrian rebels are fighting and supported" - "Syrian rebels are fighting and not supported"). In the simplest case shown here, each static environment includes one event. Modified from Pshenichny and Mouromtsev (2015).

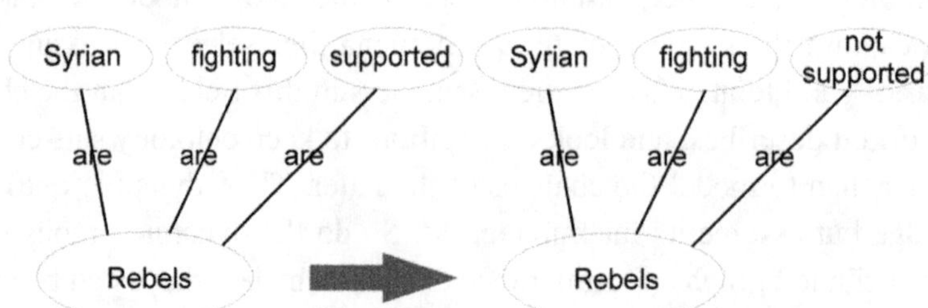

Figure 2. Classification of modeling environments. Modified from Pshenichny & Mouromtsev (2015).

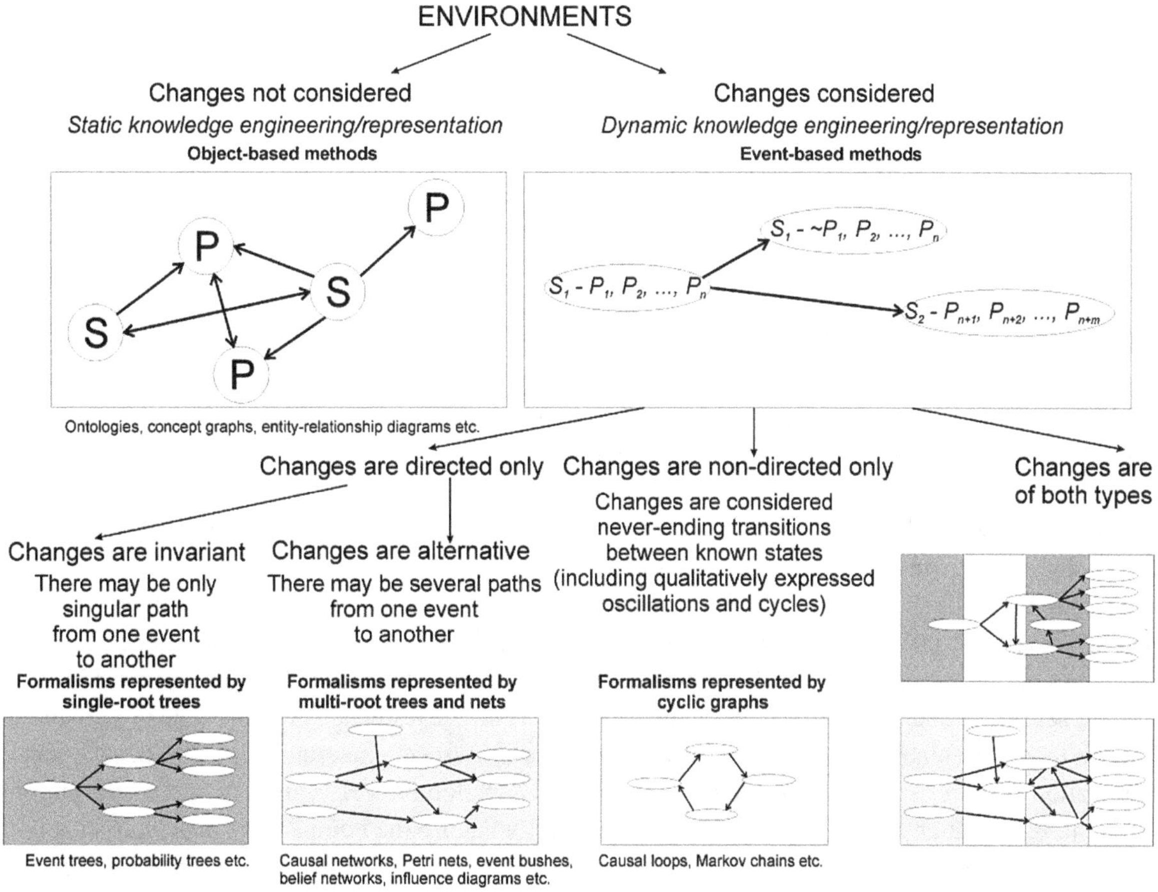

Object-based methods include a very wide range of tools (many of them were reviewed by Sowa, 2000, and broadly regarded as semantic networks). These are usually expressed as directed or non-directed, possibly cyclic graphs with each subject and each predicate occupying a separate node, and edges meaning relations "is – a", "a property of", "x acts on y" (the latter already implies an event but does not consider it), and the like. Object-based methods include ontologies, taxonomies, thesauruses, entity-relationship diagrams, relational data schemes, conceptual graphs, definitional networks, correlational nets, relational graphs, monotonic description logics, assertional networks, possibly, thematic organization packets and others.

Conversely to object-based ones, the event-based methods contain in their nodes neither separate subjects nor predicates but integer events. Some may include parameters (e.g., influence diagrams or, occasionally, belief networks). However, accurate consideration of parameters is likely to reveal hidden event-based structure (Pshenichny et al., 2013), and a virtue of knowledge representation is to make the hidden seen. Edges mean various kinds of cause-effect relation that can be well or poorly understood.

Two subtypes of changing environments can be identified, (i) those in which changes are not directed and (ii) those with directed changes, the latter being divided into two sub-subtypes. In addition, two transitional varieties are possible (Figure 2), which are not considered here.

If changes are not directed, i.e., there are no clear beginning and end of changes, there must be either (i) a cycle (a loop), or (ii) oscillations expressed qualitatively, or (iii) chaotic changes, which cannot be analyzed qualitatively (but not complete quiescence which would bring us to the no-change environment addressed by object-based methods). The formalisms that satisfy this pre-condition may include cyclic graphs (e.g., causal loops) and computational tools like Markov chains.

If changes in an environment are directed (i.e., there may be only singular path from one event to another), their flow represents a scenario (in case of invariant changes) or scenarios (if options are possible) where some causes lead to definite results, which can be somehow fixed in mind. The former type of environment gives ground to the method of event trees with ramifications (decision trees, fault trees, quantitative tools – probability trees and Bayesian trees, and so forth). Such tree has only one starting event ("root") and suggests only one path to any of its further nodes.

Finally, the most sophisticated environment type is that of directed alternative changes. This type of environment can be metaphorized as arena for various, intertangling processes, which are thought to originate from a limited number of causes or sources (but not less than two to be distinct from a tree), tend to leave a definable result and can be followed as more or less distinct scenarios.

Methods describing such environments are expressed graphically as constructions having many "roots" or net-shaped, allowing various paths at least between some of the nodes, i.e., notations strongly resembling graphs and likely interpretable as such. These methods include event bushes, sequence diagrams, activity diagrams, influence diagrams, Bayesian nets, Petri nets and, possibly, some others. Among these, the event bush method was developed specially to address this type of environment taking care of their semantics and is, in fact, the core, the motivation and the testbench of methodology of dynamic knowledge representation. Petri nets, initially designed for software engineering (Petri, 1962), have gained application from jurisprudence to biology (Rogalchuk & Solomin, 2015) and interactive storytelling (Hussein & El-Sattar, 2012). A natural question arises, what are the comparative characteristics (i.e., abilities, peculiarities and interrelation) of the two methods in description of the same environments. To answer it is the motivation and the purpose of the present paper.

For this, we will briefly narrate the fundamentals of both methods (Section 2), test them by modeling several selected environments (Section 3) and discuss the results (Section 4). The unified dynamic knowledge grammar will be used as the formal ground for comparison of the addressed methods.

METHODS IN QUESTION

Petri Nets

Petri nets are a family of mathematical models designed for modeling, analysis and development of dynamic systems in various fields of knowledge (Petri, 1962; 1973; 1977). Mathematically, Petri nets are directed labeled bipartite multigraphs (Petri, 1962). The same time, they bear semantic information in the nodes and therefore can be used for representation of qualitative knowledge about the modeled environments.

It has two types of nodes called *places* and *transitions*; directed arcs link places to transitions and transitions to places but not the nodes of similar type to each other – this is the meaning of bipartite. In some cases a place and transition may be connected by two oppositely directed arcs (thus forming a multigraph); however, this can always be substituted by a normal directed graph structure. Places represent a set of variables that define the model state. Transitions are abstractions for rules of state changes. Behavior of a system is modeled by execution (or *firing*) of transitions in the net (Figure 3a, b). Not a single node may appear in the net more than one time; in other words, all names of the nodes are unique for the net.

Thus the structure of Petri Net N is defined by a tuple (P, T, I, O), where $P = \{p_1, p_2, ..., p_n\}$ is a set of places, denoted as p_i, $T = \{t_1, t_2, ..., t_m\}$ is a set of transitions, denoted as t_j, $I: P \times T \rightarrow N$, an application of precedence, and $O: P \times T \rightarrow N$, an application of incidence, N being the set of natural numbers. $I(p_i,t_i)$ contains the number (or weights) of arcs going from p_i to t_j, and $O(p_i,t_j)$ contains the number of arcs going from t_i to p_j. Places with arcs leaving them and ending in t_i are called *input places* for t_i; Places with arcs connected to them from transition t_i are *output places* for t_i.

Figure 3. Marked Petri net with places 1-6 (p1 – p6) and transitions t1 and t2, unmarked arc is meant to have weight 1 (by default): (a) transition t1 disabled and transition t2 enabled but not executed, (b) transition t2 executed

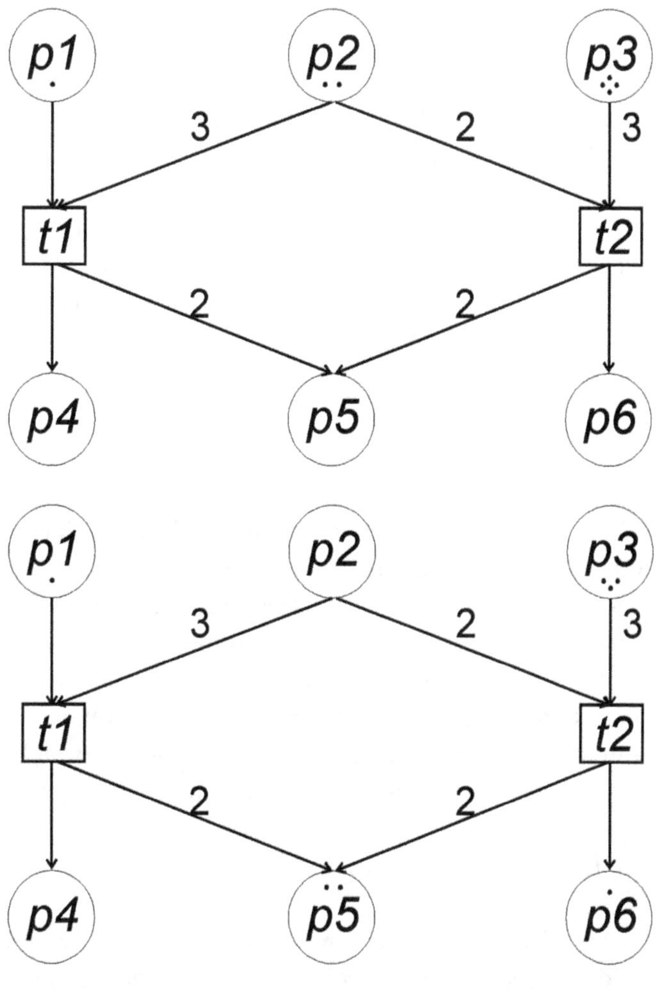

Each place can have a discrete number of 'tokens' representing some condition or resource. The token distribution in all places is called *marking* of a net. Mathematically, a marked net N is defined by a couple $\{N, M\}$, where N is a Petri net structure as stated above, and $M: P \rightarrow N$ is marking. $\mu(N)$ denotes the marking of net N, and $\mu(p_i)$ denotes the marking of a place p_i (or, a number of tokens within place p_i).

The initial distribution of tokens in the net describes the initial state of the system, and it is denoted as marking μ_0. From there, the state changes of system are modeled by executing transitions.

For each marking μ_i there is a set of transitions enabled to be fired, or executed. The transition t_i is enabled if for every input place p_j there is a greater number of tokens than the weight of arc leading to t_i $(\forall p_j \in P, \mu(p_j) \geq I(p_j, t_i))$. Places are depicted as circles with dots within representing tokens, and transitions are depicted as rectangles.

Classical Petri nets are now also called Low Level Petri Nets; many extended models have been built on top of them, linking this method to various mathematical formalisms and thus widening its modeling capabilities. Three classes of Petri Nets are now in use – discrete nets, continuous time nets and stochastic nets (Rogalchuk & Solomin, 2015).

The usefulness of Petri nets for organization of knowledge "both inside and outside computer" was emphasized yet by Petri (1977, p. 131). However, in the last twenty years, discussing the application of Petri nets to knowledge representation, researchers were inspired mainly by an ability of one particular type of Petri net, the fuzzy one, to represent one particular, however important, type of knowledge – the fuzzy knowledge (Li & Lara-Rosano, 2000; Liu et al., 2013; Xiao-zhong, 2003; Yeung, & Tsang, 1994; Wang et al., 2014; Ribarić et al., 2009). Interestingly, most of these authors state that Petri nets are especially efficient to represent the dynamic knowledge. However, virtually all of them consider the knowledge "dynamic" in the sense that this knowledge is imprecise, incomplete and may change (regardless of whether it describes fixed or changing matters), while the term "dynamic knowledge" used here means the knowledge about dynamic environments, or, in other words, the knowledge about changing events (Pshenichny & Mouromtsev, 2015). Desirably, such knowledge is precise and complete ("crisp") but includes incompatible statements corresponding to changing events. Exactly this is the case in our study.

Event Bush

The method of event bush was first suggested by Pshenichny & Khrabrykh (2002) and later developed by Pshenichny et al. (2008; 2009) to reveal the behavior of information in dynamic environments of directed alternative changes. It is gaining increasing importance in a variety of fields, from the geosciences (Behncke & Pshenichny, 2009) to marketing (Mouromtsev et al., 2014) and history (Solomin & Mandrik, 2015). At present this method is viewed as the core methodology of dynamic knowledge engineering and a working example for development of the unified grammar of dynamic knowledge.

The environment of directed alternative changes is described by the event bush as a set of events of four types arranged in certain order (defined as multiflow structure – see Figure 4) and linked by special connectives.

The events of the event bush fall into four classes, or types:

1. Primary internal events. These are primary, non-unique inputs representing the "passive conditions". These would determine any further course of events ("happenings");
2. Primary external events meaning the "invading agents". They may affect basic inputs or influence their further, indirect manifestations, thus "shaping up" different "happenings";

Figure 4. Syntax of basic blocks of event bush – a multiflow structure (from Pshenichny & Mouromtsev, 2013)

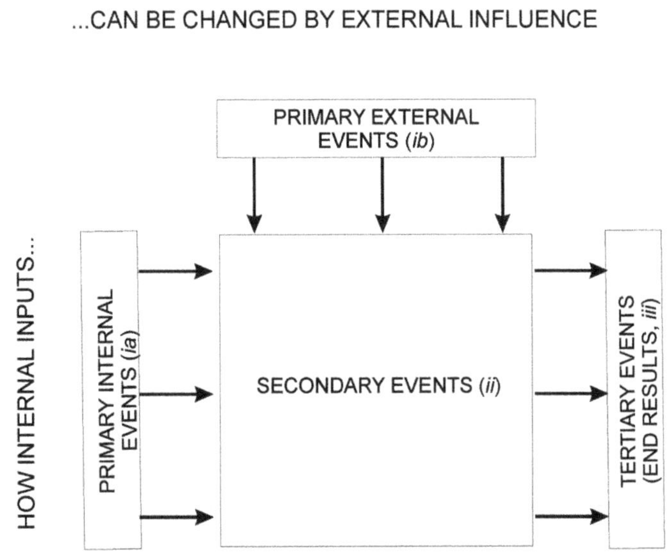

3. Secondary events that result from primary internal inputs with or without the contribution of primary external ones – the "happenings" proper formulated in a strict concise way indicating their core features determined by the causes, following the principle "one more cause – one change of event";
4. Tertiary events that denote end results, or products, generated either by primary internal or by secondary events, with or without primary external ones. Tertiary events present the completed "happenings".

Any event in event bush is uniquely characterized by the set of subjects, set of predicates and type (*ia*, *ib*, *ii* or *iii*) and, in some cases, generality. Two primary events may not have similar subject.

Other pieces of information (even not necessarily verbal) can be associated with events and/or connectives (Diviacco et al., 2014), but the presence of events of types (*i*) - (*iii*) is mandatory.

The following pairwise relations between types (*i*) - (*iii*) are set:

- Not a single event may lead to (*ia*) or (*ib*);
- Event (*ia*) without an event (*ib*) or (*ii*) may lead only to one (*ii*) or one (*iii*) event;
- Event (*ia*) may not lead to any event together with another (*ia*);
- Event (*ib*) may lead only to an event (*ii*) and only together with (*ia*) or another (*ii*);
- Event (*iii*) may not lead to any other event.

These relations are enforced by the connectives of the event bush (Figure 5).

In this subsection, for simplicity, we will consider only the events having predicates (one or more) that relate directly to the subject (i.e., no "predicate of predicate" expressions). Other cases, however interesting, will be studied later in concern of class-subclass relations in the event bush and reported elsewhere.

Figure 5. Connectives of the event bush: flux (a), influx (b), furcation (c) and conflux (d), from Pshenichny & Kanzheleva (2011)

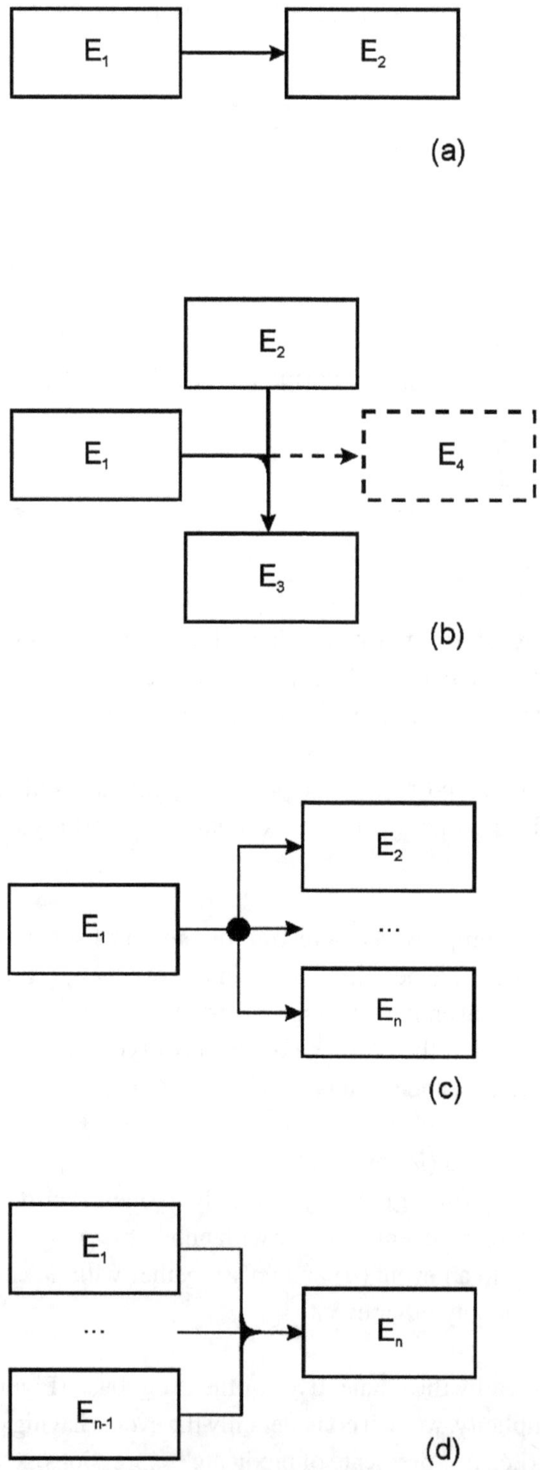

Based on the observations of behavior of information in the environments of directed alternative changes, the following connectives are introduced (Pshenichny and Kanzheleva, 2011):

1. Flux connective (Figure 5, a) describes one event (E_1) producing another (E_2):

E_1 *Flux* E_2

   e.g., "Birds fly, sing and are not scared" (E_1) *Flux* "Birds fly, sing and are not scared" (E_2).

2. Influx connective (Figure 5, b) describes two events (E_1, E_2) producing another (E_3), but playing different roles:

E_1, E_2 *Influx* E_3

   e.g., "Birds fly, sing and are not scared" (E_1), "Thunder rattles" (E_2) *Influx* "Birds fly, sing and are scared" (E_3).
   The event E_4 is not involved in the influx but, being connected by flux with E_1, is a precondition for influx.

3. Furcation connective (Figure 5, c) describes production of one of a multitude of events (E_2, ..., E_n) by one (E_1):

E_1 *Furcation* E_2, ..., E_n

   e.g., "Birds fly, sing and do not look for corn (: find corn/do not find corn)" (E_1) *Furcation* "Birds fly, sing and look for corn: find corn" (E_2), "Birds fly, sing and look for corn: do not find corn" (E_3). Importantly, this connective does not signify any classification or specification of events. It literally means that, first, one predicate changes to negation from left to right ("do not look for corn – look for corn") or, vice versa, negation to assertion, then, the predicate on the right (with or without negation) must be specified and, third, unspecified predicate makes no sense in the modeled environment. In the above example, if birds are not looking for corn, this is enough information we need, but if they are looking for it, this is not enough for us to know only this, and the predicate "look for corn" is not admitted as is. This is the reason why furcation is fundamentally different from the case of nested knowledge considered below. In this case, both unspecified predicate and its specifications are considered sense-making. Meanwhile, the issue of classification or generalization of events based on their subject-predicate structure is quite complicated and will be discussed below.

4. Conflux connective (Figure 5, d) describes production of one event (E_n) by multiple events (E_1, E_2, ..., E_{n-1}):

E_1, E_2, ..., E_{n-1} *Conflux* E_n

   e.g., "Birds fly in the sky and sing in the sky" (E_1), "Crocodiles fly in the sky and sing in the sky" (E_2) Conflux "The sky is occupied by creatures who fly and sing" (E_3).

Similar to that furcation is not classification of events, conflux does not imply their generalization. Moreover, unlike furcation, it includes modi in which not only predicates but subjects also change (Pshenichny and Kanzheleva, 2011). However, one should not take conflux for the mirror of furcation: in the conflux, the events on the left must coexist (in furcation, the events on the right may not) and, in addition, conflux may reserve an opportunity for the causal events to exist further after producing the conflowing event, while furcation, when effected, strictly prohibits further existence of the causal event.

From event to event, the following components may change:

- Subject *S*;
- Predicates P_i;
- Type (position) of event (*ia*) – (*iii*); or
- Generality of subject.

We postulate that type of event and generality of subject may change, if, and only if, the subject and predicates do not change, and vice versa, if both subjects and predicates are the same in two connected events, then either their type or generality of subject must change. If these conditions are omitted, the formalism is not an event bush.

COMPARATIVE MODELING OF SELECTED ENVIRONMENTS

To create a Petri net model of environment, first of all, places should be recognized and formulated in the modeled environment. Like in any knowledge engineering/knowledge representation procedure, this can be done either intuitively or based on a rule if available. To our mind, to formulate such rules for dynamic environments, a special language – namely, a constrained natural language – should be developed, whose grammar would be convertible into the unified dynamic knowledge grammar (Pshenichny & Mouromtsev, 2015), which, in turn, should be generic for the syntax of Petri nets and other dynamic knowledge representation methods. This is, however, a task for future research. So far we have to be guided solely by intuition when identifying basic syntactic elements in the modeled environment, i.e., places in case of Petri nets, and proposing their arrangement into a net. In other words, the ground for modeling is a hope that our own vision, when made explicit and formal, may give a consistent answer to the question that has urged us to model, e.g., when and under what provision to supply Syrian rebels with conventional weapons not to spot one's own reputation.

Likewise, practical guidelines for building an event bush were given by Pshenichny et al. (2008; 2009) and Pshenichny & Kanzheleva (2011). In particular, these guidelines prescribe to start building the bush with formulation of modeling task, then to determine primary internal and external events based on this task, then track primitive *(ia)*E_1 *Flow (iii)*E_1 successions, then introduce changes *(ia)*E_1,*(ib)*E_2 *Influx (ii)*E_3 and update predicates of *(ia)*E_1 and *(iii)*E_1, then go ahead adding connectives and events of types (*ia*), (*ib*) and (*ii*) and checking the relations between them, self-consistency of the bush, completeness of all scenarios (each must end with one (*iii*) type event) and compliance of events with the definition (the easiest error is negation of all predicates that may happen at their update).

Choosing an environment to test an application of alternative methods of dynamic knowledge representation, one should, in our opinion, consider the following prerequisites:

- Dynamic character (i.e., the environment must include changes grouped in readily tracked scenarios);
- Relevance to the society and science;
- Simplicity for understanding.

We selected four environments from vastly different domains – information technologies (IT), politics, common-sense reasoning and physics – and analyzed each of them by both methods. To emphasize the intriguing universality of these methods, we looked for environments, which, having nothing to do with each other by their contents, may be formally described by the same models.

Example From the Politics: Chemical Disarmament in Syria

Destruction of chemical weapons in Syria was thoroughly reported and widely discussed by the world's leading news agencies including CNN (Eliminating…, 2014), Al Jazeera (Syria chemical weapons…, 2014), Russia Today (The US disables diplomacy…, 2014) and even web scholarly resources such as Encyclopædia Britannica (Syrian Civil War, 2014) or Wikipedia (Destruction…, 2014). We compiled a text based on the resources quoted above (as well as many others) and used it as starting material for our query. In the future, we believe that every published text or recorded speech could be subject to dynamic knowledge extraction and representation procedures as they are now to static knowledge engineering (e.g., ontology design from texts – see, e.g., Mouromtsev et al., 2013). At present, however, the task is opposite – given the text, to test dynamic knowledge representation procedures. It is even not important therefore how true and impartial this text is, for it serves only as an example of modeled environment. However, any comments to this text are highly welcome and may be included in an updated version of it, which, in turn, again will be subject to knowledge representation procedure – this time with the purpose of creation of a real model.

The present-day version of the text is below.

Syrian rebels, fighting with the regime of President al-Assad, first got no or insignificantly few conventional weapons from their partners at the West. Later, however, they began being strongly supported by the latter including weapon supply. Later the al-Assad regime was forced to proceed to bring its chemical weapons out of the country, first from warehouses at the military bases to a seaport (Latakia) and then from the seaport by foreign vessels outside of the country for complete destruction, the foreign partners of the rebels encountered a problem: if the rebels attack Syrian army and take the chemical weapons either on the way from the bases to seaport or in the seaport before shipment, they may use for this the conventional weapons supplied by the foreign partners, and this would look not good at all. Hence, the sympathizers of rebels had to decide, when and under what provision to supply Syrian rebels with conventional weapons not to spot their own reputation.

Petri Net Model

The places for the Petri net $p_{1,\ldots,n}$ are chosen from the narration given above as determined by the unified dynamic knowledge grammar (i.e., one subject and whatever number of predicates, at least one without negation):

1. Seaports are ready to accept chemical weapons, are not occupied with chemical weapons – p_1 (i.e., chemical weapons have not been brought there);
2. Seaports are ready to accept chemical weapons, are occupied with chemical weapons – p_2 (i.e., chemical weapons are there);
3. Foreign partners sympathize Syrian rebels but neither decide nor give them conventional weapons – p_3 (indeed, if no decision is made weapons may not be delivered; however, both negations should be stated explicitly);
4. Vessels are ready to accept chemical weapons, are not occupied with chemical weapons – p_4 (i.e., weapons have not been brought aboard);
5. Foreign partners sympathize Syrian rebels, decide but do not give them conventional weapons – p_5 (i.e., the principal decision is made but still no action taken);
6. Chemical weapons are brought by vessels away from Syria – p_6;
7. Foreign partners sympathize Syrian rebels, decide to, and give them conventional weapons – p_7.

Then, these places should be united by transitions in a Petri net in the way that would reflect personal vision of the situation. This is done as shown in Figure 6.

Once the seaports are ready to accept chemical weapons (namely, one port, Latakia), this may lead via transition 1 (*t1* in Figure 6) to two places, (i) "Seaports are occupied with chemical weapons" (i.e., chemical weapons are brought to seaports) and (ii) "Foreign partners sympathize Syrian rebels but neither decide nor give them conventional weapons". These two places jointly lead to transition 2 (*t2* in Figure 6). But to enable this transition qualitatively, one more condition is necessary, that the "vessels are ready to accept chemical weapons" (but they are still not there). Then transition 2 becomes possible and leads to two more places, (i) "Chemical weapons are brought by vessels away from Syria" and (ii) "Foreign partners sympathize Syrian rebels, decide but do not give them conventional weapons". They, jointly again, lead to the next (and the last) transition, *t3* (see Figure 6), which "opens" the crucial place, "Foreign partners … give (Syrian rebels) conventional weapons", and again to the place stating that the vessels are ready to bring another portion of weapons out (though this is not necessary anymore in the modeled environment).

However, qualitative structure does not exhaust the opportunities even of low level Petri nets. An important property of this formalism is marking. In Figure 6, marked are the following places which may enable the transitions quantitatively:

1. Seaports are ready to accept chemical weapons, are not occupied with chemical weapons;
2. Seaports are ready to accept chemical weapons, are occupied with chemical weapons – by singular token equal to the number of seaports involved in the operation (Latakia); and
3. Vessels are ready to accept chemical weapons, are not occupied with chemical weapons – by *m* tokens equal to the number of vessels involved (given it does not change);
4. Chemical weapons are brought by vessels away from Syria – by *n* tokens equal to the units of Syrian chemical weapons scheduled for destruction.

Figure 6. Petri net model of chemical disarmament during civil war in. Indices of events correspond to those in the event bush model (see Figure 7).

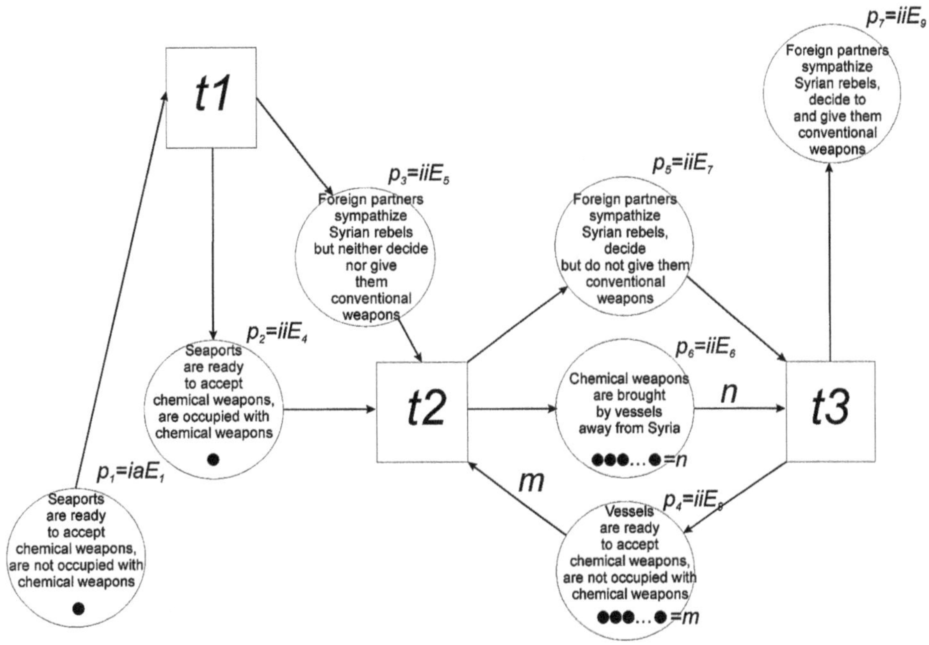

This is interpreted as follows. Until the regime of al-Assad decides to get rid of all *n* units of chemical weapons it has at its disposal, no more guns are given Syrian rebels, whatever they are in number (the place is not marked). First, the seaports, in amount of one, scheduled for the operation must be ready to accept these *n* units of chemical weapons. Then, they must physically arrive at the seaport *and* all *m* vessels must be ready to take them aboard. When all these conditions are met and *t2* is enabled ("fired"), the next two places may take place, among which one ("Foreign partners decide to give…") has no quantitative limitation, be these partners many, one or zero, and the other does have – the same very *n* units of chemical weapons that have been present in the army, have been agreed to be removed, have been transported to the seaport, now must be shipped and sent away. This "fires" *t3* and gives foreign partners a chance to put the guns in the arms of rebels without a risk to spot their own reputation. If we want to model a case that a regime is being persuaded to disarm step by step, then a number of tokens smaller than *n* may allow this transition but the whole model must be necessarily cyclic.

Event Bush Model

By the guidelines given in the beginning of this section, modeling by the event bush starts with formulation of the modeling task. We define it with a portion of irony as "How seaports and vessels are used by the al-Assad regime" (Figure 7), even though the vessels were foreign. This task only implicitly includes the question that needs to be answered, when and under what provision should mighty foreign partners supply Syrian rebels with conventional weapons not to spot their own reputation. However, the relation of this question to the modeling task is clearly seen from the structure of the bush and will be elucidated in Figure 7.

Figure 7. Event bush model of usage of seaports and vessels in civil war in Syria

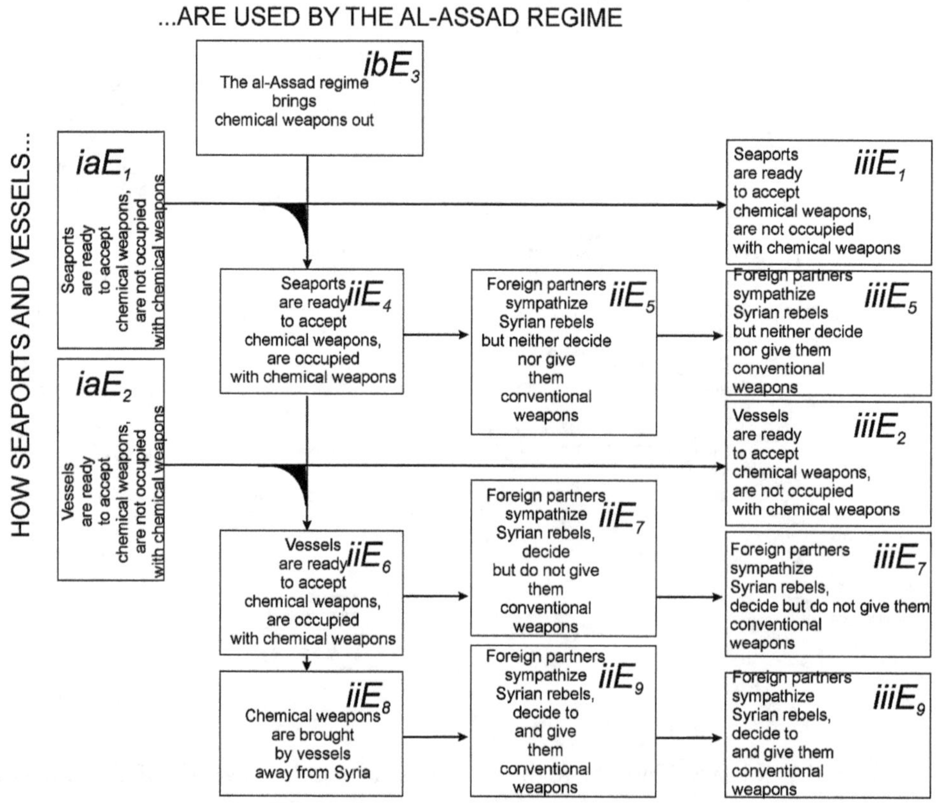

This modeling task determines the choice of primary internal and primary external events. Omitting the process of construction of event bush that was reported and discussed in several publications (see, e.g., Pshenichny et al., 2008; Mouromtsev et al., 2014), we proceed to the final result.

Primary internal events (*ia*) are:

iaE_1: Seaports (S_1) are ready to accept chemical weapons (P_1), are not occupied with chemical weapons ($\neg P_2$).

iaE_2: Vessels (S_2) are ready to accept chemical weapons (P_1), are not occupied with chemical weapons ($\neg P_2$).

In these events, ¬ is negation.

Primary external event (*ib*) is:

ibE_3: The al-Assad regime (S_3) brings chemical weapons out (P_3).

Initial changes are:

$\{(ia)\ S_1 - P_1, \neg P_2\}\ Flux\ \{(iii)\ S_1 - P_1, \neg P_2\}$

and:

$\{(ia)\ S_2 - P_1, \neg P_2\}$ *Flux* $\{(iii)\ S_2 - P_1, \neg P_2\}$

These are included in influxes:

$\{(ia)\ S_1 - P_1, \neg P_2\}, \{(ib) S_3 - P_3\}$ *Influx* $\{(ii)\ S_1 - P_1, P_2\}$

$\{(ia)\ S_2 - P_1, \neg P_2\}, \{(ii)\ S_1 - P_1, P_2\}$ *Influx* $\{(ii)\ S_2 - P_1, P_2\}$

As the theory of event bush (and unified dynamic knowledge in general) has been developed only to the simplest case of predicates that relate directly to the subject (i.e., no "predicate of predicate" expressions), while the formulations used in our model are obviously more complex (e.g., expressions like "to be occupied with chemical weapons" or "to be brought by vessels away from Syria" can be only at first approximation considered as having singular predicate), we record the events in the way that would hopefully enable further formalization. For this, the expressions are split into smaller pieces having atomic sense, e.g., "...are brought by vessels away from Syria..." (see Figure 7).

Similar to the Petri net model, this model is based on personal vision of the environment. However, the rules of composition of the event bush suggest that, having plotted the scenario that corresponds to the modeler's vision, he/she must (i) finish other scenarios that formally follow given the connectives used in this scenario and then (ii) examine the resulting bush aiming to suppose what other scenarios can bind the existing events. If this inevitably requires introduction of new events (i.e., new boxes in the event bush), these new events are also examined to find out what other, alternative relations to preexisting events they may have, and so forth. There is no formal limit for construction of the event bush. Like any other modeling, it formalizes the scientist's view and the same time strongly relies on the scientist's intuition and a will to produce an adequate and concise conceptualization.

The scenario that corresponds to the intuitive solution in the proposed bush is:

$\{iaE_1\}, \{ibE_3\}$ *Influx* $\{iiE_4\}$

$\{iaE_2\}, \{iiE_4\}$ *Influx* $\{iiE_6\}$

$\{iiE_6\}$ *Flux* $\{iiE_8\}$

$\{iiE_8\}$ *Flux* $\{iiE_9\}$

$\{iiE_9\}$ *Flux* $\{iiiE_9\}$

Along with this "desired" scenario, a few other scenarios can be seen in the model, from purely trivial, e.g., $\{iaE_1\}$ *Flux* $\{iiiE_1\}$, to more complicated ones. One may argue whether they make sense now, when the solution is already known and took place (and the situation has evolved much further, to a riot in Iraq). The answer is twofold. First, if such instrument were in hands of decision-makers before, perhaps this would have made the search of solution easier, faster and more reliable. Then, like any model, this event bush describes not only particular situation but a type of situations. If to substitute "Syrian rebels" and "regime of al-Assad" with "rebels *X*" and "regime *Y*", correspondingly, one may obtain a nice and handy decision-support tool for the foreign partners of many autocracies possessing real or imagined weapons of mass destruction.

IT Example: Direct and Inverse Data Transfer

In many information systems it is mandatory that a delivery notification, or confirmation of receipt, or another type of reverse signal, i.e., response, is sent after receiving a data packet. Importantly, this signal must be sent via the same channel. Meanwhile, it is fairly possible that another data packet is about leave through it. Collision of the two pieces of data may not take place either in the distal (i.e., closer to the receiver) or in the proximal (closer to the sender of the data packet) sections of the data transfer channel. Therefore, one should think of a way to avoid this collision – namely, not to let another data packet depart until the reverse signal related to the previous packet is received by the data packet sender.

Petri Net Model

Based on this textual description and the guidelines presented above, the following places for Petri net should be introduced:

1. Distal section of data transfer channel leads from Recipient to proximal section, is not occupied by response – p_1;
2. Distal section of data transfer channel leads from Recipient to proximal section, is occupied by response – p_2;
3. Sender has new data packet but neither prepares to, nor sends it to Recipient – p_3;
4. Proximal section of data transfer channel leads from distal section to Sender, is not occupied by response – p_4;
5. Sender has new data packet, prepares to but does not send it to Recipient – p_5;
6. Proximal section of data transfer channel leads from distal section to Sender, is occupied by response – p_6;
7. Sender has new data packet, has prepared to and sends data packet – p_7.

These places are linked by transitions as shown in Figure 8.

First, the sender sends new data packet to the recipient (p_7) and the distal part of the channel, naturally, has no reverse signal in (p_1). Then, after transition 1, the distal section must accept the response to this data packet (p_2) and the sender may have new data packet but in no way may it prepare to send it at this step (p_3). To accomplish the next step, denoted by transition 2, one more precondition is needed, that the proximal section is free (p_4). Given this, two more places are fired (transition 2), the proximal section accepts the response (p_5) and the sender is going to send new data packet (p_6). When the proximal section is free again (p_4) – or, in other words, the response has reached the sender, the latter emits new piece of data ahead (p_7). Initial marking of the net (see places 1, 4 and 7 in Figure 8) reflects the fact that there are only one sender and one data transfer channel (i.e., one proximal and one distal sections) in the system. Consequently, the recipient is also one but this does not need to be put explicitly as it explicitly follows from the above.

Figure 8. Petri net model of data transfer from recipient to sender in information system. Indices of events correspond to those in the event bush model (see Figure 9).

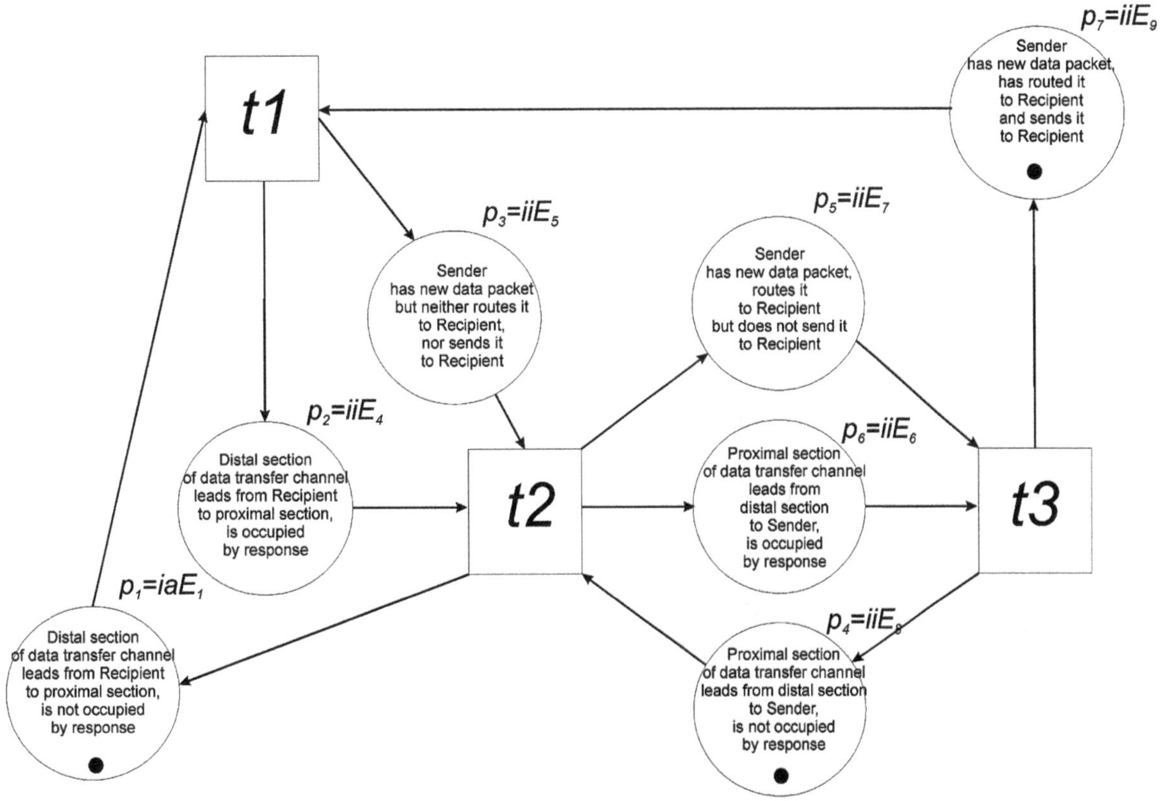

Event Bush Model

By analogy with the event bush model of military experiments in Syria presented in the subsection 3.1.2, first of all, the modeling task was defined, “How the data flow in data transfer channel is effected by the information system” (Figure 9).

Primary internal events (*ia*) are:

iaE_1: Distal section of data transfer channel (S_1) leads from Recipient to proximal section (P_1), is not occupied by response ($\neg P_2$).

iaE_2: Proximal section of data transfer channel (S_2) leads from distal section to Sender (P_3), is not occupied by response ($\neg P_2$).

In these events, $\neg$ is negation.

Primary external event (*ib*) is:

ibE_3: Recipient (S_3) sends response to Sender (P_3).

Initial changes are:

Figure 9. Event bush model of data transfer from recipient to sender in information system

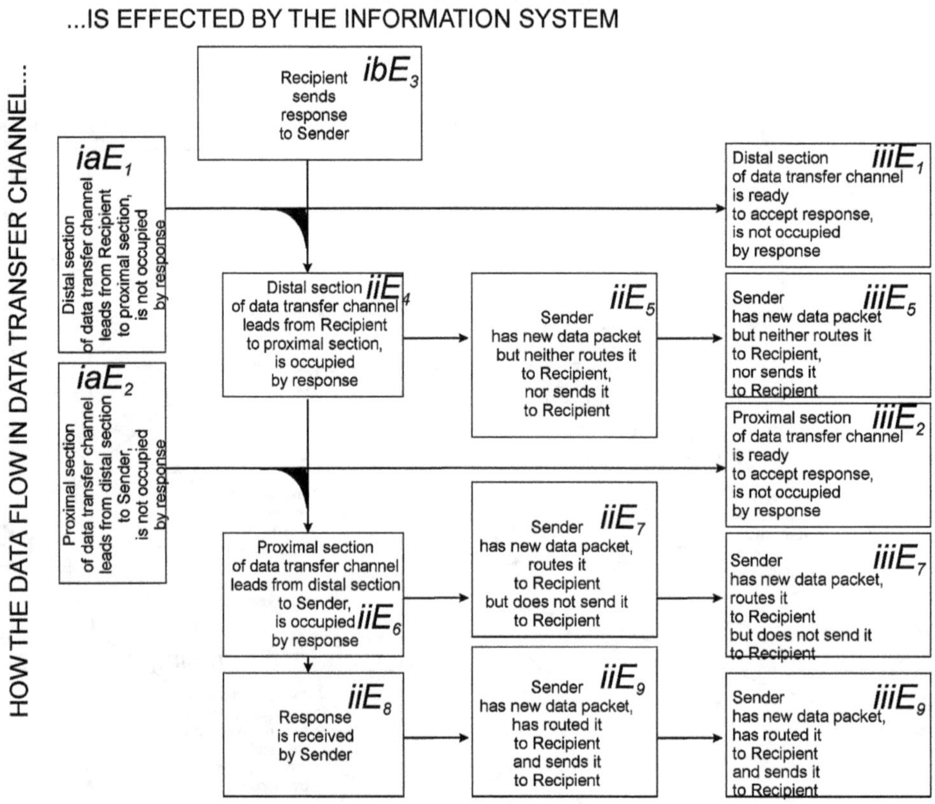

$\{(ia)\ S_1 - P_1, \neg P_2\}$ *Flux* $\{(iii)\ S_1 - P_1, \neg P_2\}$

and:

$\{(ia)\ S_2 - P_3, \neg P_2\}$ *Flux* $\{(iii)\ S_2 - P_3, \neg P_2\}$

The same primary internals join the influxes:

$\{(ia)\ S_1 - P_1, \neg P_2\}, \{(ib) S_3 - P_3\}$ *Influx* $\{(ii)\ S_1 - P_1, P_2\}$

$\{(ia)\ S_2 - P_1, \neg P_2\}, \{(ii)\ S_1 - P_1, P_2\}$ *Influx* $\{(ii)\ S_2 - P_1, P_2\}$

The scenario that corresponds to the intuitive solution in the proposed bush is:

$\{iaE_1\}, \{ibE_3\}$ *Influx* $\{iiE_4\}$

$\{iaE_2\}, \{iiE_4\}$ *Influx* $\{iiE_6\}$

$\{iiE_6\}$ *Flux* $\{iiE_8\}$

$\{iiE_8\}$ *Flux* $\{iiE_9\}$

$\{iiE_9\}$ *Flux* $\{iiiE_9\}$

All the rest paths in the bush leading to the tertiary events other than $iiiE_9$ represent either the state of repose ((ia)E_1 *Flux* (iii) E_1) and (ia)E_2 *Flux* (iii) E_2)) or possible disorders and malfunctioning of the information system, e.g.:

$\{iaE_1\}$, $\{ibE_3\}$ *Influx* $\{iiE_4\}$

$\{iaE_2\}$, $\{iiE_4\}$ *Influx* $\{iiE_6\}$

$\{iiE_6\}$ *Flux* $\{iiE_7\}$

$\{iiE_7\}$ *Flux* $\{iiiE_7\}$

Example From the Common-Sense Reasoning: Pickpocket and Petty-Dealer

Imagine a pickpocket who is heavily in debt so that either he pays back immediately or has to flee away, and a petty-dealer who is nearly in the same circumstances. Some extra earning occurs to any (or both) of them in any form – be it another sale for the dealer, another plunder for the pickpocket, or money found on street. For simplicity, let us restrict ourselves to the case when the dealer is lucky. This extra earning is not big enough to pay the debt back entirely but may help one do so if to add some more money – however, the lacking amount is not that scaring now.

Thinking so, one of them (no matter who) meets the other. In any case, with or without extra money in hands, the pickpocket would approach the dealer to steal from him. This inevitably (assumingly) leads to a scuffle. Depending on whether the dealer has got extra money or not, two variants are possible.

If both have been lucky, there takes place one of the following options: (i) the pickpocket fails to steal, and the dealer, deeming that he can "punish" the pickpocket himself by robbing him, takes money from the pickpocket, (ii) the pickpocket manages to steal and escape, (iii) none of them takes money from the other (e.g., they simply fight with no success). In two former cases, one of them becomes closer to the state of ability to pay the debt, and the other, yet farther.

If the dealer remains lucky, he can, however, spend this amount for other purposes, e.g., to purchase new goods for sale. If no one wins, they are back to the beginning. If only the dealer has been lucky, then the option (i) does not work because, even if beaten, the pickpocket has nothing to give the dealer.

Petri Net Model

Places for the Petri net were selected and arranged in similar way as in two previous models of this kind. Nevertheless, this model appears much larger. To reduce ut a bit and thus keep it displayable in a paper format, we introduced two assumptions, which are clearly seen in Figure 10:

1. Encounter of petty-dealer and pickpocket takes place only if the petty-dealer has got additional money – expressed as transitions, *t12* and *t13* without continuation (with question marks behind);
2. Pickpocket will not spend the extra money once got it – expressed as the absence of arc from p_4 to the transition *t5* that directly leads to p_2.

Weights of the arcs in this net may be attributed minimal value one, as considered are one petty-dealer, one pickpocket and a minimal, whatever small, relevant amount of extra money which also can be regarded as singular unit. This is enough to fire any transition in this model. Therefore, the values of weight are not indicated in Figure 10.

Initial marking of the net are values one of places p_1, p_2 and p_3 considered as starting for any run of the model. Two runs are possible. (i) If only the petty-dealer is lucky, places *p2* and *p3* via the transition *t2* lead to *p5*, while *p1* together with this place fire *t3* and, hence, cause *p6*. This place introduces the

Figure 10. Petri net model of street clash between a petty-dealer and pickpocket. Indices of events correspond to those in the event bush model (see Figure 11).

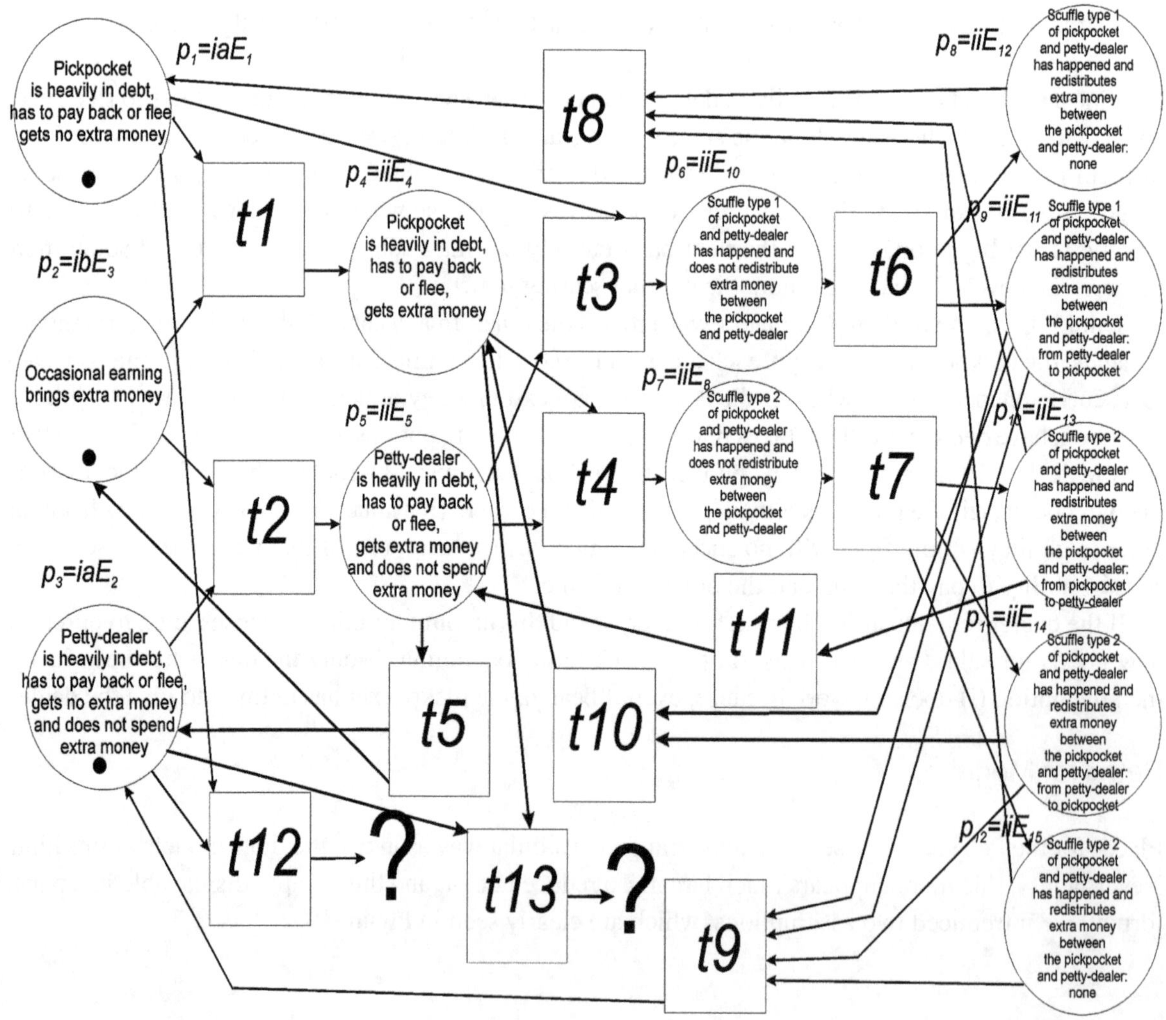

Figure 11. Event bush model of relationships in street community

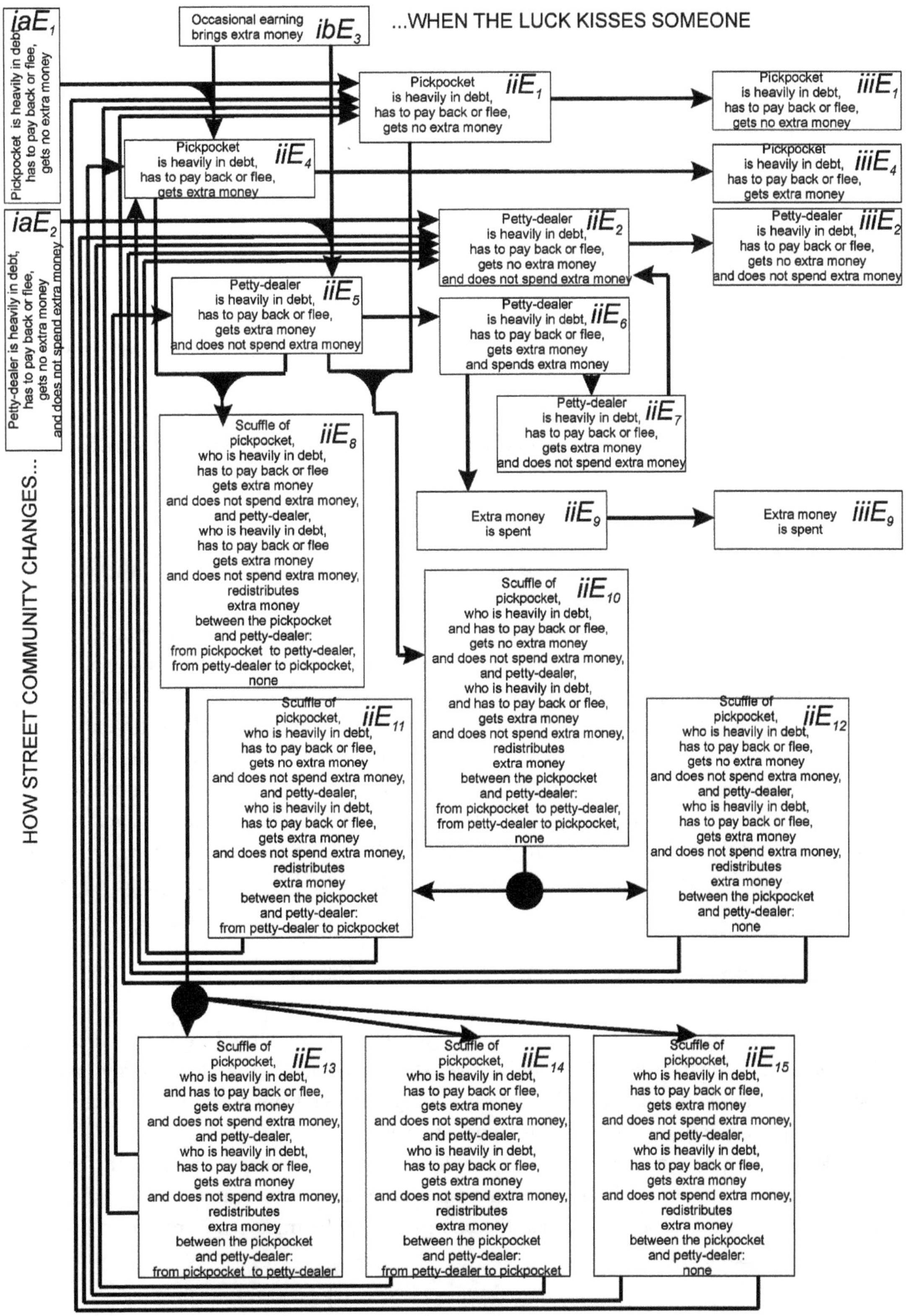

type of scuffle between the pickpocket who has no extra money and the petty-dealer who has, defined here as type 1 scuffle. This type leads, via *t6*, to two possible incompatible options described by places *p8* and *p9*. They, in turn, may lead back to *p1* and *p3* via *t8* and *t9*, correspondingly. (ii) If both the petty-dealer and pickpocket got extra money, transitions *t1* and *t2* are fired leading to *p4* and *p5*, which, in turn, fire the transition *t4*. This transition brings *p7*, which, via *t7*, to the scuffle of the second type, i.e., that of the petty-dealer and pickpocket who both have gained extra money. This type of scuffle has three possible incompatible endings, places *p10-p12* fired each by *t7* and leading via *t9-t11* back to the beginning of the story. These return options serve to show the case that either actor, the pickpocket or petty-dealer, is robbed by his counterpart.

Primary internal events describe initial states of the main actors:

iaE_1: Pickpocket: Is heavily in debt, has to pay back or flee, gets no extra money.
iaE_2: Petty-dealer: Is heavily in debt, has to pay back or flee, gets no extra money and does not spend extra money.

Primary external event describe the main change that creates the intrigue of the situation:

ibE_3: Occasional earning brings extra money.

Two important structural features of this model are repetition of primary internals as secondary events and two influxes involving the main actors.

The repetition of primary internals is necessary to model the loss of previously gained extra money by either actor as a result of robbing by the other one. Influxes are formed by events in which the actors are subjects – one, by lucky petty-dealer and unlucky pickpocket (iiE_8), and the other, by lucky petty-dealer and lucky pickpocket (iiE_{10}). They lead to events with a fundamentally different subject, the scuffle – that of type one in the former, and type two, in the latter case, in terms of the Petri net model. Correspondingly, the former has two, and the latter, three incompatible options produced by furcation.

Interestingly, this model presents only two scenarios apart from the one in question, and these two are most trivial, "technical" ones, sometimes regarded as "nothing changes" scenario: iaE_1 *Flux* iiE_1 *Flux* $iiiE_1$ and iaE_2 *Flux* iiE_2 *Flux* $iiiE_2$. This rarely happens in the event bush modeling and is a remarkable peculiarity of the presented model.

Physical Example: Electron and Atom in Upper State

Collision of an atom and electron may lead to several results depending on the energy balance between them. Except for the trivial case of the atom in lower state and electron with low energy that results in elastic scattering, let us consider a situation that an energy is added from outside either to electron or to atom (suppose for certainty that atom gets in upper state anyway).

If the electron has got higher energy too, their collision may lead either to energy transfer from electron to atom, or vice versa, or to elastic scattering again when there is essentially no transfer of energy. If the electron has got no extra energy, then the energy cannot be transferred from it to the atom and only two options remain.

Petri Net Model

The model of electron-atom collision presented in Figure 12 does not differ from the model of pickpocket-petty-dealer scuffle either qualitatively, in terms of structure of the net, or quantitatively, in terms of marking and weights. Therefore, one may straightforwardly proceed to interpret this model in terms of physics.

Considered are one electron, one atom and the minimum energy required to bring the latter in the upper state. Collision of the atom and electron implies that they interact, with or without exchange of energy. In other words, they become one entity for the moment of collision. It looks reasonable to consider them *system* sensu Bogdanov [1926], though a short-lived one. Either this is a system of an atom in upper state and electron with low energy (termed System 1 in Figure 12) causing one of the two mutually exclusive results, or a system of an atom in upper state and electron with high energy (System 2) with three possible outcomes, but in any case the system is singular. Therefore, similar to the petty dealer-pickpocket model, all arcs by default have weights equal to one and the three initial places (the same as in the said model) are also marked one.

The succession of events expressed as allowed sequences of places and transitions in Figure 12 entirely corresponds to the textual description of atom-electron collision given above but also implies an ability of the atom in upper state to emit extra energy thus passing back into a relatively lower state (*p5 – t5 – p3*) similar to spending extra money to the purposes other than the debt payback in the previous case.

Figure 12. Petri net model of collision of electron with high energy and atom in upper state. Indices of events correspond to those in the event bush model (see Figure 13).

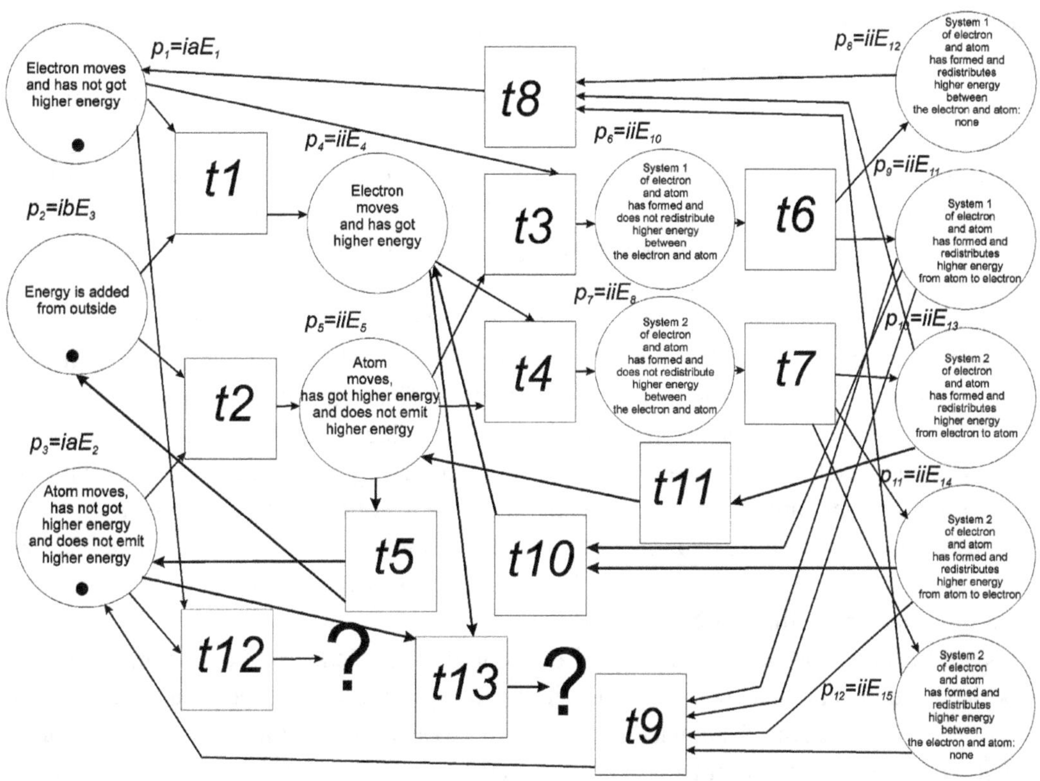

Event Bush Model

The event bush created to model this situation also looks similar to the one for the environment of street community under the kiss of luck (Figure 11). The model is presented in Figure 13.

DISCUSSION

Four environments of dynamic knowledge presented above, grouped in two pairs, were modeled each by two methods of dynamic knowledge engineering, the Petri net and the event bush. One group included relatively small environments (those of chemical disarmament and direct versus inverse signal routing) that produced obviously simpler models. The other consisted of larger environments (street history and atom-electron collision) and was mirrored by more complicated models. It should be noted herewith that in the "simple" group the similarity is not complete. While the IT example may imply cyclicity, the political one definitely does not (unless the chemical weapons are removed not from entire Syria at once but region by region, and rebel troops are being armed respectively, provided they do not migrate outside of "their" regions – however, this never happened in reality). Also, there differs the quantification of the models that is expressed in marking of the Petri nets. Models in the other group show complete qualitative and quantitative similarity of the environments.

The fact that two different environments, having nothing common in their contents, repeatedly exhibit complete or almost complete similarity when modeled by two different methods, may have numerous fruitful implications, among which the first and the most straightforward is an opportunity to develop universal qualitative knowledge-based models. These models then could be used to conceptually verify cross-disciplinary use of quantitative models (e.g., physical models in sociology – Kapitza, 1996; 2009 – or mathematical models in geology – Vistelius, 1992). Also, such models can be used for quick teaching and knowledge transfer. Ultimately, they may form a kind of library, or even continuum, allowing fast navigation between modeled environments and fields of knowledge they belong to, allocation of parametric and quantitative models, linking of datasets to the knowledge-based models and solving other tasks of academic and practical interest. Parallel modeling by two methods, from one side, allows to use the advantages of both and, from the other, implies that the coincidence of different environments is not occasional.

Still, exactly this point deserves special consideration. The first question that occurs looking at the pretty different environments modeled in the same way is how objective the approach to modeling is. Isn't there a chance that any environment at all, with some effort, can be represented in the same way? And how relevant would the revealed similarity be to the considered environments and, the more so, to the fields of knowledge to which these environments belong? E.g., in the demonstrated environments, it is enough to formulate any event in a different way – say, not "Chemical weapons are brought by vessels away from Syria" (place *p6* in Figure 6) but "The regime of Al-Assad brings chemical weapons by vessels away from Syria" – to ruin the whole harmony. Herewith, the similarity of meaning so far is entirely at the researcher's discretion.

To answer the above questions, one should examine the knowledge that constitutes the base of the events, which are then included in a dynamic environment – the relations between subjects and predicates in the corresponding domain of knowledge and plausible negations of these relations, i.e., to create a series of static knowledge environments and look at (i) how to "wrap" the same knowledge in different

Figure 13. Event bush model of collision of electron with high energy and atom in upper state

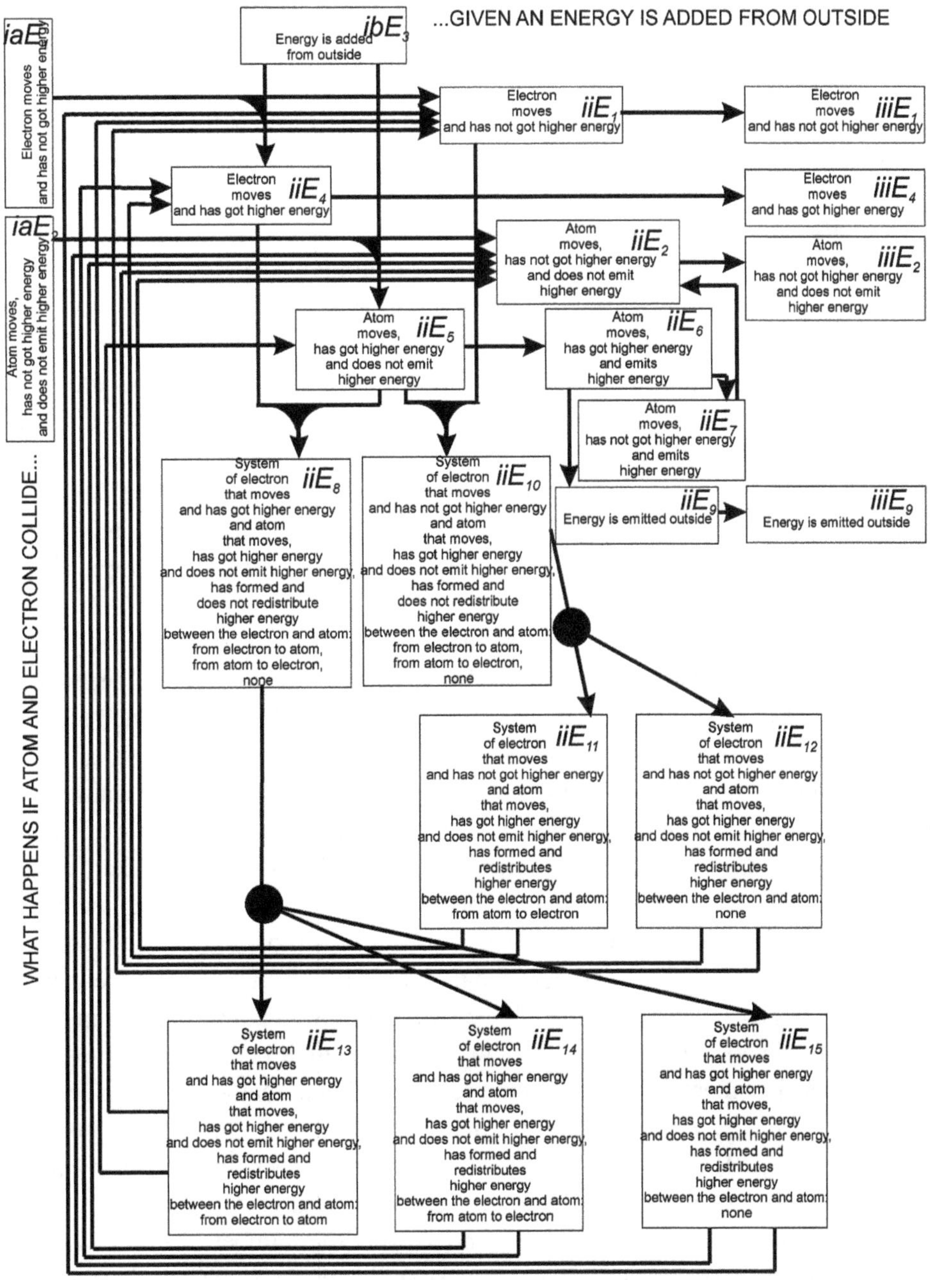

events and (ii) how much the modeled environment "covers" the domain of knowledge. The methodology for both tasks, in our view, should be semantic networks (e.g., Sowa, 1987; 2006; Bonham-Carter et al., 2003), of which the ontologies (e.g., Gruber, 2009) and conceptual graphs (Sowa, 1992) to be mentioned first of all. However, this is a problem that requires a special study and lies outside of the scope of the present paper.

Apart from this problem, however, one may query how the interpretability of the same models by strikingly different environments can be explained, once it is not occasional. For the first group of environments, this is not that obvious – and, also, the coincidence of models of different environments in this group is not full. Nonetheless, in the other group at least one metaphysical explanation can be suggested immediately. Starting the fight, a pickpocket and a petty-dealer form some new entity termed here as scuffle. Even if one is obviously stronger, his victory will always be a matter of chance. The final "decision" who wins is "made" either by the scuffle itself as an integer entity or by some supreme power. In other words, one may suppose that scuffle, as a being, may possess some degree of free will in concern of choosing the winner. The same can be said (if not to commit to the God each and every happening in the Universe) about a system of atom and electron that exists incredibly short time but might have a will to decide which of the two options, in one case, or three, in the other, to choose (see Figure 13).

From the above rationale, it can be concluded that different environments may correspond to a more or less uniformly formulated modeling task. Looking at the event bush models, this task can be derived as a generalization of the event bush modeling tasks, e.g., for the latter group, "How street community changes when the luck kisses someone" (in Figure 11) and "What happens if atom and electron collide given an energy is added from outside" (in Figure 13) accounting for the notion about free will, can be abstracted up to "What happens if two entities colliding acquire free will under external input of anything vital or valuable".

Once there may be a correspondence between the structure of the model (in terms of the event bush and Petri nets) and modeling task, it looks reasonable to expect that there can be recognized particular *types of environment* of alternative directed changes (e.g., a "free-will" type), suited for similar generic problems that may occur in various disciplines, from nuclear physics to management and from medicine to fine arts. Unlikely these types would be summarized in a classification because of their infinite diversity, but even identification and definition of each such cross-disciplinary type of environment with known formal solutions may serve for many sciences and be valuable by itself, like diagnosis of a new disease or discovery of a new mineral. Of course, ideally not only two methods but the whole armory of dynamic knowledge engineering including sequence diagrams, activity diagrams, event trees, Bayesian belief networks, influence networks and others should be tried on each such intuitively gathered type of environment. Then the key issues therein would be whether other methods behave in the same way as the event bush and Petri nets and, in any case, how to interpret their behavior.

On each type of environment the relations between the different methods are projected. Thus, it was shown that comparing the event bush and event tree, one may obtain a tree from a bush but not vice versa (Mouromtsev et al., 2014). Considering the interrelation between the even bush and Petri nets, the following observations can be made from the results of the present study.

In regard to the same environments, one may notice that the number of events in a Petri net (see Figs. 6, 8, 10 and 12) tends to be less than in the corresponding event bush (Figs. 7, 9, 11, 13). This may be due to two reasons. One is purely syntactic and applies to any pair of event bush and Petri net that include events of similar semantics, i.e., events with the same subjects and predicates. The event bush syntax requires repetition of some events in different positions – i.e., all tertiary events duplicate some secondary and possibly primary internal ones, and secondary events may duplicate primary internal and external ones.

The other reason, strictly speaking, is not mandatory. It stems from the fact that the event bush offers special rules (modi) of passage from one event to another aiming to ensure maximum accuracy in description of the environment. Thus, even without the "syntactically" repeated ones, the number of events in a well-built bush is expected to be larger than in a Petri net. This means that, from one side, the event bush provides greater conceptual thoroughness and better formal transparency, and the Petri net, from the other, may look better readable and clearer by intuition. The Petri net does not include information that is not directly related to the issue in question. Still, this is being achieved at the expense of strictness and completeness of description.

This is well seen in Figs. 10 and 12 representing Petri nets fully compliant with the event bush models. One more observation that follows from comparison of the Petri net and event bush models in Figs. 10 and 11 or Figs. 12 and 13, correspondingly, is that the four types of connectives of the event bush better describe how the consequences follow from the causes than the only type of transition in the Petri net – e.g., the influx of iiE_4 from iaE_1 and ibE_3 states that the iiE_4 takes place if, and only if, both the iaE_1 and ibE_3 occur, and their interrelation is: ibE_3 modifies iaE_1, while the transition between the same events in the Petri net means only that iiE_4 happens after both iaE_1 and ibE_3 or at least one of them spring(s) into being.

The Petri net model departs yet farther from the event bush one in the first group. There, even despite the similar semantics of the events, the scenarios are built differently and cannot be directly translated from the Petri net to event bush or vice versa. For instance, the event bush (figs. 7 and 9) defines that iiE_4 follows from the influx of iaE_1 and ibE_3 and iiE_5 follows from iiE_4, and in the Petri net both iiE_4 and iiE_5 follow directly from iaE_1 without ibE_3 at all.

Moreover, the Petri net models in Figs. 6 and 8 depict only one possible scenario, namely, the desired one, ending with $iiiE_9$ – while the event bush reconstructs and clearly presents the full set of scenarios, of which only one ends up with $iiiE_9$.

Nonetheless, the Petri net also shows some advantages compared to the event bush. Thus, the IT example represents an environment that can be transformed into cyclic, i.e., not directed change, type. The chemical disarmament of Syria, as discussed above, is definitely not a cyclic environment. This difference is fairly well reflected by the Petri net models (by presence of arc from iiE_9 to $t1$ in Figure 8 and its absence in Figure 6 – but in no way expressed in the event bush (see Figs. 7 and 9).

Also, the models notably differ quantitatively that is pronounced in the initial marking of Petri nets (compare Figs. 6 and 8). The event bush in its present form offers no such tool.

Looking at both pairs of examples, and especially to the second pair of street community and atom-electron collision – one may notice that within any intuitively gathered type of environment, patterns of modeling by each method are more or less fixed – both the patterns of formalization of semantics and setting the syntax. All subjects and relevant predicates of events of one model bijectively map those of another model of the same environment type, and the syntax of events, be it an event bush or Petri net, remains the same (compare Figures 10 and 12, Figures 11 and 13).

However, verbal formulations of environments may widely vary, and in one case it may look more straightforward to create from the existing text a Petri net, while in another, an event bush. This raises a practical issue of co-application of the two methods and, more broadly, a theoretical query of their interrelation.

As the reconstruction of scenarios in the environment in question is the primary task and challenge of the dynamic knowledge engineering, the event bush, generally speaking, has to be considered the main, while the Petri net, as subordinate, or auxiliary, means of modeling, because the former reconstructs and clearly presents the full set of scenarios by application of its rules of inference, while the latter has no such armory. Still, if the analysis of written or oral text (i.e., the environment) favors the construction of the Petri net, there is no counterindication to build it, but only to get a general idea of the modeled environment. Importantly, similar to the event tree that can be derived from event bush but cannot produce an event bush itself (Mouromtsev et al., 2014), we see no formal way to formally extend the Petri net to the event bush, for it requires information on the connectives that the Petri net may not present. Therefore, an event bush has to be built directly from the environment anyway, and perhaps building it would reveal some gaps and lapses in the description of the latter that are to be filled/removed prior to modeling. Later, if necessary or desired, another Petri net could be built from the event bush to improve the readability, representation power and common-sense clarity of the modeling – but being warranted from losing conceptual thoroughness and formal transparency if considered together with the event bush. In this case, the Petri net may be deemed as a kind of "summary" of relevant information from the event bush.

Besides, even such "subordinate" understanding of Petri nets may not disregard their main advantage to the event bushes, an opportunity of marking the events (places) with tokens and definition of "firing" (execution) of transitions by the number of tokens in particular places. At the same time, such marking is in no way dependent on the structure of Petri net. Rather, it may be introduced to the event bush method as a possible means of its quantification.

Finally, the compatibility and interrelation of the two methods, the Petri nets and event bush, as well as viability of some hybrid "Petri bush" option, can be fully justified theoretically only when both methods are sufficiently well interpreted and redefined in the framework of the unified grammar of dynamic knowledge (Pshenichny and Mouromtsev, 2015). This will allow to extend some useful properties of one method onto others – for instance, to make the syntactic primitives of any method (places in case of Petri nets) compliant with the general definition of event and, the same time, transfer some useful properties of one method to some or all of the dynamic knowledge representation tools – e.g., bring marking from Petri nets to the event bush or substitute "transitions" in Petri nets by more powerful and meaning-sensitive set of connectives borrowed from the event bush.

The interrelation and interplay of these two methods may bring fruitful revelations about how to treat dynamic knowledge, what strategy to choose to represent scenarios depending on the task in mind. This is seen as an important contribution to a new field of research, engineering and representation of dynamic knowledge.

The need for representation of dynamic knowledge obviously exists in expert system knowledge bases. An "IF… THEN…" rule may imply either (i) interrelation between properties (e.g., "IF *x* is born of woman and is not sick THEN *x* may eat apples"), or (ii) interrelation between subjects, or (iii) interrelation between events ("IF *x* is born of woman and lives on Earth now THEN *y* is looking for him to fall in love"). Two latter cases, (ii) and (iii), as seen from the example, may differ merely by the context, while semantically one of these, (ii), the interrelation of subjects (as well as that of properties), represents static knowledge along with (i), and the other, the interrelation of events, (iii) – dynamic one. Despite the fundamental semantic difference, they may be indistinguishable from one another unless there emerges a necessity to put a production rule whose left-hand and right-hand parts explicitly fit the definition of event, i.e., "IF $S - P_1, P_2$ THEN $S - \neg P_1, P_2$" or vice versa ("IF *x* lives on Earth and is brunette THEN *x* does not live on Earth anymore but so far is brunette"). Still, this is often the case in real knowledge-based systems. For instance, da Silva et al. (2014; Figure 9, p. 236), describing an expert system for design of a cogeneration power plant, quote a production rule whose antecedent includes three objects (subjects in our terms) with three facts (fact being understood as an amount of predicates relating to the same subject), and consequent has one object with one fact. Omitting two other objects and their predicates grouped into facts, one object appears both at the left and right in one rule with a fact and its negation:

IF (*Object 1* (S_1) *and related fact* (P_1, P_2)) AND (*Object 2* (S_2) *and related fact* (P_3, P_4)) AND (*Object 3* – MoverCircuit (S_3) *and related fact:* has properties: clD=engine (P_5) and selected=no ($\neg P_6$)) THEN (*Object 3* – MoverCircuit (S_3) *and related fact:* has properties: clD=engine (P_5) and selected=yes (P_6)).

To cope with the necessity of representation of events in knowledge bases, some researchers try to adapt the ontology-related approach – see, e.g., "Knowledge flow for achieving top target" by Syairudin et al. (2015, p. 67) – or create ad hoc informal solutions like behavioral model in the DANE method as the dynamic counterpart of static structural model (Goel et al., 2012). The latter option looks more methodologically pure but unlikely it can accurately encompass classical logical and most of mathematical (math *sensu stricto* and geometrical) rules, which have been the main types of rule for the majority of expert systems (La Rocca, 2012).

This urges us to look for look for a general solution for representation of dynamic knowledge in knowledge bases. One approach could be to understand the "IF… THEN…" structure as a generalization of logical implication for static knowledge and the dynamic knowledge connective for dynamic knowledge, which is further classified into four by Pshenichny and Mourotmtsev (2015). Further logical investigation of properties of this connective may lead to a new type of production rule and, probably, new class of expert systems. Among their possible applications, one of the most promising and straightforward ones should be the anaylsis of dynamic capabilities in management. By Teece et al. (1997), dynamic capability is "the firm's ability to integrate, build, and reconfigure internal and external competences to address rapidly changing environments" (p. 516). This evidences that such approach is not only of academic, but also of purely practical interest.

CONCLUSION

1. Joint application of the Petri nets and event bush is highly favorable to use the advantages of both methods in modeling of alternative directed change environments;
2. The event bush gives more general and impartial representation of an environment, while Petri net is more concise, better readable and focuses solely on the relevant information;
3. It looks promising to use both methods simultaneously in modeling dynamic change environments – first the event bush to get a general vision and understanding, and then Petri net for focusing on the particular target, or task, within the modeled environment;
4. The best co-application and synchronization of the event bush and Petri net, as well as other methods of dynamic knowledge engineering/representation, can be achieved through grounding all these methods in the unified grammar of dynamic knowledge.

ACKNOWLEDGMENT

The research was funded by the Ministry of Education of the Russian Government and the ITMO University as a part of the action aimed at maximizing the competitive advantage of the ITMO University among world's leading educational centers.

REFERENCES

Akasaka, F., Nemoto, Y., Chiba, R., & Shimomura, Y. (2012). Development of PSS Design Support System: Knowledge-based Design Support and Qualitative Evaluation. *Procedia CIRP, 3*, 239 – 244.

An, G. (2009). Dynamic Knowledge Representation Using Agent-Based Modeling: Ontology Instantiation and Verification of Conceptual Models. In I.V. Maly (Ed.), Methods in Molecular Biology. Humana Press. Doi:10.1007/978-1-59745-525-1_15

Arendt, R., & van Uden, E. (2011). A decision-making module for aiding ship system automation design: A knowledge-based approach. *Expert Systems with Applications*, *38*(1), 410–416. doi:10.1016/j.eswa.2010.06.079

Behncke, B., & Pshenichny, C. A. (2009). Modeling unusual eruptive behavior of Mt. Etna, Italy, by means of event bush. *International Journal of Volcanology and Geothermal Research*, *185*(3), 157–171. doi:10.1016/j.jvolgeores.2009.04.020

Bogdanov, A. (1926). *Allgemeine Organisationslehre*. Berlin: Alexander und Lang.

Bonham-Carter, G. F., Smyth, C., & Kerswill, J. (2003). Semantic net approach for mapping mineral potential: a feasibility study. Modeling Geohazards: IAMG 2003.

Calkins, D. E., Egging, N., & Scholz, C. (2000). Knowledge-Based Engineering (KBE) Design Methodology at the Undergraduate and Graduate Levels. *International Journal of Engineering Education*, *16*(1), 21–38.

da Silva, J. C., Matelli, J. A., & Bazzo, E. (2014). Development of a knowledge-based system for cogeneration plant design: Verification, validation and lessons learned. *Knowledge-Based Systems*, *67*, 230–243. doi:10.1016/j.knosys.2014.05.002

DANE 2.0 Users Guide. (2011). Retrieved from http://dilab.cc.gatech.edu/dane/files/DANE%202%20Users%20Guide.pdf

Destruction of Syria's chemical weapons. (2014). In *Wikipedia.* Retrieved from http://en.wikipedia.org/wiki/Destruction_of_Syria%27s_chemical_weapons

Diviacco, P., Pshenichny, C., Carniel, R., Khrabrykh, Z., Shterkhun, V., Mouromtsev, D., & Pascolo, P. et al. (2014). Organization of a geophysical information space by using an event-bush-based collaborative tool. *Earth Science Informatics*. doi:10.1007/s12145-014-0182-2

Eliminating Syria's chemical weapons. (2014). Retrieved from http://www.cnn.com/video/data/2.0/video/world/2014/07/02/id-mann-syria-chemical-gesture.cnn.html

Giarratano, J. C., & Riley, G. D. (2005). *Expert Systems - Principles and Programming* (4th ed.). Thomson Course Technology.

Goel, A. K., Vattam, S., Wiltgen, B., & Helms, M. (2012). Cognitive, collaborative, conceptual and creative — Four characteristics of the next generation of knowledge-based CAD systems: A study in biologically inspired design. *Computer Aided Design*, *44*(10), 879–900. doi:10.1016/j.cad.2011.03.010

Gruber, T. (2009). Ontology. In L. Liu & M. T. Özsu (Eds.), *Encyclopedia of Database Systems. Springer-Verlag.*

Hussein, H. K., & El-Sattar, A. (2012). State of the Art in Interactive Storytelling Technology: An Approach Based on Petri Nets. In P. Pawlewski (Ed.), Petri Nets - Manufacturing and Computer Science. InTech. doi:10.5772/48549

Kapitza, S. P. (1996). The phenomenological theory of world population growth. *Uspekhi Fizicheskikh Nauk*, *166*(1), 63–80. doi:10.3367/UFNr.0166.199601c.0063

Kapitza, S. P. (2009). Global Population Blow-up and After: The Demographic Revolution and Sustainable Development. *Bulletin of the Georgian National Academy of Sciences*, *3*, 5–12.

Kulon, J., Mynors, D. J., & Broomhead, P. (2006). A knowledge-based engineering design tool for metal forging. *Journal of Materials Processing Technology*, *177*(1-3), 331–335. doi:10.1016/j.jmatprotec.2006.04.062

La Rocca, G. (2012). Knowledge based engineering: Between AI and CAD. Review of a language based technology to support engineering design. *Advanced Engineering Informatics*, *26*(2), 159–179. doi:10.1016/j.aei.2012.02.002

Li, X., & Lara-Rosano, F. (2000). Adaptive fuzzy Petri nets for dynamic knowledge representation and inference. *Expert Systems with Applications*, *19*(3), 235–241. doi:10.1016/S0957-4174(00)00036-1

Liu, H.-C., Lin, Q.-L., Mao, L.-X. L.-X., & Zhang, Z.-Y. (2013). Dynamic Adaptive Fuzzy Petri Nets for Knowledge Representation and Reasoning. Systems, Man, and Cybernetics: IEEE Transactions, 43, 1399 – 1410, doi doi:10.1109/TSMC.2013.2256125

Mouromtsev, D., Kozlov, F., Parkhimovich, O., & Zelenina, M. (2013). Development of an Ontology-Based E-Learning System. *4th International Conference on Knowledge Engineering and Semantic Web, KESW 2013,* 394, 273-280. Doi:10.1007/978-3-642-41360-5_23

Mouromtsev, D. I., Pshenichny, C. A., & Yakovlev, A. V. (2014). Semantic and structural delineation of market scenarios by the event bush method. *Decision Support Systems and Electronic Commerce*, *57*, 34–41. doi:10.1016/j.dss.2013.07.008

Murdock, J., Buckner, C., & Allen, C. (2010). Evaluating Dynamic Ontologies. In *FCommunications in Computer and Information Science: 272 Knowledge Discovery, Knowledge Engineering and Knowledge Management. Second International Joint Conference, IC3K 2010, Valencia, Spain, October 25-28, 2010, Revised Selected Papers*. Springer.

Noy, N., Chugh, A., Liu, W., & Musen, M. (2006). A Framework for Ontology Evolution in Collaborative Environments. *Proceedings of ISWC*, 544-558. doi:10.1007/11926078_39

Petri, C. A. (1962). *Kommunication mit Automaten* (PhD dissertation). University of Bonn. (in German)

Petri, C. A. (1973). Concepts of Net Theory. *Proc. of Symposium and Summer School, High Tatras*, 137-146.

Petri, C. A. (1977). General Net Theory. Computing System Design. In B. Shaw (Ed.), *Proc. of the Joint IBM University of Newcastle upon Tyne Seminar*. University of Newcastle upon Tyne.

Pshenichny, C., Carniel, R., & Diviacco, P. (2013). Engineering of Dynamic Knowledge in Exact Sciences: First Results of Application of the Event Bush Method in Physics. In D. Mouromtsev, C. Pchenichniy, & D. Ignatov (Eds.), *MSEPS 2013 – Modeling States, Events, Processes and Scenarios; workshop proceedings*. Retrieved from http://iccs2013.hbcse.tifr.res.in/workshops/copy2_of_text.pdf

Pshenichny, C. A., & Kanzheleva, O. M. (2011). Theoretical foundations of the event bush method. Societal Challenges and Geoinformatics, 482, 139-165. doi:10.1130/2011.2482(12)

Pshenichny, C. A., & Khrabrykh, Z. V. (2002). Knowledge Base of Formation of Subaerial Eruption Unit. In S. Leroy & I. Stuart (Eds.), *Environmental Catastrophes and Recovery in the Holocene (Abstracts)*. London: Brunel University. Retrieved from http://atlas-conferences.com/cgi-bin/abstract/caiq-22

Pshenichny, C. A., & Mouromtsev, D. I. (2013). Representation of the Event Bush Approach in Terms of Directed Hypergraphs. LNAI, 7735, 287–298. doi:10.1007/978-3-642-35786-2_22

Pshenichny, C. A., & Mouromtsev, D. I. (2015). Grammar of Dynamic Knowledge for Collaborative Knowledge Engineering and Representation. In Collaborative Knowledge in Scientific Research Networks. IGI Global. doi:10.4018/978-1-4666-6567-5.ch016

Pshenichny, C. A., Nikolenko, S. I., Carniel, R., Sobissevitch, A. L., Vaganov, P. A., Khrabrykh, Z. V., . . . Gusev, E. A. (2008). The event bush as a potential complex methodology of conceptual modelling in the geosciences. *Proceedings, iEMSs – International Congress on Environmental Modelling and Software*, 2, 900-912.

Pshenichny, C. A., Nikolenko, S. I., Carniel, R., Vaganov, P. A., Khrabrykh, Z. V., Moukhachov, V. P., & Rezyapkin, A. A. et al. (2009). The Event Bush as a Semantic-based Numerical Approach to Natural Hazard Assessment (Exemplified by Volcanology). *Computers & Geosciences*, *35*(5), 1017–1034. doi:10.1016/j.cageo.2008.01.009

Quintana-Amate, S., Bermell-Garcia, P., & Tiwari, A. (2015). Transforming expertise into Knowledge-Based Engineering tools: A survey of knowledge sourcing in the context of engineering design. *Knowledge-Based Systems*, *84*, 89–97. doi:10.1016/j.knosys.2015.04.002

Ribarić, S., Pavešić, N., & Zadrija, V. (2009). Intersection Search for a Fuzzy Petri Net-Based Knowledge Representation Scheme. Lecture Notes in Computer Science, 5711, 1-10.

Rogalchuk, V., & Solomin, K. (2015). Perspectives of Use of Petri Nets in Collaborative Research. In Collaborative Knowledge in Scientific Research Networks. IGI Global. doi:10.4018/978-1-4666-6567-5.ch018

Serio Ltd. (2014). *Knowledgebase - Content is King*. Retrieved from http://www.seriosoft.com/blog/knowledgebase-content-king

Solomin, K., & Mandrik, M. (2015). Constructing historical knowledge through graphic boundary objects. In Collaborative Knowledge in Scientific Research Networks. IGI Global. doi:10.4018/978-1-4666-6567-5.ch017

Sowa, J. F. (1987). Semantic Networks. In Encyclopedia of Artificial Intelligence (2nd ed.). Academic Press.

Sowa, J. F. (Ed.). (1992). Knowledge-Based Systems, 5(3).

Sowa, J. F. (2000). *Knowledge Representation: Logical, Philosophical, and Computational Foundations*. Pacific Grove, CA: Brooks Cole Publishing Co.

Sowa, J. F. (2006). *Semantic Networks*. Retrieved from http://www.jfsowa.com/pubs/semnet.htm

Stojanovic, L. (2004). *Methods and Tools for Ontology Evolution* (PhD thesis). University of Karlsruhe.

Syairudin, B., Trisunarno, L., & Gunarta, I. K. (2015). Strategy for Achieving the Millennium Development Goals 7-10 through Knowledge-Based Sustainable Design Approach. *Procedia: Social and Behavioral Sciences*, *169*, 61–68. doi:10.1016/j.sbspro.2015.01.286

Syria chemical weapons set for destruction, Operation to destroy chemical weapons in final stages with Danish ship to transfer toxins to an American one. (2014). Retrieved from http://www.aljazeera.com/video/middleeast/2014/07/syria-chemical-weapons-set-destruction-20147225816561224.html

Syrian Civil War. (2014). In *Encyclopædia Britannica*. Retrieved from http://global.britannica.com/EBchecked/topic/1781371/Syrian-Civil-War

Teece, D. J., Pisano, G., & Shuen, A. (1997). Dynamic Capabilities and Strategic Management. *Strategic Management Journal*, *18*(7), 509–533. doi:10.1002/(SICI)1097-0266(199708)18:7<509::AID-SMJ882>3.0.CO;2-Z

The US disables diplomacy to escalate Syrian civil war, degrade Assad regime. (2014). Retrieved from http://rt.com/op-edge/us-diplomacy-escalates-syrian-war-766/

Vistelius, A. B. (1992). *Principles of mathematical geology*. Dordrecht: Kluwer Academic Publishers. doi:10.1007/978-94-011-2934-3

Wang, W.-M., Peng, X., Zhu, G., Hu, J., & Peng, Y.-H. (2014). Dynamic representation of fuzzy knowledge based on fuzzy Petri net and genetic-particle swarm optimization. *Expert Systems with Applications*, *41*(4), 1369–1376. doi:10.1016/j.eswa.2013.08.034

Xiao-zhong, LI, (2003). A Kind of Fuzzy Petri Nets and its Dynamic Characteristics. *Acta Simulata Systematica Sinica*, *2003*(S1).

Yeung, D. S., & Tsang, E. C. C. (1994). Fuzzy knowledge representation and reasoning using Petri nets. *Expert Systems with Applications*, *7*(2), 281–289. doi:10.1016/0957-4174(94)90044-2

Zablith, F. (2008). Dynamic ontology evolution. In *Proceedings of the International Semantic Web Conference (ISWC)*. Karlsruhe, Germany: Doctoral Consortium. Retrieved from http://fouad.zablith.org/docs/Dynamic_Ontology_Evolution_ISWC_2008_Fouad_Zablith.pdf

Chapter 14
The Event Bush Method in the Light of Typed Graphs Illustrated by Common Sense Reasoning

Uwe Wolter
University of Bergen, Norway

Olga Korableva
Saint Petersburg State University, Russia

Nikita Solovyov
ITMO University, Russia

ABSTRACT

The event bush method being an efficient tool for representation and engineering of dynamic knowledge still lacks a strict mathematical foundation. Many of the syntactic properties of event bushes, however, seem compliant with directed graphs and can be described by typed graphs (i.e., by homomorphisms between directed graphs). This chapter explores an opportunity to formalize the syntactic structure of event bushes by means of typed graphs and shows useful implications of this approach for knowledge engineering and representation.

INTRODUCTION

Problem Area

Traditional knowledge representation tools (e.g., ontologies and conceptual graphs) operate with objects considered either as classes (i.e., types: concept types or relation types) or as individuals (Martin, 2002). This has proven to be efficient for a wide range of tasks like building data models and search algorithms or description of those domains of knowledge in which things, their properties and relations are con-

DOI: 10.4018/978-1-5225-5261-1.ch014

sidered fixed – e.g., production units of a factory or biological species. Nevertheless, there are tasks in natural-scientific and technical domains that require consideration of changing properties and relations and claim for formal description of scenarios and patterns of evolution. The knowledge relating to such tasks has been termed dynamic, and the tasks proper, dynamic environments (Pshenichny, 2011). Correspondingly, the knowledge about fixed relations between objects is termed static.

A number of methods – influence diagrams, Petri nets, event/probability trees, event bushes, Bayesian belief networks, causal loops, activity diagrams, sequence diagrams and other approaches (Pshenichny, 2014) addresses dynamic knowledge. Some of these methods can be used in a variety of ways, from a purely qualitative conceptualization to strict mathematical models (compare, e.g., event trees and probability trees). It may seem that these methods can operate with virtually everything – quantitative data, the objects of static knowledge, mathematical variables, images, and so forth. Nevertheless, as it was argued in (Pshenichny & Mouromtsev, 2015), in case of dynamic knowledge all these can be thought as designations, or abbreviations, of one specific kind of entity, which represents a relation between the objects in form of a statement. Such entities are defined as events (Pshenichny, 2014). An event is a statement of general form $S - P_1, P_2, \ldots, P_n$, where S is subject, and $P_1, P_2, \ldots, P_n$ are predicates, with or without negation. These subject and predicate(s) are the objects of static knowledge. For the event, the following restrictions are put forth, (i) subject may be only one in a statement, (ii) subject may not be negated, (iii) all predicates may not be negated in a statement, at least one must be without negation (e.g., a statement $S - {\sim}P$, ~ being the negation, is not allowed), (iv) presence of a predicate and its negation in one event is not admitted (Pshenichny, 2014). From event to event, there may change subject, or negation of predicates, or nothing. If the subject changes, predicate may change in whatever way unless violate the above restrictions, or remain the same.

For instance:

“Grass (S_1) bursts forth (P_1) and is not mowed (${\sim}P_2$) – Grass (S_1) bursts forth (P_1) and is not mowed (${\sim}P_2$)”
(nothing changes);
“Grass (S_1) bursts forth (P_1) and is not mowed (${\sim}P_2$) – Grass (S_1) bursts forth (P_1) and is mowed (P_2)”
(change of negation of predicate);
“Grass (S_1) bursts forth (P_1) and is mowed (P_2) – Hay (S_2) is mowed (P_2) and is not burnt (${\sim}P_3$)”
(change of subject);
“Hay (S_2) is mowed (P_2) and is not burnt (${\sim}P_3$) – Hay (S_2) is mowed (P_2) and is burnt (P_3)”
(change of negation of predicate);
“Hay (S_2) is mowed (P_2) and is burnt (P_3) – Ash (S_3) is blown away (P_4)”
(change of subject).

Thus, the difference between the two types of knowledge appears essentially semantic and is rooted in the meaning of its “building blocks” – objects in one case and events in the other. Therefore, naturally, it was semantics that was mainly concerned in the abovementioned discussion on types of knowledge (static versus dynamic), while the syntax of dynamic knowledge stayed in the shade. At the same time, in the earlier discourse on the syntax of visual languages used for knowledge representation this difference was not paid attention, and the existing reviews focus mainly on the tools for static knowledge but also mention some methods that are considered now as addressing essentially dynamic knowledge (e.g., Petri nets) without accentuating any specificity of these (Kremer, 1998).

Meanwhile, different methods of dynamic knowledge engineering bring different syntax, including the rules of arrangement of statements and linking them. In general, syntax reflects the vision of the world that underlies the method in question. Therefore, understanding the syntax suggested by different methods of dynamic knowledge engineering and the relation of this syntax to the event-based semantics are essential for understanding the results these methods may bring and, hence, for the choice of the method for a particular task.

Related Work

Graph theory (Tutte, 1998) has been commonly used to visualize both static (see, e.g., Gavrilova et al., 2013) and dynamic (Pshenichny, 2014) knowledge representation tools. To represent syntax, type graphs seem to be the most appropriate option, because their vertices mean not individual entities but types of entities (see below) and therefore such graphs can efficiently mirror general rules – i.e., the rules of syntax.

However, in some methods (e.g., event trees or Bayesian/implicational networks), the syntax of resulting structures is rather simple (mainly the "parent-child" relation) and readily formalized by quite simple graphs (though exceptions exist here as well – see, e.g., Martensen & Butler, 1987).

Still, there are more complicated cases like Petri nets; here the syntax includes, in the most basic, Abstract Petri Net Notation (Bause et al., 1994), places, transitions arcs and firing rules.

Finally, the most detailed, and yet not formalized, syntax is offered by the method of event bush designed specifically for representation of dynamic knowledge. This method was first suggested by (Pshenichny & Khrabrykh, 2002) and later developed by (Pshenichny et al., 2008, 2009) to reveal the behavior of information in dynamic environments of directed alternative changes (Pshenichny & Kanzheleva, 2011). It is gaining increasing importance in a variety of fields, from the geosciences (Behncke & Pshenichny, 2009) to marketing (Mouromtsev et al., 2013) and history (Solomin & Mandrik, 2015). At present this method is viewed as the core methodology of dynamic knowledge engineering and the testbed for development of the unified grammar of dynamic knowledge (Pshenichny & Mouromtsev, 2015).

Pshenichny & Mouromtsev (2013) attempted to formalize the event bush method by directed hypergraphs. However, their results concern again mainly the semantics of the nodes (events) and of particular connectives of the bush, while the basic syntax remained uncovered. The purpose of the present work is *to give a description of the fundamental syntactic structure of event bushes by means of directed graphs and homomorphisms between them (typed graphs).*

For this, after a necessary synopsis of the event bush method (Section 2), in this paper we will:

- Perform the formalization of a singular bush by typed graphs (Section 3);
- Investigate the multi-bush structures of different types and their formalization by typed graphs (Section 4);
- Discuss the results in the context of visual language research (Section 5).

FUNDAMENTALS OF THE EVENT BUSH METHOD

Basic Syntax of the Event Bush

An environment of directed alternative changes is described by the event bush as a set of events of four types arranged in certain order (henceforth called multiflow structure – see Figure 1) and connected by special connectives (Pshenichny, Kanzheleva, 2011):

1. Primary internal events. These are primary, non-unique inputs representing the "passive conditions". These would determine any further course of events ("happenings");
2. Primary external events meaning the "invading agents". They may affect basic inputs or influence their further, indirect manifestations, thus "shaping up" different "happenings";
3. Secondary events that result from primary internal inputs with or without the contribution of primary external ones – the "happenings" proper formulated in a strict concise way indicating their core features determined by the causes, following the principle "one more cause – one change of event";
4. Tertiary events that denote end results, or products, generated either by primary internal or by secondary events, with or without primary external ones. Tertiary events document the completed "happenings".

In the event bush, an event is uniquely characterized by its subject, the set of predicates (including possible negations), position (type *ia*, *ib*, *ii* or *iii*) and, in some cases, generality. Two primary events are not allowed to have the same subject.

Other pieces of information (even not necessarily verbal) can be associated with events and/or connectives (Pshenichny et al., 2009), but the presence of events of types *(i)* - *(iii)* is mandatory.

Figure 1. Basic syntax and the multi-flow structure of the event bush (Pshenichny & Kanzheleva 2011)

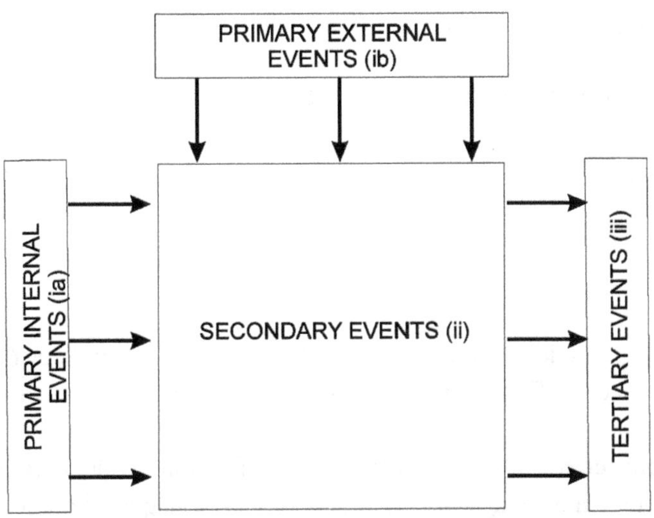

The following pairwise relations between types *(i)* - *(iii)* are set in the event bush:

- Not a single event may lead to *(ia)* or *(ib)*;
- Event *(ia)* without an event *(ib)* or *(ii)* may lead only to *(ii)* or one event *(iii)*;
- Event *(ia)* may not lead to any event together with another *(ia)*;
- Event *(ib)* may lead only to an event *(ii)* and only together with *(ia)* or another *(ii)*;
- Event *(iii)* may not lead to any other event.

These relations are enforced by the connectives of the event bush.

Connectives of the Event Bush

For simplicity, we will consider in this subsection only the events having predicates (one or more) that relate directly to the subject (i.e., no "predicate of predicate" expressions). We will study the other interesting cases later in connection with class-subclass relations in the event bush and will report them elsewhere.

Based on the observations of behavior of information in the environments of directed alternative changes, the following connectives are introduced:

1. **Flux** connective (Figure 2, a) describes one event (E_1) producing another (E_2):

E_1 *Flux* E_2

e.g., "Birds fly, sing and are not scared" (E_1) *Flux* "Birds fly, sing and are not scared" (E_2).

2. **Influx** connective (Figure 2, b) describes two events (E_1, E_2) producing another (E_3), but playing different roles (this will be described below):

E_1, E_2 *Influx* E_3

e.g., "Birds fly, sing and are not scared" (E_1), "Thunder rattles" (E_2) *Influx* "Birds fly, sing and are scared" (E_3).
The event E_4 is not involved in the influx but, being connected by flux with E_1, is a precondition for influx.

3. **Furcation** connective (Figure 2, c) describes production of one of a multitude of events (E_2, …, E_n) by one (E_1):

E_1 *Furcation* E_2, …, E_n

e.g., "Birds fly, sing and do not look for corn (: find corn/do not find corn)" (E_1) *Furcation* "Birds fly, sing and look for corn: find corn" (E_2), "Birds fly, sing and look for corn: do not find corn" (E_3).

Importantly, this connective does not signify any classification or specification of events. It literally means that, first, one predicate changes to negation from left to right ("do not look for corn – look for corn") or, vice versa, negation to assertion, then, the predicate on the right (with or without negation) must be specified and, third, unspecified predicates make no sense in the modeled environment. In the above example, if birds are not looking for corn, this is enough information we need, but if they are looking for it, this is not enough for us to know only this, and the predicate "look for corn" is not admitted as is. This is the reason why furcation is fundamentally different from the case of nested knowledge considered below. In this case, both unspecified predicate and its specifications are considered sense-making. Meanwhile, the issue of classification or generalization of events based on their subject-predicate structure is quite complicated and will be discussed below.

4. **Conflux** connective (Figure 2, d) describes production of one event (E_n) by multiple events (E_1, E_2, …, E_{n-1}):

E_1, E_2, …, E_{n-1} *Conflux* E_n

e.g., "Birds fly in the sky and sing in the sky" (E_1), "Crocodiles fly in the sky and sing in the sky" (E_2) *Conflux* "Sky is occupied by creatures who fly and sing" (E_3).

Similar to furcation, conflux is not a classification of events. Conflux does not imply their generalization. Moreover, unlike the furcation, it includes modi in which not only predicates but subjects also change (Pshenichny & Kanzheleva, 2011). However, one should not take conflux for the mirror of furcation: in the conflux, the events on the left must coexist (in furcation, the events on the right may not) and, in addition, conflux may reserve an opportunity for the causal events to exist further after producing the conflowing event, while furcation, when effected, strictly prohibits further existence of the causal event.

From event to event, the following components may change:

- Subject S;
- Predicates P;
- Type (position) of event; or
- Generality of subject.

We postulate that type of event and generality of subject may change, if, and only if, the subject and predicates do not change, and vice versa, if both subjects and predicates are the same in two connected events, then either their type or generality of subject must change. If these conditions are omitted, the formalism is not an event bush.

All the changes listed above are governed by the connectives. For detailed consideration of connectives and their modi the reader is referred to (Pshenichny & Kanzheleva, 2011), and for their formalization by means of directed hypergraphs, to (Pshenichny & Mouromtsev, 2013).

An example of an event bush is given in Figure 3. It illustrates the multiflow structure of the event bush as well as the semantics of flux, influx and furcation connectives.

A graphic notation was adopted for the event bush. It is based on the multiflow structure (Figure 1) placing events *(ia)* on the left, *(ib)*, on the top, *(ii)*, in the center, and *(iii)*, on the right. Events are plotted as rectangular boxes. The connectives are marked, flux, as ordinary arrow from cause to effect (Figure 2, a), influx, as an arrow from the modifier to the effect with a "right turn" from the modified event (Figure 2, b), conflux, as "double right turn" (Figure 2, c), and furcation, as a circle with ramification to effects (Figure 2, d). This visualizes the bush and makes it possible to "drive" along the flows following the "road painting" of connectives.

Figure 2. Connectives of the event bush: flux (a), influx (b), furcation (c) and conflux (d), from (Pshenichny & Kanzheleva 2011)

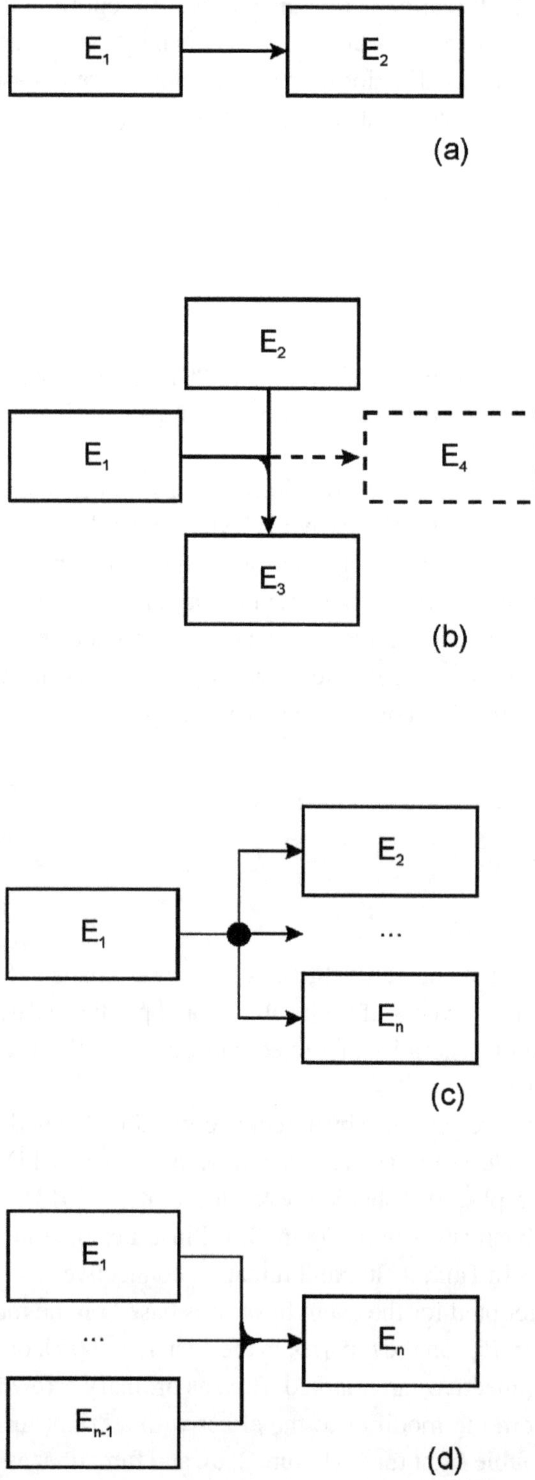

Figure 3. An example of event bush. Modeled is the evolution of hard rock and heavy metal rock bands in the 1970s-1990s (modified from Mouromtsev et al., 2014)

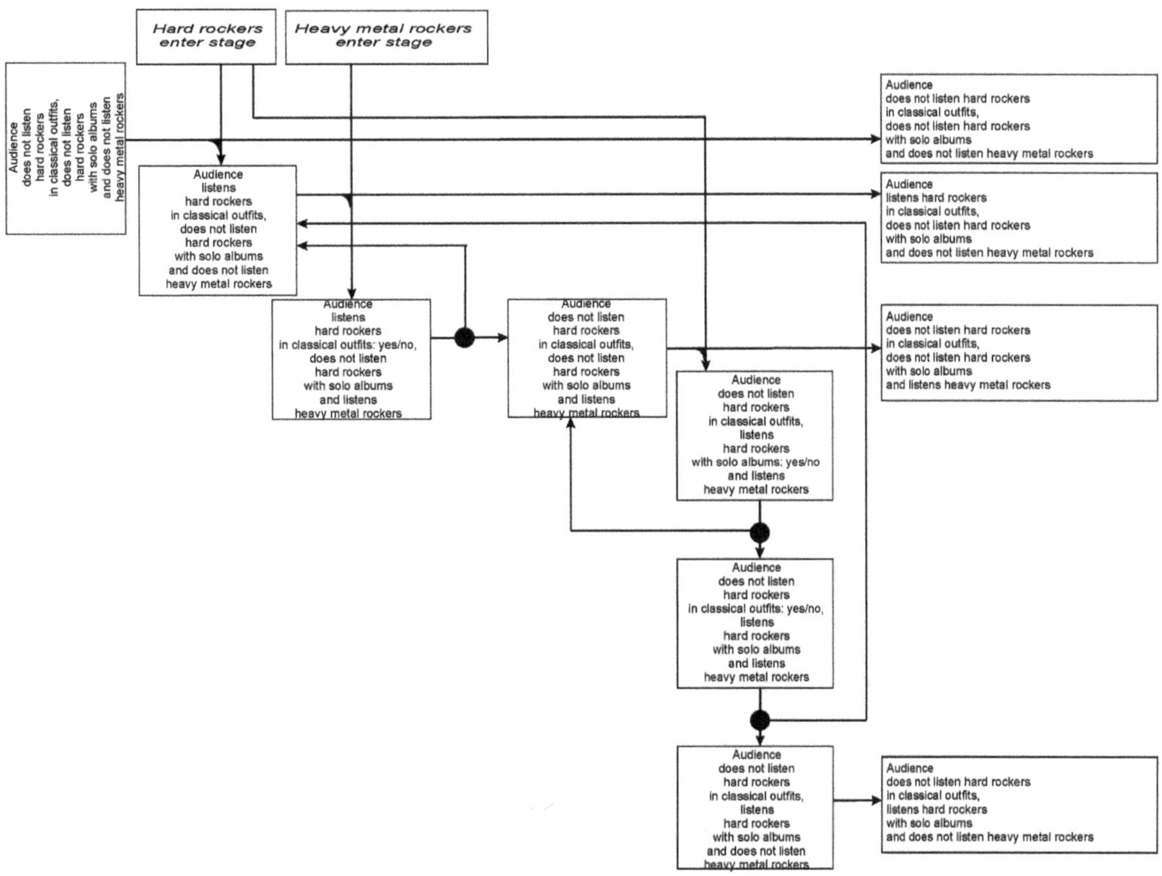

The conceptual framework put forth above brings us to a definition of the event bush.

The event bush:

- Is a method of dynamic knowledge engineering;
- Aims at construction of scenarios in environments of directed alternative change;
- Is based on the multiflow structure; and
- Must include the connectives of flux and influx and may include the connectives of conflux and furcation.

This definition, still being far from formal, fixes the event bush in a state in which it can be, and has been, applied in various knowledge representation tasks.

Concomitant Definitions, Rules and Corollaries

The corollaries and ultimate consequences of the definitions of event bush and the connectives need to be explored further to optimize their use and avoid possible ambiguity. Though, some observations can be made right now. For instance, events *(ii)* can form cyclic patterns but either with an entrance from *(ia)* and exit to another *(ii)*, or with an entrance from another *(ii)* and exit to yet another *(ii)* or *(iii)*. However, such patterns are not considered in the present work. Also, for the obvious reason that the resulting events of furcation are incompatible, neither they nor their consequences (provided they have no other causes) may be united by conflux.

Another point stated in the definitions of all the connectives can be summarized as follows. If a predicate *P* appears for the first time in relation to subject *S* in event *E*, this automatically means that in all events throughout the bush, which have the same subject but do not follow directly or indirectly from *E*, there must be the negation of *P*, if not explicitly put otherwise.

Some more definitions would enable reasoning and inference in the event bush framework.

Change in the event bush is a correctly built expression including one (only one!) connective of any type. General formulae for these expressions were listed in the subsection 2.2.

Change has one causal part (that includes the events, one or more, left of the connective) and one effect part (the events to the right of the connective).

Flow is a finite totally ordered set (Schröder, 2003) of changes, in which the left part of the first change is an *(ia)* event, the right part of the last change is a *(iii)* event, and the right part of every preceding change is the left part of the following change.

Flow is always followed from *(ia)* to *(iii)*. If there is a change with influx in it, the modifier of this influx (*(ib)* or *(ii)*) is included, but not the events that caused it, unless there are successive influx changes. There must be no events in a bush not included into a single flow.

Ramifying (furcating or influxing), confluxing, or coming to common effect by different fluxes, flows form a *multiflow*, which is, by definition, the structure of the event bush.

FORMALIZATION OF SINGULAR EVENT BUSHES BY TYPED GRAPHS

In this section we are going to formalize the syntax of event bushes by means of graphs and graph homomorphisms. Not all relevant aspects of event bushes will be covered but we address the most principal features:

- An event bush comprises events and connectives;
- Events and connectives are related by directed edges;
- An event bush has a multiflow structure as it is visualized in Figure 1.

The discussion in the last section shows that an event bush can be described, in first approximation, as a directed graph with vertices that are either events or connectives. The graph is simple in the sense that there is at most one edge between two vertices in the graph. There must be no vertices (events or connectives) in a bush unconnected by edges.

Definition 1. (Directed graph (Ehrig et al., 2006)). A (simple) directed graph $G = (G_V, G_E)$ consists of a set G_V of vertices and a set $G_E \subseteq G_V \times G_V$ of edges where each edge is an ordered pair of vertices $(x,y) \in G_V \times G_V$. We call x the source of the edge (x,y) and y the target of (x,y), respectively.

Figure 4. Representation of connectives of the event bush as directed graphs

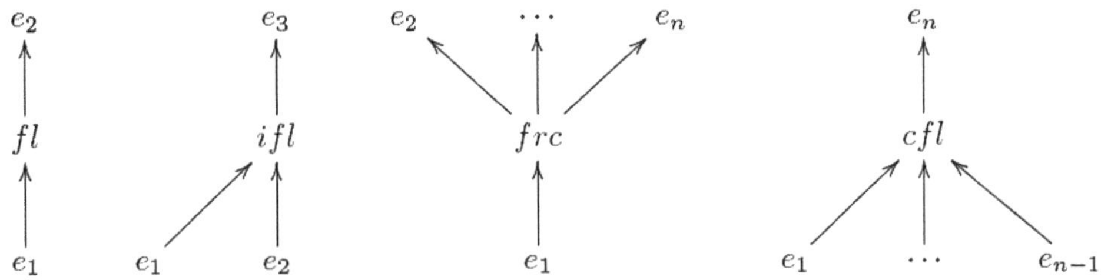

Figure 5. A type graph B for event bushes mirroring that event bushes have two types of nodes and edges only between different types of nodes

B: *Event* ⇄ *Connective*

The presence of an edge *(x,y)* in a graph *G* is usually visualized by an arrow $x \to y$. Thus, the connectives of the event bush can be presented by the four graphs in Figure 4.

In a graph that mirrors an event bush only two types of vertices are present, event vertices and connective vertices. Moreover, edges go only from an event vertex to a connective vertex or the other way around. There are no edges between event vertices and no edges between connective vertices, respectively. These syntactic requirements can be represented by graph *B* (where *B* stands for bipartite) as shown in Figure 5.

We call the graph *B* a "type graph" (Ehrig et al., 2006) since its vertices are meant to represent the two types of vertices present in event bushes.

The information what vertices in an event bush represent events or connectives, respectively, can be described by a graph homomorphism into the type graph *B*. Graph homomorphisms are defined as follows.

Definition 2. (Graph homomorphism (Ehrig et al., 2006)). A graph homomorphism $\phi: G \to H$ between two directed graphs *G* and *H* is given by a function $\phi_V: G_V \to H_V$ from the vertex set of *G* to the vertex set of *H* such that for each edge $(x, y) \in G_E$ there is an edge $(\phi_V(x), \phi_V(y)) \in H_E$.

Similar to functions between sets, graph homomorphisms allow different interpretations and may serve quite different purposes. One possible interpretation is to consider the target graph *H* of a graph homomorphism $\phi: G \to H$ as a "type graph" and to refer to ϕ, in such a way, as a "typing morphism" since ϕ assigns to each vertex or edge in *G* a vertex (type) or edge (type), respectively, in *H*. This interpretation can be formulated equivalently by the statement that *G* is a typed graph, namely a graph typed over *H* via the typing morphism ϕ (Ehrig et al., 2006).

Relying on this interpretation, we can say now that an event bush is given by an underlying directed graph $\mathcal{E}$ typed over graph *B* via a typing morphism $\tau_{\mathcal{E}}: \mathcal{E} \to B$. The events are all vertices in $\mathcal{E}$ that are mapped by $\tau_{\mathcal{E}}$ to the type *Event*, and the connectives are all the vertices mapped to the type *Connective*, respectively.

Figure 6. Graph S, typed over graph B, reflecting the multiflow structure of event bushes (see comments in the text)

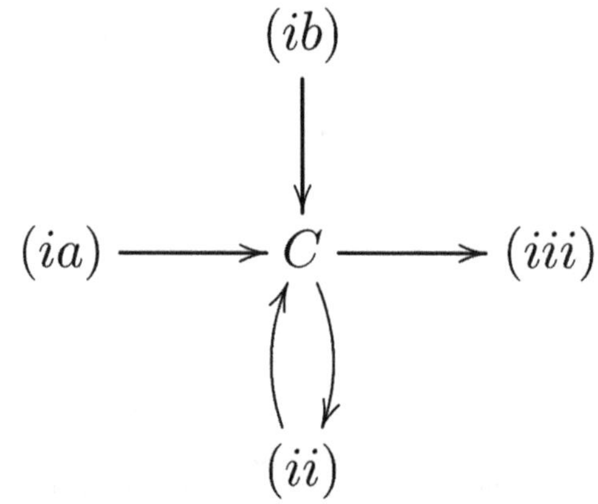

Note that the homomorphism requirement that $\tau_{\mathcal{E}}$ has to preserve edges implies that there cannot be an edge in $\mathcal{E}$ between events or connectives, respectively. If there would be such an edge in $\mathcal{E}$ then this edge would be mapped to an edge (*Event,Event*) or (*Connective,Connective*) in *B*. There are, however, no such loops in *B*.

From now on we consider also each of the four graphs in Figure 4 as a graph typed over *B* where all the vertices e_i are typed as *Event* and the vertices *fl*, *ifl*, *frc*, or *cfl*, respectively, are typed as *Connective*.

Finally, the assumed multiflow structure of event bushes, as visualized in Figure 1, can be also naturally formalized by a graph. There are four kinds of event vertices occupying four boxes, or fields, in Figure 1, and all the connective vertices can be also placed in the central box that hosts secondary events. This "multiflow structure" can be formally precisely reflected by a graph *S* (where *S* stands for multiflow structure) as shown in Figure 6.

S is, in the obvious naturally way, typed over *B* via the typing morphism τ_S:$S \rightarrow B$ with $\tau_S(C)=Connective$ and $\tau_S(ia)=\tau_S(ib)=\tau_S(ii)=\tau_S(iii)=Event$.

The assumption that the events in an event bush are classified as primary internal *(ia)*, primary external *(ib)*, secondary *(ii)*, and tertiary *(iii)* can be formalized now by assuming a graph homomorphism $\sigma_{\mathcal{E}}$: $\mathcal{E} \rightarrow S$ representing exactly this classification (σ stands for "structure"). Obviously, the connectives in the event bush should be mapped by $\sigma_{\mathcal{E}}$ to the type *C* and any event should be mapped to either of the types *(ia)*, *(ib)*, *(ii)*, or *(iii)*. Formally, this means that the composition $\sigma_{\mathcal{E}};\tau_S:\mathcal{E} \rightarrow B$ of the two graph homomorphisms $\sigma_{\mathcal{E}}$: $\mathcal{E} \rightarrow S$ and $\tau_S:S \rightarrow B$ should coincide with the typing morphism $\tau_{\mathcal{E}}:\mathcal{E} \rightarrow B$.

This consideration rounds up our analysis of the three main syntactic features of event bushes and we close this section by summarizing the outcome of this analysis.

Syntax of the event bush is represented as follows. An event bush is given by:

1. A directed graph $\mathcal{E}$ that is;
2. Typed over the type graph *B* via a typing morphism $\tau_{\mathcal{E}}:\mathcal{E} \rightarrow B$; and
3. Has a multiflow structure given by a graph homomorphism $\sigma_{\mathcal{E}}$: $\mathcal{E} \rightarrow S$ that is compatible with typing, i.e., $\sigma_{\mathcal{E}};\tau_S=\tau_{\mathcal{E}}$ (Figure 7).

MULTI-BUSH ENSEMBLES FOR INFORMATION MODELING AND KNOWLEDGE REPRESENTATION

Individual event bushes may make use in information modeling of many fields of knowledge; however, it is well known that the best performance of most of information and knowledge-based systems can be achieved by their integration (compare the use of individual computer against the Internet). Regarding the event bushes, one may have in mind an integration either "on plane" (connecting the event bushes into chains or networks) or "in depth" (bursting the information stored in one bush and thus making several bushes out of one, or, conversely, merging information from several bushes into one). In some sense, this can be considered analogous to the types of links on the Internet – the links to other websites and those to other documents within the same website. The complexity of such multi-bush ensembles increases manifold; however, so does also their importance for the information modeling and knowledge representation.

Types of Multi-Bush Structures and Associated Cognitive Problems

Cross-Domain Reasoning

Human reasoning is a continuous process and, despite that every task and every modeled environment has their beginning and end, usually it is very important, and often required, to track the next task to which the considered one contributes or in what new environment(s) the results of evolution of given one may occur and be relevant.

To do this, the event bushes should be combined into chains or nets (Figure 8, a-c). While every singular bush is assumed to represent knowledge and reasoning in one domain of knowledge (or information domain, by (Pshenichny & Kanzheleva, 2011), a succession of bushes, correspondingly, must represent a cross-domain reasoning.

The simplest case (Figure 8a) looks trivial and unlikely may have wide application. More complicated cases of networks of event bushes (Figure 8, b, c) are rather hard to comprehend but, in our opinion, they resemble real structures of human knowledge. The difference between the cases (b) and (c) is purely operational and appears only assuming the structure of knowledge is growing. Otherwise, if we

Figure 7. Compatibility of multiflow structure and typing

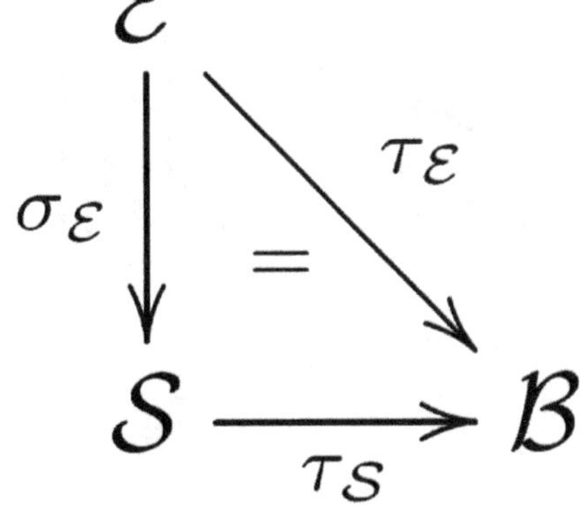

Figure 8. Planar structures of event bushes: (a) trivial chain, in which tertiary events of the previous bush and primary internal events of the next bush are the same; (b) and (c): scheme of complex network of event bushes in which any of (ia)-(iii) events of any number of preceding bushes may become any of (ia)-(iii) events of the given bush (b) and any of (ia)-(iii) events of the given bush may become any of (ia)-(iii) events of any number of the next bushes (c)

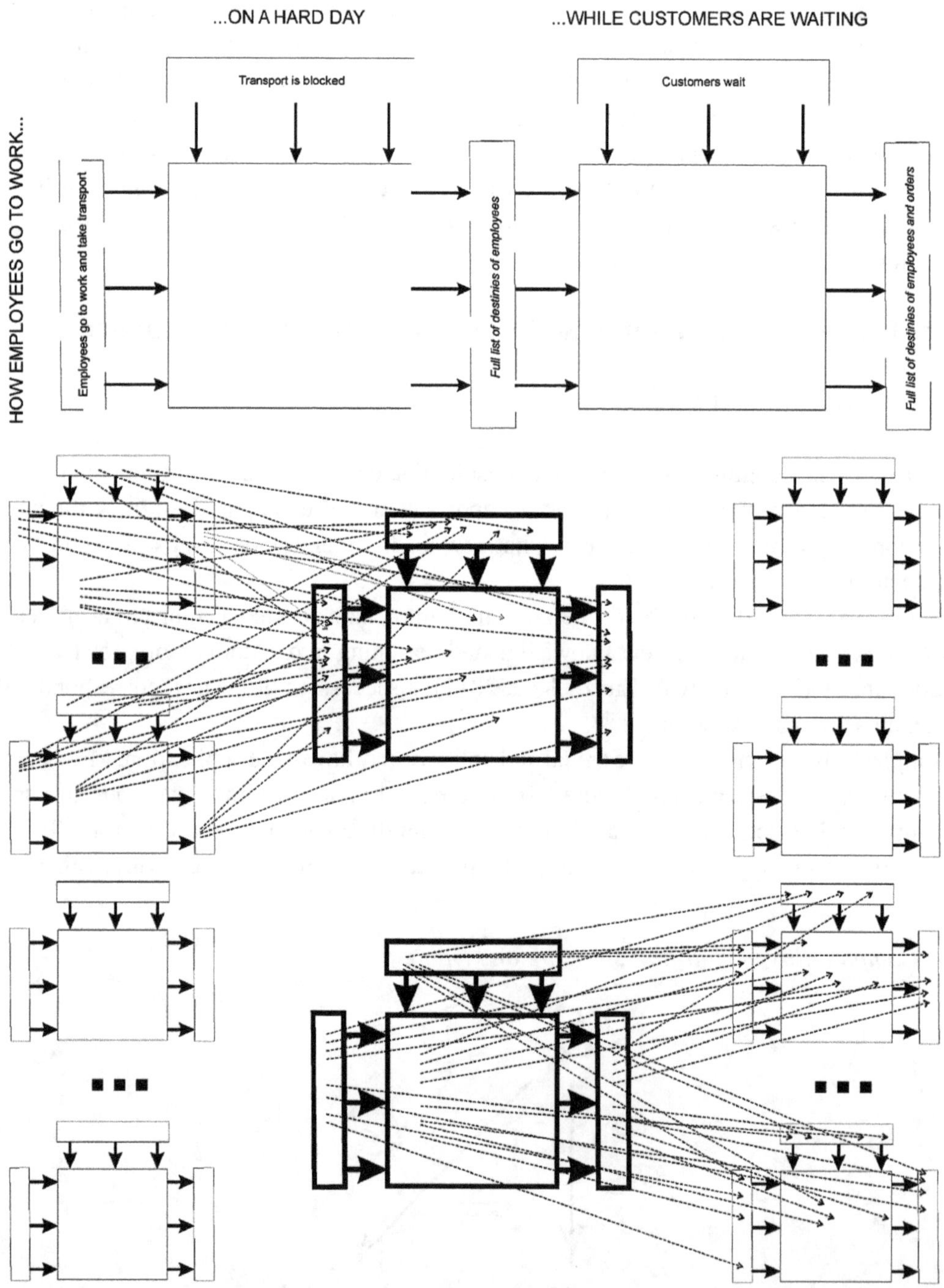

investigate an existing structure, given to us entirely, it makes no difference for us what is "preceding" and what is "following".

Still, operationally, case (b) reflects such mental activities as explanation, reconstruction or comparison, while in the other case (c), forecasting, constructing, imagination, and the like. This can be parallelized, respectively, with the inverse and direct problems in exact and technical sciences. However, here the list of disciplines is much wider: from psychology and fiction to physics and technical disciplines.

For instance, in economy, (Korableva & Litun, 2014) noticed that an unexpected type of innovation had been developed in Russian pharmaceutical industry – the so-called disruptive innovation. Disruptive innovations are those that aim at creation of new markets, which are competition-free and have high rate of return (Christensen, 2014). For a number of reasons, this rarely happens in the pharmaceutics. Still, a new segment of market of pharmaceutical technologies, that of immediately applicable microbial solutions, appeared in Russia (Figure 9, see the larger event bush in the center). Examination of this phenomenon revealed that this happened due to activities of one company who conducted research to improve microbial technologies (see smaller bush on top of Figure 9). The primary external event in the following bush placed in the center of the figure, "Microbial production exists and is improved to allow immediate application", appears to be one of the tertiary events of the preceding bush (put on top) that describes the activity of this company in the field of microbial technologies.

This pathway of thinking is an example of the case (b) in Figure 8 and represents an inverse problem sensu lato, in which the effect is known and the cause is in question.

In contrast, the direct problem depicted in Figure 8c is a passage from causes to possible effects. This passage requires one to unbind imagination, and thus it may lead far away to complicated, intriguing,

Figure 9. Example of an ensemble of two event bushes, in which one tertiary event of the former bush is primary external of the latter. See comments in the text.

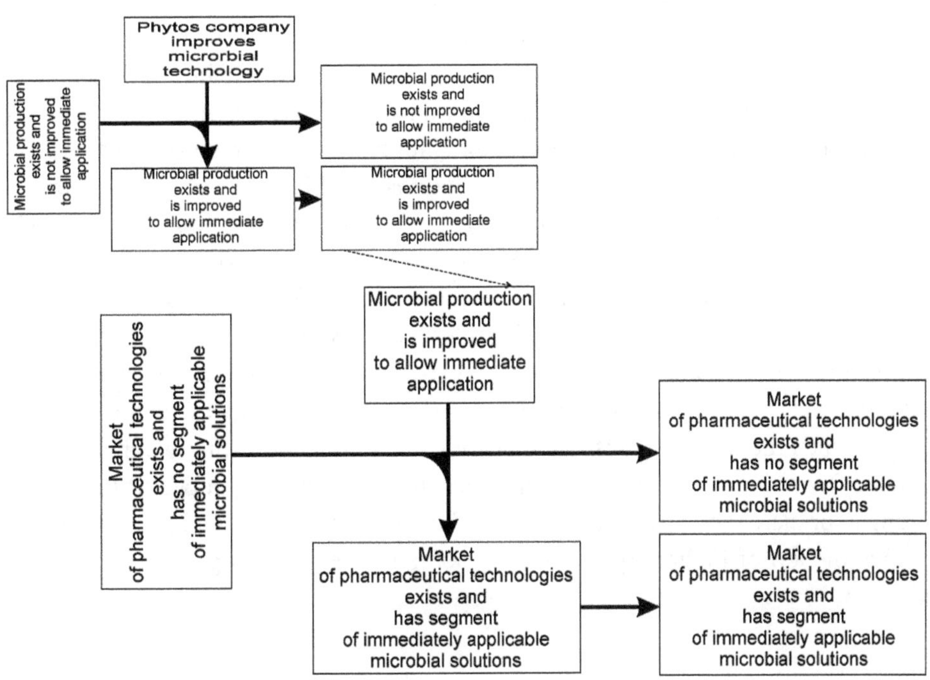

unlikely and even unrealistic scenarios arising from "projecting" of events from an existing domain onto new, real or imaginary, circumstances.

Naturally, in all possible applications, from fiction to technical design, there is a need to constrain and rationalize this activity. Classical logic, as was construed above, cannot provide enough assistance in this due to its static nature.

Since the 1980s a research is conducted aiming at formalization of the process of storytelling (including interactive storytelling) for computer game design and other related purposes (Meehan, 1981). Several formalisms have been proposed including, in particular, Petri nets (Abd El-Sattar, 2012) and an ad-hoc semi-formal graph-based approach to define discourse strategies based on their semantics (McKeown, 1985). These approaches are methodologically very close to the event bush, and one may reasonably expect that a chain of event bushes similar to that plotted in Figure 10 is a promising candidate for development of algorithms of interactive scenario generation for computer-aided storytelling.

Figure 10a describes a simple situation that a thief is heavily in debt and has got into serious trouble with his creditors so that he has either to pay them back or to run. In addition, he has no "free" money at all. Meanwhile, an occasional earning (of any nature) may occur, and in this case he would have some "free" money – still having the debt. All events involved in this situation are included in the following bush (Figure 10b) describing how the situation may evolve with appearance of another actor – a street merchant having similar problem (a threatening debt) and the same dilemma, to pay back or to run. An interference of such circumstance as occasional earning creates an option of having some free money for the merchant (similar to the thief). Furthermore, merchant may either spend this money or keep it. In the considered context, with no other circumstances involved, this is entirely an issue of his free will. In the event bush notation, this is expressed as furcation. It is noteworthy that furcation may imply either insufficient information (and then, with new events coming, will likely be replaced by influx) or the case of free choice, or probabilistic occurrence, and in two latter cases it will be never substituted by any other connective.

Yet more interesting plots arise due to interference of each of these two secondary events, "Thief is heavily in debt..., gets no extra money" and "Thief is heavily in debt..., gets extra money", into the merchant's destiny. This is expressed as confluxes of events in which the thief and merchant participate as subjects. New events resulting from each conflux have a different subject, "scuffle". This literally means that, in relation to the situation described, the thief and the merchant become one for a moment, and this single entity is their fight, in which they are indivisible. The result of this fight may be threefold if the thief has extra money and twofold, if not. Again, we are having here a furcation, which may be either due to lack of information (e.g., we do not know that one of the fighters has a pistol), or due to pure probability, or – again, due to the fact that the fight, being an entity that exists for a while, may also have its own will – in other words, the scuffle "decides" itself who takes whose money. This interpretation, looking bizarre for the common sense, may appear consistent with the systems theory, assuming the thief and the merchant make up an integral system when one assaults the other.

Interestingly, when the "decision" is made and either the thief robs the merchant (thus getting another *extra* money), or vice versa, or no one manages, the collision may repeat if the same thief meets *another* merchant, or merchant meets another thief (these options are shown by a bundle of arrows coming bottom-up at the left of Figure 10b). Thus a hypothetic interactively generated story may continue as long as desired.

Figure 10. Example of building alternative fictious plots with one (a) and two (b) primary internal events. Mind how much complicated the plots become with adding just one primary internal event.

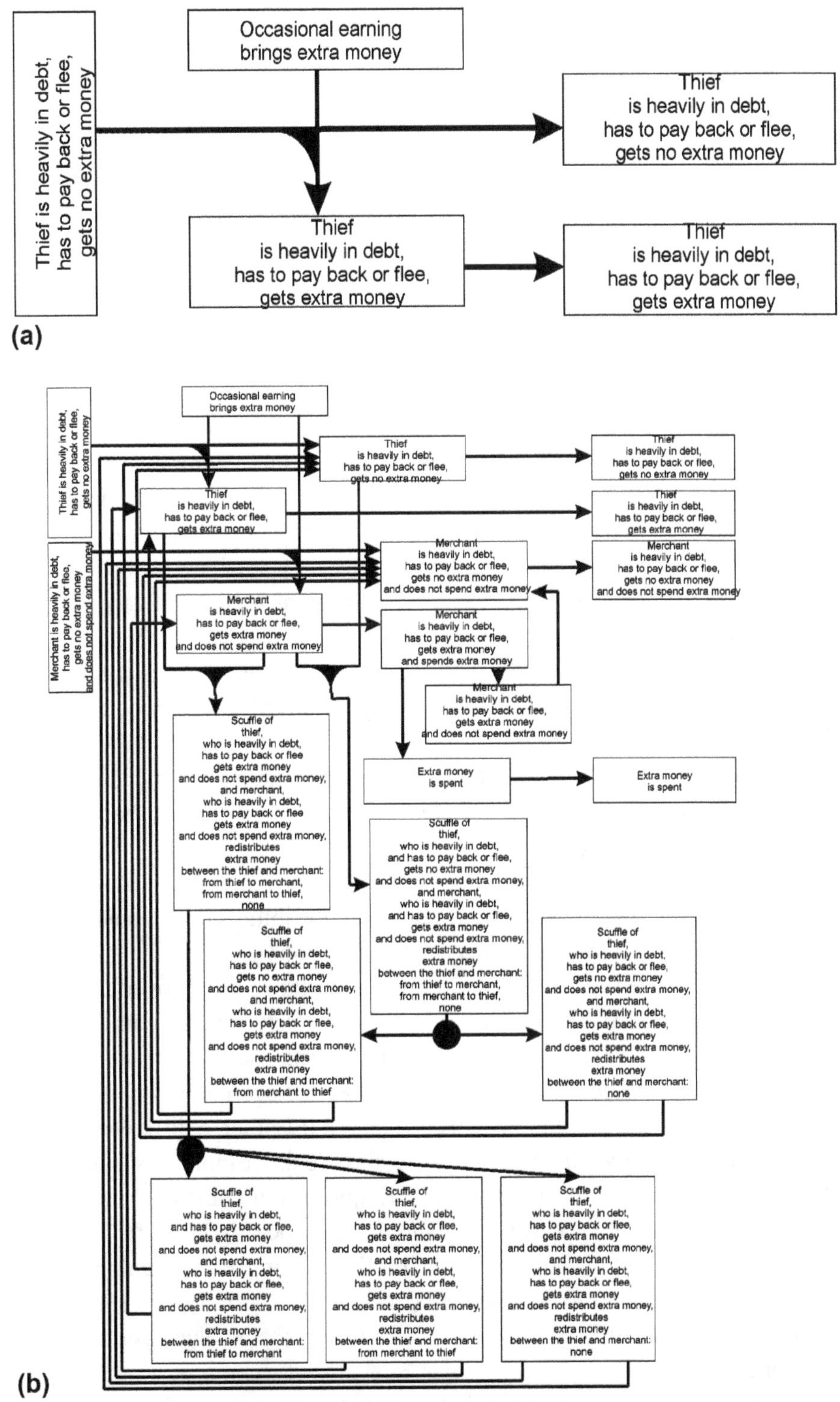

Problem of Nested Structures in Dynamic Knowledge Representation

An immanent part of knowledge representation is the consideration of class-subclass relations in a given domain. For the static knowledge, it gives ground, first, for building hierarchies of objects (considered either as classes, i.e., types – concept types or relation types – or as individuals (Ph. Martin, 2002), and then, for inference of relation of an instance to a class and possessing/not possessing by a class or instance of given properties and relations. Visualization of the class-subclass relation commonly gives either nested or tree-shaped patterns. Both these patterns fascinate the eye of a mathematician or computer scientist by their transparency, simplicity and formal beauty.

However, things become much more opaque and complicated if we proceed from static to dynamic knowledge. This is due to the structure of events that represents a statement with one subject and *n* predicates (with or without negation), each participating in class-subclass relations. Thus, in concern of events, these relations superimpose manifold and become hardly tractable. Indeed, taking simple relations "class: employees – subclass: workers" and "class: work – subclass: factory", we can specify the general event "Employees go for a work" in three ways: "Workers go for a work", "Employees go for a factory" and "Workers go for a factory", and there is no formal rule saying which one to prefer.

However, one should not skip considering these relations in dynamic knowledge because here they produce a very important pattern of thinking, which can be metaphorized as *nested events*. By intuition, it is clearly felt that all three "particular" events given above are somehow "nested" in the general one. Moreover, like in static knowledge, this gives ground – again, at least intuitively – to proceed from a general event to some particular ones or, vice versa, generalize particular ones by a general. In the event bush, this may take place either in the field(s) of primary or secondary events (because tertiary events only repeat some secondary or primary internal ones). In Figure 11, a-c, this pattern is shown for primary internal events. For simplicity, only the class-subclass relation between the subjects of events will be considered in this work, and set of predicates (including negations) will be considered the same for any two events whose subjects are in a class-subclass relation – e.g., "Employees go to a factory" and "Workers go to a factory".

In Figure 11a, the subject of event "Employees go to a factory", "employees", has two non-overlapping subclasses, "workers" and "clerks". As shown in Figure 11a, the initial general bush is decomposed into two particular ones, which have common primary external event(s).

If we specify subclasses in a subject class of the general event that would overlap, the result would be more complicated (see Figure 11b). Two particular bushes will have not only common primary external event(s) but also an overlap of subjects of some primary internal, secondary and tertiary events (and, hence, overlap in scenarios) painted gray in Figure 11b, and this overlap will not be explicit, as in one bush the events will have one subject, e.g., "Workers", and in the other, a different one ("Employees of company X)". To avoid this, a division of the generic class (e.g., "Employees") into non-overlapping subclasses ("Workers not employed by company X", "Employees of company X who are workers" and "Employees of company X who are not workers") should be made (Figure 11c).

Structurally this is a so-called pushout construction that generalizes the union construction in the sense that some shared structure has to be respected (Ehrig et al., 2006). Figure 11c represents one possibility how to construct a pushout in case the relations between the involved event bushes are given by injective graph homomorphisms.

From purely formal point of view, the "gray" class born as intersection of classes may be empty, i.e., include no instances in real world. This is the virtue of knowledge, which can be absolutely imaginary

Figure 11. Transformation of an event bush with substitution of subject of primary internal event(s) from generic ("employees") to particular ("workers", "clerks", "employees of company X", "workers who are not employees of company X", "workers who are employees of company X", "employees of company X who are not workers"). See comments in the text.

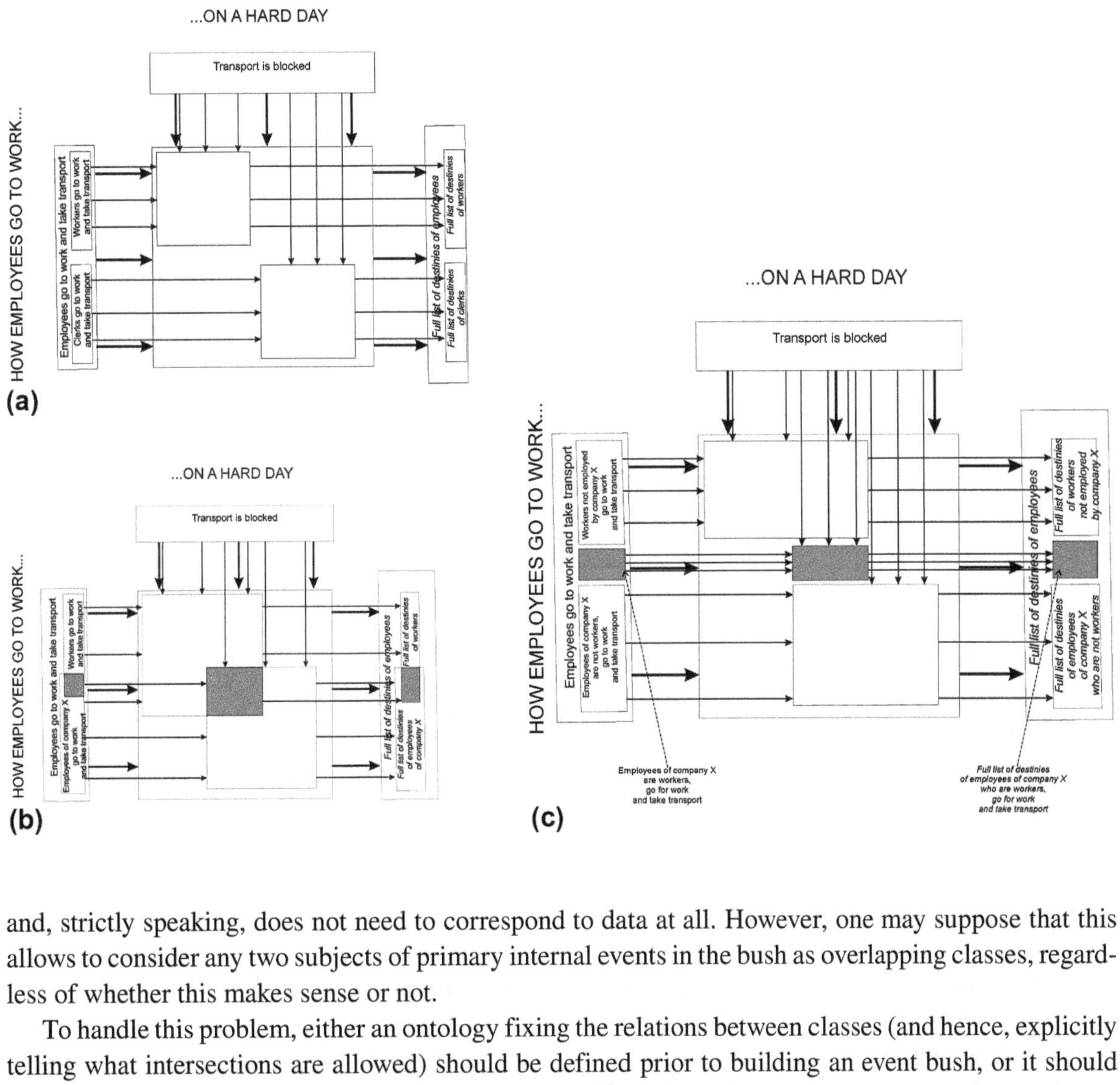

and, strictly speaking, does not need to correspond to data at all. However, one may suppose that this allows to consider any two subjects of primary internal events in the bush as overlapping classes, regardless of whether this makes sense or not.

To handle this problem, either an ontology fixing the relations between classes (and hence, explicitly telling what intersections are allowed) should be defined prior to building an event bush, or it should be accepted that all subjects of primary internal events are formulated in the most generic way (more or less like individual constants in the predicate logic) – e.g., for Figure 11a this could be, "*People* are employed, go to work, take transport", "*People* are employed as workers, go to work, take transport", "*People* are employed as clerks, go to work, take transport". This moves the problem of nested events entirely to the area of predicates but creates, at the same time, a new problem of compound predicates: "to be employed", "to be employed as worker/clerk", "to be employed by company X" and so forth – that requires stating class-subclass relations between the predicates (i.e., again, an ontology or the like). The issue of optimal formulation of events will be addressed in a separate study, while here the concern is mostly the structure and structural changes of event bushes.

Figure 12a. Some of the possible relationships between the specifications of the subject of primary internal event and the structure of the bush: the initial bush with the primary internal event having generic subject; the general structure is changed by adding an (ii) event between an (ii) event resulting from the furcation and the (iii) event, and the set of subjects is changed by a new subject that this newly introduced event has, "streets" (instead of "employees")

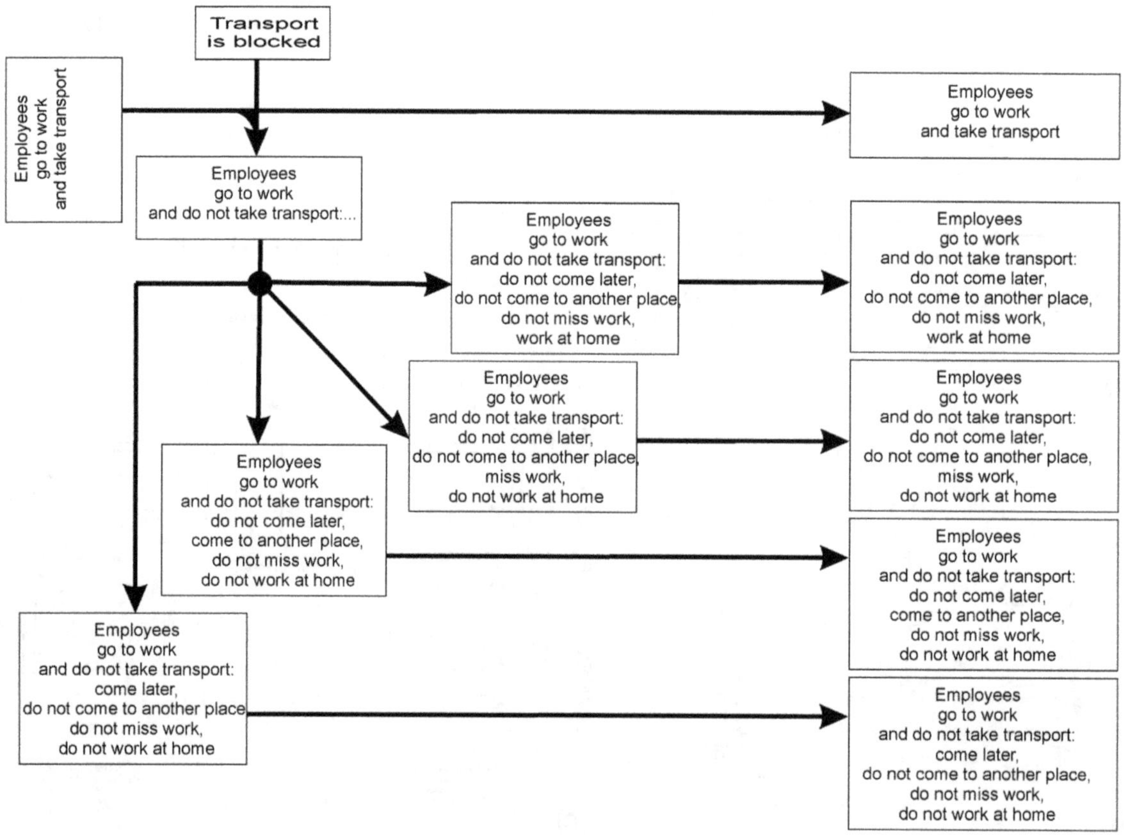

In this respect, it looks important to analyze whether the specification or generalization of subjects of *(ia)* events changes the structure of the bush – in other words, whether there appear new events and connectives (or only connectives), disappear old ones or change the subject-predicate structure of existing events when the subject of a primary internal event is specified or generalized. Full classification of possible structural and compositional changes still needs to be constructed, and this issue will be approached in the following section. Figure 12, a-d, only illustrates this issue by the example of a simple bush with one primary internal and one primary external events, one influx and one furcation connectives. For this purpose, the specification of the term "employee" is made somewhat different from that in Figure 11.

In the general case (primary internal event subject is "employees"; Figure 12a), the furcation implies four options. Mind that furcation here may resemble classification but should not be taken as such, because it means that, if (when) the transport is blocked, *first* employees realize that they do not take transport and *then*, in some time (and possibly under some additional conditions like a phone call to the boss; these conditions are not stated in Figure 12 in order to keep the example simple) they take one of the following options: come to working place later, come to a different place (e.g., to do some

Figure 12b. Some of the possible relationships between the specifications of the subject of primary internal event and the structure of the bush: specification of the subjects causes no change in the structure of the bush; the general structure is changed by adding an (ii) event between an (ii) event resulting from the furcation and the (iii) event, and the set of subjects is changed by a new subject that this newly introduced event has, "streets" (instead of "employees")

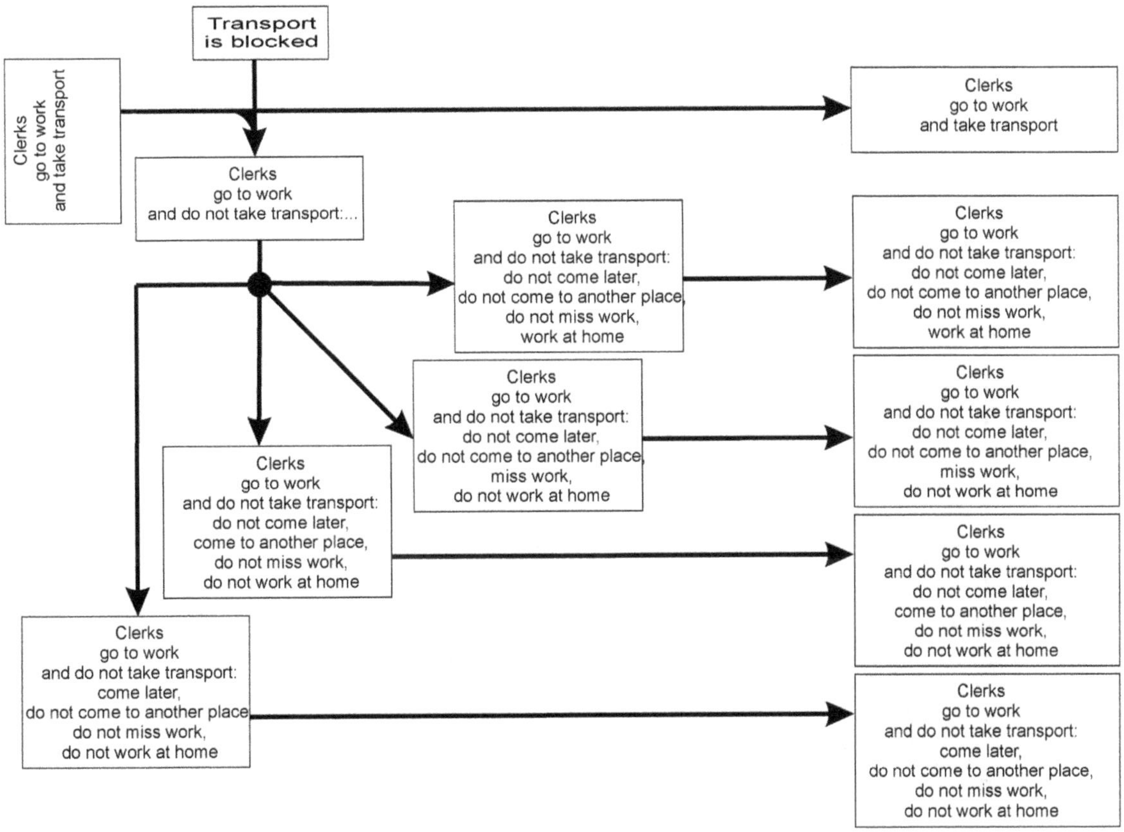

paperwork in an office of a partner company to save time), to return home and stay working there, or not to work at all (to miss work). Also, there is no contradiction in a formulation like "Employees go to work… do not come to work" or "Employees go to work… work at home". Time in the event bush is understood as a succession of events, and each event, as a set of predicates relating to the same subject. Once employees *were* going to work (regardless of whether they reached it and if not, whether they are going to work now), the predicate "to go to work" is there. However, if, after they were already going to work they decide to return home (e.g., because they are unable to reach work without transport), both predicates ("go to work" and "come home") are listed in an event in a narrative style, "Employees go to work, do not take transport, and come home". However, time interval that may be attributed to this composite event should be the sum of the intervals that correspond to all subject-predicate pairs where the predicate is not negated, "Employees go to work", "Employees come home".

Possible structural changes of event bush with specification of a subject of primary internal event deserve special consideration. For instance, in the considered example, no change occurs if to substitute

Figure 12c. Some of the possible relationships between the specifications of the subject of primary internal event and the structure of the bush: specification changes the overall number of events and connectives but does not change the general structure and set of subjects of events in the bush; the general structure is changed by adding an (ii) event between an (ii) event resulting from the furcation and the (iii) event, and the set of subjects is changed by a new subject that this newly introduced event has, "streets" (instead of "employees")

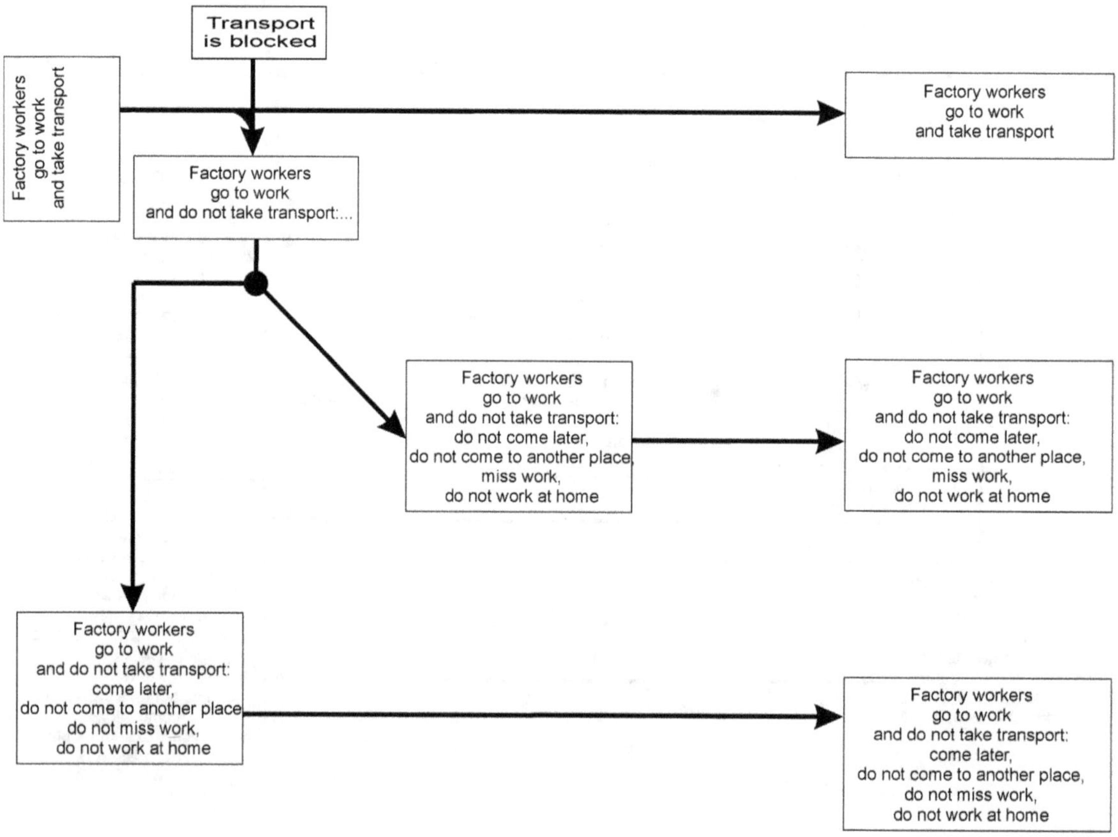

"employees" with "clerks" (Figure 12b); the number of options obviously decreases for factory workers (neither they can come to do their job to a different place but their factory nor can work at home – Figure 12c), and in case of another type of employee, policemen, one more secondary and tertiary events with a new subject appear ("Streets are patrolled by policemen"). As is seen yet in this simple example, structural and compositional changes in nested event bushes can be highly diverse.

Similar changes that arise from specification of subjects of primary external and secondary events need to be addressed in a separate study, including the possible phenomenon of cross-nesting of secondary events (theoretically, one may think of a pair of bushes in which one bush includes a secondary event whose subject is generic for those of some secondary events of the other bush, while the other also includes such event that has a subject generic for those of some secondary events of the former bush).

Finally, specified can be not only subjects but predicates of events, and this also leads to various and still unexplored patterns of nesting.

Figure 12d. Some of the possible relationships between the specifications of the subject of primary internal event and the structure of the bush: specification changes the overall number of events and connectives changes the general structure and set of subjects of events in the bush; the general structure is changed by adding an (ii) event between an (ii) event resulting from the furcation and the (iii) event, and the set of subjects is changed by a new subject that this newly introduced event has, "streets" (instead of "employees")

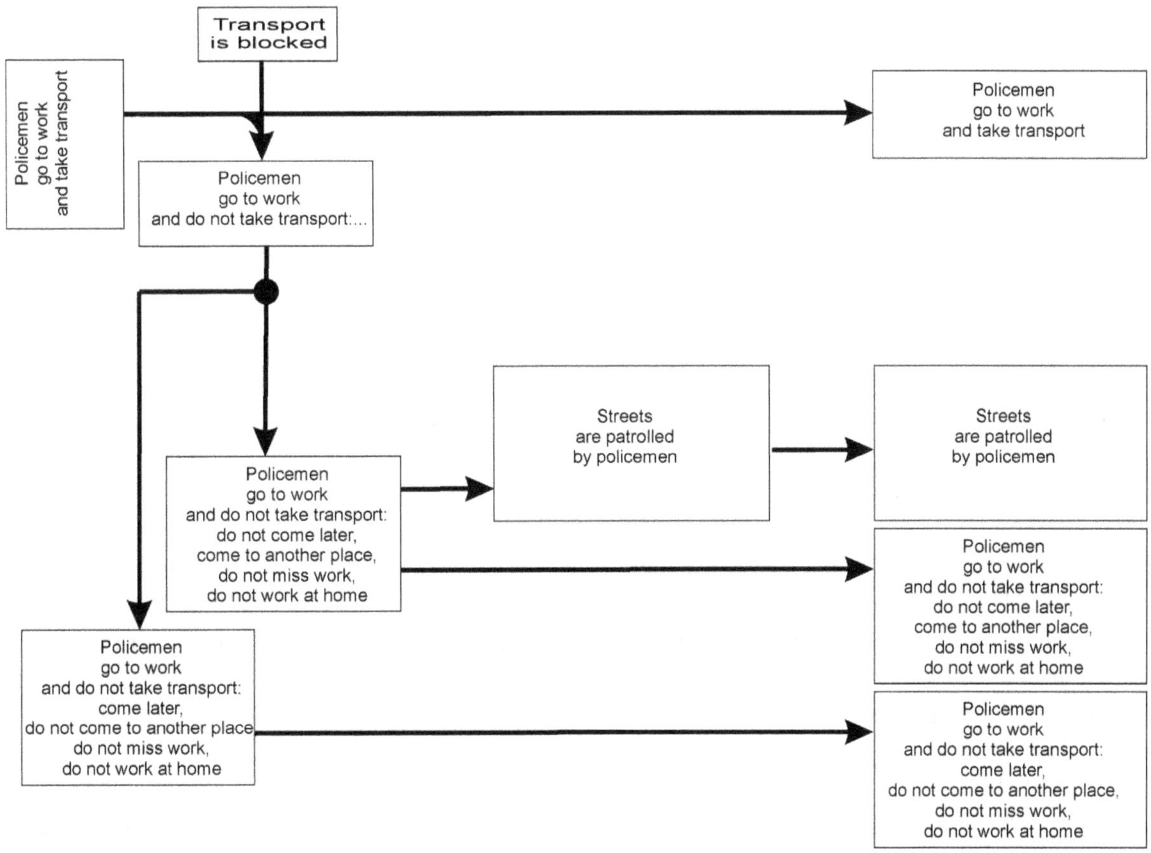

If not entirely, then at least partially these issues can be formalized by describing the event bush in terms of typed graphs.

MULTI-BUSH STRUCTURES IN THE LIGHT OF TYPED GRAPHS

In this section, we show how the description of the syntax of event bushes by means of typed graphs that we developed in section Formalization of singular event bushes by typed graphs gives us naturally the opportunity to model, to formalize and to reason about the multi-bush structures discussed in section. Types of multi-bush structures and associated cognitive problems. We discuss in detail the chaining of two event bushes (see Figure 8, a) and the union of two bushes (see Figure 11a). This demonstrates how the approach will work for larger structures. The formalization of these structures, not to become too abstract and lift-up, should proceed in relation to particular practical applications. However, the

Figure 13. Graph S'|S" reflecting the multiflow structure of a chain of two event bushes where S' and S" are copies of the graph S in Figure 6

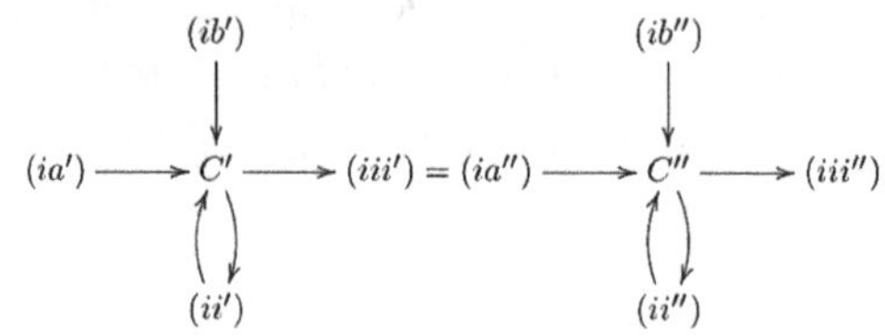

Figure 14. Relation between the multiflow structure S'|S" of a chain of two event bushes depicted in Figure 13 and the multiflow structures S' and S" of the two single event bushes that are chained

$$S' \xrightarrow{in_1} S' \mid S'' \xleftarrow{in_2} S''$$

description of any of such applications here will well exceed the paper format and therefore has to be addressed in a subsequent paper.

Chaining of Event Bushes

The idea of chaining two event bushes by identifying the tertiary events of the first bush with the primary internal events of the next bush can be abstractly described, on the abstraction level of multiflow structures, by a graph *S'|S"* (see Figure 13) obtained by chaining two copies, *S'* and *S"*, of the graph *S* that we introduced in section 3 to reflect the multiflow structure of event bushes.

That *S'|S"* is constructed out of *S'* and *S"* is formally reflected by the existence of two injective graph homomorphisms (Figure 14).

The graph homomorphisms in_1 and in_2 are jointly surjective in the sense that we have $in_1(S')\cup in_2(S'')=S'|S''$. Note, that the image $\phi(G)$ of a graph *G* w.r.t. a graph homomorphism $\phi{:}G\rightarrow H$ is a subgraph of *H* defined by:

$$\phi(G)_V = \{\phi_V(x) \mid x\in G_V\}$$

and:

$$\phi(G)_E = \{(\phi_V(x), \phi_V(y)) \mid (x,y)\in G_E\}$$

Moreover, there is a graph homomorphism $\alpha{:}S'|S''\rightarrow S$ that is compatible with the obvious typing morphism $\tau_{S'|S''}{:}S'|S''\rightarrow B$. We have $\alpha(ia') = (ia)$, $\alpha(ib')= \alpha(ib'')=(ib)$, $\alpha(ii')=\alpha((iii')=(ia''))=\alpha(ii'')=(ii)$, $\alpha(iii'')=(iii)$, and $\alpha(C')=\alpha(C'')=C$. (From now on we assume always implicitly that graph homomorphisms are compatible with typing and we will leave out the typing morphisms from the figures.)

That an event bush $\mathcal{E}$ is obtained by (or can be described as) a chain of two event bushes means that there exists a type-compatible surjective graph homomorphism $\gamma{:}\ \mathcal{E}\rightarrow S'|S''$ such that $\gamma;\alpha=\sigma_{\varepsilon}$ (Figure

Figure 15. Condition that an event bush $\mathcal{E}$ possesses the multiflow structure S'|S" of a chain of two event bushes

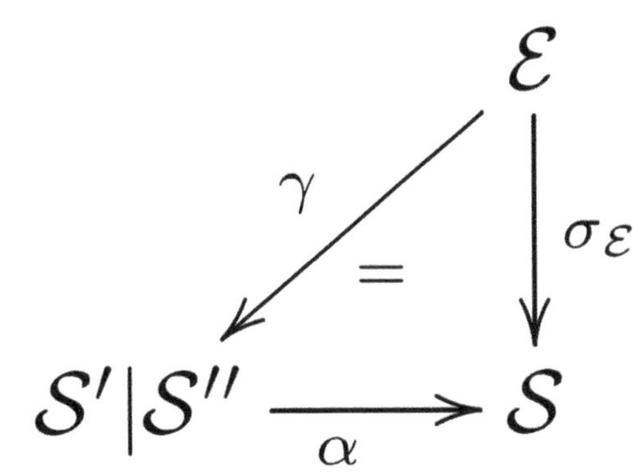

Figure 16. Decomposition of an event bush $\mathcal{E}$ into a chain of two single event bushes $\gamma^{-1}(S')$ and $\gamma^{-1}(S'')$

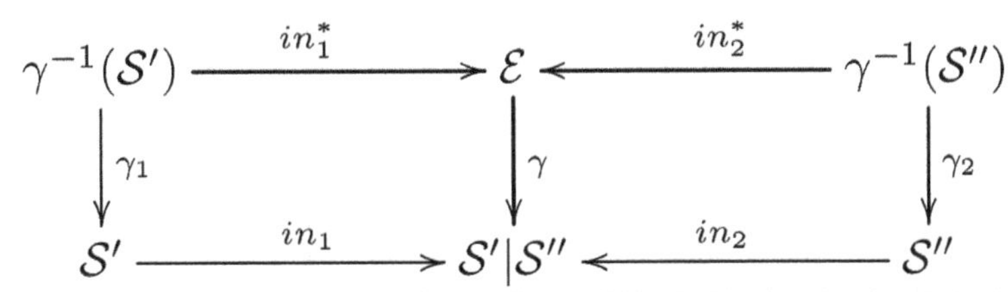

15) where the first event bush in the chain is given by the pre-image of *S'* and the second one is given by the pre-image of *S"* (Figure 16).

Given an injective graph homomorphism *in: K→ H* the pre-image $\phi^{-1}(K)$ of *K* w.r.t. a graph homomorphism$\phi: G\rightarrow H$ is a subgraph of *G* with vertices defined by:

$$\phi^{-1}(K)_V = \{x \in G_V \mid \text{exists } z \in K_V \text{ with } in_V(z) = \phi_V(x)\}$$

Since the function $in_V: K_V \rightarrow H_V$ is assumed to be injective, there is for any $x \in \phi^{-1}(K)_V$ exactly one $z \in K_V$ with $in_V(z) = \phi_V(x)$. We denote this unique vertex by z_x. In such a way, we can describe the edges in $\phi^{-1}(K)$ by:

$$\phi^{-1}(K)_E = \{(x,y) \in \phi^{-1}(K)_V \times \phi^{-1}(K)_V \mid (z_x, z_y) \in K_E\}$$

The construction ensures that the inclusion $\phi^{-1}(K)_V \subseteq G_V$ establishes a graph homomorphism $in^*: \phi^{-1}(K) \rightarrow G$ and that the assignments $x \rightarrow z_x$ define a graph homomorphism $\phi^*: \phi^{-1}(K) \rightarrow K$ (Figure 17).

Union of Event Bushes

The idea to build a general event bush by putting together two particular event bushes (Figure 11a) can be abstractly reflected by a graph *S'+S"* that is the union of two copies *S'* and *S"* of *S* (see Figure 18).

Figure 17. A diagram visualizing the formation of the pre-image $\phi^{-1}(K)$ of a graph K w.r.t. a graph homomorphism ϕ:G→H

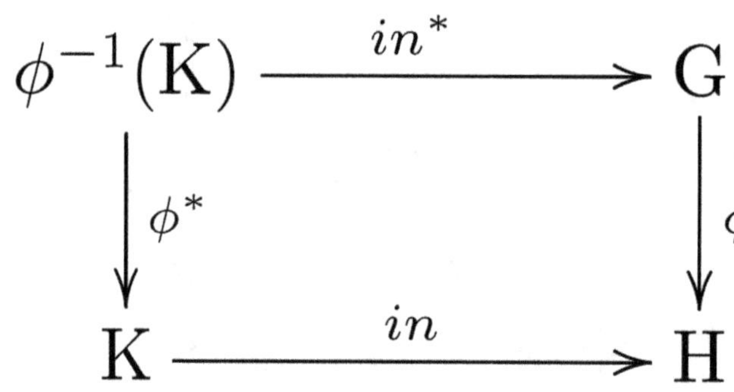

Figure 18. Graph S'+S" reflecting the multiflow structure of the union of two event bushes where S' and S" are copies of the graph S in Figure 6. C' and C'' are connectives

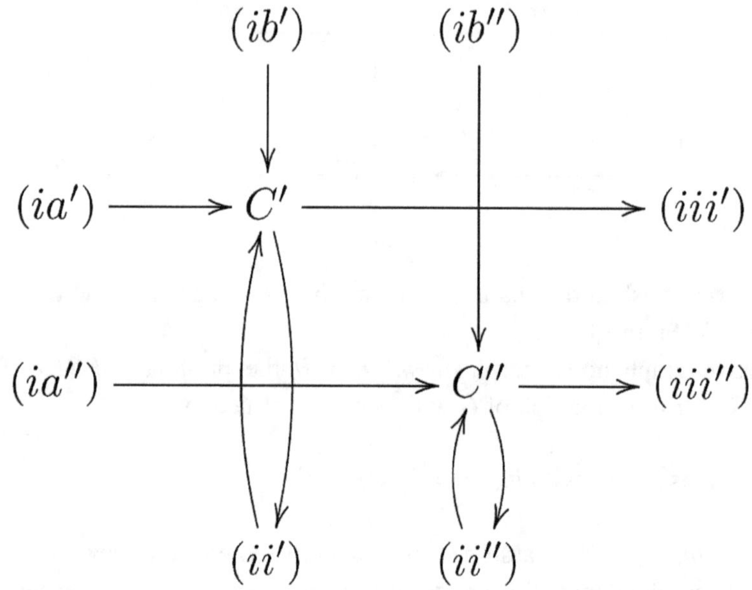

Figure 19. Relation between the multiflow structure S'+S" of the union of two event bushes and the multiflow structures S' and S" of the two single event bushes that are put together

$$S' \xrightarrow{in_1} S' + S'' \xleftarrow{in_2} S''$$

That $S'+S''$ is constructed out of S' and S'' is formally reflected by the existence of two injective graph homomorphisms (Figure 19) that are actually inclusions.

As in the case of chaining, the graph homomorphisms in_1 and in_2 are jointly surjective in the sense that we have $in_1(S')\cup in_2(S'')=S'+S''$. In contrast to chaining, in_1 and in_2 are, in addition, also disjoint in the sense that $in_1(S')\cap in_2(S'')=\varnothing$. That is, $S'+S''$ is actually the disjoint union, i.e., the sum, of S' and S''.

There is a graph homomorphism $\beta:S'+S''\rightarrow S$ that is compatible with the obvious typing morphism $\tau_{S'+S''}:S'+S''\rightarrow B$. We do have $\beta(ia')=\beta(ia'')=(ia)$, $\beta(ib')=\beta(ib'')=(ib)$, $\beta(ii')=\beta(ii'')=(ii)$, $\beta(iii')=\beta(iii'')=(iii)$, and $\beta(C')=\beta(C'')=C$. Be aware that this graph homomorphism β mirrors a class-subclass relation on an abstract level. The event nodes in S represent general events while the event nodes in S' and S'', respectively, represent particular events. Moreover, β describes nothing but the relation between particular and general events.

That an event bush $\mathcal{E}$ is obtained (or can be described) as the disjoint union of two particular event bushes means that there exists a type-compatible surjective graph homomorphism γ: $\varepsilon\rightarrow S'+S''$ such that $\gamma;\beta=\sigma_\varepsilon$ (Figure 20) where the left event bush in the sum is given by the pre-image of S' and the right event bush is given by the pre-image of S'' (Figure 21).

By general results about directed graphs and graph homomorphisms it follows that in_1* and in_2* are as well jointly surjective as disjoint. That is, ε can be seen indeed as the disjoint union of $\gamma^{-1}(S')$ and $\gamma^{-1}(S'')$.

The attentive reader may object that in Figure 11a only the primary internal and the tertiary events are decomposed while the primary external events are common. To reflect this situation we have to do nothing but just to replace the disjoint union of $S'+S''$ by another non-disjoint union $S'\pm S''$ of the two copies S' and S'' (Figure 22).

Figure 20. Condition that an event bush $\mathcal{E}$ possesses the multiflow structure $S'+S''$ of the union of two event bushes

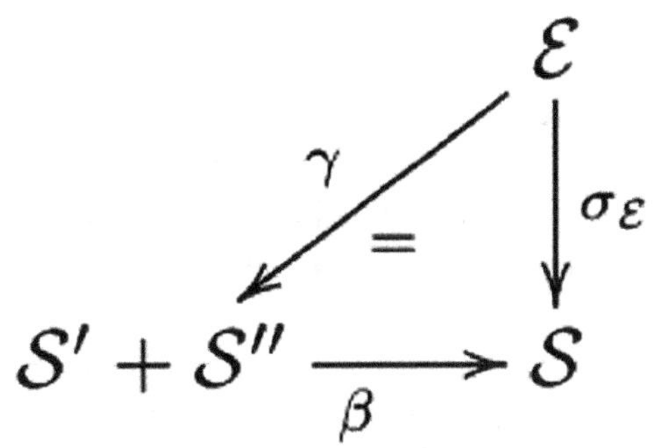

Figure 21. Decomposition of an event bush $\mathcal{E}$ into the union of two single event bushes $\gamma^{1}(S')$ and $\gamma^{1}(S'')$

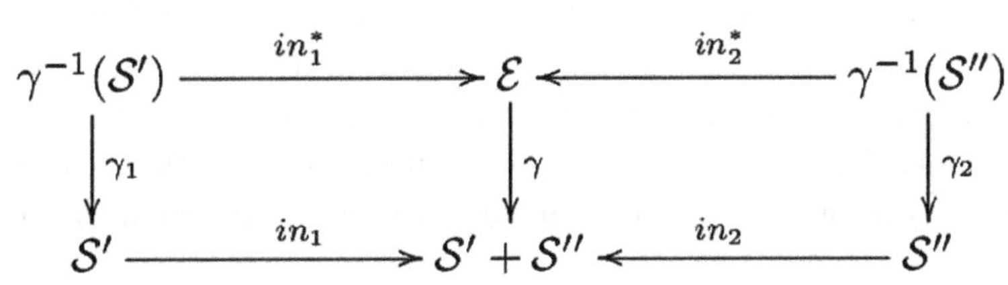

Figure 22. Graph S'+S" reflecting the multiflow structure of the non-disjoint union of two particular event bushes with common primary external events

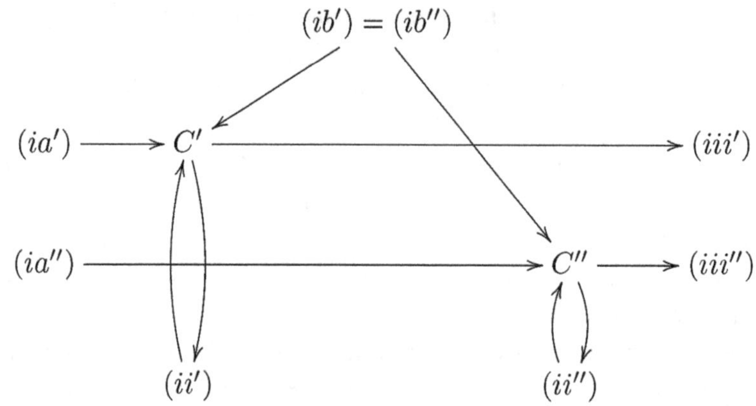

The decomposition of an event bush into two particular event bushes with common primary external events can be described fully analogously to the disjoint union case.

A detailed discussion of situations where subclasses overlap (Figure 11, b+c) goes beyond the scope of this paper. We want to mention, however, that we are convinced that those situations can be treated most appropriately by adapting the so-called pushout construction for (typed) graphs (Ehrig et.al.,2006) to event bushes.

DISCUSSION: EVENT BUSH AS A VISUAL LANGUAGE

The formalization of event bushes by typed graphs can be straightforwardly applied in a number of existing applications of the event bush method in various fields of knowledge. Though most of these applications are still rather demonstrative, in some areas, e.g., in the geosciences, the progress has been enough to approach solutions of practical problems. (Diviacco et al., 2014) report on the event-bush qualitative modeling of eruption activity of Etna Volcano, Sicily, Italy, based on the results of (Behncke & Pshenichny, 2009), and suggest a method of quantitative assessment of similarity of various eruptions from the event bushes built in the same semantics, provided each eruption is described by a separate bush. The measure of similarity between any two eruptions can be expressed as a function of the number of cancelled and introduced vertices of both types – connectives and events – required to proceed from one event bush to the other. Obviously, this method can be applied to any kind of environment of directed alternative changes modeled as a series of event bushes with the same semantics.

Then, the event bush is used in an Internet-based information system for support of collaborative studies in seismology that is being built in Istituto Nazionale di Oceanografia e di Geofisica Sperimentale (OGS), Trieste, Italy, in the COLLA environment, currently available online after registration at http://colla.ogs.trieste.it (Diviacco et al., 2014). The system aims at bringing together all concepts, qualitative standpoints expressed in any natural language, quantitative models using any mathematical formalism and data in any format that describe the phenomenon of site effects. Site effect is a rapid amplification of seismic waves in loose sediments that leads to increased local damage at earthquakes (Luzón et al., 2002).

The system should organize the information domain that mirrors site effects, including (i) the corresponding field of geophysical and geological knowledge and (ii) related data. Nested event bushes are thought to be the "skeleton" of the information domain and, hence, the information system. This is illustrated in Figure 23.

It is noteworthy that structural nesting is accompanied here by setting semantic equivalences of some subjects and predicates, e.g., "Source" in one bush is meant equal to "Earthquake-generating fault" in the other. Other equivalences include: "Path" - "Hard rock bed", "Site" - "Sedimentary rock bed". This allows one to give double definitions (listed below via "*or*") to those events (E_1 to E_5) occurring in both bushes.

The events are defined as follows:

E_1: Source seismic wave propagates unaffected by path, unaffected by site or Seismic wave propagates from earthquake-generating fault unaffected by hard rock bed, unaffected by sedimentary rock bed.
E_2: Path affects or Hard rock bed affects.
E_3: Source seismic wave propagates affected by path, unaffected by site or Seismic wave propagates from earthquake-generating fault affected by hard rock bed, unaffected by sedimentary rock bed.
E_4: Site affects or Sedimentary rock bed affects.
E_5: Source seismic wave propagates affected by path, affected by site or Seismic wave propagates from earthquake-generating fault affected by hard rock bed, affected by sedimentary rock bed.
E_6: Seismic wave propagates from earthquake-generating fault unaffected by hard rock bed, affected by sedimentary rock bed.

Although basic, even these event bushes serve well to organize the information domain of site effect studies. For this purpose, the event bushes originally plotted in the Corel Draw software were uploaded then as scalable vector graphics (SVG) files to COLLA, thus giving start to a new COLLA project. Textual evidence, graphic and tabular data were then linked to different parts of the bushes, such as: (i) particular subject or predicate occurring in a number of events throughout a bush, (ii) particular event, (iii) several events linked by connectives, (iv) particular scenario or scenarios, (v) the bush in general.

The presented system takes considerable advantage of building multi-bush structures. If an information resource is linked to a part of or the entire event bush, and the bush itself appears to have some relationship to another bush, this may mean, inter alia, that the information linked to one bush appears to be somehow linkable to (and therefore, accessible from) the other. In other words, a scientist who builds an event bush suiting his concept or model would immediately relate the information he links to "his" bush to a number of other bushes describing the same information domain, and vice versa, remaining within "his" bush, he would receive updates from others. This actually means that, while retaining his own language, terminology and modeling approach, a researcher will be able to successfully communicate and collaborate with others.

A perspective goal of development of this technology is automation of this relinking. However, this automation could not be performed because the relationships between the bushes and their parts in a multi-bush structure have not been formalized. The results of the present paper open an opportunity to do this in terms of typed graphs.

To underline that typed graphs are indeed a flexible tool to formalize different scenarios of overlapping of event bushes we mention that situations, as in Figure 23a, b, where two particular event bushes do have common primary internal and common primary external events can be described by a graph as in Figure 24.

Figure 23. Nested event bushes representing a part of the COLLA information system on site effects. The bush in Figure 23a is nested in the bush in Figure 23b. See comments in the text.

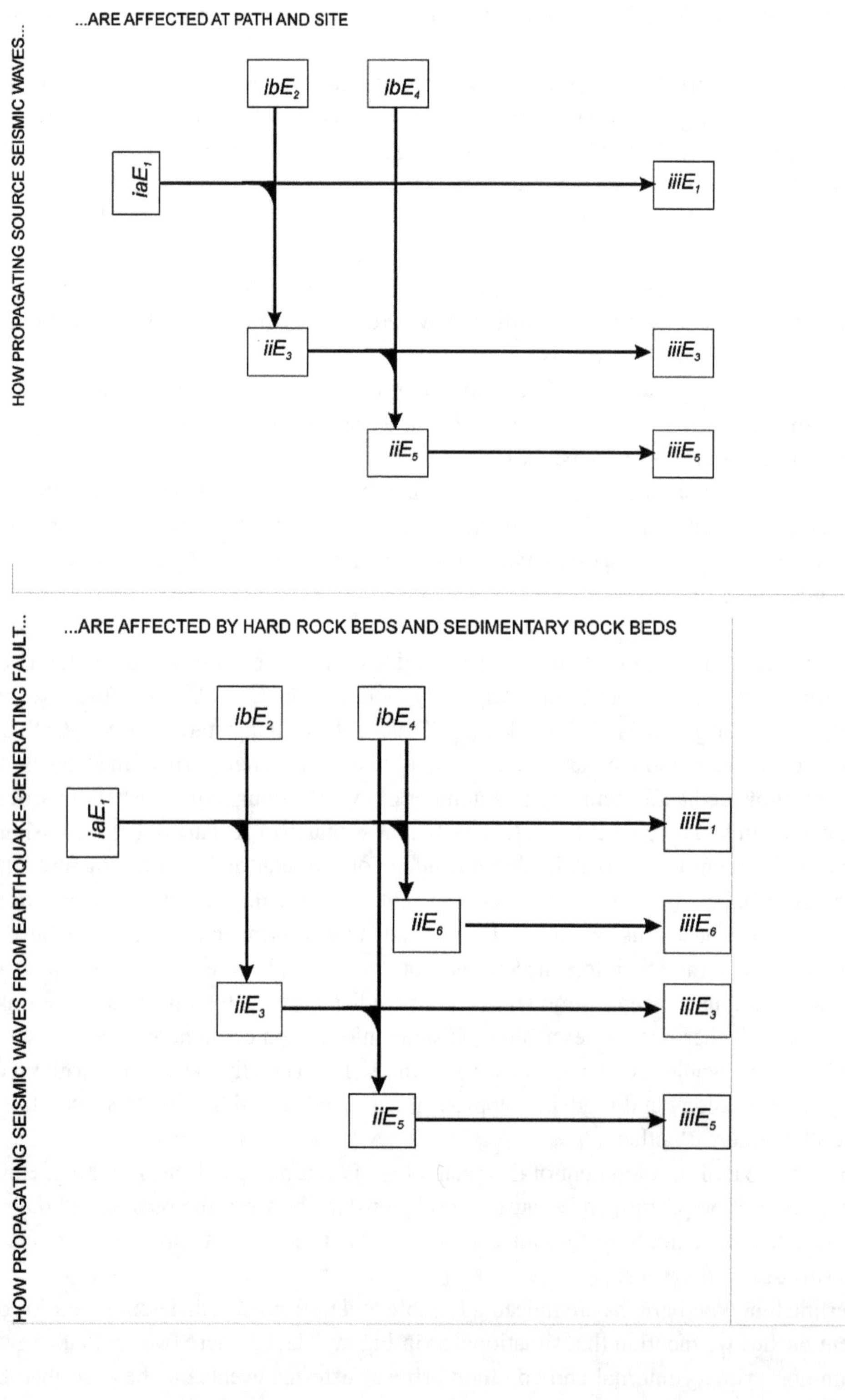

Figure 24. Graph S'+S" reflecting the multiflow structure of the non-disjoint union of two particular event bushes presented in Figure 23a, b with common primary internal and common primary external events

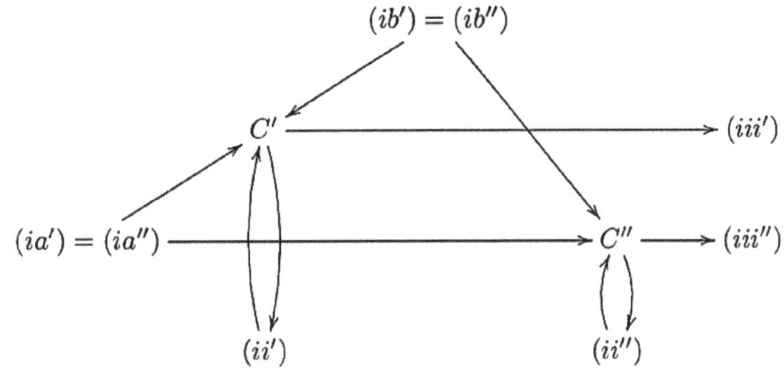

However, for a more comprehensive formalization of multi-bush structures we have to develop more general concepts for relating event bushes. Moreover, we have to develop concepts and techniques to specify properties of those relationships between event bushes.

Our results show that there are both practical necessity and theoretical opportunity to develop a visual language of event bushes based on their typed graph formalization, with account of the subject-predicate structure of events. Such language is needed to produce special software for creation, edition, and exchange of event bushes, because the general-purpose products used so far (Corel Draw, MS Visio and some others) do not support the control of correctness of semantics and syntax. However, assuming this language is developed, a question arises how it is related to existing visual languages used in knowledge engineering and representation.

In the available literature, we can identify three kinds of visual (graph-based) languages and formalisms relevant to knowledge engineering – concept-centered, state-centered and event-centered.

Concept-centered formalisms (see, e.g., (Kremer, 1998) and references therein) provide tools for engineering and representation of static knowledge as discussed in the introduction.

State-centered formalisms, in contrast, aim to model systems of changing entities, i.e., dynamic knowledge. Those formalisms are based on the concept of state (configuration) of the system in question, and the dynamics of the system is described as transitions between states. There is a variety of transition systems in this state-centered sense, and some efforts have been made to bring an order into this variety (see the comprehensive analysis of Abdulla et al. (2007).

Visual variants of state-centered formalisms include mainly finite state machines, of which DEVS (Discrete Event Systems Specification) is perhaps the most universal (Zeigler et al., 2000). Many formalisms used or usable for dynamic knowledge representation (e.g., Petri nets or state charts) can be described as state-centered (Zeigler, 2009). However, in many finite state machines the states are just enumerated and have no inner structure. State charts go a bit further and focus on the hierarchical structure of states as we find it often in human-made systems like airplanes, computers, and organizations. Still, not a single formalism of the state-centered type, to our knowledge, is concerned with the internal subject-predicate structure of events as do the event bush and the dynamic knowledge grammar in general.

At the same time, Petri nets (place/transition nets – Reisig, 1985) can be seen as a visual formalism that combines, according to our classification, states and events. Similar to event bushes, Petri nets are bipartite graphs given by a set of places, a set of transitions and edges from places to transitions or vice versa. States of a Petri net are defined as markings (of places by a number of tokens). Firing an enabled transition defines then a transition between two states of the Petri net and can be seen as an event. One of our ideas for future research is to construct Petri net for any event bush. The events in the bush are interpreted as transitions and the connective nodes are taken as places (Rogalchuk & Solomin, 2015), where we have to add some auxiliary places to mimic the semantics of the connectives. The intuition is that a token on a place indicates that a certain event happened. Once we have abstracted from an event bush a corresponding Petri net we intend to use available Petri net tools to analyze reachability of critical events in the event bush.

Event-based approach sensu stricto has been developed mostly by Glynn Winskel and includes, to our knowledge, one formalism, the so-called event structures (Winskel, 1987). An event structure is given by a set of events, a partial order $\leq$ on the events, and an irreflexive and symmetric incompatibility relation # on events. Obviously, we can define for any event bush a corresponding event structure. We define $E_1 \leq E_2$ if, and only if, there is a sequence of edges from E_1 to E_2 in the bush. Moreover, we declare the resulting events of furcation connective, $E_2,...,E_n$, as incompatible. The transition from an event bush to an event structure implies, however, a severe abstraction. On one side, we lose essential information by abstracting away from the multiflow structure and from the connectives. On the other side, to make $\leq$ into a partial order, we are forced to introduce identity loops on events and we are forced to collapse any cycle of events into a single event to achieve anti-symmetry. Especially, the second step of collapsing cycles is inadequate since it does not reflect a property of the process we are modelling by the event bush.

Creation of big and powerful knowledge-based systems inevitably requires integration of static and dynamic knowledge engineering approaches, and among the latter, strict organization of different methods based on formal relations between them. Our formalization of the event bush method by typed graphs is based on category theory, which is seen as one of the most universal mathematic languages (Barr & Wells, 2012).

Therefore, one may reasonably expect that this formalization would not only optimize the event bush method itself but also lead to a clear definition of its relation with the static-knowledge (concept-centered) and other dynamic-knowledge (state- and event-centered) methods and probably, alongside, even between the dynamic-knowledge methods in general.

CONCLUSION

The fundamentals of the event bush method were interpreted and formalized in terms of graph theory, namely typed graphs and graph homomorphisms.

We have shown that the description of the syntactic structure of event bushes by means of typed graphs provides us appropriate conceptual tools to model, to formalize and to reason about the problem of multi-bush structures. We discussed in detail the chaining of event bushes and the decomposition of a general event bush into two particular event bushes.

In general, the proposed formalization enables us to access, to adapt and to develop further the wide range of constructions, results and methods offered, for example, by the theory of graph transformations (Ehrig et al., 2006) and the Diagram Predicate Framework (Rutle et al., 2012).

As next steps in this direction we like to identify the following tasks and problems:

- **Composition and Decomposition of Event Bushes:** As mentioned in the section 4.2, we will need a more sophisticated arsenal of composition and corresponding decomposition constructions to describe, in a formal and precise way, the multi-bush structures discussed in the section 4.1. Especially, we have to develop an appropriate variant of the so-called pushout construction;
- **Nested Event Bushes:** In the present paper we addressed the issue of "nested event bushes" rather in the sense of "structured event bushes" along the line of class-subclass relationships. This is reflected especially by the fact that the graph homomorphisms between event bushes that appear in the section 4.2 are injective ones. To address "nested" along the line of class-object relationships we have, however, to take into consideration non-injective homomorphisms between event bushes. Until now non-injective homomorphisms arise only in form of the homomorphisms $\sigma_{\mathcal{E}}$: $\mathcal{E}\rightarrow S$, γ:$\mathcal{E}\rightarrow S'|S''$ and γ:$\mathcal{E}\rightarrow S'+S''$. A natural idea is now to consider the typed graphs S, $S'|S''$, and $S'+S''$ as abstract event bushes with new kinds of abstract connectives. This idea opens up a promising line of formalizing and of reasoning about nested event bushes;
- **Semantics of Event Bushes:** In our formalization of event bushes as typed graphs $\mathcal{E}$ with a multiflow structure we treat the nodes in $\mathcal{E}$ as abstract nodes without any structure and semantics. To formalize also the semantics of events (as statements of general form $S - P_1, P_2, \ldots, P_i$, where S is subject, and $P_1, P_2, \ldots, P_i$ are predicates) we have to move from directed graphs to an appropriate variant of attributed graphs (compare, (Ehrig et al., 2006)). That is, we have to extend our formalization of an event bush by a map from the set of event nodes in $\mathcal{E}$ into an appropriate "universe of statements" that is defined prior to event bushes.

Note, that only such a separation of syntax/structure on one side and semantics on the other side allows us to formalize adequately the presence of the same event at different positions in an event bush (see, e.g., Figure 3). "Different positions" mean different nodes in $\mathcal{E}$ but these different nodes can be of course mapped to the same statement.

- **Typed Graph Grammar:** Based on the general theory of graph transformations (compare, (Ehrig et al., 2006)) it is quite straightforward to define typed graph grammar rules for event bushes that allow, for example, to generate event bushes. Especially, the syntactic restrictions for connectives, presented in subsection 2.3, can be naturally reflected by typed graph grammar rules.

REFERENCES

Abd El-Sattar, H. K. H. (2012). State of the Art in Interactive Storytelling Technology: An Approach Based on Petri Nets. In Petri Nets – Manufacturing and Computer Science. InTech. doi:10.5772/48549

Abdulla, P., Delzanno, G., & Van Begin, L. (2007). *A Classification of the Expressive Power of Well-structured Transition Systems.* Technical Report 95. Centre Federe en Verification, Université libre de Bruxelles. Retrieved from http://www.ulb.ac.be/di/ssd/cfv

Barr, M., & Wells, C. (2012). Category Theory for Computing Science. *Reprints in Theory and Applications of Categories*, *22*, 1–538.

Bause, F., Kemper, P., & Kritzinger, P. (1994). *Abstract Petri Net Notation.* Universitaet Dortmund.

Behncke, B., & Pshenichny, C. A. (2009). Modeling unusual eruptive behavior of Mt. Etna, Italy, by means of event bush. *Journal of Volcanology and Geothermal Research, 185*(3), 157–171. doi:10.1016/j.jvolgeores.2009.04.020

Christensen, C. M. (2014). *The Innovator's Dilemma: When New Technologies Cause Great Firms to Fail.* Alpina Publishers.

Diviacco, P., Pshenichny, C., Behncke, B., & Kanzheleva, O. (2014). Systems for support of collaborative studies in the COLLA environment based on the event bush. *Scientific and Technical Journal of Information Technologies Mechanics and Optics*, *5*(93), 83–91. (in Russian)

Diviacco, P., Pshenichny, C., Carniel, R., Khrabrykh, Z., Shterkhun, V., Mouromtsev, D., & Pascolo, P. et al. (2014). Organization of a geophysical information space by using an event-bush-based collaborative tool. *Earth Science Informatics*, *4*(3), 139–160. doi:10.1007/s12145-014-0182-2

Ehrig, H., Ehrig, K., Prange, U., & Taentzer, G. (2006). *Fundamentals of Algebraic Graph Transformations.* Berlin: Springer.

Gavrilova, T., Bolotnikova, E., Leshcheva, I., Blagov, E., & Yanson, A. (2013). Measuring Psychological Impact on Group Ontology Design and Development: An Empirical Approach. Knowledge Engineering and the Semantic Web. Communications in Computer and Information Science, 394, 29-43.

Korableva, O., & Litun, V. (2014). The potential of transitive economies' growth based on innovative strategy. *WSEAS Transactions on Business and Economics*, *11*, 725–736.

Kremer, R. (1998). Visual Languages for Knowledge Representation. *KAW'98: Eleventh Workshop on Knowledge Acquisition, Modeling and Management.* Retrieved from http://tinyurl.com/l4sc9ze

Luzón, F., Palencia, V. J., Morales, J., Sánchez-Sesma, F. J., & García, J. M. (2002). Evaluation of Site effects in sedimentary basins. *Fisica della Tierra*, *14*, 183–214.

Martensen, A. L., & Butler, R. W. (1987). *The fault-tree compiler.* NASA Technical Memorandum 89098.

Martin, Ph. (2002). Knowledge representation in CGLF, CGIF, KIF, Frame-CG and Formalized-English. In *Proceedings of ICCS'02, 10th International Conference on Conceptual Structures*. Borovets, Bulgaria: Springer Verlag. doi:10.1007/3-540-45483-7_7

McKeown, K. R. (1985). Discourse strategies for generating natural-language text. *Artificial Intelligence*, *27*(1), 1–41. doi:10.1016/0004-3702(85)90082-7

Meehan, J. (1981). TALE_SPIN. In Inside Computer Understanding: Five programs plus Miniatures. Erlbaum.

Mouromtsev, D. I., Pshenichny, C. A., & Yakovlev, A. V. (2014). Semantic and structural delineation of market scenarios by the event bush method. *Decision Support Systems and Electronic Commerce*, *57*, 34–41. doi:10.1016/j.dss.2013.07.008

Pshenichny, C. A. (2014). Knowledge engineering in volcanology: Practical claims and general approach. *Journal of Volcanology and Geothermal Research*, *286*, 78–92. doi:10.1016/j.jvolgeores.2014.07.012

Pshenichny, C. A., & Kanzheleva, O. M. (2011). Theoretical foundations of the event bush method. Societal Challenges and Geoinformatics, 482, 139-165. doi:10.1130/2011.2482(12)

Pshenichny, C. A., & Khrabrykh, Z. V. (2002). Knowledge Base of Formation of Subaerial Eruption Unit. In S. Leroy & I. Stuart (Eds.), *Environmental Catastrophes and Recovery in the Holocene (Abstracts)*. London: Brunel University. Retrieved from http://atlas-conferences.com/cgi-bin/abstract/caiq-22

Pshenichny, C. A., & Mouromtsev, D. I. (2013) Representation of the Event Bush Approach in Terms of Directed Hypergraphs. In *13th International Conference on Conceptual Structures*. Mumbai, India: Springer-Verlag. doi:10.1007/978-3-642-35786-2_22

Pshenichny, C. A., & Mouromtsev, D. I. (2015). Grammar of Dynamic Knowledge for Collaborative Knowledge Engineering and Representation. In Collaborative Knowledge in Scientific Research Networks. IGI Global. doi:10.4018/978-1-4666-6567-5.ch016

Pshenichny, C. A., Nikolenko, S. I., Carniel, R., Sobissevitch, A. L., Vaganov, P. A., Khrabrykh, Z. V., . . . Gusev, E. A. (2008). The event bush as a potential complex methodology of conceptual modelling in the geosciences. *Proceedings, iEMSs - International Congress on Environmental Modelling and Software, 2*, 900-912.

Pshenichny, C. A., Nikolenko, S. I., Carniel, R., Vaganov, P. A., Khrabrykh, Z. V., Moukhachov, V. P., & Rezyapkin, A. A. et al. (2009). The event bush as a semantic-based numerical approach to natural hazard assessment (exemplified by volcanology). *Computers & Geosciences, 35*(5), 1017–1034. doi:10.1016/j.cageo.2008.01.009

Reisig, W. (1985). *Petri Nets*. Springer-Verlag. doi:10.1007/978-3-642-69968-9

Rogalchuk, V., & Solomin, K. (2015). Perspectives of Use of Petri Nets in Collaborative Research. In Collaborative Knowledge in Scientific Research Networks. IGI Global. doi:10.4018/978-1-4666-6567-5.ch018

Rutle, A., Rossini, A., Lamo, Y., & Wolter, U. (2012). A formal approach to the specification and transformation of constraints in MDE. *Journal of Logic and Algebraic Programming, 81*(4), 422–457. doi:10.1016/j.jlap.2012.03.006

Schröder, B. S. W. (2003). *Ordered Sets: An Introduction*. Boston: Birkhäuser Boston, Inc. doi:10.1007/978-1-4612-0053-6

Solomin, K., & Mandrik, M. (2015). Constructing historical knowledge through graphic boundary objects. In Collaborative Knowledge in Scientific Research Networks. IGI Global. doi:10.4018/978-1-4666-6567-5.ch017

Tutte, W. T. (1998). *Graph theory as I have known it*. New York: Oxford University Press.

Winskel, G. (1987). Event Structures. Lecture Notes in Computer Science, 255.

Zeigler, B. P. (2009). *DEVS-Centered Modeling and Simulation: Core Concepts for Engineering Education*. Retrieved from http://acims.asu.edu/publications/presentations

Zeigler, B. P., Praehofer, H., & Kim, T. G. (2000). *Theory of Modeling and Simulation. Integrating Discrete Event and Continuous Complex Dynamic Systems* (2nd ed.). Academic Press.

Chapter 15
First Results of Formalization of Drilling Knowledge by the Event Bush Method

Pavel Komar
Donetsk National Technical University, Donetsk People's Republic, formerly Ukraine

Alexander Khokhulia
Donetsk National Technical University, Donetsk People's Republic, formerly Ukraine

Yuri Gatchin
ITMO University, Russia

ABSTRACT

Geological drilling, being one of the key methods of investigation of the earth's interior and extraction of hydrocarbons and groundwater, represents a highly promising, urgent, yet virtually uncovered area for dynamic knowledge representation. In one sense, it favors some straightforward solution; in another, it brings serious challenges – such as interdisciplinarity and difference in reasoning by formulae and by concepts. Most importantly, existing conceptual models developed in the static knowledge representation paradigm generally do not reflect the dynamic character of the process of drilling. This chapter aims only to start paving this road by examining whether, and how, the principal method of dynamic knowledge representation, the event bush, may work in the interdisciplinary domain of drilling.

INTRODUCTION

Drilling, being one of the key methods of investigation of the earth's interior and extraction of hydrocarbons and groundwater, represents an interdisciplinary domain, which brings together geological and technical, qualitative and quantitative, well-understood and purely intuitive knowledge. Specialists from different fields (engineers, geologists, geophysicists, economists) perceive the issues of drilling differently, and understanding may be readily hampered by discordance of personal perceptions of collaborators or a teacher and student. To efficiently plan and manage drilling campaigns, support teamwork, handle

DOI: 10.4018/978-1-5225-5261-1.ch015

problems and adequately interpret and process the drilling data, a unified framework is desired, which could reconcile different perceptions.

This entirely falls in the scope of the field of knowledge engineering and representation. Having appeared in the 1970s to operationalize knowledge-based systems and help to create large systems for industrial and commercial use, knowledge engineering aimed to turn the process of constructing such systems from an art into an engineering discipline (Studer et al., 1998). It has been traditionally focused on medicine, economics, technical writing and technical design, some fields of chemistry and biology and such particular fields of the geoscience as mineral exploration or hydrology. Knowledge engineering quickly evolved from transferring expert knowledge to a knowledge-based system to modeling of the field of knowledge by a knowledge-based system.

In recent years, quite a few studies appeared on application of knowledge engineering and representation in hydrocarbon industry, which largely deal with drilling (e.g., Xu, 2012). However, following the mainstream of knowledge engineering – the ontology design, these studies, as pointed by Norheim & Fjellheim (2006), come to a challenge that the knowledge processes are not well connected to existing core working processes. In our opinion, this challenge cannot be met sufficiently well by the said approach because of the static nature of ontology (Gruber, 2009) and clearly dynamic character of the modeled environment of drilling that implies appearance and disappearance of events in a single scenario. Representation of a drilling scenario as a labeled directional graph that consists of nodes (vertices) indicating concepts and edges (links) representing relations between concepts (Skalle & Aamodt, 2005), does not capture this change either, because exactly these links may appear and fade, while the graph structure should be fixed. The solution is deemed to be found in the newly-emerged subfield of *dynamic knowledge engineering and representation*, the key method of which is the *event bush* (Pshenichny & Kanzheleva, 2011).

The event bush conceptualization of the entire field of drilling is a large and comprehensive task. It may be approached differently – from geological, technical, physical, economic ends and so forth. Pshenichny (2017, Chapter One of this volume) shows that various approaches may be brought together in one context. The visualization of this context suggested by this author is termed predmetka. Still, the purpose of this work is much less pretentious: to probe different approaches to this conceptualization. Herewith, we shall examine, at this stage at the most basic level, two pathways: formalization of underlying geological and physical knowledge.

GEOLOGICAL KNOWLEDGE: DRILLING WITH AND WITHOUT CORE OUTPUT

The first thing that occurs to one's mind about drilling is consequent penetration of layers of rock. This idea, however trivial, is the basement of all our knowledge about the geological aspect of drilling. Hence, exactly this should be represented first of all in terms of the event bush (Figure 1).

The event bush in Figure 1 is very simple in structure. It has only one primary internal and one primary external event and implies only two "actors", rock sequence and drilling (event predmets, in terms of the theory of multitudes, Pshenichny, 2017). The latter is involved in the primary external event, which has also one, very simple, primeta ("to be on"), and all other events of the bush have the predmet "rock sequence". The universum primeta for this event predmet is "to be in area X", and other primetas are composite and have structure: "to have layer … disturbed/not disturbed by the drill". Importantly, the layers of the sequence are not considered independent predmets, though in the predmetka depicting the

Figure 1. Event bush describing the drilling of a rock sequence without core output

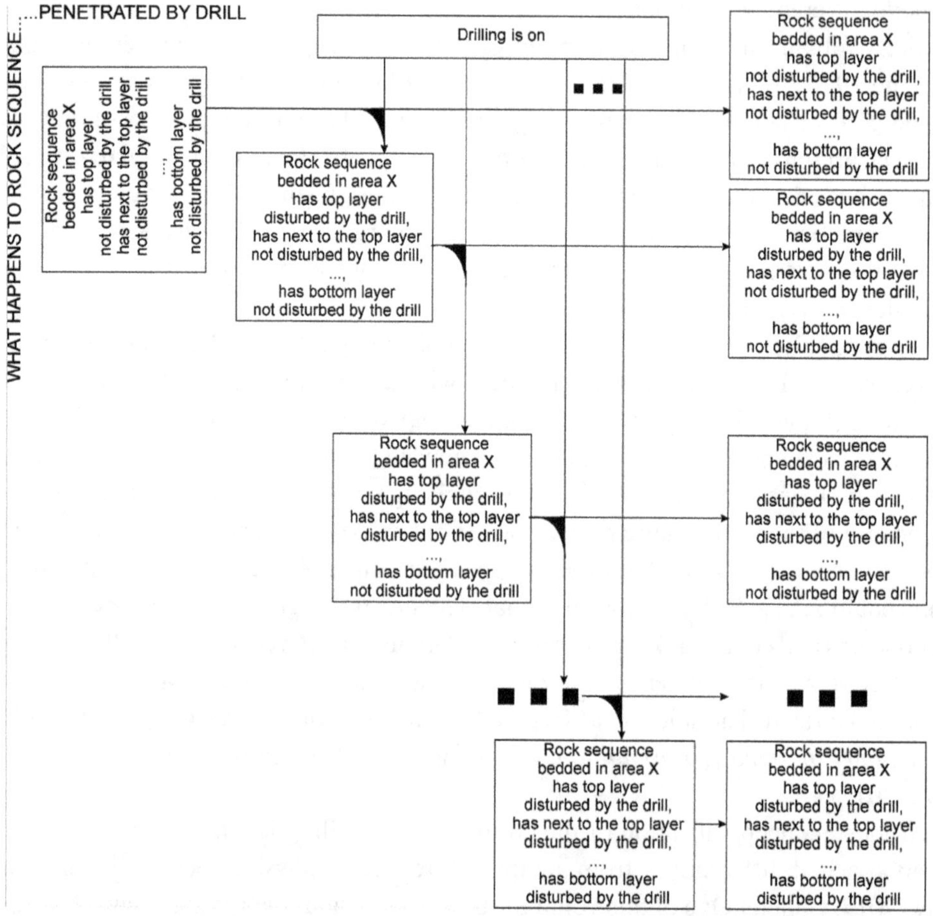

entire context that has given birth to the event bush, corresponding multitudes must be present as event resulting from intersection of the predmet "Layer" (or event "Rock layer") with primetas "to be top", "to be next to top", …, "to be bottom".

Scenarios of this event bush are built in a cascade form, starting with the "nothing happens" scenario (drilling is not on) and the first "drilling" scenario implying penetration of the top layer only. Once the top layer is penetrated, the next to the top one may be penetrated, and drilling may stop at it or go further, and so on. The model does not specify explicitly whether a layer is penetrated entirely or only partially, but, obviously, to touch the next layer, all upper ones must be penetrated entirely.

Just visually, the structure of the bush strongly resembles the spatial position of modeled entities: events describing the rock sequence are put horizontally, similar to the bedding in real geoenvironment, and the primary external event denotes the drilling instrument, which, indeed, is placed on the top of rock sequence (i.e., on the earth surface) and acts downward, as downward point the arrows of influx connectives of the event bush visualizing the results of work of this instrument.

Still, this model describes the case of "bare drilling", without producing a core. Drilling without core output is widely used in various types of geologic and construction project. Still, not less important option is drilling with core output. Core is sampled continuously from each penetrated level, thus modeling the entire interval of the sequence penetrated by drill. This mode of drilling is shown in Figure 2.

This event bush has somewhat more complicated structure. In contrast to the previous model, each event resulting from influx connective ("Rock sequence in area X has… layer *N* disturbed by the drill") itself produces another event by flux: "Core of layer *N* is taken", which immediately leads to the tertiary event of the same formulation. Collecting all such tertiary events, one may readily "build up" a core log of a drillhole.

One of possible developments of models presented in Figs. 1 and 2 could be adding quantitative parameters to primetas, e.g., values of thickness and density (drillability or the like) to layers and parameters indicating its power, to the drill. This will be surely done in future publications, but, importantly, these parameters would give an overall picture but tell us little or nothing on what happens inside a particular layer entered by the drill. To take a closer look at this, a physical model is needed.

Figure 2. Event bush describing the drilling of a rock sequence with core output

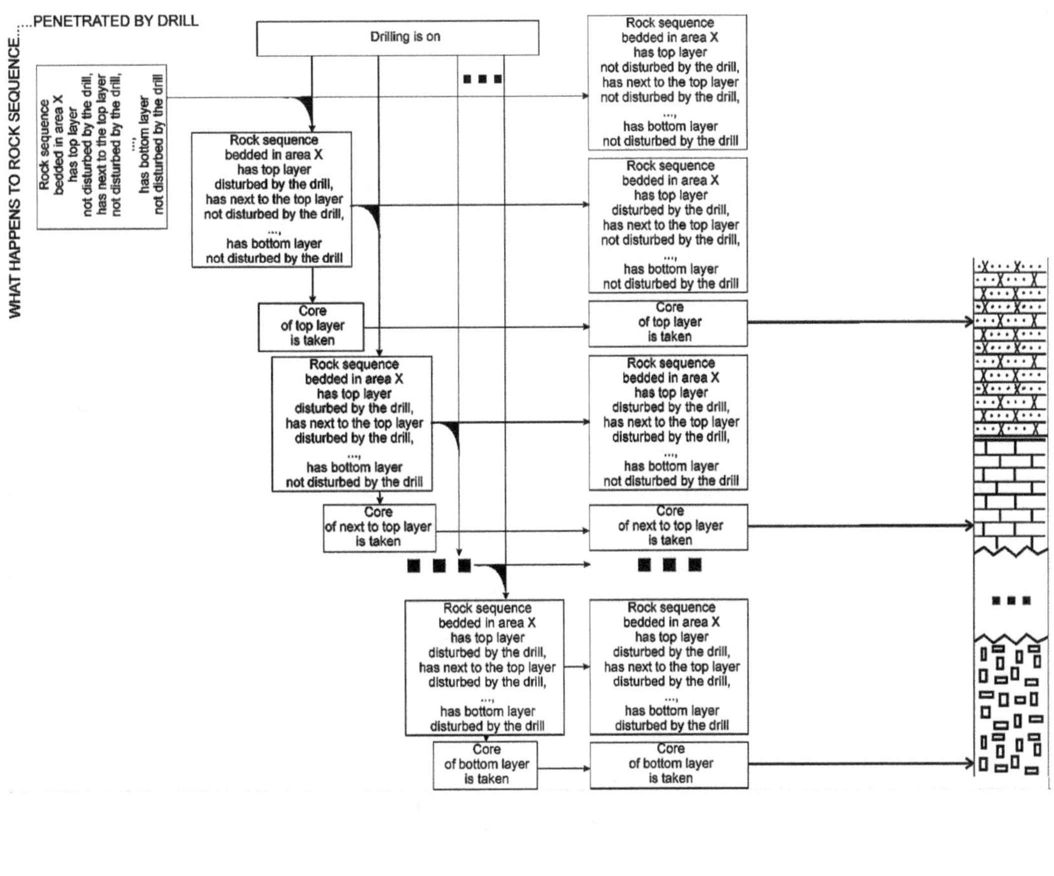

PHYSICAL KNOWLEDGE: PARSING A MODEL OF DISRUPTION OF MATERIAL BY SINGULAR CHISEL WITHOUT ROTATION, STRIKE AND BLUNTING

To perform physical modeling by means of event bush, one should, as suggested by Pshenichny et al. (2009); Pshenichny (2013) and Diviacco et al. (2014), either construct a physical model based on the event bush or parse an existing model by means of the event bush.

In the considered case, physical models for explanation and numerical assessment of the process in question have been suggested in abundance, mainly in the basic study courses for drilling engineers, and represent rather trivial task in the field of solid mechanics. For instance, in the course by Filimonenko (2008) taught at the Mining Faculty of Donetsk National Technical University, one of the largest ex-Soviet mining educational establishments, physical modeling of drilling is split into subsequent stages, of which the simplest is modeling of disruption of material by singular chisel without rotation, strike and blunting of the tool (Figure 3).

The model suggested by is Filimonenko (2008) is as follows.

Specific axial load on the chisel C_0 is defined as:

$$C_0 = \frac{C}{m}$$

where C is the axial load on the drilling instrument and m is the number of chisels in it.

Let N_1, N_2 be the resistant forces to chisel (see Figure 3) F_1, F_2, friction forces; friction angle; cutting angle; h, depth of penetration; f, friction coefficient defined as $f = tg\phi$.

Projecting the forces affecting the chisel onto the horizontal plane, we have:

Figure 3. Disruption of material by singular chisel without rotation, strike and blunting of the tool (modified from Filimonenko, 2006-2016). See comments in the text.

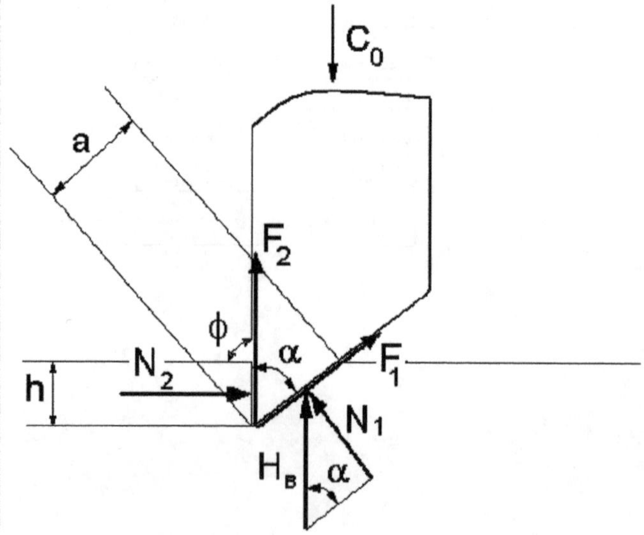

$$N_2 - N_1 \cos\alpha + F_1 \sin\alpha = 0$$

Then, $F_1 = N_1 f = N_1 tg\phi$ and hence, $N_2 - N_1 \cos\alpha + N_1 \sin\alpha tg\phi = 0$. Consequently:

$$N_2 = N_1 \left(\cos\alpha - \sin\alpha tg\phi\right) = N_1 \left(\cos\alpha \frac{\cos\phi}{\cos\phi} - \frac{\sin\phi}{\cos\phi} \sin\alpha \right) = N_1 \frac{\cos\left(\alpha + \phi\right)}{\cos\phi}$$

which leads to:

$$N_2 = N_1 \frac{\cos\left(\alpha + \phi\right)}{\cos\phi}$$

Projecting the same forces onto the vertical plane, obtain:

$$C_0 - F_2 - F_1 \cos\alpha - N_1 \sin\alpha = 0$$

The formula for C_0 is inferred as follows:

$$C_0 - N_2 tg\phi - N_1 tg\phi \cos\alpha - N_1 \sin\alpha = 0$$

$$C_0 = N_2 tg\phi + N_1 \left(\frac{\sin\phi}{\cos\phi} \cos\alpha + \sin\alpha \right)$$

$$C_0 = N_2 tg\phi + N_1 \left(\frac{\sin\phi \cos\alpha + \sin\alpha \cos\phi}{\cos\phi} \right) = N_2 tg\phi + N_1 \frac{\sin\left(\alpha + \phi\right)}{\cos\phi}$$

$$C_0 = N_1 \frac{\cos\left(\alpha + \phi\right)}{\cos^2\phi} + N_1 \frac{\sin\left(\alpha + \phi\right)}{\cos\phi} = N_1 \frac{\sin\left(\alpha + 2\phi\right)}{\cos^2\phi}$$

Finally:

$$C_0 = N_1 \frac{\sin\left(\alpha + 2\phi\right)}{\cos^2\phi}$$

Let us denote $H_в$ is the breakdown point of the top layer material at incision of a chisel. Specific pressure at the back side of the chisel is $H_H = H_в \sin\alpha$.

Given the contact surface of chisel with the material equal to $a \cdot b$ (where *b* is the chisel width) the following can be put forth:

$$N_1 = H_{=}ab = H_2 ab \sin\alpha$$

and:

$$h = a\cos\alpha$$

Hence:

$$a = \frac{h}{\cos\alpha}$$

and:

$$N_1 = H_2 \frac{h}{\cos\alpha} b \sin\alpha = H_2 hbtg\alpha$$

Therefore, the resistant force N_1 appears in a relation with the breakdown point, depth of penetration, chisel width and cutting angle:

$$N_1 = H_2 hbtg\alpha$$

Substituting N_1 in the above equation for C_0, one yields:

$$C_0 = \frac{H_2 hbtg\alpha \sin(\alpha + 2\phi)}{\cos^2\phi}$$

From this, the depth of penetration h can be expressed as:

$$h = \frac{C_0 \cos^2\phi}{H_2 btg\alpha \sin(\alpha + 2\phi)}$$

Denote:

$$\eta = \frac{\cos^2\phi}{\sin(\alpha + 2\phi)}$$

Then:

$$h = \frac{C_0\eta}{H_2 btg\alpha}$$

This shows that the depth of penetration h of a singular chisel without rotation can be computed as:

$$h = \frac{C_0\eta}{H_2 btg\alpha}$$

From this formula, it is evident that the depth of penetration is positively related to the axial load and inversely related to the breakdown point, chisel width and cutting angle.

In this inference, as commonly occurs in physical modeling, mathematical expressions are used to set the relations between physical parameters and to substitute one parameter by another. Although mathematically perfect, this approach, however, is not guaranteed to match physical reality at least at some steps. At times, to proceed from one seemingly reasonable formula to another, one needs to make a few steps that look formally correct but can hardly be interpreted in terms of physics or look even misleading, and these transformations are allowed only because "at the end" we obtain what we need; however, the "end" in inference is set quite arbitrarily and, given we stop, say, at the step:

$$a = \frac{h}{\cos\alpha}$$

we would need to admit that the length of the part of the chisel chamfer contacting the penetrated material is determined by the vertical depth of penetration into this material, which unlikely can be accepted as physically reasonable, as these two things, vertical and oblique penetration, happen simultaneously and have common cause, which is completely outside of this formula.

Also, it should be noted that incision of chisel into a solid material may be considered as having three distinct, however short, stages: first touch, pressure on the surface and incision. Perhaps this process could be best modeled by differential equations, but such model, due to its much higher complexity, would be less valuable practically, and the existing model can also be rethought and "parsed" with the account of these qualitative constraints.

This parsing can be performed by an event bush built to conceptualize what happens in the system "chisel – material" when the drilling instrument is put to work (Figure 4).

As was done in related studies in other fields (Pshenichny et al., 2009; Pshenichny, 2013, Diviacco et al., 2014), first an event bush is built that has no parameters and no additional computational meaning of arcs (connectives). The composition of this event bush reflects the three stages of interaction between the chisel and material, of which the second one (pressure on the surface of the material) is assumed to inevitably lead to the third one (breaking the surface and incision into the interior of the material). Indeed, one can hardly imagine a drilling tool endlessly pressing a material without penetration, from one side, and blunting (as set by the definition of the bush), from the other.

Two event predmets are involved in the bush, the top rock layer and chisel. A few primetas relate to each, and all of them appear valuable with further parametrization.

Parameters are introduced as extension of primetas when the event bush is built and checked for completeness and self-consistence. First of all, to understand what happens when a drilling instrument equipped with n chisels cuts the rock, it looks reasonable first to analyze the case of single chisel doing the same. Therefore, we consider reasonable not to define the axial load of a single chisel C_0 as the ratio of total load and the number of chisels but take it for initial parameter.

Figure 4. Event bush model of disruption of material by singular chisel without rotation, strike and blunting of the tool

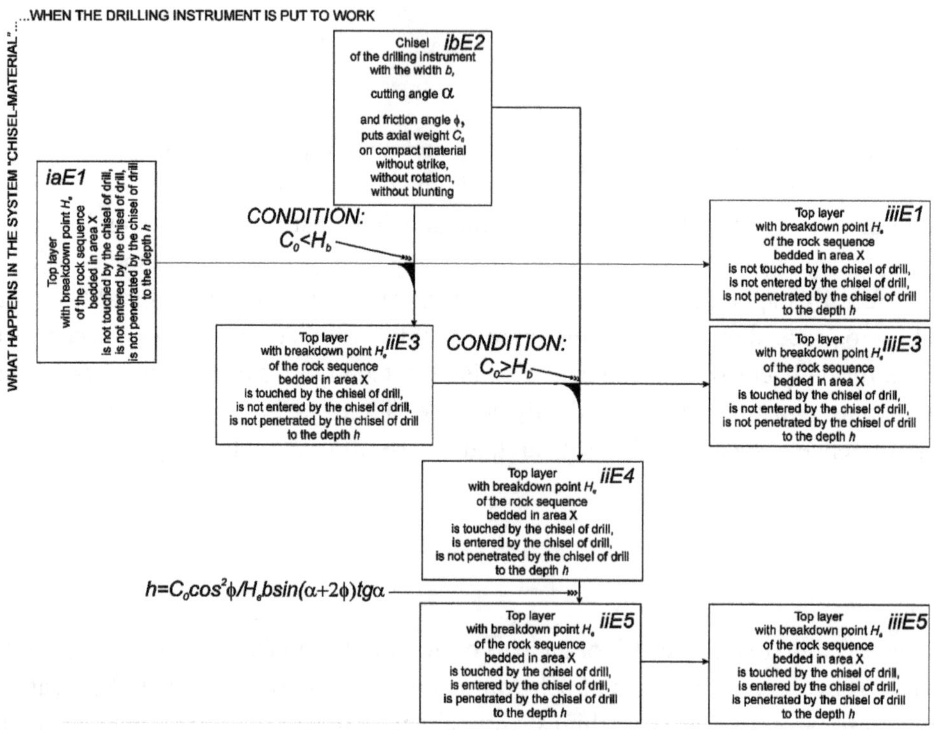

Then, each primeta used in the event bush acquires value for the physical modeling, either putting on one parameter or indicating constraints.

Primetas of the event predmet "Top layer":

"with breakdown point" – introduces the parameter $H_в$, meaning the breakdown point;
"of the rock sequence bedded in area X" – universum primeta that, in addition, relates this event bush to the bushes describing the geological aspect of drilling;
"to be touched by the chisel of drill" – allows comparison of C_0 and $H_в$;
"to be entered by the chisel of drill" – allows further primeta, "to be penetrated by the chisel of drill" (in other words, the event without this primeta but with the further one is excluded, by the rules of the multitude theory);
"to be penetrated by the chisel of drill" – introduces the parameter *h*, the depth of penetration.

Primetas of the event predmet "Chisel":

"to be of the drilling instrument" – universum primeta;
"to have the width" – introduces parameter *b*, the width;
"to have cutting angle" – introduces parameter α, the cutting angle;
"to have friction angle" – introduces parameter φ, the friction angle;
"to put axial weight on compact material" – introduces parameter C_0, the axial load;

Figure 5. Disruption of material by singular chisel without rotation, strike and blunting of the tool as seen after the event bush modeling

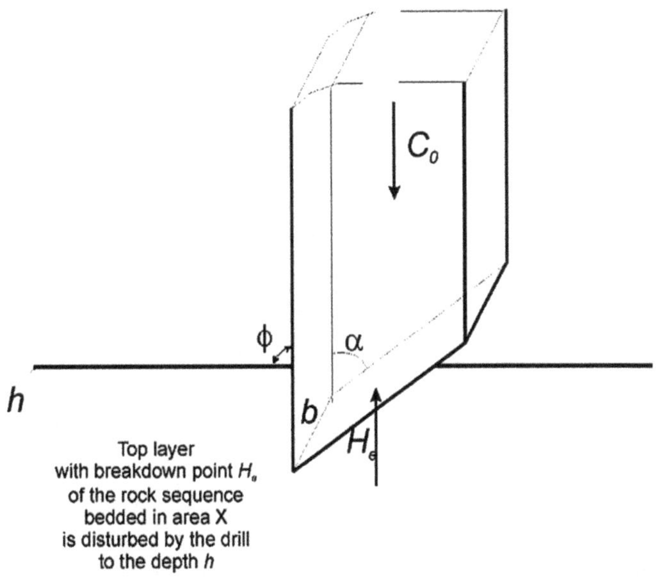

"to be struck" – negation of this primeta puts a constraint on the bush composition ("no strike");
"to rotate" – negation of this primeta puts a constraint on the bush composition ("no rotation");
"to blunt" – negation of this primeta puts a constraint on the bush composition ("no blunting").

Now, each connective of the event bush except those leading to tertiary events may be attributed a computational sense:

iaE1, ibE2 Influx iiE3 – condition: $C_0<H_{в}$;
iiE3, ibE2 Influx iiE4 – condition: $C_0 \geq H_{в}$;
iiE4 Flux iiE5 – computation: $h= C_0 cos^2\varphi /H_b bsin(\alpha+2\phi))tg\alpha$.

Putting this model back to the traditional sketch, we will see it is drastically reduced compared to the initial one (Figure 5).

DISCUSSION

The question arises, whether it is good or bad, to see the initial inference thus simplified (compare Figure 5 with Figure 3). The event-bush-based model takes only few parameters (axial load, breakdown point, cutting angle and friction angle, chisel width, depth of penetration) and omits others (friction forces, resistant forces, friction coefficient, specific pressure on the back side of the chisel, the length of the part of the chisel chamfer contacting the penetrated material). The selected parameters appear to be enough to describe the whole process in question (i.e., form the "skeleton" of the model readily parsed quantitatively), while others may be used just as auxiliary, to simplify computation given the data. Still, another question is whether the choice of the "skeleton" parameters is unique. Can there be taken other primetas for the same predmets (event predmets) entailing different parameters in the existing event bush, or could even a different event bush be built with other predmets and primetas to describe the same phenomenon (e.g., disruption of material by singular chisel without rotation, strike and blunting of the tool)? Even for this simple case, we have not enough ground to strictly prove the uniqueness of choice of the "skeleton" parameters.

Still, a definite advantage of this representation is clear "flow" of the model, in which the succession of constraints and inferred formulae corresponds to the change of events, i.e., scenario (chisel touches the material – chisel presses the material – chisel incises in the material). The same can be said about the "geologic" model of drilling with core output, where the core intervals would be "stacked" in the event bush exactly in the same order as rock layers are penetrated, if to introduce a general rule that once layer of rock is penetrated, it produces a core of the said rock. This seems to be exactly the link between the knowledge processes and core working processes claimed by Norheim & Fjellheim (2006).

Notwithstanding, in contrast to the "geological" model where the composition of core is obviously determined by that of the penetrated layers, in the physical model it is far from obvious why the depth of penetration h is positively related to the square of the cosine of friction angle, $cos^2\varphi$, and negatively, say, to the sine of an angle defined as the sum of cutting angle and double friction angle, $sin(\alpha+2\phi)$. Its meaning is absolutely unclear from the event bush.

To clarify it, obviously an inference is needed within event, which is not implied by the event bush, nor by any other tool of dynamic knowledge engineering (see, e.g., Naumov et al., 2017, or Rogalchuk et al., 2017 – Chapters 12 and 13 of this book). Hence, to be fully visualized and represented in terms of the theory of multitudes, this physical model also needs involvement of traditional, "static" tools of knowledge engineering sensu Pshenichny (2014) or Pshenichny & Mouromtsev (2015) – ontology, conceptual graphs or the like – but this, in turn, puts a theoretical task of reformulation of these methods in the predmet-primeta language of the theory of multitudes. This task is being only approached now for the IDEF notations and ontology design (Mishin, personal communication).

Another important issue is bringing together different aspects of drilling knowledge; in the present case – the geological and physical models. The predmet of the primary initial event of the "physical" event bush in Figure 4 is exactly the predmet that forms the first refutable (non-universum) primeta of the primary internal event in the "geological" event bushes in Figs. 1 and 2. To explain this relation, one should construct the entire context of drilling knowledge and plot a special graph suggested by Pshenichny (2017), termed predmetka in the Chapter Two of this book. This task should be considered as an issue of a separate study, in which, alongside, the following dimensions should be covered:

- Link to the knowledge domain of solid mechanics (as the physical model presented here applies not only to "Top layer … of the rock sequence bedded in area X", and not only to any other layer in any location, but also to any kind of solid material, not necessarily in geology) – more or less in the way proposed by Wolter et al. (2017) discussing the nested structures of reasoning (see Chapter 14 of this book);
- Development of both physical and geological aspects to capture various details of drilling (e.g., penetration by n chisels, with a strike, with rotation, with blunting, accounting for drilling parameters such as drillability, modified core recovery, rock fragmentation in core, based on rock hardness, abrasiveness, core diameter, drill angle relative to fracture surface and so forth, as well as drilling with a slurry, drilling of sea/lake floor substrate, drilling of frozen or partially molten rock and other options);

- Account of various aspects of drilling – economic, engineering, environmental and others.

As a result, a comprehensive framework of drilling knowledge in terms of the theory of multitudes will be created. This framework may hopefully serve as the conceptual ground for various industrial and academic drilling projects and drilling instrument design and be free of the shortcomings of existing knowledge models used in this field. It should include existing quantitative models suggested to compute drilling parameters and allow one to plot various scenarios, favorable and unfavorable, which may occur at drilling, and figure out the critical points in terms of technology and penetrated geological structure, being extendable specifiable to address particular drilling tasks, particular types of geological framework, particular types of equipment, and other particularities and peculiarities of drilling activity.

Importantly, it can be brought into an Internet-based system for support of collaborative research (Diviacco et al., 2014) and thus facilitate joint and distributed work on drilling projects.

Finally, the vocabulary of the bush, deliberately kept as simple and descriptive as possible (Pshenichny & Kanzheleva, 2011), opens an opportunity of binding the event bushes plotted in the domain of drilling activities, from one side, with upper-level bushes describing physical experimentation and other issues that constrain drillwork, and from the other, with existing ontologies (including the ontological description of drilling scenarios of Skalle & Aamodt, 2005), case-based reasoning patterns and other methods of knowledge engineering and representation successfully implemented in the domain of drilling, thus leading to creation of a fullfledged unified framework for engineering and representation of knowledge about drilling activities.

CONCLUSION

In the most trivial cases, the knowledge of geologic and physical aspects of drilling is readily represented by the event bush method. However, even at this stage, the relationship between the basic terms of geological and physical models appears not straightforward, though explicable in terms of the theory of multitudes. Handling physical knowledge reveals a substantial difference in understanding of the drilling phenomenon by formulae and by concepts. "Parsing" of formulae by conceptual framework shows different meanings wrapped in the same mathematical form.

Despite all the peculiarities, further development of research is expected to lead to a comprehensive framework of drilling knowledge in terms of the theory of multitudes that would serve as the conceptual ground for various industrial and academic drilling projects and drilling instrument design.

REFERENCES

Diviacco, P., Pshenichny, C., Carniel, R., Khrabrykh, Z., Shterkhun, V., Mouromtsev, D., & Pascolo, P. et al. (2014). Organization of a geophysical information space by using an event-bush-based collaborative tool. *Earth Science Informatics*, *4*(3), 139–160. doi:10.1007/s12145-014-0182-2

Filimonenko, N. T. (2008). *Lectures on Disruption of Rocks*. Donetsk National Technical University. (unpublished; in Russian)

Gruber, T. (2009). Ontology. In L. Liu & M. T. Özsu (Eds.), *Encyclopedia of Database Systems. Springer-Verlag*.

Norheim, D., & Fjellheim, R. (2006). AKSIO - Active Knowledge management in the petroleum industry. In A. Leger, A. Kulas, L. Nixon, & R. Meersman (Eds.), *How Business application challenges Semantic Web Research (ESWC'06 Industry Forum)* (pp. 22–26). Available from http://ceur-ws.org/Vol-194/complete.pdf

Pshenichny, C., Carniel, R., & Diviacco, P. (2013). Engineering of Dynamic Knowledge in Exact Sciences: First Results of Application of the Event Bush Method in Physics. *MSEPS 2013 – Modeling States, Events, Processes and Scenarios; Workshop Proceedings*, 60-73. Retrieved from http://iccs2013.hbcse.tifr.res.in/workshops/copy2_of_text.pdf

Pshenichny, C. A. (2014). Knowledge engineering in volcanology: Practical claims and general approach. *Journal of Volcanology and Geothermal Research, 286*, 78–92. doi:10.1016/j.jvolgeores.2014.07.012

Pshenichny, C. A., & Kanzheleva, O. M. (2011). Theoretical Foundations of the Event Bush. In K. Sinha, L. Gundersen, J. Jackson, & D. Arctur (Eds.), Geological Society of America Special Paper: Vol. 482. Societal Challenges and Geoinformatics (pp. 139–165). Academic Press. doi:10.1130/2011.2482(12)

Pshenichny, C. A., & Mouromtsev, D. I. (2015). Grammar of Dynamic Knowledge for Collaborative Knowledge Engineering and Representation. In Collaborative Knowledge in Scientific Research Networks. IGI Global, 323-354. doi:10.4018/978-1-4666-6567-5.ch016

Pshenichny, C.A., Nikolenko, S.I., Carniel, R., Vaganov, P.A., Khrabrykh, Z.V.., Moukhachov, V.P., … Rezyapkin, A.A. (2009). The Event Bush as a Semantic-based Numerical Approach to Natural Hazard Assessment (Exemplified by Volcanology). *Computers and Geosciences, 35*(5), 1017-1034. doi 10.1016/j.cageo.2008.01.009

Rogalchuk, V. V., Arustamov, S. A., Solovyov, N. A., & Malisheva, A. V. (2017). Petri Net versus Event Bush in Description of Environments of Alternative Directed Changes. In C. A. Pshenichny, P. Diviacco, & D. I. Mouromtsev (Eds.), *Representation of Dynamic Knowledge in Scientific Domains*. IGI Global.

Skalle, P., & Aamodt, A. (2005). Knowledge-based decision support in oil well drilling: combining general and case-specific knowledge for problem solving. In Z. Shi & Q. He (Eds.), *Intelligent information processing II* (pp. 443–455). London, UK: Springer-Verlag. doi:10.1007/0-387-23152-8_56

Studer, R., Benjamins, V. R., & Fensel, D. (1998). Knowledge Engineering: Principles and methods. *Data & Knowledge Engineering, 25*(1-2), 161–197. doi:10.1016/S0169-023X(97)00056-6

Xu, Y. (2012). Research on Method of Ontology-Based Knowledge Representation and Integration for Drilling Risk Decision. *International Journal of Digital Content Technology and Its Applications, 6*(16). doi:10.4156/jdcta.vol6.issue16.33

Compilation of References

Abdulla, P., Delzanno, G., & Van Begin, L. (2007). *A Classification of the Expressive Power of Well-structured Transition Systems*. Technical Report 95. Centre Federe en Verification, Université libre de Bruxelles. Retrieved from http://www.ulb.ac.be/di/ssd/cfv

Acocella, V., & Neri, M. (2009). Dike propagation in volcanic edifices: Overview and possible developments. *Tectonophysics*, *471*(1-2), 67–77. doi:10.1016/j.tecto.2008.10.002

Acocella, V., Neri, M., & Norini, G. (2013). An overview of analogue models to understand a complex volcanic instability: Application to Etna, Italy. *Journal of Volcanology and Geothermal Research*, *251*, 98–111. doi:10.1016/j.jvolgeores.2012.06.003

Acocella, V., Neri, M., & Scarlato, P. (2006a). Understanding shallow magma emplacement at volcanoes: Orthogonal feeder dikes during the 2002–2003 Stromboli (Italy) eruption. *Geophysical Research Letters*, *33*(17), L17310. doi:10.1029/2006GL026862

Acocella, V., Neri, M., & Sulpizio, R. (2009). Dike propagation within active central volcanic edifices: Constraints from Somma-Vesuvius, Etna and analogue models. *Bulletin of Volcanology*, *71*(2), 219–223. doi:10.1007/s00445-008-0258-2

Acocella, V., Porreca, M., Neri, M., Massimi, E., & Mattei, M. (2006b). Propagation of dikes at Vesuvio (Italy) and the effect of Mt. Somma. *Geophysical Research Letters*, *33*(8), L08301. doi:10.1029/2005GL025590

Aguirre-Díaz, G. (2008). Types of collapse calderas. *IOP Conference Series: Earth and Environmental Science*, 3. doi:10.1088/1755-1307/3/1/012021

Akasaka, F., Nemoto, Y., Chiba, R., & Shimomura, Y. (2012). Development of PSS Design Support System: Knowledge-based Design Support and Qualitative Evaluation. *Procedia CIRP*, *3*, 239 – 244.

Akimova-Shterkhun, V. L., & Pshenichny, C. A. (2007). Quick Extraction of Multi-Format Relevant Volcanological Information by Means of Semantic and Smysl Networks. *Abstracts, IUGG General Assembly XXIV2007*. Retrieved from http://www.iugg2007perugia.it/webbook/

Allen, J. F. (1981). An interval-based representation of temporal knowledge. In *International Conference of Artificial Intelligence*, Morgan Kaufmann.

Allen, J. F. (1983). Maintaining knowledge about temporal intervals. *Communications of the ACM*, *26*(11), 832–843. doi:10.1145/182.358434

Allen, J. F. (1984). Towards a general theory of action and time. *Artificial Intelligence*, *23*(2), 123–154. doi:10.1016/0004-3702(84)90008-0

Allen, J. F., & Ferguson, G. (1994). Actions and events in interval temporal logic. *Journal of Logic and Computation*, *4*(5), 531–579. doi:10.1093/logcom/4.5.531

Alley, R. B., Clark, P. U., Huybrechts, P., & Joughin, I. (2005). Ice-sheet and sea-level changes. *Science*, *310*(5747), 456–460. doi:10.1126/science.1114613 PMID:16239468

Alley, R. B., Spencer, M. K., & Anandakrishnan, S. (2007, October). Ice-sheet mass balance: Assessment, attribution and prognosis. *Annals of Glaciology*, *46*(1), 1–7. doi:10.3189/172756407782871738

AMAP. (2011). Snow, Water, Ice and Permafrost in the Arctic (SWIPA): Climate Change and the Cryosphere. Arctic Monitoring and Assessment Programme (AMAP).

An, G. (2009). Dynamic Knowledge Representation Using Agent-Based Modeling: Ontology Instantiation and Verification of Conceptual Models. In I.V. Maly (Ed.), Methods in Molecular Biology. Humana Press. Doi:10.1007/978-1-59745-525-1_15

Anoprienko, A. Ya., & Ivanitsa, S. V. (2001). Postbinarnyi kompiuting i interval'nye vychisleniya v kontekste kodologicheskoi evolutsii [Post-Binary Computing and Interval Computation in the Context of Code-Logical Evolution]. Donetsk National Technical University – UNITEX Publishers.

Aramayo, A., Guzmán, S., Hongn, F., del Papa, C., Montero López, C., & Sudo, M. (2017). A Middle Miocene (13.5–12 Ma) deformational event constrained by volcanism along the Puna – Eastern Cordillera border, NW Argentina. *Tectonophysics*, *703-704*, 9–22. doi:10.1016/j.tecto.2017.02.018

Arendt, R., & van Uden, E. (2011). A decision-making module for aiding ship system automation design: A knowledge-based approach. *Expert Systems with Applications*, *38*(1), 410–416. doi:10.1016/j.eswa.2010.06.079

Armstrong-Altrin, J. S., Madhavaraju, J., Asir, N. G. G., & Ramasamy, S. (2005). *Provenance and depositional history of sandstones from the Upper Miocene Kudankulam Formation*. Tamil Nadu.

Armstrong-Altrin, J. S., Ramasamy, S., & Makhnach, A. (2001). Stable Isotope Geochemistry and Evidence for Meteoric Diagenesis in Kudankulam Formation,Tamil Nadu. *Journal of the Geological Society of India*, *57*, 39–48.

Armstrong-Altrin, J. S., Verma, S. P., Lee, Y. I. I., & Ramasamy, S. (2004). Geochemistry of sandstones from the upper Miocene Kudankulam Formation, southern India: Implications for provenance, weathering, and tectonic setting. *Journal of Sedimentary Research*, *74*(2), 285–297. doi:10.1306/082803740285

Armstrong-Altrin, J. S., Verma, S. P., Madhavaraju, J., Lee, Y. I. I., & Ramasamy, S. (2003). Geochemistry of Upper Miocene Kudankulam Limestones, Southern India. *International Geology Review*, *45*(1), 16–26. doi:10.2747/0020-6814.45.1.16

Baillie, P. W., Shaw, R. D., Liyanaarachchi, D. T. P., & Jayaratne, M. G. (2002). *A New Mesozoic Sedimentary Basin, Offshore Sri Lanka*. EAGA 64th Conference & Exhibition, Florence, Italy.

Barr, M., & Wells, C. (2012). Category Theory for Computing Science. *Reprints in Theory and Applications of Categories*, *22*, 1–538.

Bartholomew, I., Nienow, P., Mair, D., Hubbard, A., King, M. A., & Sole, A. (2010). Seasonal evolution of subglacial drainage and acceleration in a Greenland outlet glacier. *Nature Geoscience*, *3*(6), 408–411. doi:10.1038/ngeo863

Battaglia, M., Di Bari, M., Acocella, V., & Neri, M. (2011). Dike emplacement and flank instability at Mount Etna: Constraints from a poro-elastic model of flank collapse. *Journal of Volcanology and Geothermal Research*, *199*(1-2), 153–164. doi:10.1016/j.jvolgeores.2010.11.005

Bause, F., Kemper, P., & Kritzinger, P. (1994). *Abstract Petri Net Notation*. Universitaet Dortmund.

Becher, T., & Trowler, P. R. (2001). *Academic tribes and territories intellectual enquiry and culture of disciplines*. Buckingham, UK: The Society for Research into Higher Education and Open University Press.

Behncke, B. (2009). Hazards from pyroclastic density currents at Mt. Etna (Italy). *Journal of Volcanology and Geothermal Research*, *180*(2), 148–160. doi:10.1016/j.jvolgeores.2008.09.021

Behncke, B., Fornaciai, A., Neri, M., Favalli, M., Ganci, G., & Mazzarini, F. (2016). LiDAR surveys reveal eruptive volumes and rates at Etna, 2007-2010. *Geophysical Research Letters*, *43*(9), 4270–4278. doi:10.1002/2016GL068495

Behncke, B., Neri, M., & Carniel, R. (2003). An exceptional case of lava dome growth spawning pyroclastic avalanches at Mt. Etna (Italy): The 1999 Bocca Nuova eruption. *Journal of Volcanology and Geothermal Research*, *124*, 15–128. doi:10.1016/S0377-0273(03)00072-6

Behncke, B., & Pshenichny, C. A. (2009). Modeling unusual eruptive behavior of Mt. Etna, Italy, by means of event bush. *Journal of Volcanology and Geothermal Research*, *185*(3), 157–171. doi:10.1016/j.jvolgeores.2009.04.020

Bennett, B., & Galton, A. P. (2004). A unifying semantics for time and events. *Artificial Intelligence*, *153*(1–2), 13–48. doi:10.1016/j.artint.2003.02.001

Bishop, M. A. (1992). Theory-Ladenness of Perception Arguments. In *Proceedings of the Biennial Meeting of the Philosophy of Science Association* (pp. 287-299). Academic Press.

Bittner, T. (2002). Approximate qualitative temporal reasoning. *Annals of Mathematics and Artificial Intelligence*, *36*(1–2), 39–80. doi:10.1023/A:1015899702951

Blizard, W. (1989). Multiset Theory. *Notre Dame Journal of Formal Logic*, *30*(1), 36–66. doi:10.1305/ndjfl/1093634995

Bogdanov, A. (1926). *Allgemeine Organisationslehre*. Berlin: Alexander und Lang.

Bonham-Carter, G. F., Smyth, C., & Kerswill, J. (2003). Semantic net approach for mapping mineral potential: a feasibility study. Modeling Geohazards: IAMG 2003.

Booch, G. (1999). *The Unified Modeling Language User Guide*. Pearson Education.

Box, P., Jenkins, G., Reinsel, G., & Ljung, G. M. (2012). *Time Series Analysis Forecasting and Control*. Wiley Series in Probability and Statistics.

Boyd, S., Zowghi, D., & Farroukh, A. (2005). Measuring the expressiveness of a constrained natural language: an empirical study. *13th IEEE International Conference on Requirements Engineering*, 339-349. doi:10.1109/RE.2005.39

Branca, S., Coltelli, M., & Groppelli, G. (2015). Carta Geologica del Vulcano Etna. *Mem Descr Carta Geol D'Italia, 98*.

Branney, M. J., & Kokelaar, P. (2002). Pyroclastic Density Currents and the Sedimentation of Ignimbrites. Geological Society Memoir, 27.

Braun, I., & Kriegsman, L. M. (2003). Proterozoic crustal evolution of southernmost India and Sri Lanka. In M. Yoshida & B. F. Windley (Eds.), Geological Society, London, Special Publications: Vol. 206. Proterozoic East Gondwana: Supercontinents Assembly and Breakup (pp. 169–203). Academic Press. doi:10.1144/GSL.SP.2003.206.01.10

Bruno, P. P., & Rapolla, A. (1999). Study of the sub-surface structure of Somma-Vesuvius (Italy) by seismic reflection data. *Journal of Volcanology and Geothermal Research*, *92*(3-4), 373–387. doi:10.1016/S0377-0273(99)00093-1

Bursik, M. (2009). A general model for tectonic control of magmatism: Examples from Long Valley Caldera (USA) and El Chichón (México). *Geofísica Internacional*, *48*(1), 171–183.

Calkins, D. E., Egging, N., & Scholz, C. (2000). Knowledge-Based Engineering (KBE) Design Methodology at the Undergraduate and Graduate Levels. *International Journal of Engineering Education*, *16*(1), 21–38.

Callender, C., & Cohen, J. (2005). There Is No Special Problem About Scientific Representation. *Theoria*, *55*, 67–81.

Calvari, S., Tanner, L. H., Groppelli, G., & Norini, G. (2004). A comprehensive model for the opening of the Valle del Bove depression and hazard evaluation for the eastern flank of Etna volcano. AGU (Geophysical Monograph), 143, 65–75.

Campelo, C. E. C. (2013). *Representing and Reasoning about Changing Spatial Extensions of Geographic Features* (Doctoral thesis). University of Leeds.

Campelo, C. E. C., Bennett, A. P., & Dimitrova, V. (2011). Identifying geographical processes from time-stamped data. *Lecture Notes in Computer Science*, *6631*(5), 70–87. doi:10.1007/978-3-642-20630-6_5

Campelo, C. E. C., Bennett, B., & Dimitrova, V. (2012). From polygons and timestamps to dynamic geographic features: Grounding a spatio-temporal geo-ontology. *Lecture Notes in Computer Science*, *7518*(10), 251–260. doi:10.1007/978-3-642-33999-8_31

Campelo, C. E. C., Bennett, B., & Dimitrova, V. (2013). Identifying geographic events and processes in spatio-temporal data. *Frontiers of Artificial Intelligence and Applications.*, *241*(8), 47–58.

Cappello, A., Neri, M., Acocella, V., Gallo, G., Vicari, A., & Del Negro, C. (2012). Spatial vent opening probability map of Mt. Etna volcano (Sicily, Italy). *Bulletin of Volcanology*, *74*(9), 2083–2094. doi:10.1007/s00445-012-0647-4

Carniel, R., Pshenichny, C., Khrabrykh, Z., Shterkhun, V., & Pascolo, P. (2011). Modeling Models: Understanding of Structure of Geophysical Knowledge by Means of the Event Bush Method. *Mathematical geosciences at the crossroads of theory and practice, IAMG Proceedings*, 1336-1350. doi doi:10.5242/iamg.2011.0074

Carniel, R., Guzmán, S., & Neri, M. (2017). FIERCE: FInding volcanic ERuptive CEnters by a grid-searching algorithm in R. *Bulletin of Volcanology*, *79*(2), 19. doi:10.1007/s00445-017-1102-3

Cas, R. A. F., Wright, H. M. N., Folkes, C. B., Lesti, C., Porreca, M., Giordano, G., & Viramonte, J. G. (2011). The flow dynamics of an extremely large volume pyroclastic flow, the 2.08-Ma Cerro Galán Ignimbrite, NW Argentina, and comparison with other flow types. *Bulletin of Volcanology*, *73*(10), 1583–1609. doi:10.1007/s00445-011-0564-y

Cas, R. A. F., & Wright, J. V. (Eds.). (1987). Volcanic Successions. Modern and Ancient. *Geological Magazine*, *125*(2), 189. doi:10.1017/S0016756800009602

Chadwick, W. W. Jr, & Howard, K. A. (1991). The pattern of circumferential and radial eruptive fissures on the volcanoes of Fernandina and Isabela islands, Galapagos. *Bulletin of Volcanology*, *53*(4), 259–275. doi:10.1007/BF00414523

Chakravartty, A. (1995). Truth and Representa- tion in Science: Two Inspirations from Art. In R. Frigg & M. Hunter (Eds.), *Beyond Mimesis and Convention: Representation in Art and Science* (pp. 33-50). Berlin: Springer.

Chandra, S., Gideon, N. D., Chaturvedi, C., & Rana, M. S. (2006). *6th International Conference & Exposition on Petroleum Geophysics*. Retrieved from https://www.spgindia.org/conference/6thconf_kolkata06/161.pdf

Chauhan, T., Longhinos, B., & Seralathan, P. (2008). Characterization of carbonate substratum in and around Chettikulam-Shrirangnarayanapuram villages, Radhapuram Taluk, Tirunelveli, Tamilnadu, India. *Proceedings of the XXV Annual Convention of Indian Association of Sedimentologists and National Seminar on Sedimentary basins of India - Their geological significance and Economic Prospects,* 49-50.

Chen, J. L., Wilson, C. R., & Tapley, B. D. (2006). Satellite gravity measurements confirm accelerated melting of Greenland ice sheet. *Science*, *313*(5795), 1958–1960. doi:10.1126/science.1129007 PMID:16902089

Chen, P. P.-S. (1976). The Entity-Relationship Model: Toward a Unified View of Data. *ACM Transactions on Database Systems*, *1*(1), 9–36. doi:10.1145/320434.320440

Christensen, C. M. (2014). *The Innovator's Dilemma: When New Technologies Cause Great Firms to Fail*. Alpina Publishers.

Cioni, R., Santacroce, R., & Sbrana, A. (1999). Pyroclastic deposits as a guide for reconstructing the multi-stage evolution of the Somma-Vesuvius caldera. *Bulletin of Volcanology*, *60*(4), 207–222. doi:10.1007/s004450050272

Claramunt, C., & Theriault, M. (1995). Managing time in GIS: An event-oriented approach. *Proceedings Recent Advances in Temporal Databases*, *1*(9), 23–42.

Claramunt, C., & Theriault, M. (1996). *Towards semantics for modelling spatio-temporal processes within GIS. In Advances in GIS Research* (pp. 27–43). Taylor and Francis.

Clarke, B. L. (1981). A calculus of individuals based on connection. *Notre Dame Journal of Formal Logic*, *22*(3), 204–218. doi:10.1305/ndjfl/1093883455

Clarke, B. L. (1985). Individuals and points. *Notre Dame Journal of Formal Logic*, *26*(1), 61–75. doi:10.1305/ndjfl/1093870761

Clark, P. U., Shakun, J. D., Marcott, S. A., Mix, A. C., Eby, M., Kulp, S., & Plattner, G.-K. et al. (2016). Consequences of twenty-first-century policy for multi-millennial climate and sea-level change. *Nature Climate Change*, *6*(4), 360–369. doi:10.1038/nclimate2923

Codd, E. F. (1970). A relational model of data for large shared data banks. *Communications of the ACM*, *13*(6), 377–387. doi:10.1145/362384.362685

Coelho, C., Silva, R., Gomes, F. V., & Pinto, F. T. (2006). A Vulnerability Analysis Approach for the Portuguese West Coast. *Transactions of the Wessex Institute*, *12*. doi:10.2495/RISK060241

Cohn, A. G., Gooday, J. M., & Bennett, B. (1994). A comparison of structures in spatial and temporal logics. Proceedings of Philosophy and the Cognitive Sciences, 1–14.

Cohn, A. G., Bennett, B., Gooday, J., & Gotts, N. (1997). Qualitative spatial representation and reasoning with the region connection calculus. *GeoInformatica*, *1*(3), 275–316. doi:10.1023/A:1009712514511

Coltelli, M., Garduño, V. H., Neri, M., Pasquarè, G., & Pompilio, M. (1994). Geology of northern wall of Valle del Bove, Etna (Sicily). *Acta Vulcanologica*, *5*, 55–68.

Cooray, P. G. (1984). Geology with special reference to the Precambrian. In C. H. Fernando (Ed.), *Ecology and Biogeography of Sri Lanka* (pp. 1–34). The Hague, The Netherlands: Dr. W. Junk B.V. Publishers. doi:10.1007/978-94-009-6545-4_1

Corazzato, C., Francalanci, L., Menna, M., Petrone, C., Renzulli, A., Tibaldi, A., & Vezzoli, L. (2008). What controls sheet intrusion in volcanoes? Petrological and structural characters of the Stromboli sheet complex, Italy. *Journal of Volcanology and Geothermal Research*, *173*, 26–54. doi:10.1016/j.jvolgeores.2008.01.006

Corbi, F., Rivalta, E., Pinel, V., Maccaferri, F., Bagnardi, M., & Acocella, V. (2015). How caldera collapse shapes the shallow emplacement and transfer of magma in active volcanoes. *Earth and Planetary Science Letters*, *431*, 287–293. doi:10.1016/j.epsl.2015.09.028

Corsaro, R. A., Neri, M., & Pompilio, M. (2002). Paleo-environmental and volcano-tectonic evolution of the south-eastern flank of Mt. Etna during the last 225 ka inferred from volcanic succession of the «Timpe», Acireale, Sicily. *Journal of Volcanology and Geothermal Research*, *113*(1-2), 289–306. doi:10.1016/S0377-0273(01)00262-1

Courts, H. R., & Keith, G. B. (1997). *Method for preventing overwriting cache buffer transaction entries until corresponding log buffer entries have been copied to log partition of the disk.* U.S. Patent No. 5,636,360. 3.

Cramer, J. G. (1986). The transactional interpretation of quantum mechanics. *Reviews of Modern Physics*, *58*(3), 647–688. doi:10.1103/RevModPhys.58.647

Cuffey, K. M., & Paterson, W. S. B. (2010). *The physics of glaciers*. Burlington, MA: Butterworth-Heinemann/Elsevier.

Cui, Z., Cohn, A., & Randell, D. A. (1993). Qualitative and topological relationships in spatial databases. Lect. Notes Comput. Sci., Advances in Spatial Databases, 692(5), 296–315. doi:10.1007/3-540-56869-7_17

Curray, J. R. (1984). Sri Lanka: Is it a mid-plate platelet? *J. Nara*, *31*, 30–50.

da Silva, J. C., Matelli, J. A., & Bazzo, E. (2014). Development of a knowledge-based system for cogeneration plant design: Verification, validation and lessons learned. *Knowledge-Based Systems*, *67*, 230–243. doi:10.1016/j.knosys.2014.05.002

Damodaran, T., Srinivasa Rao, G., Radhakrishna, M., & Murthy, K. S. R. (2016). Crustal structure and rift tectonics across the Cauvery–Palar basin, Eastern Continental Margin of India based on seismic and potential field modeling. *Journal of Earth System Science*, *125*(2), 329–342. doi:10.1007/s12040-016-0669-y

DANE 2.0 Users Guide. (2011). Retrieved from http://dilab.cc.gatech.edu/dane/files/DANE%202%20Users%20Guide.pdf

Davey, B. A., & Priestley, H. A. (2002). *Introduction to Lattices and Order* (2nd ed.). Cambridge University Press. doi:10.1017/CBO9780511809088

David, P. A., & Bunn, J. A. (1988). The economics of gateway technologies and network evolution. *Information Economics and Policy*, *3*(2), 165–202. doi:10.1016/0167-6245(88)90024-8

DeConto, R. M., & Pollard, D. (2016). Contribution of Antarctica to past and future sea-level rise. *Nature*, *531*(7596), 591–597. doi:10.1038/nature17145 PMID:27029274

Destruction of Syria's chemical weapons. (2014). In *Wikipedia.* Retrieved from http://en.wikipedia.org/wiki/Destruction_of_Syria%27s_chemical_weapons

Diskin, Z., & Wolter, U. (2008). A Diagrammatic Logic for Object-Oriented Visual Modeling. *ENTCS*, *203*(6), 19–41.

Diviacco, P. (2012). Addressing Conflicting Cognitive Models in Collaborative E-Research: A Case Study in Exploration Geophysics in Collaborative and Distributed E-Research. In Collaborative and Distributed E-Research: Innovations in Technologies, Strategies and Applications. IGI Global. doi:10.4018/978-1-4666-0125-3.ch012

Diviacco, P. (2015) Reconciling Knowledge and Collaborative e-Research. In Collaborative Knowledge in Scientific Research Networks. Hershey, PA: IGI Global. doi:10.4018/978-1-4666-6567-5.ch001

Diviacco, P., Juan, A. A., Daradoumis, T., Roca, M., Grasman, S. E., & Faulin, J. (2012). Addressing Conflicting Cognitive Models in Collaborative E-Research: A Case Study in Exploration Geophysics in Collaborative and Distributed E-Research. In Collaborative and Distributed E-Research: Innovations in Technologies, Strategies and Applications. IGI Global.

Diviacco, P. (2005). An open source, web based, simple solution for seismic data dissemination and collaborative research. *Computers & Geosciences*, *31*(5), 599–605. doi:10.1016/j.cageo.2004.11.008

Diviacco, P. (2013). Addressing Conflicting Cognitive Models in Collaborative e-Research: A Case Study in Exploration Geophysics. In *Collaborative and Distributed e-Research: Innovations in Technologies, Strategies and Applications*. Hershey, PA: IGI Global.

Diviacco, P. (2017). On Formalization and Representation in Collaborative Research. In C. A. Pshenichny, P. Diviacco, & D. I. Mouromtsev (Eds.), *Dynamic Knowledge Representation in Scientific Domains*. IGI Global.

Diviacco, P., & Busato, A. (2013). The Geo-Seas Seismic data viewer: A tool to facilitate control of data access. *Bollettino di Geofisica Teorica ed Applicata*, *54*(3), 257–270.

Diviacco, P., & Busato, A. (2015). Maps, graphs and annotations as Boundary Objects in Knowledge Networks, Distributed Cognition and Collaborative e-Research. In P. Diviacco, P. Fox, C. Pshenichny, & A. Leadbetter (Eds.), *Collaborative Knowledge in Scientific Research Networks*. Hershey, PA: IGI Global. doi:10.4018/978-1-4666-6567-5.ch019

Diviacco, P., DeCawer, K., Leadbetter, A., Sorribas, J., Stojanov, Y., Busato, A., & Cova, A. (2014). *Bridging semantically different paradigms in the field of marine acquisition event logging, Earth Science Informatics*. Springer. doi:10.1007/s12145-014-0192-0

Diviacco, P., & Pshenichny, C. A. (2010). Concept-referenced spaces in computer-supported collaborative work. *Geophysical Research Abstracts*, (12).

Diviacco, P., & Pshenichny, C. A. (2011). *A case study on the use of event bushes as a formal representation for Computer Supported Collaborative Work in the GeoSciences. European Geosciences Union*. EGU.

Diviacco, P., Pshenichny, C. A., Behncke, B., & Kanzheleva, O. M. (2014b). Sozdanie sistem podderzhki sovmestnykh nauchnykh issledovanii v naukah o Zemle v srede COLLA na osnove metoda kusta sobytii [in Russian]. *Vestnik ITMO*, *5*(93), 83–91.

Diviacco, P., Pshenichny, C., Behncke, B., & Kanzheleva, O. (2014). Systems for support of collaborative studies in the COLLA environment based on the event bush. *Scientific and Technical Journal of Information Technologies Mechanics and Optics*, *5*(93), 83–91. (in Russian)

Diviacco, P., Pshenichny, C., Carniel, R., Khrabrykh, Z., Shterkhun, V., Mouromtsev, D., & Pascolo, P. et al. (2014). Organization of a geophysical information space by using an event-bush-based collaborative tool. *Earth Science Informatics*, *4*(3), 139–160. doi:10.1007/s12145-014-0182-2

Drury, S. A., Harris, N. B. W., Holt, R. W., Reeves-Smith, G. J., & Wightman, R. T. (1984). Pre-Cambrian Tectonics and Crustal Evolution in South India. *Journal of Geophysics*, *92*, 3–20.

Eco, U. (1981). Guessing: From Aristotle to Sherlock Holmes. *Versus*, *30*.

Egenhofer, M. J., Mark, D. M., & Herring, J. (1994b). *The 9-intersection: Formalism and Its Use for Natural-Language Spatial Predicates*. Technical Report 94-1, National Center for Geographic Information and Analysis, Santa Barbara, CA.

Egenhofer, M. J. (1991). Reasoning about binary topological relations. *Advanced Spatial Databases*, *525*(8), 141–160. doi:10.1007/3-540-54414-3_36

Egenhofer, M. J., Clementini, E., & Di Felice, P. (1994a). Topological relations between regions with holes. *International Journal of Geographical Information Science*, *8*(2), 129–142. doi:10.1080/02693799408901990

Egenhofer, M. J., & Franzosa, R. D. (1991). Point-set topological spatial relations. *International Journal of Geographical Information Systems*, *5*(2), 161–174. doi:10.1080/02693799108927841

Ehrig, H., Ehrig, K., Prange, U., & Taentzer, G. (2006). *Fundamentals of Algebraic Graph Transformations*. Berlin: Springer.

Ehrig, H., Ehrig, K., Prange, U., & Taentzer, G. (2006). *Fundamentals of Algebraic Graph Transformations. EATCS Monographs on Theoretical Computer Science*. Berlin: Springer.

Eliminating Syria's chemical weapons. (2014). Retrieved from http://www.cnn.com/video/data/2.0/video/world/2014/07/02/id-mann-syria-chemical-gesture.cnn.html

Elsukov, M.Yu., Prozorova, V.B., Pshenichny, C.A., & Cherkashin, K.V. (2015). Smysl i prednaznachenie Novorossii (Smysl and Predestination of Novorossia). *MIA Novorossia*. Retrieved from mianews.ru/ru/2015/12/23/smysl-i-prednaznachenie-novorossii

Encoded Archival Description. (2017). Library of Congress. Retrieved from http://loc.gov/ead/

Engelhardt, W., & Zimmerman, J. (1982). *Theory of earth science*. Cambridge, UK: Cambridge University Press.

European Commission. (2004). *EUROSION: Living with coastal erosion in Europe: Sediment and Space for Sustainability, Part IV – A guide to coastal erosion management practices in Europe: Lessons Learned.* Technical Report 27. Author.

Falsaperla, S., & Neri, M. (2015). Seismic footprints of shallow dyke propagation at Etna, Italy. *Scientific Reports*, *5*(1), 11908. doi:10.1038/srep11908 PMID:26173557

Felpeto, A. & Marti, J. (2005). GIS-based automatic system for volcanic hazard assessment. In *Geophysical Research Abstracts* (vol. 7). European Geosciences Union (EGU) 2nd General Assembly, Vienna, Austria, EGU05-A-04020.

Ferrari, L., Garduño, V. H., & Neri, M. (1991). I dicchi dellaValle del Bove, Etna: Un metodo per stimare le dilatazioni di un apparato vulcanico [in Italian]. *Memorie della Societa Geologica Italiana*, *47*, 495–508.

Filimonenko, N. T. (2008). *Lectures on Disruption of Rocks*. Donetsk National Technical University. (unpublished; in Russian)

Fiske, R. S., & Jackson, E. D. (1972). Orientation and growth of Hawaiian volcanic rifts: The effect of regional structure and gravitational stresses. *Proceedings of the Royal Society of London*, *329*(1578), 299–326. doi:10.1098/rspa.1972.0115

Foster, G. L., Royer, D. L., & Lunt, D. J. (2017). Future climate forcing potentially without precedent in the last 420 million years. *Nature Communications*, 8. PMID:28375201

Francalanci, L., Manetti, P., Peccerillo, A., & Keller, J. (1993). Magmatological evolution of the Stromboli volcano (Aeolian Arc, Italy): Inferences from major and trace element and Sr-isotopic composition of lavas and pyroclastic rocks. *Acta Vulcanologica*, *3*, 127–151.

Frank, A. U. (1992). Theories and methods of spatio-temporal reasoning in geographic space. *International Conference GIS – From Space to Territory, Proceedings Lecture Notes Computer Science*, 639. doi:10.1007/3-540-55966-3

Frank, A. U. (1991). Qualitative spatial reasoning about cardinal directions. *International Journal of Geographical Information Systems*, (10): 1–10.

Frank, A. U. (1996). Qualitative spatial reasoning: Cardinal directions as an example. *International Journal of Geographical Information Systems*, *10*(3), 269–290. doi:10.1080/02693799608902079

Freksa, C. (1991). Qualitative spatial reasoning. *Cognitive and Linguist Aspect of Geographic Space*, *63*, 361–372. doi:10.1007/978-94-011-2606-9_20

Freksa, C. (1992). Temporal reasoning based on semi-intervals. *Artificial Intelligence*, *54*(2), 199–227. doi:10.1016/0004-3702(92)90090-K

Frodeman, R. (1995). Geological reasoning: Geology as an interpretive and historical science. *Geological Society of America Bulletin*, *107*(8), 960–968. doi:10.1130/0016-7606(1995)107<0960:GRGAAI>2.3.CO;2

Gai, K., Qiu, M., & Elnagdy, S. A. (2016). Security-Aware Information Classifications Using Supervised Learning for Cloud-Based Cyber Risk Management in Financial Big Data. *IEEE 2nd International Conference on Big Data Security on Cloud (BigDataSecurity), IEEE International Conference on High Performance and Smart Computing (HPSC), and IEEE International Conference on Intelligent Data and Security (IDS).* New York: IEEE. Doi:10.1109/BigDataSecurity-HPSC-IDS.2016.66

Galton, A., & Worboys, M. (2005). Processes and events in dynamic geo-networks. Lecture Notes in Computer Science, 3799(11), 45–59. doi:10.1007/11586180_4

Galton, A. (2003). Desiderata for a spatio-temporal geo-ontology. Spatial Information Theory. *Foundation of Geographic Information Science*, *2825*(2), 1–12.

Galton, A. (2004). Fields and objects in space, time and space-time. *Spatial Cognition and Computation*, *4*(1), 15–37. doi:10.1207/s15427633scc0401_4

Galton, A. (2008). Experience and history: Processes and their relation to events. *Journal of Logic and Computation*, *18*(3), 323–340. doi:10.1093/logcom/exm079

Galton, A. (2009). Spatial and temporal knowledge representation. *Earth Science Informatics*, *2*(3), 169–187. doi:10.1007/s12145-009-0027-6

Garduño, V. H., Neri, M., Pasquarè, G., Borgia, A., & Tibaldi, A. (1997). Geology of NE rift of Mount Etna, Sicily (Italy). *Acta Vulcanologica*, *9*(1/2), 91–100.

Gavrilova, T., Bolotnikova, E., Leshcheva, I., Blagov, E., & Yanson, A. (2013). Measuring Psychological Impact on Group Ontology Design and Development: An Empirical Approach. Knowledge Engineering and the Semantic Web. Communications in Computer and Information Science, 394, 29-43.

Gentzen, G. (1934). Untersuchungen über das logische Schliessen [Investigations into logical deduction]. In The Collected Papers of Gerhard Gentzen (pp. 68-128). Amsterdam: North-Holland.

Geourghiou, L. (1998, September). 1998, Global cooperation in research. *Research Policy*, *27*(6), 611–626. doi:10.1016/S0048-7333(98)00054-7

Geshi, N., & Neri, M. (2014). Dynamic feeder dyke systems in basaltic volcanoes: The exceptional example of the 1809 Etna eruption (Italy). *Frontiers of Earth Science*, *2*, 13. doi:10.3389/feart.2014.00013

Ghosh, J. G., de Wit, M. J., & Zartman, R. E. (2004). Age and Tectonic Evolution of Neoproterozoic Ductile Shear Zones in the Southern Granulite Terrain of India, with implications for Gondwana studies. *Tectonics*, 23.

Giammanco, S., Melián, G., Neri, M., Hernández, P. A., Sortino, F., Barrancos, J., & Perez, N. M. et al. (2016). Active tectonic features and structural dynamics of the summit area of Mt. Etna (Italy) revealed by soil CO2 and soil temperature surveying. *Journal of Volcanology and Geothermal Research*, *311*, 79–98. doi:10.1016/j.jvolgeores.2016.01.004

Giarratano, J. C., & Riley, G. D. (1998). *Expert Systems: Principles and Programming*. Thomson Course Technology.

Giarratano, J. C., & Riley, G. D. (2005). *Expert Systems - Principles and Programming* (4th ed.). Thomson Course Technology.

Gladkaya, E. A., Egorova, E. N., & Torsunova, E. R. (2017). *Using Economic and Mathematical Methods During Formation of Construction Cluster*. Integration and Clustering for Sustainable Economic Growth. doi:10.1007/978-3-319-45462-7_26

Goel, A. K., Vattam, S., Wiltgen, B., & Helms, M. (2012). Cognitive, collaborative, conceptual and creative — Four characteristics of the next generation of knowledge-based CAD systems: A study in biologically inspired design. *Computer Aided Design*, *44*(10), 879–900. doi:10.1016/j.cad.2011.03.010

Goodday, J. M., & Cohn, A. G. (1994). *Conceptual neighborhoods in temporal and spatial reasoning. ECAI-94 Spatial Temporal Reasoning Workshop.*

Gopalakrishnan, K., & Varadan, G. N. (1996). Geothermal Manifestations and their relation to Geotectonic features along Tamil Nadu-Pondicherry Coast. Geothermal Energy in India. Geological Survey of India, 45, 25-37.

Gornitz, V. M., Daniels, R. C., White, T. W., & Birdwell, K. R. (1994). The Development of a Coastal Risk Assessment Database: Vulnerability to Sea-Level Rise in the U.S. Southeast. *Journal of Coastal Research*, *12*, 327–338.

Gray, J. (1981). The Transaction Concept: Virtues and Limitations. In *Seventh International Conference on Very Large Databases*. Tandem Computers Incorporated.

Gruber, T. (2009). Ontology. In L. Liu & M. T. Özsu (Eds.), *Encyclopedia of Database Systems. Springer-Verlag.*

Gruber, T. R. (1993). Toward Principles for the Design of Ontologies Used for Knowledge Sharing. *International Journal Human-Computer Studies, 43,* 907-928.

Gruber, T. R. (1993). Toward Principles for the Design of Ontologies Used for Knowledge Sharing. *International Journal of Human-Computer Studies*, *43*(5-6), 907–928. doi:10.1006/ijhc.1995.1081

Guru-Rajesh, K., & Chetty, T. R. K. (2006). Structure and tectonics of the Achankovil shear zone, Southern India. *Gondwana Research, 10*(1-2), 86–98. doi:10.1016/j.gr.2005.11.018

Gutierrez, B. T., Plant, N. G., & Thieler, E. R. (2011). A Bayesian network to predict coastal vulnerability to sea level rise. *Journal of Geophysical Research*, *116*(F2), F02009. doi:10.1029/2010JF001891

Guzmán, S., Strecker, M. R., Martí, J., Petrinovic, I., Schildgen, T. F., Grosse, P., & Sudo, M. et al. (2017). Construction and degradation of a broad volcanic massif: The Vicuña Pampa Volcanic Complex, southern Central Andes, NW Argentina. *Geological Society of America Bulletin*, *129*(5-6), 750–766. doi:10.1130/B31631.1

Haerder, T., & Reuter, A. (1983). Principles of transaction-oriented database recovery. *ACM Computing Surveys, 15*(4), 287. doi:10.1145/289.291

Hancock, P. L. (1994). *Continental Deformation*. Pergamon Press.

Hanson, N. R. (1958). *Patterns of Discovery*. Cambridge, UK: Cambridge University Press.

Hapke, C., & Plant, N. (2010). Predicting coastal cliff erosion using a Bayesian probabilistic model. *Marine Geology*, *278*(1-4), 140–149. doi:10.1016/j.margeo.2010.10.001

Harris, M. (2009). The ten most expensive online crimes ever. *CNET*. Retrieved from https://www.cnet.com/news/the-ten-most-expensive-online-crimes-ever/

Harris, N. B. W., & Santosh, M. (1993). Geological Society of India Memoir: Vol. 25. Chronological constraints on granulite formation in southern India and Sri Lanka. Academic Press.

Harris, N. B. W., Santosh, M., & Taylor, P. N. (1994). Crustal Evolution in South India: Constraints from Nd Isotopes. *The Journal of Geology*, *102*(2), 139–150. doi:10.1086/629659

Hazarika, S. M. (2005). *Qualitative Spatial Change: Space-Time Histories and Continuity*. Thesis.

Hazarika, S. M., & Cohn, A. G. (2001). *Qualitative spatio-temporal continuity*. Spatial Information Theory.

He, J. (2010) Transaction calculus. In *UTP'08 Proceedings of the 2nd international conference on Unifying theories of programming*. Dublin, Ireland: Springer-Verlag.

Health and Safety Executive (HSE). (1996-2017). *HSE Offshore: Offshore topics. Statistics, Health and Safety Executive*. Retrieved from http://www.hse.gov.uk/offshore/statistics.htm

Hodkinson, I., & Reynolds, M. (2007). Temporal Logic. In Handbook of Modal Logic. Elsevier.

Holdsworth, R. E., Strachan, R. A., & Dewey, J. F. (1998). Continental Transpressional and Transtensional Tectonics. *Geological Society of London, Special Publications*, 135.

Hornsby, K., & Egenhofer, M. J. (2000). Identity-based change: A foundation for spatio-temporal knowledge representation. *International Journal of Geographical Information Science*, *14*(3), 207–224. doi:10.1080/136588100240813

Hume, D. (1910). *[1748]. An enquiry concerning human understanding*. New York, NY: P. F. Collier & Son.

Hurkmans, R. T. W. L., Bamber, J. L., Davis, C. H., Joughin, I. R., Khvorostovsky, K. S., Smith, B. S., & Schoen, N. (2014). Time-evolving mass loss of the Greenland Ice Sheet from satellite altimetry. *The Cryosphere*, *8*(5), 1725–1740. doi:10.5194/tc-8-1725-2014

Hussein, H. K., & El-Sattar, A. (2012). State of the Art in Interactive Storytelling Technology: An Approach Based on Petri Nets. In P. Pawlewski (Ed.), Petri Nets - Manufacturing and Computer Science. InTech. doi:10.5772/48549

IPCC. (2013). Retrieved from http://www.ipcc.ch/report/ar5/wg1/

Jagannathan, S., Vitek, J., Welc, A., & Hosking, A. (2005). A transactional object calculus. *Science of Computer Programming*, *57*(2), 164–186. doi:10.1016/j.scico.2005.03.001

Jaquet, O., & Carniel, R. (2003). Multivariate stochastic modelling: Towards forecasts of paroxysmal phases at Stromboli. *Journal of Volcanology and Geothermal Research*, *311*(1-3), 261–271. doi:10.1016/S0377-0273(03)00259-2

Jayanthi, G, Uma, V. (2016a). Event Attributed Spatial Entity Knowledge (EASE) Based Spatio-Temporal Reasoning to Infer Geographic Processes. *Proceeding of IEEE(ICGHPC.16).*

Jayanthi, G., Uma, V.(2016d). Automated Planning Considering Event Based Spatial Object Interactions Using Spatio-Temporal Reasoning. *Proceeding of IEEE(ICRTIT.16).*

Jayanthi, G., & Uma, V. (2016b). Reasoning About Events Causing Geographic Process Using Spatial and Temporal Relations. *Proceedings of the IEEE, ICCCI*, 16.

Jayanthi, G., & Uma, V. (2017). A qualitative inference method for prediction of geographic process using spatial and temporal relations. *International Journal of Artificial Intelligence and Soft Computing*, *6*(1), 1–23. doi:10.1504/IJAISC.2017.084215

Jayanthi, G., & Uma, V. (under review). submitted on June 2016c). Reasoning Temporal Description of Spatial Entity Knowledge Towards Qualitative Geographic Process Prediction. *International Journal of Knowledge and Information Systems.*

Jensen, F. V. (2001). *Bayesian Networks and Decision Graphs*. New York: UCL Press. doi:10.1007/978-1-4757-3502-4

Johannessen, O. M., Khvorostovsky, K., Miles, M. W., & Bobylev, L. P. (2005). Recent ice-sheet growth in the interior of Greenland. *Science*, *310*(5750), 1013–1016. doi:10.1126/science.1115356 PMID:16239440

Joughin, I., Das, S. B., King, M. A., Smith, B. E., Howat, I. M., & Moon, T. (2008). Seasonal speedup along the western flank of the Greenland Ice Sheet. *Science*, *320*(5877), 781–783. doi:10.1126/science.1153288 PMID:18420901

Kanishkan, B., & Renganathan, M. (1990). *Study of the Tertiary formations to the south of Coleroon river to locate possible lignite bearing areas in Tamil Nadu.* Geological Survey of India, Unpubished Report.

Kapitza, S. P. (1996). The phenomenological theory of world population growth. *Uspekhi Fizicheskikh Nauk, 166*(1), 63–80. doi:10.3367/UFNr.0166.199601c.0063

Kapitza, S. P. (2009). Global Population Blow-up and After: The Demographic Revolution and Sustainable Development. *Bulletin of the Georgian National Academy of Sciences, 3*, 5–12.

Karunarathna, H., & Reeve, D. (2007). Predicting morphodynamic response of a coastal plain estuary using a Boolean model. *Proceedings of the 5th IAHR Symposium on River, Coastal and Estuarine Morphodynamics*, 1109–1115. doi:10.1201/NOE0415453639-c139

Karunarathna, H., & Reeve, D. (2008). A Boolean approach to prediction of long-term evolution of estuary morphology. *Journal of Coastal Research, 24*(2B), 51–61. doi:10.2112/05-0542.1

Katz, M. B. (1978). Sri Lanka in Gondwanaland and the Evolution of Indian Ocean. *Geological Magazine, 115*(4), 237–244. doi:10.1017/S0016756800037146

Kechris, A. S. (1995). *Classical Descriptive Set Theory.* Springer. doi:10.1007/978-1-4612-4190-4

Keet, C. M. (2006). *Introduction to part-whole relations: mereology, conceptual modelling and mathematical aspects.* KRDB Research Centre Technical Report KRDB06-3, Faculty of Computer Science, Free University of Bozen-Bolzano, Italy.

Khvorostovsky, K. (2012). Merging and analysis of elevation time series over Greenland ice sheet from satellite radar altimetry. *IEEE Transactions on Geoscience and Remote Sensing, 50*(1), 1. doi:10.1109/TGRS.2011.2160071

Kim, W. (1990). *Introduction to Object-Oriented Databases.* The MIT Press.

Korableva, O., & Litun, V. (2014). The potential of transitive economies' growth based on innovative strategy. *WSEAS Transactions on Business and Economics, 11*, 725–736.

Kowalski, R., & Sergot, M. (1986). A logic based calculus of events. *New Generation Computing, 4*(1), 67–95. doi:10.1007/BF03037383

Krabill, W. & 12 others (2004). Greenland Ice Sheet: increased coastal thinning. *Geophysical Research Letters, 31*(2), L24402. 10.1029/2004GL021533

Kremer, R. (1998). Visual Languages for Knowledge Representation. *KAW'98: Eleventh Workshop on Knowledge Acquisition, Modeling and Management.* Retrieved from http://tinyurl.com/l4sc9ze

Krisak, O.S., & Kupenko, V.I. (2014). Usloviya formirovania kvartza s beloi polosoi Seleznyovskogo uglenosnogo raiona Donbassa [Conditions of formation of quartz with white tape in Seleznyovsky Coal Basin, Donbass]. In *Problemi Teoretichnoi I Prikladnoi Mineralogii, Geologii, Metallogenii Girnichnodobuval'nikh Regioniv.* Vid. Krivoriz'kogo Nazional'nogo Universitetu, Krivoi Rog. (in Russian)

Kuhn, T. S. (1962). *The structure of scientific revolutions.* Chicago, IL: University of Chicago Press.

Kulon, J., Mynors, D. J., & Broomhead, P. (2006). A knowledge-based engineering design tool for metal forging. *Journal of Materials Processing Technology, 177*(1-3), 331–335. doi:10.1016/j.jmatprotec.2006.04.062

La Rocca, G. (2012). Knowledge based engineering: Between AI and CAD. Review of a language based technology to support engineering design. *Advanced Engineering Informatics, 26*(2), 159–179. doi:10.1016/j.aei.2012.02.002

Lakatos, I. (1970). Falsification and the methodology of scientific research programmes. In Criticism and the Growth of Knowledge. Cambridge, UK: Cambridge University Press. doi:10.1017/CBO9781139171434.009

Latour, B., & Woolgar, S. (1979). *Laboratory life: The construction of scientific facts*. Princeton, NJ: Princeton University Press.

Lawlor, D. A., Davey Smith, G., & Ebrahim, S. (2004). Commentary: The hormone replace- ment-coronary heart disease conundrum: Is this the death of observational epidemiology? *International Journal of Epidemiology*, *33*(3), 464–467. doi:10.1093/ije/dyh124 PMID:15166201

Le Cozannet, G., Garcin, M., & Pshenichny, C. A. (2012). Reconsidering data models in coastal databases. *KEMG-2012 Proceedings*. Retrieved from geognosis.ifmo.ru

Lemmlein, G. G. (1946). *O proiskhozhdenii ploskikh kvartzev s beloi polosoi [On the origin of flat quartz with white tape]. Voprosy Mineralogii, Geohimii i Petrografii*. Moscow: Academy of Sciences of the USSR. (in Russian)

Lemmlein, G. G. (1973). *Morfologia i Genesis Kristallov [Morphology and Genesis of Crystals]*. Moscow: Nauka.

Leyrit, H., Zylberman, W., Lutz, P., Jaillard, A., & Lavina, P. (2016). Characterization of Phreatomagmatic Deposits from the Eruption of the Pavin Maar (France). In Lake Pavin: History, geology, biogeochemistry, and sedimentology of a deep meromictic maar lake. Springer.

Lipman, P. W. (1968), Geology of the Summer Coon volcanic center, eastern San Juan Mountains, Colorado. Colorado School of Mines Quarterly, 63, 211–236.

Liu, H.-C., Lin, Q.-L., Mao, L.-X. L.-X., & Zhang, Z.-Y. (2013). Dynamic Adaptive Fuzzy Petri Nets for Knowledge Representation and Reasoning. Systems, Man, and Cybernetics: IEEE Transactions, 43, 1399 – 1410, doi doi:10.1109/TSMC.2013.2256125

Li, X., & Lara-Rosano, F. (2000). Adaptive fuzzy Petri nets for dynamic knowledge representation and inference. *Expert Systems with Applications*, *19*(3), 235–241. doi:10.1016/S0957-4174(00)00036-1

Longhinos, B., Anand, S. P., & Rajaram, M. (2010). Physical Geology of Subvolcanic Systems: Laccolith, Sills and Dykes. *Proceedings, LASI 4 Conference*, 22-26.

Longhinos, B., & Ramasarma, M. (2008). *Seismo-tectonic signatures in and around Kudankulam, Tirunelveli, India. In Proceedings in Nuclear Power safety*. Kudangulam.

Loudon, T. V. (2000). Geoscience after IT. *Computers & Geosciences*, 26.

Lucchi, F., Keller, J., De Astis, G., Francalanci, L., & Tranne, C. A. (2013). *Geological Map of Stromboli (Aeolian Archipelago)*. Università di Bologna, Universitat Freiburg, Istituto Nazionale di Geofisica e Vulcanologia, Università di Firenze.

Luthcke, S. B., Zwally, H. J., Abdalati, W., Rowlands, D. D., Ray, R. D., Nerem, R. S., & Chinn, D. S. et al. (2006). Recent Greenland ice mass loss by drainage system from satellite gravity observations. *Science*, *314*(5803), 1286–1289. doi:10.1126/science.1130776 PMID:17053112

Luzón, F., Palencia, V. J., Morales, J., Sánchez-Sesma, F. J., & García, J. M. (2002). Evaluation of Site effects in sedimentary basins. *Fisica della Tierra*, *14*, 183–214.

Maitra, A., Chakrobarty, Chandan, Rana, M.S. & Chandra, S. (2007). Tectono-sedimentation model and hydrocarbon prospectivity of Gulf of Mannar, Cauvery Basin. *KDMIPE, ONGC Unpub. Report*, 17.

Mandrik, M. V., & Arustamov, S. A. (2017). Biography as Scenario: Representing Historical Knowledge by the Event Bush. In Dynamic Knowledge Representation in Scientific Domains. IGI Global.

Martensen, A. L., & Butler, R. W. (1987). *The fault-tree compiler.* NASA Technical Memorandum 89098.

Martí, J., Aspinall, W. P., Sobradelo, R., Felpeto, A., Geyer, A., Ortiz, R., & Lopez, C. et al. (2008). A long-term volcanic hazard event tree for Teide-Pico Viejo stratovolcanoes (Tenerife, Canary Islands). *Journal of Volcanology and Geothermal Research*, *178*(3), 543–552. doi:10.1016/j.jvolgeores.2008.09.023

Martin, Ph. (2002). Knowledge representation in CGLF, CGIF, KIF, Frame-CG and Formalized-English. In *Proceedings of ICCS'02, 10th International Conference on Conceptual Structures*. Borovets, Bulgaria: Springer Verlag. doi:10.1007/3-540-45483-7_7

Marzocchi, W., Sandri, L., & Selva, J. (2008). BET_EF: A probabilistic tool for long- and short-term eruption forecasting. *Bulletin of Volcanology*, *70*(5), 623–632. doi:10.1007/s00445-007-0157-y

McCarthy, J., & Hayes, P. J. (1969). *Some philosophical problems from the standpoint of artificial intelligence. In Readings in Nonmonotonic Reasoning* (pp. 26–45). Morgan Kaufmann Publishers Inc.

McGuire, W. J., & Pullen, A. D. (1989). Location and orientation of eruptive fissures and feeder-dykes at Mount Etna: Influence of gravitational and regional stress regimes. *Journal of Volcanology and Geothermal Research*, *38*(3-4), 325–344. doi:10.1016/0377-0273(89)90046-2

McKeown, K. R. (1985). Discourse strategies for generating natural-language text. *Artificial Intelligence*, *27*(1), 1–41. doi:10.1016/0004-3702(85)90082-7

Meehan, J. (1981). TALE_SPIN. In Inside Computer Understanding: Five programs plus Miniatures. Erlbaum.

Moats, W. P. (1990). *Geometry and emplacement of silicic dikes at Summer Coon Volcano, Rio Grande and Saguache counties, Colorado* (MS Thesis). Arizona State University, Tempe, AZ.

Mondo, D., Stell, J. G., Claramunt, C., & Thibaud, R. (2010). A Graph Model for Spatio-Temporal Evolution. *Journal of Universal Computer Science*, *16*(11), 1452–1477.

Moon, T., Joughin, I., Smith, B., & Howat, I. (2012). 21st-Century Evolution of Greenland Outlet Glacier Velocities. *Science*, *336*(6081), 576–578. doi:10.1126/science.1219985 PMID:22556249

Moores, E. M., & Twiss, R. J. (1995). *Tectonics*. Freeman.

Morris, C. (1971). *Foundations of the theory of signs*. Chicago, IL: University of Chicago Press.

Mouromtsev, D. I., Pshenichny, C. A., & Yakovlev, A. V. (2014). Semantic and structural delineation of market scenarios by the event bush method. *Decision Support Systems and Electronic Commerce*, *57*, 34–41. doi:10.1016/j.dss.2013.07.008

Mouromtsev, D., Kozlov, F., Parkhimovich, O., & Zelenina, M. (2013). Development of an Ontology-Based E-Learning System. *4th International Conference on Knowledge Engineering and Semantic Web, KESW 2013,* 394, 273-280. Doi:10.1007/978-3-642-41360-5_23

Muller, P. (1998a). A qualitative theory of motion based on spatio-temporal primitives. Principles of Knowledge Representation and Reasoning, 131–143.

Muller, P. (1998b). Space-time as a primitive for space and motion. *Proceedings of FOIS*, 63–76.

Muller, P. (2002). Topological spatio-temporal reasoning and representation. *Computational Intelligence*, *18*(3), 420–450. doi:10.1111/1467-8640.00196

Murdock, J., Buckner, C., & Allen, C. (2010). Evaluating Dynamic Ontologies. In *FCommunications in Computer and Information Science: 272 Knowledge Discovery, Knowledge Engineering and Knowledge Management. Second International Joint Conference, IC3K 2010, Valencia, Spain, October 25-28, 2010, Revised Selected Papers*. Springer.

Napoleon Bonaparte. (2014). Retrieved from http://www.biography.com/people/napoleon-9420291

Narayanaswami, S. (1947). *The geology and mineral resources of the Tiruchendur and southern and eastern portions of the Nanguneri taluk, Tirunelvelli District.* Unpublished Report, GSI P.R.

Naumov, A., Popov, I., Bondarenko, I., Krylov, B., Timonin, R., & Ofitserov, I. (2018). Modeling of Transaction Scripts Using Dynamic Knowledge Representation. In C. A. Pshenichny, P. Diviacco, & D. I. Mouromtsev (Eds.), *Representation of Dynamic Knowledge in Scientific Domains*. IGI Global.

Neri, A., Aspinall, W. P., Cioni, R., Bertagnini, A., Baxter, P. J., Zuccaro, G., & Woo, G. et al. (2008). Developing an Event Tree for probabilistic hazard and risk assessment at Vesuvius. *Journal of Volcanology and Geothermal Research, 178*(3), 397–415. doi:10.1016/j.jvolgeores.2008.05.014

Neri, M., Acocella, V., Behncke, B., Giammanco, S., Mazzarini, F., & Rust, D. (2011). Structural analysis of the eruptive fissures at Mount Etna (Italy). *Annales Geophysicae, 54*(5), 464–479. doi:10.4401/ag-5332

Neri, M., Lanzafame, G., & Acocella, V. (2008). Dike emplacement and related hazard in volcanoes with sector collapse: The 2007 Stromboli eruption. *Journal of the Geological Society, 165*(5), 883–886. doi:10.1144/0016-76492008-002

Newhall, C. G., & Hoblitt, R. P. (2002). Constructing event trees for volcanic crises. *Bulletin of Volcanology, 64*(1), 3–20. doi:10.1007/s004450100173

Nick, F. M., Vieli, A., Howat, I. M., & Joughin, I. (2009). Large-scale changes in Greenland outlet glacier dynamics triggered at the terminus. *Nature Geoscience, 2*(2), 110–114. doi:10.1038/ngeo394

Norheim, D., & Fjellheim, R. (2006). AKSIO - Active Knowledge management in the petroleum industry. In A. Leger, A. Kulas, L. Nixon, & R. Meersman (Eds.), *How Business application challenges Semantic Web Research (ESWC'06 Industry Forum)* (pp. 22–26). Available from http://ceur-ws.org/Vol-194/complete.pdf

Noy, N., Chugh, A., Liu, W., & Musen, M. (2006). A Framework for Ontology Evolution in Collaborative Environments. *Proceedings of ISWC*, 544-558. doi:10.1007/11926078_39

ONGC. (1993a). *Lithostratigraphy of the Indian Petroliferous Basins; document VII-Cauvery Basin. KDM Instt. Petro. Expl.* Dehra Dun, India: Oil and Natural Gas Commission.

ONGC. (1993b). *Fault systems of part of coastal belt of Tamil Nadu; document VII-Cauvery Basin. KDM Instt. Petro. Expl.* Dehra Dun, India: Oil and Natural Gas Commission.

ONGC. (1993c). *Composite Bouguer Gravity Map of Cauvery basin, Tamil Nadu; document VII-Cauvery Basin. KDM Instt. Petro. Expl.* Dehra Dun, India: Oil and Natural Gas Commission.

Pani, A. K., & Bhattacharjee, G. P. (2001). Temporal representation and reasoning in artificial intelligence: A review. *Mathematical and Computer Modelling, 34*(1–2), 55–80. doi:10.1016/S0895-7177(01)00049-8

Paramasivam, T. S., & Srinivasan, R. (1980). Sedimentary Limestones of Tisaiyanvilai Area, Tirunelveli District, Tamil Nadu. *Oil And Natural Gas Commission of India Bulletin, 17*, 183–186.

Parfitt, E., & Wilson, L. (2008). *Fundamentals of Physical Volcanology*. Wiley-Blackwell.

Pasquarè, G., Francalanci, L., Garduño, V. H., & Tibaldi, A. (1993). Structure and geologic evolution of the Stromboli volcano, Aeolian islands, Italy. *Acta Vulcanologica, 3*, 79–89.

Patterson, D. W. (2005). *Introduction to Artificial Intelligence and Expert Systems*. PHI learning.

Pearl, J. (1988). *Probabilistic Reasoning in Intelligent Systems: Networks of Plausible Inference*. San Mateo, CA: Morgan Kaufmann.

Peirce, C. S. (1958). *Collected Writings* (Vols. 1-8; C. Hartshorne, P. Weiss, & A. W. Burks, Eds.). Cambridge, MA: Harvard University Press.

Petri, C. A. (1962). *Kommunication mit Automaten* (PhD dissertation). University of Bonn. (in German)

Petri, C. A. (1973). Concepts of Net Theory. Mathematical Foundations of Computer Science. *Proc. of Symposium and Summer School*, 137-146.

Petri, C. A. (1977). General Net Theory. Computing System Design. In B. Shaw (Ed.), *Proc. of the Joint IBM University of Newcastle upon Tyne Seminar*. University of Newcastle upon Tyne.

Petri, C. A. (1973). Concepts of Net Theory. *Proc. of Symposium and Summer School, High Tatras*, 137-146.

Petrinovich, I. A., Marti, J., Aguirre-Diaz, G. J., Guzman, S., Geyer, A., & Salado Paz, N. (2010). The Cerro Aguas Calientes caldera, NW Argentina: An example of a tectonically controlled polygenetic collapse caldera, and its regional significance. *Journal of Volcanology and Geothermal Research*, *194*(1-3), 15–26. doi:10.1016/j.jvolgeores.2010.04.012

Pin, P., & Chen, S. (1976). The entity-relationship model—toward a unified view of data. *ACM Transactions on Database Systems, 1*(1), 9-36.

Plant, N. G., & Holland, K. T. (2011). Prediction and assimilation of surfzone processes using a Bayesian network, Part I: Forward models. *Coastal Engineering*, *58*, 119–130. doi:10.1016/j.coastaleng.2010.09.003

Poland, M. P., Moats, M. P., & Fink, J. H. (2008). A model for radial dike emplacement in composite cones based on observations from Summer Coon volcano, Colorado, USA. *Bulletin of Volcanology*, *70*(7), 861–875. doi:10.1007/s00445-007-0175-9

Polanyi, M. (1966). *The tacit dimension*. New York, NY: Anchor Day Books.

Popper, K. (1963). *Conjectures and refutations*. London, UK: Routledge.

Porreca, M., Acocella, V., Massimi, E., Mattei, M., Funiciello, R., & De Benedetti, A. A. (2006). Geometric and kinematic features of the dike complex at Mt. Somma, Vesuvio (Italy). *Earth and Planetary Science Letters*, *245*(1-2), 389–407. doi:10.1016/j.epsl.2006.02.027

Project Vasundara, GSI. (1994). *Seismotectonic map*. Geological Survey of India. T-VI.

Pschenichny, C., & Mouromtsev, D. (n.d.). Grammar of Dynamic Knowledge for Collaborative Knowledge Engineering and Representation. In Collaborative Knowledge in Scientific Research Networks. Hershey, PA: IGI Global.

Pshenichny, C. A. (2004). Classical logic and the problem of uncertainty. Geological Society, 239, 111-126.

Pshenichny, C. A., & Kanzheleva, O. M. (2011). Theoretical Foundations of the Event Bush. In K. Sinha, L. Gundersen, J. Jackson, & D. Arctur (Eds.), Geological Society of America Special Paper: Vol. 482. Societal Challenges and Geoinformatics (pp. 139–165). Academic Press. doi:10.1130/2011.2482(12)

Pshenichny, C. A., & Mouromtsev, D. I. (2013). Representation of the Event Bush Approach in Terms of Directed Hypergraphs. LNAI, 7735, 287–298. doi:10.1007/978-3-642-35786-2_22

Pshenichny, C. A., & Mouromtsev, D. I. (2015). Grammar of Dynamic Knowledge for Collaborative Knowledge Engineering and Representation. In Collaborative Knowledge in Scientific Research Networks. IGI Global. doi:10.4018/978-1-4666-6567-5.ch016

Pshenichny, C. A., Carniel, R., & Diviacco, P. (2013). Engineering of Dynamic Knowledge inExact Sciences: First Results of Application of the Event Bush Method in Physics. MSEPS 2013 - Modeling States, Events, Processes and Scenarios, Proceedings, 60–73.

Pshenichny, C. A., Nikolenko, S. I., Carniel, R., Sobissevitch, A. L., Vaganov, P. A., Khrabrykh, Z. V., . . . Gusev, E. A. (2008). The event bush as a potential complex methodology of conceptual modelling in the geosciences. *Proceedings, iEMSs - International Congress on Environmental Modelling and Software, 2*, 900-912.

Pshenichny, C. A., Nikolenko, S. I., Carniel, R., Sobissevitch, A. L., Vaganov, P. A., Khrabrykh, Z. V., . . . Gusev, E. A. (2008). The event bush as a potential complex methodology of conceptual modelling in the geosciences. *Proceedings, iEMSs – International Congress on Environmental Modelling and Software*, *2*, 900-912.

Pshenichny, C., Carniel, R., & Diviacco, P. (2013). Engineering of Dynamic Knowledge in Exact Sciences: First Results of Application of the Event Bush Method in Physics. In D. Mouromtsev, C. Pchenichniy, & D. Ignatov (Eds.), *MSEPS 2013 – Modeling States, Events, Processes and Scenarios; workshop proceedings*. Retrieved from http://iccs2013.hbcse.tifr.res.in/workshops/copy2_of_text.pdf

Pshenichny, C., Carniel, R., & Diviacco, P. (2013). Engineering of Dynamic Knowledge in Exact Sciences: First Results of Application of the Event Bush Method in Physics. *MSEPS 2013 – Modeling States, Events, Processes and Scenarios: workshop proceedings*, 60-73. Retrieved from http://iccs2013.hbcse.tifr.res.in/workshops/copy2_of_text.pdf, http://ceur-ws.org/Vol-1223

Pshenichny, C., Carniel, R., & Diviacco, P. (2013). Engineering of Dynamic Knowledge in Exact Sciences: First Results of Application of the Event Bush Method in Physics. *MSEPS 2013 – Modeling States, Events, Processes and Scenarios; Workshop Proceedings*, 60-73. Retrieved from http://iccs2013.hbcse.tifr.res.in/workshops/copy2_of_text.pdf

Pshenichny, C.A., Nikolenko, S.I., Carniel, R., Vaganov, P.A., Khrabrykh, Z.V., Moukhachov, V.P., … Rezyapkin, A.A. (2009). The Event Bush as a Semantic-based Numerical Approach to Natural Hazard Assessment (Exemplified by Volcanology). *Computers and Geosciences, 35*(5), 1017-1034. doi: 10.1016/j.cageo.2008.01.009

Pshenichny, C. A. (2014). Knowledge engineering in volcanology: Practical claims and general approach. *Journal of Volcanology and Geothermal Research*, *286*, 78–92. doi:10.1016/j.jvolgeores.2014.07.012

Pshenichny, C. A. (2018). Theory of Multitudes as an Alternative to the Set Theory. In C. A. Pshenichny, P. Diviacco, & D. I. Mouromtsev (Eds.), *Representation of Dynamic Knowledge in Scientific Domains*. IGI Global.

Pshenichny, C. A. (2018b). Qualitative and Quantitative Calculi in the Theory of Multitudes. In C. A. Pshenichny, P. Diviacco, & D. I. Mouromtsev (Eds.), *Representation of Dynamic Knowledge in Scientific Domains*. IGI Global.

Pshenichny, C. A., & Khrabrykh, Z. V. (2002). Knowledge Base of Formation of Subaerial Eruption Unit. In S. Leroy & I. Stuart (Eds.), *Environmental Catastrophes and Recovery in the Holocene (Abstracts)*. London: Brunel University. Retrieved from http://atlas-conferences.com/cgi-bin/abstract/caiq-22

Quintana-Amate, S., Bermell-Garcia, P., & Tiwari, A. (2015). Transforming expertise into Knowledge-Based Engineering tools: A survey of knowledge sourcing in the context of engineering design. *Knowledge-Based Systems*, *84*, 89–97. doi:10.1016/j.knosys.2015.04.002

Raiverman, V. (1992). Trans-Asiatic lineaments and Himalayan orogeny. In A. K. Sinha (Ed.), Himalayalan orogen and global tectonics. The International Lithosphere Programme.

Ramalingeswara Rao, B. (1992). Seismicity and Geodynamics of the low- to high-grade transition zone of Peninsular India. *Tectonophysics*, *201*(1-2), 175–185. doi:10.1016/0040-1951(92)90182-6

Ramanathan, S. (1979). Tertiary Formations of South India. *Miscellaneous Publication - Geological Survey of India*, *45*, 165–180.

Ramasamy, S., Armstrong-Altrin, J. S., & Chandrasekar, K. (1996). Heavy Mineral Distribution around Manavalakurichi, Kanyakumari District, Tamil Nadu, India. Indian Association of Sedimentologists.

Ramasamy, S. (1995). Petrography and Major Element Geochemistry of Sedimentary Rocks exposed around Kudankulam, Tamil Nadu. In *Proceedings, XIth Convention of Indian Association of Sedimentologists*. Roorkee University.

Ramasamy, S., & Armstrong-Altrin, J. S. (1998). Inferences on Rhodoids from Neogene arbonates of Kudankulam, Tamil Nadu. *Journal of the Geological Society of India*, *52*, 341–344.

Ramasamy, S., Paul, B. A. J., & Madhavaraju, J. (1994). Stratigraphy, petrography and sedimentation history of Kudankulam formation along the south Tamil Nadu coast, India. *Bulletin of Pure and Applied Sciences*, *13*, 43–58.

Ramillien, G., Lombart, A., Cazenave, A., Ivins, E. R., Llubes, M., Remy, F., & Biancale, R. (2006). Interannual variations of the mass balance of the Antarctica and Greenland ice sheets from GRACE. *Global and Planetary Change*, *53*(3), 198–208. doi:10.1016/j.gloplacha.2006.06.003

Rana, M. S., Chakraborty, C., Sharma, R., & Giridhar, M. (2008). Mannar Volcanics - Implications for Madagascar Breakup. *Proceedings, 7th International Conference & Exposition on Petroleum Geophysics.*

Randell, D. A., Cui, Z., & Cohn, A. G. (1992). *A spatial logic based on regions and connection.* Proceedings Knowledge Representation and Reasoning.

Rathore, S. S., Bansal, M., Vijan, A. R., & Pangtey, P. S. (2007). Ar-Ar dating of intrusives from Gulf of Mannar sub-basin of Cauvery Basin. *KDMIPE, ONGC Unpublished Report*, 17.

Records in Contexts. A Conceptual Model for Archival Description. (2016). International Council on Archives, Experts Group on Archival Description. Retrieved from https://www.ica.org/sites/default/files/RiC-CM-0.1.pdf

Reddy, A. G. B. (1995). A Geophysical approach to the problems of seismicity in the Indian shield. *Journal of the Geological Society of India*, *45*, 5–17.

Reisig, W. (1985). *Petri Nets.* Springer-Verlag. doi:10.1007/978-3-642-69968-9

Ribarić, S., Pavešić, N., & Zadrija, V. (2009). Intersection Search for a Fuzzy Petri Net-Based Knowledge Representation Scheme. Lecture Notes in Computer Science, 5711, 1-10.

Rich, Knight, & Nair. (2009). *Artificial Intelligence.* Tata McGraw Hill.

Richards, R. P. (1990). The origin of faden quartz. The Mineralogical Records, 21.

Ridley, J., Gregory, J. M., Huybrechts, P., & Lowe, J. (2010). Thresholds for irreversible decline of the Greenland ice sheet. *Climate Dynamics*, *35*(6), 1049–1057. doi:10.1007/s00382-009-0646-0

Rignot, E., & Kanagaratnam, P. (2006). Changes in the velocity structure of the Greenland Ice Sheet. *Science*, *311*(5673), 986–990. doi:10.1126/science.1121381 PMID:16484490

Rignot, E., Velicogna, I., van den Broeke, M. R., Monaghan, A., & Lenaerts, J. T. M. (2011). Acceleration of the contribution of the Greenland and Antarctic ice sheets to sea level rise. *Geophysical Research Letters*, *38*(5), L05503. doi:10.1029/2011GL046583

Ritov, M. Y., Gorlov, A. P., & Eryomenko, V. T. (2017). Mathematical modeling of the assessing process the of security level of the complex organizational and technical systems of industrial enterprises. *2nd International Conference on Industrial Engineering, Applications and Manufacturing (ICIEAM)*. Doi:10.1109/ICIEAM.2016.7911697

Ritz, C., Edwards, T. L., Durand, G., Payne, A. J., Peyaud, V., & Hindmarsh, R. C. (2015). Potential sea-level rise from Antarctic ice-sheet instability constrained by observations. *Nature*, *528*(7580), 115. PMID:26580020

Rogalchuk, V., & Solomin, K. (2015). Perspectives of Use of Petri Nets in Collaborative Research. In Collaborative Knowledge in Scientific Research Networks. IGI Global. doi:10.4018/978-1-4666-6567-5.ch018

Rogalchuk, V. V., Arustamov, S. A., Solovyov, N. A., & Malisheva, A. V. (2017). Petri Net versus Event Bush in Description of Environments of Alternative Directed Changes. In C. A. Pshenichny, P. Diviacco, & D. I. Mouromtsev (Eds.), *Representation of Dynamic Knowledge in Scientific Domains*. IGI Global.

Rossini, A., de Lara, J., Guerra, E., Rutle, A., & Wolter, U. (2014). A formalisation of deep metamodelling. *Formal Aspects of Computing*, 1–38.

Rutle, A., Rossini, A., Lamo, Y., & Wolter, U. (2012). A formal approach to the specification and transformation of constraints in MDE. *Journal of Logic and Algebraic Programming*, *81*(4), 422–457. doi:10.1016/j.jlap.2012.03.006

Rykart, R. (1977). Zum Wachstum plattiger Quarze mit "Faden". Schweizer Strahler, 4.

Sahani, A. (1979). Miocene Vertebrates from the Coastal Tertiary Rocks of Peninsular India and Sri Lanka. *Miscellaneous Publication - Geological Survey of India*, *45*, 197–205.

Santacroce, R., & Sbrana, A. (2003). Carta Geologica del Vesuvio. *Scala*, *1*, 15000.

Santosh, M., Kagami, H., Yoshida, M., & Nanda-Kumar, V. (1992). Pan-African charnockite formation in East Gondwana: Geochronologic (Sm-Nd and Rb-Sr) and petrogenetic constraints. *Bulletin of Indian Geological Association*, *25*, 1–10.

Santosh, M., Tanaka, K., Yokoyama, K., & Collins, A. S. (2005). Late Neoproterozoic–Cambrian feisic magnetism along transcrustal shear zones in southern India: U–Pb electron microprobe ages and implications for the amalgamation of the Gondwana supercontinent. *Gondwana Research*, *8*(1), 31–42. doi:10.1016/S1342-937X(05)70260-4

Scandone, R., Giacomelli, L., & Gasparini, P. (1993). Mount Vesuvius: 2000 years of volcanological observations. *Journal of Volcanology and Geothermal Research*, *58*(1-4), 4–25. doi:10.1016/0377-0273(93)90099-D

Scardamalia, M., & Bereiter, C. (2006). Knowledge building: Theory, pedagogy, and technology. In R. K. Sawyer (Ed.), *The Cambridge handbook of the learning sciences* (pp. 97–115). New York, NY: Cambridge University Press.

Schröder, B. S. W. (2003). *Ordered Sets: An Introduction*. Boston: Birkhäuser. doi:10.1007/978-1-4612-0053-6

Schukin, I. S. (1960). *Obschaya geomorphologia [General Geomorphology]* (Vols. 1–3). Moscow State University Publishers. (in Russian)

Scientific method. (n.d.). In *Oxford dictionary*. Retrieved from https://en.oxforddictionaries.com/definition/scientific_method

Serio Ltd. (2014). *Knowledgebase - Content is King*. Retrieved from http://www.seriosoft.com/blog/knowledgebase-content-king

Shafer, G. (1976). *A Mathematical Theory of Evidence*. Princeton, NJ: Princeton University Press.

Shannon, C. E., & Weaver, W. (1949). *A mathematical model of communication*. Chicago, IL: University of Illinois Press.

Shaw, J., Taylor, R. B., Solomon, S., Christian, H. A., & Forbes, D. L. (1998). Potential Impacts of Global Sea-Level Rise on Canadian Coasts. *The Canadian Geographer*, *42*(4), 365–379. doi:10.1111/j.1541-0064.1998.tb01352.x

Shepherd, A., Ivins, E. R., A, G., Barletta, V. R., Bentley, M. J., Bettadpur, S., & Zwally, H. J. et al. (2012). A Reconciled Estimate of Ice-Sheet Mass Balance. *Science*, *338*(6111), 1183–1189. doi:10.1126/science.1228102 PMID:23197528

Simons, P. (2005). Against Set Theory. *Proceedings of the 2004 Wittgenstein Symposium*, 143–152.

Skalle, P., & Aamodt, A. (2005). Knowledge-based decision support in oil well drilling: combining general and case-specific knowledge for problem solving. In Z. Shi & Q. He (Eds.), *Intelligent information processing II* (pp. 443–455). London, UK: Springer-Verlag. doi:10.1007/0-387-23152-8_56

Smith, B. (2005). Against Fantology. *Proceedings of the 2004 Wittgenstein Symposium*, 153–170.

Solomin, K., & Mandrik, M. (2015). Constructing historical knowledge through graphic boundary objects. In Collaborative Knowledge in Scientific Research Networks. IGI Global. doi:10.4018/978-1-4666-6567-5.ch017

Sørensen, L. S., Simonsen, S. B., Nielsen, K., Lucas-Picher, P., Spada, G., Adalgeirsdottir, G., & Hvidberg, C. S. et al. (2011). Mass balance of the Greenland ice sheet (2003–2008) from ICESat data – the impact of interpolation, sampling and firn density. *The Cryosphere*, *5*(1), 173–186. doi:10.5194/tc-5-173-2011

Sowa, J. (2006). *Semantic Networks*. Retrieved from http://www.jfsowa.com/pubs/semnet.htm

Sowa, J. F. (1987). Semantic Networks. In Encyclopedia of Artificial Intelligence (2nd ed.). Academic Press.

Sowa, J. F. (2006). *Semantic Networks*. Retrieved from http://www.jfsowa.com/pubs/semnet.htm

Sowa, J. F. (Ed.). (1992). Knowledge-Based Systems, 5(3).

Sowa, J. F. (2000). *Knowledge Representation: Logical, Philosophical, and Computational Foundations*. Pacific Grove, CA: Brooks Cole Publishing Co.

Speaks, J. (2016). Theories of Meaning. In *The Stanford Encyclopedia of Philosophy*. Retrieved from http://plato.stanford.edu/archives/spr2016/entries/meaning

Sreejith, C., & Ravindra Kumar, G. R. (2013). Petrogenesis of high-K metagranites in the Kerala Khondalite Belt, southern India: A possible magmatic-arc link between India, Sri Lanka, and Madagascar. *Journal of Geodynamics*, *63*, 69–82. doi:10.1016/j.jog.2012.10.002

Star, S. L., & Griesemer, J. R. (1989). Institutional ecology, translations and boundary objects: Ama- teurs and professionals in Berkeley's museum of vertebrate zoology. *Social Studies of Science*, *19*(4), 387–420. doi:10.1177/030631289019003001

Stasiuk, A. I., Hryshchuk, R. V., & Goncharova, L. L. (2017, May). A Mathematical Cybersecurity Model of a Computer Network for the Control of Power Supply of Traction Substations. *Cybernetics and Systems Analysis*, *53*(3), 476–484. doi:10.1007/s10559-017-9949-z

Stell, J. G. (2000). Boolean connection algebras: A new approach to the region-connection calculus. *Artificial Intelligence*, *122*(1), 111–136. doi:10.1016/S0004-3702(00)00045-X

Stell, J. G. (2004). Part and complement: Fundamental concepts in spatial relations. *Annals of Mathematics and Artificial Intelligence*, *41*(1), 1–17. doi:10.1023/B:AMAI.0000018575.09226.ff

Stell, J. G., Mondo, D., Thibaud, R., & Claramunt, C. (2011). Spatio-Temporal Evolution as Bigraph Dynamism. *COSIT, LNCS.*, *6899*, 148–167.

Stell, J., & Worboys, M. (2008). A theory of change for attributed spatial entities. *Lecture Notes in Computer Science, 5266*(9), 308–319. doi:10.1007/978-3-540-87473-7_20

Stojanovic, L. (2004). *Methods and Tools for Ontology Evolution* (PhD thesis). University of Karlsruhe.

Storetvedt, K. M., & Longhinos, B. (2012) The Atlantic and its bordering continents – a wrench tectonic analysis: lithospheric deformation, basin histories and major hydrocarbon provinces. *New Concepts in Global Tectonics Newsletter, 64.*

Storetvedt, K. M. (2003). *Global wrench tectonics: theory of earth evolution.* Bergen: Fagbokforlaget.

Studer, R., Benjamins, V. R., & Fensel, D. (1998). Knowledge Engineering: Principles and methods. *Data & Knowledge Engineering, 25*(1-2), 161–197. doi:10.1016/S0169-023X(97)00056-6

Subramanian, K. S., & Gopalakrishnan, K. (2002). Earth tremors in the coastal belt of Tamil Nadu and Pondicherry in September, 2001 – A Geological Perspective. *Journal of the Geological Society of India, 60*, 691–694.

Swanson, B., Yamangil, E., & Charniak, E. (2014). Natural Language Generation and Vocabulary Constraints. *Proceedings of the Ninth Workshop on Innovative Use of NLP for Building Educational Applications*, 124-133. doi:10.3115/v1/W14-1815

Syairudin, B., Trisunarno, L., & Gunarta, I. K. (2015). Strategy for Achieving the Millennium Development Goals 7-10 through Knowledge-Based Sustainable Design Approach. *Procedia: Social and Behavioral Sciences, 169*, 61–68. doi:10.1016/j.sbspro.2015.01.286

Sylvester, A. G. (1958). Strike-slip Faults. *Geological Society of America Bulletin, 100*(11), 1666–1703. doi:10.1130/0016-7606(1988)100<1666:SSF>2.3.CO;2

Syria chemical weapons set for destruction, Operation to destroy chemical weapons in final stages with Danish ship to transfer toxins to an American one. (2014). Retrieved from http://www.aljazeera.com/video/middleeast/2014/07/syria-chemical-weapons-set-destruction-20147225816561224.html

Syrian Civil War. (2014). In *Encyclopædia Britannica.* Retrieved from http://global.britannica.com/EBchecked/topic/1781371/Syrian-Civil-War

Tadini, A., Bevilacqua, A., Neri, A., Cioni, R., Aspinall, W. P., Bisson, M., & Pistolesi, M. et al. (2017). Assessing future vent opening locations at the Somma-Vesuvio volcanic complex: 2. Probability maps of the caldera for a future Plinian/sub-Plinian event with uncertainty quantification. *Journal of Geophysical Research. Solid Earth, 122*(6), 4357–4376. doi:10.1002/2016JB013860

Takada, A. (1997). Cyclic flank-vent and central-vent eruption patterns. *Bulletin of Volcanology, 58*(7), 539–556. doi:10.1007/s004450050161

Teece, D. J., Pisano, G., & Shuen, A. (1997). Dynamic Capabilities and Strategic Management. *Strategic Management Journal, 18*(7), 509–533. doi:10.1002/(SICI)1097-0266(199708)18:7<509::AID-SMJ882>3.0.CO;2-Z

The US disables diplomacy to escalate Syrian civil war, degrade Assad regime. (2014). Retrieved from http://rt.com/op-edge/us-diplomacy-escalates-syrian-war-766/

Thieler, E. R., & Hammar-Klose, E. S. (1999). *National assessment of coastal vulnerability to future sea-level rise: Preliminary results for the U.S. Atlantic Coast.* Technical Report, Open-File Report 99-953, US Geological Survey.

Thomas, R., Frederick, E., Krabill, W., Manizade, S., & Martin, C. (2006). Progressive increase in ice loss from Greenland. *Geophysical Research Letters, 33*(10), L10503. 10.1029/2006GL026075

Thomas, R. H., Rignot, E. J., Kanagaratnam, K., Krabill, W. B., & Casassa, G. (2004). Force-perturbation analysis of Pine Glacier, Antarctica, suggests cause for recent acceleration. *Annals of Glaciology*, *39*, 133–138. doi:10.3189/172756404781814429

Tibaldi, A. (2001). Multiple sector collapses at Stromboli volcano: How they work. *Bulletin of Volcanology*, *63*(2-3), 112–125. doi:10.1007/s004450100129

Tibaldi, A., Corazzato, C., Marani, M., & Gamberi, F. (2009). Subaerial and submarine evidence of structures feeding magma to Stromboli Volcano, Italy, and relations with edifice flank failure and creep. *Tectonophysics*, *469*(1-4), 112–136. doi:10.1016/j.tecto.2009.01.031

Tutte, W. T. (1998). *Graph theory as I have known it*. New York: Oxford University Press.

Van de Wal, R. S. W., Boot, W., van den Broeke, M. R., Smeets, C. J. P. P., Reijmer, C. H., Donker, J. J. A., & Oerlemans, J. (2008). Large and rapid meltinduced velocity changes in the ablation zone of the Greenland Ice Sheet. *Science*, *321*(5885), 111–113. doi:10.1126/science.1158540 PMID:18599784

Van den Broeke, M. R., Bamber, J., Ettema, J., Rignot, E., Schrama, E., van de Berg, W. J., & Wouters, B. et al. (2009). Partitioning recent Greenland mass loss. *Science*, *2434*(326), 984–986. doi:10.1126/science.1178176 PMID:19965509

van der Pluijm, B. (2016). Wrench Tectonics: Strike-slip Boundaries. *Processes in Structural Geology & Tectonics*. Retrieved from http://psgt.earth.lsa.umich.edu/PDF/14_WrenchTectonics_2016.pdf

Van der Pluijm, B. A., & Marshak, S. (1997). *Earth Structure: An introduction to structural geology and tectonics*. McGraw-Hill.

Vassiliou, Y., & Lochovsky, F. H. (1980). DBMS Transaction Translation. *Information Systems Working Papers Series*.

Vaz, G., Subbarao, V., & Ravikumar, V. (2006). Thermal Springs in Indian Coastal Areas of the Palk Bay: Their implication in Relation to Lineaments, Coastal Morphology and Seismicity. *Journal of the Geological Society of India*, *68*, 593–596.

Velicogna, I. (2009). Increasing rates of ice mass loss from the Greenland and Antarctic ice sheets revealed by GRACE. *Geophysical Research Letters*, *36*(19), L19503. doi:10.1029/2009GL040222

Venema, T. (2001). Temporal Logic. In L. Goble (Ed.), *The Blackwell Guide to Philosophical Logic* (pp. 203–223). Malden, MA: Blackwell Publishers.

Vistelius, A. B. (1992). *Principles of mathematical geology*. Dordrecht: Kluwer Academic Publishers. doi:10.1007/978-94-011-2934-3

Wang, W.-M., Peng, X., Zhu, G., Hu, J., & Peng, Y.-H. (2014). Dynamic representation of fuzzy knowledge based on fuzzy Petri net and genetic-particle swarm optimization. *Expert Systems with Applications*, *41*(4), 1369–1376. doi:10.1016/j.eswa.2013.08.034

Weikum, G., & Vossen, G. (2002). *Transactional Information Systems: Theory, Algorithms, and the Practice of Concurrency Control and Recovery*. Morgan Kaufmann Publishers.

White, S.A. (2004). Process Modeling Notations and Workflow Patterns. *BPTrends,* 1-20.

Whitehead, A. N. (1929). Process and Reality: An Essay in Cosmology. Macmillan.

Whitley, R. (2000). *The intellectual and social organization of the sciences*. Oxford, UK: Clarendon Press.

Winskel, G. (1987). Event Structures. Lecture Notes in Computer Science, 255.

Wittgenstein, L. (1921). Logisch-philosophiche Abhandlung. Annalen der Naturphilosophie, 14, 185–262.

Wolter, U., Korableva, O. N., & Solovyov, N. A. (2018). The event bush method in the light of typed graphs illustrated by common sense reasoning. In C. A. Pshenichny, P. Diviacco, & D. I. Mouromtsev (Eds.), Representation of Dynamic Knowledge in Scientific Domains. IGI Global.

Wolter, U., Korableva, O., & Solovyov, N. (2018). The Event Bush Method in the Light of Typed Graphs. In C. A. Pshenichny, P. Diviacco, & D. I. Mouromtsev (Eds.), *Representation of Dynamic Knowledge in Scientific Domains*. IGI Global.

Worboys, M. (2005). Event oriented approaches to geographic phenomena. *International Journal of Geographical Information Science*, *19*(1), 1–28. doi:10.1080/13658810412331280167

Worboys, M., & Hornsby, K. (2004). From objects to events: GEM, the geospatial event model. *GIScience Lecture Notes in Computer Science*, *3234*(10), 1–17.

Wouters, B., Chambers, D., & Schrama, E. J. O. (2008). GRACE observes small-scale mass loss in Greenland. *Geophysical Research Letters*, *35*(20), L20501. doi:10.1029/2008GL034816

Wu, X., Heflin, M. B., Schotman, Y., Vermeersen, B. L. A., Dong, D., Gross, R. S., & Owen, S. E. et al. (2010). Simultaneous estimation of global present-day water transport and glacial isostatic adjustment. *Nature Geoscience*, *3*(9), 642–646. doi:10.1038/ngeo938

Xiao-zhong, LI, (2003). A Kind of Fuzzy Petri Nets and its Dynamic Characteristics. *Acta Simulata Systematica Sinica*, *2003*(S1).

Xu, Y. (2012). Research on Method of Ontology-Based Knowledge Representation and Integration for Drilling Risk Decision. *International Journal of Digital Content Technology and Its Applications*, *6*(16). doi:10.4156/jdcta.vol6.issue16.33

Yates, M. L., & Le Cozannet, G. (2012). Brief communication 'Evaluating European Coastal Evolution using Bayesian Networks. *Natural Hazards and Earth System Sciences*, *12*(4), 1173–1177. doi:10.5194/nhess-12-1173-2012

Yeung, D. S., & Tsang, E. C. C. (1994). Fuzzy knowledge representation and reasoning using Petri nets. *Expert Systems with Applications*, *7*(2), 281–289. doi:10.1016/0957-4174(94)90044-2

Zablith, F. (2008). Dynamic ontology evolution. In *Proceedings of the International Semantic Web Conference (ISWC)*. Karlsruhe, Germany: Doctoral Consortium. Retrieved from http://fouad.zablith.org/docs/Dynamic_Ontology_Evolution_ISWC_2008_Fouad_Zablith.pdf

Zadeh, L. (1965). Fuzzy sets. *Information and Control*, *8*(3), 338–353. doi:10.1016/S0019-9958(65)90241-X

Zeigler, B. P. (2009). *DEVS-Centered Modeling and Simulation: Core Concepts for Engineering Education*. Retrieved from http://acims.asu.edu/publications/presentations

Zeigler, B. P., Praehofer, H., & Kim, T. G. (2000). *Theory of Modeling and Simulation. Integrating Discrete Event and Continuous Complex Dynamic Systems* (2nd ed.). Academic Press.

Zwally, H. J., Abdalati, W., Herring, T., Larson, K., Saba, J., & Steffen, K. (2002). Surface melt-induced acceleration of Greenland ice-sheet flow. *Science*, *297*(5579), 218–222. doi:10.1126/science.1072708 PMID:12052902

Zwally, H. J., Giovinetto, M. B., Li, J., Cornejo, H. G., Beckley, M. A., Brenner, A. C., & Yi, D. et al. (2005). Mass changes of the Greenland and Antarctic ice sheets and shelves ... to sea level rise: 1992-2002. *Journal of Glaciology*, *51*(175), 509–527. doi:10.3189/172756505781829007

Zwally, H. J., Li, J., Brenner, A. C., Beckley, M., Cornejo, H. G., DiMarzio, J., & Wang, W. et al. (2011). Greenland ice sheet mass balance: Distribution of increased mass loss with climate warming; 2003-07 versus 1992–2002. *Journal of Glaciology*, *57*(201), 88–102. doi:10.3189/002214311795306682

About the Contributors

Cyril Pshenichny graduated from Faculty of Geography, St. Petersburg State University, Russia, in 1995, and obtained a PhD in volcanology and petrology from Faculty of Geology of the same university in 1998. Cyril Pshenichny worked on different volcanoes of the world and studied the problems of argumentation and reasoning in geohazard assessment. This led him to the domains of logic and knowledge representation, and brought him to current employment at ITMO University, St. Petersburg, Russia. Cyril Pshenichny has authored and coauthored more than 70 scientific publications and edited two monographs. His research interests include knowledge engineering and representation, logic, methodology of science, volcanology and history.

Paolo Diviacco is a geophysicist at Istituto Nazionale di Oceanografia e di Geofisica Sperimentale (OGS). He holds a PhD in exploration geophysics, and has 20 years of experience in seismic data processing. He maintained always an interest in computer science and programming, developing software in the field of geophysics and Web-based data management. At the same time, he has always been interested in the philosophical and sociological aspects of scientific production in general and in the geo-sciences in particular. He has been active in developing ideas and Web-based systems that could support scientists while collaborating and foster the collaborative attitude among research institutions. In this context, he worked within several international projects, such as, for example, FP7 ODIP, FP7 Geo-Seas, FP7 Emodnet, FP7 Eurofleets, and the Antarctic Seismic Data Library.

* * *

Vladimir Anokhin was born on June 7, 1956, in Mariupol, USSR. Obtained a degree in geology from Leningrad State Mining Institute in 1981, a PhD in geology (fracture tectonics) in 1997, and DSc in geography (lineament analysis and geomorphology), in 2011. In 1981 – 2016, he was continuously employed as marine geologist (working in all oceans of the Earth) and involved in geological and environmental projects on land. Since 2015 is employed as a chief researcher at Institute of Limnology, Russian Academy of Sciences, St. Petersburg, and professor at Herzen St. Petersburg State Pedagogical University. Vladimir Anokhin has authored and coauthored more than 100 publications including three monographs and three sets of Russian State Geological Map. His research interests span from the deep structure of the Earth and planets to lacustrine ecosystems, geohazards and formalization of geological reasoning.

Sergey Arustamov is a professor at Computer Systems Design Department of ITMO University, St. Petersburg, Russia, since February 2008. His research activities are focused around computer-aided design system in microelectronics, system –on-chip design, hardware and software tools for information security, management of information at the corporate level. The same time, he conducts or supervises the study courses on information security and simulation of complex systems. Prof. Arustamov also has an experience in R&D activities in university and business environments.

Vyacheslav Bezrukov obtained a degree in optical electronics in 1975, and a PhD at Leningrad Institute of Fine Mechanics and Optics (now ITMO University), USSR, in 1986. Since 1997 he is employed as an assistant professor at ITMO, has five patents of invention and 42 published papers.

Igor Bondarenko obtained a degree in software engineering at Leningrad Institute of Fine Mechanics and Optics (now ITMO University), USSR, in 1995, and PhD, in 2003 in ITMO University. Since 2005, he is an assistant professor at the Computer Systems Design Department of ITMO University. He has more than 90 published papers, seven course manuals and textbooks and one monograph. His research interests include information technologies, automated design, information security, multiagent and hierarchical systems, artificial neural networks, evolution algorithms and data science.

Teena Chauhan received her Ph.D. Degree in Arctic marine geology from the University Centre in Svalbard (UNIS) at Longyearbyen and UiT The Arctic University of Norway at Tromso in 2015. She has worked at the National Institute of Oceanography (NIO), Goa-India, and, since 2017, at the Department of Earth Science, University of Bergen, Norway. Her research interests are in the field of paleoclimate and paleo-oceanography, particularly the reconstruction of past climates of the Arctic based on the microfossils and marine sediments. Teena is also involved in deciphering the Quaternary landscape evolution of the south-eastern coast of India.

Yuri Gatchin graduated from Leningrad Institute of Fine Mechanics and Optics (now ITMO University), USSR, in 1975, is employed as a lecturer in the same institute since 1980, obtained PhD and DSc in optical electronics, has more than 100 publications. Research interests include optical electronics, information system design, hardware/software interaction and information security.

G. Jayanthi has completed M. Tech Computer Science and Engineering in the year 2016 from Pondicherry Central University, Puducherry, India. She was awarded Pondicherry University Gold Medal in the year 2017. She has an interdisciplinary postgraduate degree in Remote Sensing completed in the year 2007 from Anna University, India. She is the university rank holder both in graduate and under graduate engineering studies. She has 9 years of teaching experience in engineering education (Computer Science and Engineering and Civil Engineering). Presently, she is pursuing research at Anna University, Chennai, India under faculty of Information and Communication Engineering. She has presented papers in international (IEEE) conference. She has authored a paper published in International Journal of Artificial Intelligence and Soft Computing, Inderscience, indexed in ACM digital library. Her research interests include Artificial Intelligence, Knowledge Engineering, Statistical & Machine Learning, Spatial Analytics, Spatio temporal Network Modeling and Analysis.

Jijoy Joseph did his masters in Mathematics from University of Kerala in 2003. He did his doctoral research in Mathematical Biology under Dr. Roschen Sasikumar, Scientist in CSIR-India. The Ph.D. dissertation was entitled "Chaos Game Representation for genome sequence analysis". One of his research publications as a Ph.D. scholar, an algorithm to compare genome sequences, has more than 50 citations to its credit. He was awarded the degree from Cochin University of Science and Technology in May 2011. He is currently working as Assistant Professor in Mathematics at University College, Thiruvananthapuram. His current research interests include signal processing, digital image analysis and mathematical geosciences.

Alexander Khokhulia graduated from Donetsk National Technical University, Ukraine, in 2005, as a drilling engineer. Since 2006 he is involved in numerous drilling projects, primarily in sea, and teaches in the same university, and defended a PhD in 2016. Dr. Khokhulia has authored and coauthored about 30 papers and monograph chapters. His research interest is drilling in marine environments.

Kirill S. Khvorostovsky received the Ph.D. Degree in Climatology and Meteorology in St. Petersburg State University and Nansen International Environmental and Remote Sensing Centre in 2004. Since 2006, he has worked at the Nansen Environmental and Remote Sensing Center, Bergen, Norway and since 2017 at Satellite Oceanography Laboratory of Russian State Hydrometeorological University, Saint Petersburg, Russia. His research interests are in the area of remote sensing, particularly satellite altimetry, of the Greenland ice sheet and Arctic sea ice. He has more than a decade expertise in deriving ice sheet elevation changes from satellite radar altimetry including developing the methodology of merging the data from different satellites as well as detailed analysis of time series for accounting of geophysical corrections. He is involved in the developing of the Greenland ice sheet surface elevation change retrievals in the frame of the ESA's CCI (Climate Change Initiative) Greenland ice sheet project. He has also experience in assessment and validation of sea ice thickness retrieved by satellite altimeter observations, and he is participating in the ESA CCI Sea ice project. He has 13 peer-reviewed publications.

Pavel Komar obtained an engineering degree from Donetsk Polytechnic Institute (now Donetsk National Technical University), USSR, in 1990. Since that, for 27 years, he develops drilling equipment, performs drilling and educates students in Donetsk National Technical University. He has coauthored several research papers, has certificates of invention and submitted dozens of technical reports. His research interests include various aspects of drilling and design of drilling equipment.

Sophie Koneva graduated from Geomorphology Dept., Earth Science Institute, St. Petersburg State University, in 2016, and is employed as geologist at Karpinsky All-Russian Research Geological Institute (VSEGEI), St. Petersburg, Russian Federation. She participated in several research projects in geomorphology and photogrammetry and composed a new geomorphological map of Iceland. The scope of her research interests includes volcanology, geomorphology, Quaternary geology and knowledge engineering in the geosciences.

Olga Korableva is a Doctor of Economics, Professor, Department of Economics of Research and Development, Saint Petersburg State University and professor of Saint Petersburg University of Economics; Member of the Association of Enterprise Architecture, Managing Partner of the Cluster of High-

Tech and Engineering. Research interests: knowledge-based economy, digital transformation, enterprise architecture, business analytics, competitive intelligence.

Oleg Krisak stared his research career and publication activity in the field of mineralogy in early studentship. Since 2008, he regularly reported his results at student conferences in Ukraine and abroad. He obtained a degree in geology at Donetsk National Technical University in 2016. Since that, he is a lecturer at the department of Geology and Exploration of Mineral Deposits of the same university.

Boris Krylov obtained a degree in hardware and software engineering in 1980 from Leningrad Institute of Fine Mechanics and Optics (now ITMO University), USSR, and a PhD from the same university in 2000. Currently he is employed there as assistant professor. Dr. Krylov has about 37 published papers, 3 student textbooks and 15 registered software products. His research scope includes all aspects of software engineering.

Biju Longhinos is an active academician and researcher in Earth sciences. He received degree of 'Philosophiae Doctor' in Marine Geology in Cochin University of Science and Technology, India, in 2009. Since 2000, he worked at the Department of Geology, University College, Trivandrum City, India, and since 2016 at Geo-Technology and Geology Division, College of Engineering, Trivandrum City. His research focuses on Global tectonics and landscape evolution since Pliocene. As an extensive field worker, he has come across a number of geological imprints that are either overlooked or suppressed by fellow researchers, which led him to convene an international meeting in 2011 at Kanyakumari, enabling scientists of different geological schools of thought to assemble and to discuss the possibilities on the evolution of earth, at planetary scale. He has worked on the tectonic evolution of North Atlantic, especially Iceland and the Arctic. He is currently involved in project looking into the planetary scale fractures and their geometry in south west Indian Ocean in relation with the geomechanics of the same exposed in land across India and south-east Africa. He is a member of high level core committee on river basin management constituted by the Government of the state Kerala, India.

Pavel Lunev graduated in geology from the State University of Saint-Petersburg, Russia in 2007. Since 2005, he is an employee of the Polar Marine Exploration Expedition (Lomonosov, St. Petersburg) performing research on geological structure and evolution of the Earth's crust and the ice sheet of Antarctica. He is mainly engaged in the interpretation of seismic data and ground-based radar. His current activity relates to the seismic research of the subglacial Lake Vostok area (Central Antarctica) and sedimentary basins of the Antarctic shelf.

Aleksandra Malisheva obtained her Bachelor degree in linguistics from Saint Petersburg State University of Economics and Finance in 2016 and enrolled in the MD program at Universite d'Angers, France, in 2017.

Maria Mandrik graduated from Faculty of History, Saint Petersburg State University, Russia, in 1996. Since that, she is specialized in Russian history and archivistics. Working in education and research in a few educational establishments in St. Petersburg, she defended in 2000 her PhD degree, and from 2002 on she is employed as researcher and head of department in St. Petersburg Branch of Archive of the Russian Academy of Sciences. She was among those who inspired and launched the computer-based information system in this archive in 2008 and since that is actively involved in its maintenance and development. Research interests of Dr. Mandrik include historiography of Russia in late 19th to early 20th century, biographies of Russian scientists, Russian-Mongolian relations in the first half of the 20th century and, in the last ten years, the issues of database design and knowledge representation related to the archivistics.

Andrei Naumov graduated from ITMO University, St. Petersburg, Russia, in 2015 and currently is a PhD student in the same university. His research interests include information security, integrated security systems, personal data protection, cryptography, cloud computing, cloud security.

Ivan Ofitserov graduated from ITMO University, St. Petersburg, Russia, in 2015 with a degree in information security. From then on, he is employed at state service in St. Petersburg and continues research in the field of his education.

Ilya Popov studied at ITMO University in 2010-2015, enrolled in PhD course in 2015 and began teaching in the same university the same year. Up to now, he has authored and coauthored more than 20 scientific publications. His research interests include information technologies, personal data security and information security sensu lato.

Vyacheslav V. Rogalchuk is a part-time senior lecturer at Information Technology Systems department of St. Petersburg State Transport University and software engineer at Dell EMC. His main research interests relate to knowledge management, semantic web technologies and software engineering processes. He has more than a decade experience in web systems development and maintenance, database integration and performance analysis. As development team leader, he has produced a radio network optimization tools suite on top of geographic information systems for railroads.

Victoria Shterkhun graduated from Faculty of Geography, St. Petersburg State University, in 2002, and is employed as geomorphologist and Quaternary geologist at Karpinsky All-Russian Research Geological Institute (VSEGEI), St. Petersburg, Russian Federation. She has about 20 research papers. Her research interests span from volcanology, Quaternary geology and palaeontology to modern morphodynamics and lithodynamics, geohazards, and knowledge engineering in geology.

Nikita Solovyov graduated from the Leningrad University majoring in physics. Researched physical processes in high temperature plasmas for laser fusion. Was engaged in consulting in the field of industrial safety. He is currently researching the metaphysical problem of free will and decision-making under uncertainty with the use of quantum analogies.

Roman Timonin received a degree in information security from ITMO University, St. Petersburg, Russia, in 2015 with a degree. Since that, he is employed in private business.

V. Uma received the M.Tech, and PhD degrees in computer science from Pondicherry University in 2007 and 2014 respectively. She was awarded the Pondicherry University gold medal for M.Tech. degree in Distributed Computing Systems. She has more than 10 years of teaching experience at PG level. Her research interest includes Data mining, knowledge representation and reasoning (spatial and temporal knowledge) and sentiment analysis. She has authored and co-authored more than 20 peer-reviewed journal papers, which includes publications in Springer and Inderscience. She received the Best Paper Award in International Conference on Digital Factory in the year 2008.

Uwe E. Wolter received the Ph.D. degree in Mathematics from the Technical University Magdeburg, Germany in 1989. His Ph.D. thesis on Partial Algebraic Specifications was awarded with the medal of honor of the Mathematical Society of East-Germany. Since then his research is devoted to application oriented further developments of a broader spectrum of formal specifications techniques. He is interested to combine different specification techniques and has been working in the areas Universal Algebra, Process Algebra, Universal Coalgebra, Graph Transformation, Logic and Abstract Model Theory. He is dedicated to the application and adaption of concepts and results from Category Theory. In the last decade his research has been focused on the mathematical foundation of diagrammatic specification techniques related to Model Driven Engineering and Knowledge Engineering. He has more than 50 peer-review publications.

Index

K

L

M

N

P

Q

S

T

V

W

www.ingramcontent.com/pod-product-compliance
Lightning Source LLC
LaVergne TN
LVHW060403300926
849162LV00006B/1335